Second Edition

Insect Diets

Second Edition

Insect Diets

Science and Technology

Allen Carson Cohen

NORTH CAROLINA STATE UNIVERSITY

INSECT REARING EDUCATION & RESEARCH PROGRAM

RALEIGH NC

CRC Press
Taylor & Francis Group
Boca Raton London New York

CRC Press is an imprint of the
Taylor & Francis Group, an **informa** business

Cover credits (photos used with permission): Clockwise starting with upper left: Instron Universal Testing Machine (from Chris Daubert and Chris Pernell, NCSU Department of Food Science); Mile-a-minute weed predator, *Rhinoncomimus latipes* (Curculionidae: Coleoptera), feeding on Cohen and Cheah artificial diet; ovarioles from healthy *Heliothis virescens* (Noctuidae: Lepidoptera); *Sasajiscymnus tsugae* (Coccinelidae: Coleoptera), a hemlock woolly adelgid predator, feeding on slurry-type artificial diet (image from Dr. Carole Cheah, Connecticut Experimental Station); *Arundo donax* (giant reed) parasite, *Tetramesa latipes* (Eurytomidae: Hymenoptera) eating *Arundo* callus from tissue culture developed by Dr. Sergei F. Krasnyanski of the Plant Transformation Laboratory in the Department of Horticulture at North Carolina State University; mandibles from termite, showing highly sclerotised dentition.

CRC Press
Taylor & Francis Group
6000 Broken Sound Parkway NW, Suite 300
Boca Raton, FL 33487-2742

First issued in paperback 2020

ISBN 13: 978-0-367-57569-4 (pbk)
ISBN 13: 978-1-4665-9194-3 (hbk)

Library of Congress Cataloging-in-Publication Data

Cohen, Allen Carson.
Insect diets : science and technology / Allen Carson Cohen. -- Second edition.
pages cm
Includes bibliographical references and index.
ISBN 978-1-4665-9194-3
1. Insects--Feeding and feeds. 2. Insect rearing. I. Title.

SF518.C64 2015
638'.5--dc23 2015000203

Visit the Taylor & Francis Web site at
http://www.taylorandfrancis.com

and the CRC Press Web site at
http://www.crcpress.com

Contents

Preface to the Second Edition

Since finishing the first edition of *Insect Diets: Science and Technology*, I had spent a short time in a private business dedicated to insect rearing education and research, and I joined the Entomology Department at North Carolina State University (NCSU), where I started a program in insect rearing education and research. In these two endeavors, I have had a chance to expand my background with many species of insects that were new to me and with many techniques that helped shape my understanding of and humility toward the complexities of diets and rearing systems. The exposure to new insects and new techniques and equipment was influential, but the exposure to students and colleagues in varied disciplines has been the most powerful force in my opportunity to learn and reexamine rearing and its greater interdisciplinary context. Some of the insects that I have worked with recently to develop diets and rearing systems include yellow jackets (bait development); house crickets; houseflies; mealworms; hemlock woolly adelgids; balsam woolly adelgids; predators of the adelgids (coccinellids and derodontids); painted lady and monarch butterflies; western bean cutworms; cactus moths; various weed-eating insects, especially beetles such as *Rhinoncomimus latipes*, a predator of mile-a-minute weeds; brown lacewings; hymenopteran gall-forming parasites of *Arundo*; silkworms; drosophilid flies; honeybees; buprestid and long-horned beetles; stinkbugs; western thrips; and, recently, termites and cockroaches. In teaching several rearing courses and workshops, I was exposed to many other species of insects (mosquitoes, mayflies, turf beetles, and many others), causing me to have to expand my thinking about the biology (especially feeding systems) of a broad range of insects with varied lifestyles.

With these research and teaching experiences, and with my consultation work with private companies and government rearing facilities, I have become deeply convinced that statistical process control (SPC) is a tremendously useful tool and approach to insectary system development and management. Using a data-driven approach to developing and producing diets and the other components of rearing systems is the most powerful way of assuring quality and reliability of the insects we work with. This push toward SPC is evident in the second edition of *Insect Diets: Science and Technology*. In my observations of production systems, we can divide the approaches into being either predominantly academic or industrial. The academic approach to production of any entity, including insects, has advantages such as being science driven and cognizant of rationale-based efforts, but industry is more dominated by a culture of customer satisfaction and as such it is involved with economics, efficiency, and quality. I think that systems such as diet-based insect rearing can have both the academic understanding and the industrially oriented quality control behind their operations. In my work on rearing systems and diets over the past decade, I have become convinced that the strongest feature that we can incorporate into rearing systems (for both research and practical applications) is a strong SPC (Chapter 11) basis for all aspects of production. Also, because my teaching experiences taught me that most people who work in insect rearing need specific guidelines about how to perform process and quality control measurements of diets and diet components, I included a chapter (Chapter 17) on how to perform the analytical techniques—mainly chemical measurements of proteins, lipids, antioxidants, and peroxidation.

A second new aspect of my understanding of diets is my increased exposure to food science, especially texture analysis with rheological techniques. It seems inescapable that texture features (mouth feel, penetration forces, viscosity or flow characteristics, bite force required for ingestion, and the many other characteristics of materials subjected to forces of stress) are important to the insect's response to food materials (natural or artificial diets)—just as the chemical stimuli are important. Equally important is the fact that the choice of components and processing steps will be manifested in the diet's texture so that using rheological techniques adopted from food science can be tremendously valuable in diet assessments for SPC. These two realizations about insect feeding choices being influenced by texture and textural changes being related to better evaluation of diet-making processes led to my invitation of two food rheologists (Professor Christopher Daubert and Christopher Pernell) to coauthor the new chapter (Chapter 15) on diet texture.

Thanks to my experiences in my own or my students' efforts to develop or modify insect diets and also in my roles as editor or reviewer for original research papers, I had the opportunity to study hundreds of papers on diets. Through these studies, I have modified my understanding of how diets are developed (presented in Chapters 4, 7, 9, and especially 10), including heuristic vs. analytical approaches, and also how and why certain materials (such as beans, wheat germ, yeast, and complex protein sources) are used in the vast majority of diets. In this vein, I have also developed an increased grasp of the additives from host plants used in many diets—plant extracts, for example. These studies are reflected throughout the second edition where I track the historical progress of certain materials, casein and yeast products, for example, and the rationale behind the success of these materials is explained further. In all these discussions of materials and processing steps in diets, I have tried to examine—mechanistically—how or why the components behave as they do in terms of palatability, nutritional value, bioavailability, or stability that they contribute to diets. These discussions are also presented, whenever possible, as a basis for further research on diet development. Also, in this new version, I have included many case studies—not just as formulations but as stories of why some aspect of the insect's feeding biology is served by an incipient diet. This is again part of the effort of this text to clarify for the reader the relationship of understanding the insect's biology to the increased likelihood of the success of the diet (Chapter 4).

This raises the questions of microbial involvement in insect diets (natural and artificial), and there is an expanded treatment, in the chapter on microbial relations and in several other chapters, on various kinds of beneficial relationships between insects and their microbial associates (microbiomes). The amazing relationships between insects such as the drosophilid flies and their yeasts, the ambrosia beetles and their mycangial-borne fungal associates, or between termites and their cornucopia of gut microbes are treated in greater depth in the second edition (Chapter 13). The realization that the microbial associations are an incredibly powerful influence on insect feeding processes has struck me with the awesome responsibility that we have to try to understand these relationships to help foster them or at least to try not to derail them.

Also with so much of my newly found understanding of food science and technology equipment and practices, more microbiology experience (including experiences with fermentation systems and biomanufacturing with bioreactors as basic tools), and more background with the diversity of uses of insects such as the uses of insects as foods for other animals including humans, I have expanded the chapter (Chapter 16) on new directions for the future of insect diets. One of the most lively topics in this chapter is the worldwide awakening (some would say reawakening) of the drive to use insects more extensively as foods for humans. In this context, the many suggestions that insects produced by mass-rearing technology will be prominent in the supply of nutritious, reliable, safe, and economical foods for the world's population, which is thought (by UN thinkers) to be poised to expand from seven billion in 2015 to nine billion by 2050. If this is going to be a reality, there is certainly a need to greatly improve our understanding and capacity to control the fundamentals of mass rearing, and this lofty goal is a major theme and intention in this book. The precept that I teach (preach?) to my students, "know your insect," is the underlying premise of this book as is the expansion of this concept, "know your rearing/diet system." This understanding will serve our multifaceted goals of using insect rearing for research, teaching, and sterile release programs (or other genetic pest management strategies that are rapidly emerging), as food for other organisms, as biological control agents, as protein production "factories" (or other biomanufacturing systems), as subjects of conservation programs, and for many other purposes.

Acknowledgments to the Second Edition

Between the time when the first edition of this book was written and now (October 2014), I have had a logarithmic learning curve of my rearing and diet experiences. These experiences came in part from consulting contracts with many private companies and government organizations. These projects gave me experience with dozens of species of insects that were new to me and that gave me whole new perspectives about insect diets and the rearing context of these diets. My assignments included baits for yellow jackets and diets and rearing systems for house crickets; houseflies; mealworms; hemlock woolly adelgids; balsam woolly adelgids; predators of the adelgids (coccinellids and derodontids); painted lady and monarch butterflies; western bean cutworms; cactus moths; various weed-eating insects, especially beetles such as *Rhinoncomimus latipes*, a predator of mile-a-minute weeds; brown lacewings; hymenopteran gall-forming parasites of *Arundo*; silkworms; drosophilid flies; honeybees; buprestid and long-horned beetles; stinkbugs; and, through interactions with workshop participants, students, and professional colleagues, many other insects (and a few mites).

In my more recent experiences at the NCSU, I have had the enriching and growth-promoting experiences of working with people from the Entomology Department and from other departments across the campus, including the Department of Food, Bioprocessing, and Nutritional Science, the Department of Industrial and Systems Engineering, Plant Sciences, Biomanufacturing Training and Education Center (BTEC), Forestry Biomaterials, and the Department of Biological Sciences.

Some of the richest learning experiences have come from teaching on-site and online insect rearing courses. Preparing to teach courses in rearing forced me to clarify my perspective and understanding to meet the scrutiny of an intelligent and critical-thinking audience. Furthermore, the students in these courses (and in the rearing workshops that I have taught over the past decade) were stimulating in the questions that they asked from a fresh viewpoint, and in the answers to test questions and especially in their term projects, students caused me to examine my perspectives of diets and rearing from a 360° perspective. Included in my teaching experiences were contacts with rearing professionals from private industry and public sector rearing labs. Often, these people had perspectives and techniques that I had never experienced, so interaction with these people was incredibly enriching. Also, my opportunities to do on-site visits to industrial and government rearing laboratories were beyond an eye-opening, mind-expanding experience. I got to see firsthand how FMC, Monsanto, BASF, DuPont Pioneer, Beneficial Insectary, S.C. Johnson, and other industrial labs performed their rearing operations, often with great competence, good equipment, and remarkable personnel. Likewise, during my visits, state laboratories such as the PABIL facility in New Jersey, the State of Connecticut Experimental Station, the Florida DPI, and the North Carolina Laboratory of Biological Control conduct their excellent and highly professional operations. And of course, the leadership provided by the USDA-sponsored laboratories is underappreciated in my opinion in terms of the remarkable advancements that they made in rearing and diet technology, including the APHIS laboratories at Otis Air Base (Massachusetts), the Mission, Texas Mexican Fruit Fly Laboratory, and of course the Pink Bollworm Rearing Facility, in Phoenix, Arizona, where they kindly hosted visits from my workshop participants and showed all aspects of their rearing accomplishments. The United States Department of Agriculture, Agricultural Research Service (USDA, ARS, where I got my start for the first 23 years of my rearing career) has excellent facilities and professional personnel whose mission is to advance insect rearing in ways that are of service to the people of the United States and the world.

A very important part of my recent learning experiences and being introduced into a new realm was the considerable time that I spent with forest entomology, where Dr. Fred Hain included me in his group of students and colleagues. Thanks to Fred's inclusiveness, I got to serve on several graduate students' committees, and I found the programs in forest entomology a kind of second home.

Through my association with the forest entomology group, I worked with a wonderful group of students who contributed to my learning about the basics of insect feeding systems, including Dr. Kelly F. Oten, Dr. David Bednar, Micah Gardner, and Mary Talley. During my interactions with the forest entomology group at NCSU, I was most privileged to work with John Strider, whose constant curiosity and energy fueled many efforts to study one of the most challenging insects that I have ever encountered, the adelgids. At the same time (starting in 2004), I became a research partner of Dr. Carole Cheah (from the Connecticut Experimental Station), and with my interactions with Fred and Carole, we were supported (especially by the USDA, Forest Service and the USDA, APHIS) to do research on hemlock woolly adelgids, balsam woolly adelgids, and the predators of these two species. With further support from the USDA, APHIS, I worked with other forest insects, including the emerald ash borer in a project with Dr. Jason Hansen, who is currently a postdoctoral research associate working out of the Otis Airbase rearing facility in Cape Cod, MA. Through this project sponsored by the USDA, APHIS, I have been able to work with Dr. Hannah Nadel and David Lance. During this period, I also had the support and productive learning experiences from Dr. Kenneth Bloem (USDA, APHIS), who brought me into the cactus moth project and the emerald ash borer project. Ken's rearing background was a tremendous asset to my learning, and his further support for the USDA, APHIS, webinar series on insect rearing was a further opportunity to work with many of the top-notch rearing experts in the world: the staffs of the rearing labs run by APHIS.

PEOPLE WHO WERE PART OF MY LEARNING EXPERIENCES

Over the time frame when I started and ended my business venture (Insect Diet and Rearing Research, LLC), I had been funded by several companies (Clearwater Butterflies, The Bug Company, MGK, Beneficial Insectary, FMC, DuPont Pioneer, and Monsanto) and several government organizations (USDA, APHIS; USDA, ARS; USDA, Forest Service; USDA, National Institute of Food and Agriculture [NIFA] grants, or SBIR grants). I have also gotten support in terms of facility use, provision of insects, and scientific advice from the state of Connecticut Experimental Station, the state of New Jersey Beneficial Insect Rearing Lab, the state of Florida Department of Plant Industry, and the state of North Carolina Department of Agriculture and Consumer Services (Plant Industry—Plant Protection Section Biological Control Laboratory).

Also, I offer my special thanks to Dr. Sergei F. Krasnyanski of the Plant Transformation Laboratory in the Department of Horticulture at NCSU; José M. Bruno-Bárcena and James W. Brown (Department of Biological Sciences and formerly Department of Microbiology); Tom Dorsey, Amy Dierks, Kathleen Kidd, and Rebecca Norris (quarantine at the Biological Control Program of the North Carolina Department of Agriculture and Consumer Services); Thom Hodgson (Department of Industrial and Systems Engineering); Roger Thompson (retired, USDA, ARS); Den Truong (USDA, ARS, NC State); Richard Lawless and Lucas Vann (BTEC, NCSU); James Carpenter and Susan Dawdry (USDA, ARS in Tifton, Georgia); Erica Simbro and Leigh Christensen (DuPont Pioneer); Gordon Vadis and Hong Ji (The Bug Company); Sinthya Penn (Beneficial Insectary); Robert Suryani (MGK); Chris Daubert, Chris Pernell, and Sharon Ramsey (NCSU, Department of Food Science); Jenny DeSio, Tom Dorsey, Amy Dierks, and Cynthia Detweiler (of the State of New Jersey PABIL facility); Peter Ebling (Great Lakes Forestry Centre, Sault Ste. Marie, Ontario); Kenneth Johnson and Mary Wallin (FMC Corporation); and Hugh Conway and Donald Vacek (from the USDA, APHIS). Two dear friends, Dr. Ann Black and Brad Heidinger from ImmunoReagents, Inc. (our former neighbors on the NCSU Centennial Campus), added greatly to my learning about potential immunological and biochemical tests that could support rearing and diet quality measurements; plus, they were always there to help us with students and workshops.

Several of my colleagues from the Entomology Department have been instrumental and supportive of my role at NCSU, and in serving on graduate student committees, having research partnerships, and other interactions with these people, I have expanded my thinking about many ideas in this book: James Harper, George Kennedy, Ed Vargo, Wes Watson, Fred Gould, David Orr, Clyde Sorenson, Michael Roe, Charles Apperson, David Tarpy, and Coby Schal.

Many of my richest learning experiences have come from interactions with students, especially the graduate students at the University of Arizona (where I began my insect diet and rearing career), Mississippi State University (where I worked for several of my last years with the USDA, ARS, and where I had the chance to work with my friend Dr. Frank Davis and with many other outstanding scientists and rearing staff), and most recently and currently at NCSU (where I came to pursue efforts to make valuable rearing accomplishments through the private business sector from 2005 until now). Both at Mississippi State University and here at NCSU, I have cotaught workshops in insect rearing, and I have taught several courses in insect rearing at NCSU (as both in-person and online courses). Because of sheer numbers and logistic constraints, I cannot mention all the students who influenced me and helped me be a better diet and rearing specialist, but there is no substitute for the experience that teaching these people has provided to me and most especially learning from them. One of the long-standing supporters of my research and someone from whom I learned a great deal about insect physiology, insect/plant interactions, and some wonderful insights into rearing is W. S. Bowers. He has not only been a mentor to me but also a great advocate of insect rearing and artificial diets. He helped advance the field in ways that even he may not appreciate, especially with his pioneering work on tobacco hornworms.

Special mention is in order for the late Professor Henry Hagedorn. Henry was not only a great insect biochemist and teacher (and leader of the Center for Insect Science for many years), but he founded the open access *Journal of Insect Science* (*JIS*). His vision was to establish a journal that would give free access to all aspects of entomology to scientists and students from around the world, including emerging nations where funds for publishing and access to insect research were limited. He was inclusive to a greater degree than anyone can imagine—helping authors with their writing and science to make *JIS* a teaching system, as well as an outlet for new research. He invited me to become editor of the section in *JIS* dealing with insect diets, rearing systems, and feeding biology, and in that function, he encouraged researchers to produce papers on all aspects of insect rearing. Henry truly manifested the spirit of what I speak of in this book about establishing a home for insect rearing in a reputable journal with worldwide circulation. Everyone in entomology will miss him, especially the people in the rearing community who need someone of his stature and vision to help advance the field.

Finally, I could never have gotten done one-fourth of what I have done in my career without the support, encouragement, and help from my wife, Jackie. She does research, artwork, IT work, office management, and teaching and is the most important sounding board that I have for all my ideas, which always profit from her organizational genius.

Preface to the First Edition

A large part of the success of entomology over the past century has come from the ability to rear insects on artificial diets. Probably, much of future entomology will continue to depend on diet-based programs. This reliance points to the need to better understand how and why diets work and how and why they fail. In more than three decades of research in entomology, I have found that insect diets make up one of the most complex, misunderstood, and underappreciated aspects of entomology. This book is written to help explain these complexities and dynamics.

Unlike the handful of other texts on this subject, this book is not a compendium of diet formulations. Instead, it is an effort to explain what the various ingredients and processing steps do to make diets work. It explains the nutrient classes and how foods and diet components meet the insects' nutritional and other feeding needs. The book explains diets in terms of overall insect feeding biology (feeding stimuli, digestion and absorption, and metabolic frameworks). It explains the effects of various processing steps used in the preparation of components and complete diets, including refinement of foods, size reduction, heat and cold processing, prevention of microbial contamination, and removal of antinutrients. It deals with the chemical and physical interactions of components, explaining how insect diets are matrices or dispersions with complex organization that predetermines the diets' food value and stability. This book offers perspective on how diets are developed and how a program of quality assessment can be applied to rearing systems.

The book draws heavily from food science and technology because the base of knowledge of these fields is highly advanced in developing a base of understanding of virtually every aspect of foods—their chemistry, physics, microbiology, and the effects of processing techniques. My personal *discovery* of food science was an epiphany that was like a biologist who had squinted at specimens for years trying to see minute structures and then discovered the existence of microscopes! What I have found in the food science community is an energetic quest to understand foods, and between the vast resources behind such studies and an atmosphere of open-minded inquiry, there is a wealth of information and methods for all of us dedicated to insect diets.

In the movie *Inherit the Wind* about the Scopes trial, a sarcastic reporter quipped that religion's purpose was "...to comfort the afflicted and to afflict the comfortable." In a very real sense, that is also the purpose of this book. I have found that the complexity and difficulty in developing and using artificial diets properly have been greatly underestimated, and those who perform these practices competently have been underappreciated. In this light, I have tried to fill in the gaps in understanding for those who work with insect diets and to illustrate for everyone connected with insect diets how complex and special these tools actually are. To "afflict the comfortable," I tried to explain the many pitfalls that result from complacency and oversimplification of the complex dynamics of diets, and to "comfort the afflicted," I provided explanations of why we use the specific ingredients and processing steps called for in diet formulations and how to anticipate and troubleshoot problems with diets.

The driving force behind this book is the demystification of insect diets as *black boxes* whose mechanisms and modes of action had been obscure. Hopefully, once the scientific and mechanistic basis is clear as to how diets work or fail to work, the community of rearing and diet specialists will be better equipped to develop new diets and to improve their efficiency in handling established diets. Such improvements will serve the entomology community as a whole by making available increased numbers of various species of insects produced under conditions that at once are quality enhancing and economical. Most hopefully, this book will be a bridge for rearing specialists and their stakeholders to use artificial diets as ever-improving tools to better manipulate insects in ways that benefit humanity and our environment.

Acknowledgments to the First Edition

Many authors say that their works could not have been completed without certain kinds of help. Writing this book has taught me what they meant and how such statements are not exaggerations. First and foremost, I thank my wife, Jackie, who has given me limitless encouragement and countless patient hours of listening and discussing virtually every idea in the entire body of my work, including those in this book. There are others who were mentors, teachers, and reviewers who helped me develop and improve my knowledge and ideas. But I am especially indebted to some outstanding workers, without whose diligence and capability I would not have had the successes in developing insect diets (or I would not have known that the diets were successful without their competent bioassays and culture handling). Those whose contributions most profoundly affected my work are Nina M. Urias, Lisa Smith, Gay McCain, Brenda Woods, and Patrick Crittenden. Nina, Lisa, Gay, and Brenda helped me build a grasp of how the bioassay fits inextricably into the diet development/diet assessment paradigm. Patrick has tirelessly worked at the analytical aspects of my work in taking apart diets so that we could understand what the components do individually and collectively. Amanda Lawrence and Bill Monroe contributed to my microscopic studies of diet and insect feeding systems and the nature of diet/insect interactions. Discussions with Gerald Baker changed my approach to looking at diets visually (via microscopy), which led to looking at diets as dispersions, a concept that so permeates this work. Many fruitful and enjoyable discussions with G. Doug Inglis expanded my understanding of diet contaminants, symbionts, and were supportive of my progress in the field. Jack Debolt's work was a model for diet development techniques. Nelson Thompson was a mentor and early influence on my thinking about insect nutrition. The late Ken Hagen was an intellectual stimulant, model of top-notch research, and the kindest of encouragers and supporters for my work. And Margaret Connor, as the patent advisor for my patents on diets and diet delivery systems, forced me to think in the most analytical and mechanistic terms how and why diets worked or did not work; her rigorous approach forced me to develop a deeper understanding of what we do when we feed insects artificial diets. Finally, the pioneers of insect diet development and related rearing system accomplishments laid down the foundation of all the successes that are taken for granted by insect diet studies by the pioneers in the fields of insect nutrition and dietetics. Robert T. Staten encouraged my efforts over nearly two decades, and his own work has been an impetus to my grasp of technological approaches to scaling up diet-based mass-rearing systems. Two reviewers made substantial improvements in an earlier version of this book. Finally, I gratefully acknowledge the pioneers in the field of insect feeding on artificial diets for having built the foundation upon which stand the many successful programs that are based on insect mass rearing.

Author

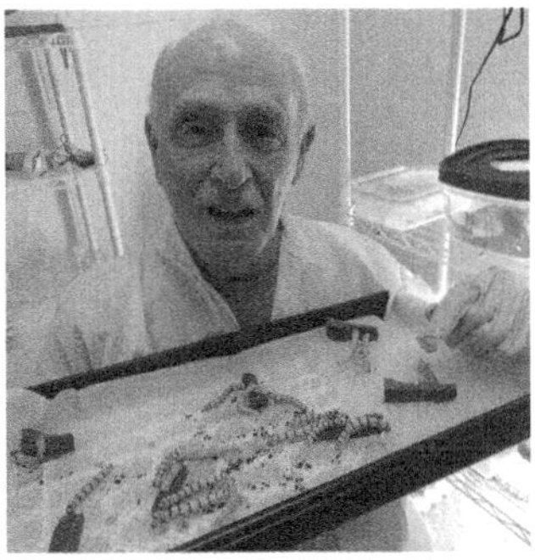

Allen Carson Cohen completed his high school education in Chicago, Illinois. He served four years in the U.S. Marine Corps and then went to California for a BA and MA in English and a teaching credential. He taught high school and college English for four years, then returned to college for an MA in biology, specializing in physiological ecology of rattlesnakes, and taught biology for nine years at Cypress College. During this time, he completed his PhD in entomology at the University of California, Riverside (1978). In 1979, he became a postdoctoral associate at the University of Arizona, and in 1980, he became a research entomologist with the USDA, Agricultural Research Service (ARS). In his ARS career, he advanced from GS-11 to GS-15 and held positions as research leader of the Biological Control of Insects Laboratory (in Tucson, Arizona) and of the Biological Control and Mass-Rearing Research Unit (at Mississippi State, Mississippi). He served several times as a panelist of the Research Panel Evaluation System, including advancement to panel chair. He was a member of the Patent Committee for ARS Pacific West and Mid-South Areas. During his ARS career, Dr. Cohen also served as adjunct professor at the University of Arizona, Mississippi State University, and Clemson University. He served as outside examiner at the University of Montreal (Quebec) and the Australian National University.

After his 23-year career in ARS, Dr. Cohen started a private research and consultation company (Insect Diet and Rearing Research, LLC), locating the company first in Tucson, Arizona, then at North Carolina State University (NCSU) on the Centennial Campus. Realizing that his central, lifetime goal was to advance entomology by helping to develop an infrastructure in insect rearing, in 2009, he accepted the position of research professor in the Department of Entomology and Coordinator of the Insect Rearing Education and Research Program at NCSU. During his current tenure as research professor and program coordinator, he has served the university and the field of entomology by teaching classes and workshops in insect rearing (online and face-to-face classes), by serving as graduate student advisor and student committee member, and has maintained an active research program on the development of diets and other aspects of rearing systems. He has served for seven years as editor for insect diets, insect rearing, and insect feeding biology for the *Journal of Insect Science*, and he serves as mentor for postdoctoral research associates and other students and professionals who seek his guidance in diet and rearing science and technology. He holds six U.S. patents, has published more than 125 refereed papers, and has received several technology transfer awards for his work on insect diets.

At NCSU, Dr. Cohen has developed a network of scientists, technicians, and students from various departments, including food science, engineering, microbiology, statistics, plant sciences, biomanufacturing, and, of course, entomology. In this network, the idea is developed of insect rearing being a multidisciplinary process, which demands that all aspects of the insects' well-being be addressed as only an integrated, multidisciplinary approach can accomplish. This multidisciplinary approach and efforts to establish rearing as a science and rationale-based technology are reflected in the rearing program and in the current text. Hopefully, this book and the other aspects of the rearing program will support the exhortation: "know your insect."

Contributing Authors

Chris Daubert
Department of Food Science
North Carolina State University
Raleigh, North Carolina

Chris Pernell
Department of Food Science
North Carolina State University
Raleigh, North Carolina

Humane Treatment of Organisms Statement

The Insect Rearing Education and Research Program coordinated by Allen C. Cohen subscribes to a policy of humane and ethical treatment of organisms. The values in this policy include these features: (1) Recognizing the diversity and complexity of life systems, we value all living things. (2) In light of this value, we try to design our work to help us understand the nature of life and living systems with a minimum of harm to or discomfort of living organisms. (3) We treat the subjects of our studies with respect, dignity, and with efforts to inflict a minimum of pain, trauma, or damage. This includes using, whenever possible, physical models instead of living systems to conduct our experiments. (4) We first consider ways that can inflict a minimum of potential harm to organisms and we ask the question: is the experiment necessary to give us the information we seek? Is the information of sufficient value to merit such sacrifices? (5) Furthermore, we subscribe to the tenet that the information derived from studies of organisms can be no better than the health, homeostasis, and minimization of stress of the organisms involved. Therefore, we strive to make our rearing systems and every other aspect of our handling of insects compatible with insect well-being.

Translation of these tenets into practice:

1. We use the concept of insect homeostasis or "comfort" in the sense that sensitive and reasonable thought about our insect handling can suggest guidelines about issues like food quality, space, management of waste, access to avoidance of adverse stimuli, sufficient gas exchange, protection from contaminants and pathogens, and all other aspects of insects' well-being as it would pertain to the populations of insects in the wild.
2. We subscribe to the model that teaches that well-treated insects are closer to "normal" insects from the wild. Such insects are of higher quality than ones that have been reared under suboptimal conditions. Therefore, we believe that our science is better when all aspects of our inquiries are conducted with the highest possible quality of components.
3. Therefore, our research into development and improvement of rearing systems keeps at its forefront the homeostasis of the insect subjects in our studies, and we further strive continuously to improve the base of knowledge for the scientific community to improve the conditions of rearing systems and the quality and well-being of the subjects.

CHAPTER 1

Scope of Insect Diet Science and Technology

1.1 INTRODUCTION

Insects that are reared on artificial diets are used in many programs—as agents of biological control and sterile insect technologies (Knipling 1979), as feed for other animals (Versoi and French 1992), as bioreactors for the production of pharmaceuticals and other recombinant proteins (Hughes and Wood 1998), and as foods for people (DeFoliart 1999)—and one of their most important uses is in research on virtually all areas of entomology and of other biological sciences. Thousands of papers written over the past century deal with artificial diets for insects. Although the topic of most of these papers is a subject other than artificial diets, it is evident that high-quality insects are essential to the assurance of meaningful studies, and the quality of the insect diets is, in turn, essential to the acquisition of healthy laboratory insects.

In fact, with the exception of a few subdisciplines such as field ecology and systematics, most insect studies rely upon laboratory-reared insects, and most of these studies incorporate insects reared on artificial diets. The reliability of all of these programs depends upon the insects' health, which depends upon the quality of the diets (Cohen 2001). While these successes testify to the value of artificial diet technology, there remain many problems in the existing programs and the potential to develop new programs based on applications of artificial diets. Most of the barriers to much-needed successes stem from the lack of a thorough understanding of the complexities of artificial diets, both on the part of those who develop diets and those who use them—collectively referred to in this book as *insect diet professionals.*

The accomplishments in development, improvement, and application of artificial diets have come from direct efforts to suit the needs of insects by studying the target insects and less directly by application of a knowledge base of various aspects of food sciences and their related disciplines (e.g., nutrition, microbiology, and biochemistry). Although the accomplishments associated with and stemming from applications of insect diets are noteworthy, with a better understanding of insect diets, progress in entomology could be much further accelerated and amplified. Review of the literature on insect diets reveals that many of the most noteworthy advancements have come from using information from the food sciences. Examples of this point are the breakthroughs discussed in Chapters 2 and 4. The point that insect dietetics has profited from and would continue to be improved by tapping into the pool of information from food science does not detract from the marvelous discoveries that are insect specific and that could only be made in the context of direct experimentation with insects (such as the uniqueness and universality of insects' requirements for dietary sterols discovered by Hobson in 1935). However, a wealth of information on various aspects of foods exists and, if properly utilized, could greatly enhance efforts of specialists to improve insect diets and diet processing.

1.2 FOOD SCIENCE AND FOOD TECHNOLOGY AND INSECT DIET PROGRAMS

In contrast with the enormous base of resources invested in research on human and livestock foods, research on artificial diets for insects is meager, and support for this research has been modest. For example, in the USDA Agricultural Research Service (ARS), there are several major research centers dedicated solely or extensively to programs on food science and nutrition. The extent of the USDA ARS research programs related to food science includes a scope of dozens of laboratories with major centers in Beltsville, MD; Wyndmoor, PA; Albany, CA; and Grand Forks, ND, and other specialty laboratories in nearly every U.S. state. Every aspect of human food is explored in these world-class laboratories where every aspect of nutrition is explored: food processing, food safety, obesity, novel food materials, and nearly countless other projects. Several laboratories including a major group in Beltsville are dedicated to the detailed analysis of the nutritional and biochemical composition of thousands of foods. Many of the approaches and findings on these subjects are presented in several places in this book.

This is in contrast with a handful of centers where research on insect diets is a fairly minor part of other programs in insect management and miniscule in comparison with the foods of humans. This differential between studies of human diets versus insect diets is not unique to the USDA ARS but rather is typical of the research profile around the world. While the difference in the foundation of knowledge is understandable considering the vast economic, social, and health importance of human and livestock foods, the shortcomings in our understanding of many of the basics of insect diets are a hindrance to progress in many entomology programs. This knowledge gap can be bridged by using the knowledge base and approaches of the food science/food technology community. Such a shift is not incongruous because the fundamentals of the insect diet and the human food domains share many commonalities.

Insect diets must serve insects in much the same way that human foods serve people. They must fulfill sensory requirements, be nutritious, and be reasonably stable—all within a framework of economic feasibility. Research on human and livestock foods has targeted virtually every aspect of foods and food processing. Food characterizations include analysis of nutrient and antinutrient profiles, sensory qualities, microbial populations, various additives, and components that are nutritionally inert. Food processing studies focus on every aspect of preparation and storage, including the effects of sorting, size reduction, various heat treatments, as well as preservation and storage strategies in the contexts of nutritional and sensory qualities, as well as economic impacts of various processing strategies.

Food science and food technology are characterized by a base of literature built upon well-defined approaches and standards. The advancements in these disciplines are documented in dozens of books, journals, and popular press articles, listed at the end of this chapter to suggest a valuable base of literature. This literature reveals a pursuit of questions on nearly every conceivable aspect of foods, including their nutritional content, sensory qualities, and the effects of various kinds of food preparation techniques (such as pasteurization, extrusion, flash sterilization, size reduction of food components, mixing, and packaging) and food preservation techniques (cold storage, dry storage, and chemical preservation). In summation, the body of information on food science and technology is more extensive, detailed, robust, and thorough than the base of information on diets for insects.

Besides the extensive resources that have gone into studies of human foods, the other key factor that has stimulated rapid progress in food science and technology is the approach to research questions and what is considered as publishable material. Insect diet research, in contrast with food science research, has been traditionally limited to direct studies of effects of diet components on target insects. In contrast, food researchers have had a more open domain in which they could conduct research on virtually any question about a wide array of topics. In the food science community, research is considered appropriate as long as it advances our understanding of the nature of food.

A summary of the following food science papers demonstrates the approaches and base of knowledge of the food science community. These studies seek to explain the nature of foods in various contexts without the use of human or other animal subjects to validate the findings. Once it had become established, for example, that ascorbic acid is a vital human nutrient, exploration of the stability, preservation, biochemical interactions, or other aspects of ascorbic acid in the context of any of several foods could be undertaken without human or animal subjects. It is tacitly accepted that it is of inherent value to know whether or not a given process (such as heating a product in boiling water or processing in a high-pressure twin screw extruder) is going to reduce the content or availability of antioxidants or a labile protein or amino acid or a delicate lipid or vitamin. In all of the studies that are summarized here, the nature of the foods and food components is explored. Such exploration of insect diet components and interactions of those components would serve the entomology community in a way that is parallel to the benefits derived from these studies in the food science and technology community. Each of these studies expands our body of knowledge of foods with respect to the nutritional value, sensory characteristics, safety, and availability of foods, the characteristics explained by Fennema (1996) as the salient features of foods and food processing that serve as the basis for improvement of food quality, economics of foods, and the resultant improvement of the human condition.

1.2.1 Representative Case Study: Antioxidants

The first example is a series of papers that deal with the antioxidant content of various foods. The presence of several classes of phyto-antioxidants, including ascorbic acid, phenolic compounds, lipid-soluble components with antioxidant properties, and a profile of the total antioxidant capacity in rose hip extracts were revealed by Gao et al. (2000). This study and several others like it are predicated upon the well-established principle that antioxidant quality is an important value of a food. Such studies are appearing in increasing numbers to show that many substances besides ascorbic acid, α-tocopherol, and β-carotene (three of the most popularly recognized antioxidants) are natural antioxidants. Other antioxidants are becoming recognized as important in reducing the destructive effects of oxidation. This work is valuable because it opens doors to viewing this supplement in a much broader context of antioxidant qualities than simply the ascorbic acid content.

In another study of antioxidants, Cao and Prior (1999) present a method for determining the overall oxygen radical absorbance capacity (called ORAC values) of biological materials. Cao and Prior also make the point that it is important to look beyond the handful of well-recognized antioxidants to discover other agents that confer protection against oxidative degradation of foods and oxidative stress within organisms that ingest these foods, emphasizing the importance of the total food (or other biological material) *matrix* as a complex that works simultaneously and synergistically to scavenge free radicals and other agents of oxidative stress.

These methods were applied to insect diets to examine components (cryptic antioxidants) that were not deliberately added as antioxidants, but that did confer antioxidant capacity to the diets (Cohen and Crittenden 2004). It would be useful to know how much of such components were present and their contribution to the antioxidant capacity for every insect diet. The application of this information to insect diets is potentially far-reaching in light of a growing realization of the direct value of antioxidants to insects (discussed in detail in Chapters 3 and 5) and the indirect value of these substances in the preservation of diet.

1.2.2 Representative Case Study: Antibiotics

The next example is a study that demonstrates that chitosan additives confer antimicrobial capacity to foods such as tofu (No et al. 2002). Chitosans, biopolymers that are derivatives from the exoskeleton from crustacean shellfish, have been shown to have health benefits when added to foods or as pharmaceutical supplements (No et al. 2002). In addition to these putative benefits, chitosans have been shown to have antimicrobial activity when added to soy products such as tofu to reduce spoilage, as well as to add a desirable texture to this important soy product (Chun et al. 1997). Several species of bacteria from the genus *Bacillus* and *Enterobacter sakazakii* (all known as spoilage factors of tofu) were reduced by 3–4 log cycles (i.e., 1,000–10,000-fold) by the presence of chitosans (No et al. 2002). This paper provides a model for testing putative antimicrobial substances to reduce or prevent spoilage of insect diets, to be studied on a case-by-case, diet-by-diet series of studies. It would be useful in improving insect diets to have a greater knowledge of inexpensive, nontoxic, but effective antimicrobial additives such as the chitosan derivatives.

1.2.3 Representative Case Study: Sensory Qualities and Storage

Another instructive model derived from the literature on food science and food technology is an approach typified by a study of nonenzymatic browning (discussed further in Chapter 5 as the Maillard reaction) and chemical changes in grape juice as a result of prolonged storage (Buglione and Lozano 2002). One of the most important issues throughout the history of food science and technology is that of maintenance of nutritional and sensory quality and safety during storage of foods, especially after prolonged storage. A parallel problem is the fact that insect diets must often be kept at elevated temperatures with prolonged exposure to degradation-inducing conditions making storage even more challenging in insect diet domains than it is in human foods. It is noteworthy, the insect diets with water activities of more than 0.95 are conventionally held at rearing room temperatures of 25°C–30°C for weeks, where human foods that are at comparable water activity would never be held outside of a refrigerator or freezer, unless there is a special container to prevent microbial growth or oxidative degradation. This disparity between handling human foods and insect diets places a great burden on rearing personnel and insect diet specialists to use special measures such as antimicrobial compounds and other preservation techniques to reduce losses or deterioration of insect diets.

Stored juice samples from three varieties of grape at temperatures including 10°C, 20°C, and 30°C for 20 weeks were sampled at weekly intervals measuring changes in the pigment color, amino acid and sugar concentrations, and accumulation of a palatability-degrading contaminant known as hydroxymethyl-furfural (Buglione and Lozano 2002). As would be expected, the degradation of all factors took place much more rapidly at the two higher temperatures than they did at 10°C, but the extent of degradation was not linear with the linear increase in temperature. This emphasizes the importance of temperature in storage systems (a point further discussed in Chapter 12 on food processing and Chapter 13 on microbial aspects of diets).

1.2.4 Representative Case Study: Twin Screw Extrusion

Another aspect that attracts considerable attention in the literature of the food science and technology community is the effect of food processing techniques on the nutritional quality, stability, and sensory qualities of various foods. One such study reports the effects of extrusion cooking and sodium bicarbonate on the carbohydrate composition of black bean flours (Berrios et al. 2002). The use of extrusion has grown in the food community, and the twin screw extruder has become a central tool for the processing of countless foods (covered more extensively in Chapter 12 on food processing techniques).

The extrusion process was shown to cause an increase in the concentration of total sugars, while the concentration of oligosaccharides was unaffected as were the various concentrations of sodium bicarbonate (Berrios et al. 2002). Previous studies indicated that high-temperature extrusion processing caused a marked decrease in gas-inducing sugars from pinto beans (Borejszo and Khan 1992).

Later studies, showing no such decrease, attributes the disparity to differences in the types of twin screw extruders used in each study (Berrios et al. 2002). The extruder in the Borejszo and Khan study had a higher rate of turning (300 rpm) compared to the rate of turning of screws in the Berrios study (200 rpm). The differences between the carbohydrates processed in legumes in these two studies are possibly the *tip of the iceberg* as far as nuances that result from different processing techniques are concerned. The processing program (including temperature profile, rate of turning, types of screw configuration, point of introduction of different components) plays a profound role in the outcome in terms of texture, nutritional content, sensory characteristics, and preservation qualities of foods processed by extruders. Although the many other characteristics of extruded foods compared with their preextruded counterparts are too numerous to review here, one other important feature of extrusion has been demonstrated and deserves mention: decrease in the concentration of the antinutrient compound, phytic acid, as well as denaturation of proteins that act as protease inhibitors, amylase inhibitors, and toxic lection (Singkhornart et al. 2013). This is an important improvement that exemplifies how special kinds of processing of diets and diet components can contribute to the improvement of insect diets in ways that had not been anticipated in previous diet processing conventions. The details on how twin screw extruders were adopted for pink bollworm diets and later for other species will be discussed in the section on extruders in Chapter 12.

Subtle differences in the processing can affect profound differences in the outcome of the final product in the application of extruder technology to insect diets. This point is so well demonstrated in the food science and technology literature that one more example is presented here in the following section.

1.2.5 Representative Case Study: Assessing *Cryptic Phytosterols*

The cholesterol content of most commercially available food used by humans is well known, and the importance of this subject is well accepted because of its relationship to public health. In contrast, although plant sterols are becoming increasingly well reputed for reducing blood serum cholesterol levels, the profiles of plant sterols of most foods are only sparsely known, especially those foods commonly used in insect diets. Toivo et al. (2001) report novel methods of analyzing plant sterols that are associated with various functional groups that could disguise them and that are present in a variety of foods of plant origin. Toivo et al. (2001) describe these poorly characterized sterols as *cryptic nutrients*. Figure 1.1 shows the free sterol form and three combined forms, steryl esters, steryl glycosides, and acylated steryl glycosides—forms that commonly occur in a wide variety of plant materials. They characterized sterols from soy flour, wheat flour, rapeseed oil, corn meal, sunflower kernel, and onion, showing that their method worked for phytosterols from a variety of matrices and plant sources. The authors showed that their method modification—using an initial acid hydrolysis prior to a saponification step—proved to be far more reliable for determining the glycoside-bound sterols. As a result of using this method, the underestimation of phytosterols (cryptic nutrients) in several foods such as cornmeal and dried onions could be averted. This method and approach can be used to develop artificial diets for insects and in understanding the composition of foods and the contribution of *cryptic nutrients*.

1.2.6 Representative Case Study: Fine Structure of Foods

Finally, there have been so many studies of the microscopic characteristics and matrix organization of foods that a specialty journal, *Food Structure*, was established. As of 1994, the journal is no longer published, but papers in this subject area are now published in *Food Science and Technology*. And in fact, several websites on microscopic characteristics of food are available online. Two papers typify the microscopic approaches to understanding foods, Heertje and Leunis (1997) and Heertje et al. (1996). The authors of these works used confocal microscopy and electron microscopy to

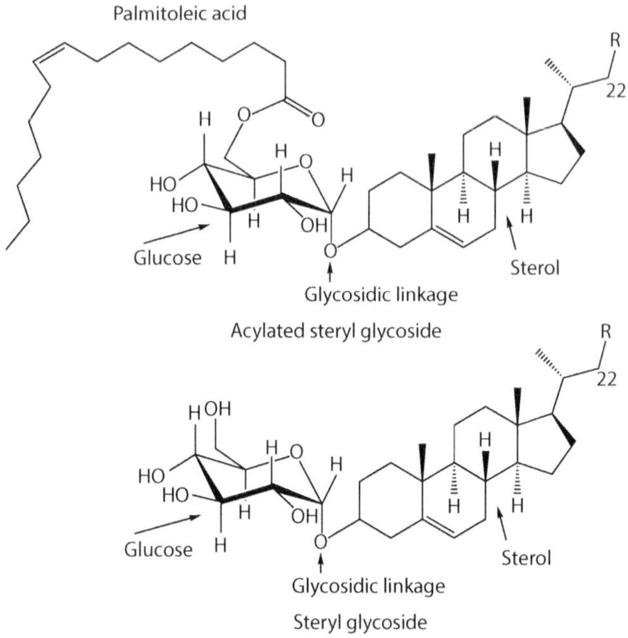

Figure 1.1 Various forms of modified or combined sterols, steryl glycosides, and acylated steryl glycosides.

examine the matrix (dispersion) interaction of oil and water as influenced by emulsifiers and the size and shapes of fat crystals in various foods. Various microscopic techniques are useful for the characterization of the organizational matrix of foods, to show the structural relationship of components such as lipids, proteins, and carbohydrates. Direct visualization of how such components are distributed, the size of subunits, and the stability of these complexes would be as useful for studies of insect diets as they have been for understanding foods (Chapter 4). Such approaches have not been applied to insect diets, but they could be useful in diagnosing why and how diets work or may fail to work.

1.2.7 Summary of Potential Application to Insect Diets

The papers that are summarized earlier were selected for two reasons: (1) They represent the food science and food technology literature as being typical samples of what are considered fit topics for exploration and (2) they present information and procedures that are directly applicable and useful to the insect diet science and technology community. The first reason for the selection of these papers is further discussed in the next section, which is an effort to offer a structure for the kind of studies and accompanying publications that would advance the insect diet science and technology community. The second reason illustrates how much the food science and technology literature has to offer to the insect-rearing community and why insect diet professionals will profit from careful attention to the literature and methodology of these fields.

1.3 SUBDISCIPLINES OF FOOD SCIENCE AND TECHNOLOGY

Food science and technology studies are conventionally divided into three main domains: (1) food chemistry and food physics, (2) food microbiology, and (3) food processing technology. A recurring theme in the literature on foods is that these domains and their subsets are interrelated and are best understood in terms of interdisciplinary approaches. As is the case for human and livestock foods, many aspects of insect diets are interrelated and are best understood through interdisciplinary studies. The chemical and physical character and interaction of diet components are related to their nutritional role, and these factors are intimately related to the processing by which the components were combined. The preservation of the intact diet and the inevitable changes that take place after the diet is completely synthesized include microbe–diet, insect–diet, and component–component interactions. The literature on insect diet development explains little of why certain diet components were selected. Without such statements of rationale, diet development emerges as an intuitive or gestalt process. A mechanistic (cause-and-effect), hypothesis-driven approach to diets will help in our understanding of how and why diets work or fail (Cohen 2001). Good models for such an approach are found throughout the literature of food sciences and food technology.

1.3.1 Food Chemistry and Physics as Models for Insect Diets

Good examples of this cause-and-effect approach are the food chemistry/food physics models of Fennema (1996) and Damodaran et al. (2008). As has been done in food science, insect diet chemistry and physics could be further divided logically into these subtopics: the nature of water and its role in diets, insect diets as dispersed systems, and treatment of the individual chemical classes, including carbohydrates, lipids, nitrogenous nutrients (amino acids, peptides, and proteins), enzymes, vitamins, minerals, and food additives. Although these topics are treated in several places throughout this book, they are explored in depth in Chapter 5.

A comment is in order about the expression, *insect diet chemistry and physics*. The physical aspects of foods and insect diets, including texture, viscosity, homogeneity, specific heat capacity, and a great array of other qualities, are related to the chemistry of foods and insect diets and their components in the most intricate and intimate ways. For example, water and carrageenan (a gelling agent derived from seaweed composed of sulfated polysaccharides) are commonly used diet ingredients. The overall (gross) water and carrageenan content of a diet are virtually identical in an unheated versus a heated mixture. However, after the diet has been heated above *activation temperature* and then cooled, what had been a free-flowing liquid becomes a gel. The physical property known as viscosity is directly related to the chemical interactions of the water and the carrageenan. The viscosity is a principal determinant of several rheological properties (effects of distortion energy on form and flow of matter), including gel strength, solute mobility, and shear strength, all of these being aspects of a diet's sensory qualities, stability, and numerous other functions. It would be valuable to understand the interrelationships between the physical–chemical properties of the gelling agent and water (such as the heat required to fully activate the gel former, the requirements for calcium to assure cross-linking, and the properties of water that lend themselves to gel characteristics). The complex details of gel chemistry and physics are treated in depth in Chapter 5. Also, the study of texture and its chemical and physical measurement, known as rheology, will be presented in a chapter (Chapter 15) that deals exclusively with this topic.

1.3.2 Food Microbiology and Microbial Relations in Insect Diets

Insect diet professionals share many of the same challenges faced by food microbiologists. A consideration of the table of contents of a food microbiology text such as Jay (2000) will substantiate this point. Both groups must perform a kind of balancing act of reducing or eliminating microbial contaminants without lowering the palatability or nutritional quality of foods. Both groups are concerned with diet or food safety, and they are also charged with solving their problems within the constraints of cost. Both groups are concerned with preservation or shelf life of their target foods. There are ever-present problems of dealing with the fine line between devising treatments that are too harsh and those that are gentle to the point of being ineffective. As noted in several places in this book (especially Chapter 13), insect diets contain many of the same contaminants as do human and livestock foods, and therefore, the knowledge base and techniques that have been developed in the food science community can be profitably and almost seamlessly applied to insect diets.

1.3.3 Food Processing Technology and Insect Diet Processing

Foods for people and their livestock are processed often on a large scale, and the processing is done with highly specialized equipment. The food processing equipment and the theory behind the various facets of processing are described and explained in numerous articles summarized by Fellows (2000). This includes the properties of foods such as density, specific gravity, viscosity, rheology, texture, material transfer, fluid flow, heat transfer, water activity, sensory characteristics, nutritional properties, quality assurance, and safety. Fellows survey the processing activities from the preliminaries of cleaning, sorting, peeling through intermediary processes of size reduction, mixing and forming, separation and concentration of food components, finishing with thermal processing (heat treating and cold preservation), and finally packaging.

Anyone who has worked with artificial diets for insects, especially in larger-scale production systems, recognizes the relevance of most of these topics. First, many of the foods and food components intended for use as human or domestic animal foods are also the materials of insect diets, including meals and flours of various seeds (soy, cottonseed, wheat, rice, and numerous others), oils, meat and dairy products, and more purified components such as sugar, proteins, starches, and finally multipurpose additives such as gelling and thickening agents. If a given process removes the fats from soy flour or changes the protein structure in that flour, making the proteins more (or less) nutritious, that information is certainly of importance to the diet professional who is using soy flour in his/her insect diet. Besides potential nutritional changes, a further complication can stem from changes in diet components that lead to changes in the other functional properties of insect diets: palatability, stability, and/or bioavailability.

Figure 1.2 shows the four major characteristics of successful diets. The first factor is that the diet must have sensory qualities that make it meet the insect's feeding biology requirements—taste, odor, and texture. The factors that stimulate feeding have been topics of extensive study and are excellently explained by Dethier (1962, 1976). The complexities of feeding responses are enormous, but the interplay between the internal state of the insect, the insect's sense organs, and the chemical/physical food matrix is the driving force. As might be expected, the adaptations that assure finding and accepting foods that are appropriate to the insects' needs are finely tuned because they are driven by survival and reproductive potential, which are products of extensive selection and evolution. The complexity of these nuances of feeding stimuli and responses makes the labor of developing an artificial diet that sustains palatability an understandably onerous task.

Once the diet passes the first hurdle, being eaten, it must then meet the requirement of being digested and absorbed.

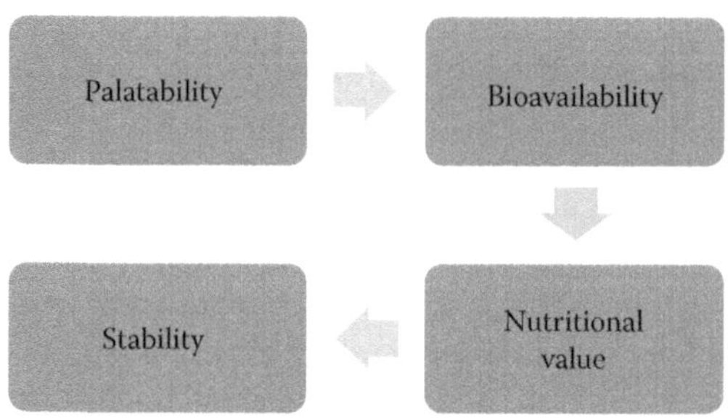

Figure 1.2 The four requirements for successful diets.

Besides the importance of having a comprehensive grasp of the nature of their diets' raw materials, a thorough knowledge of food processing equipment as it applies to preparation of materials specific to insect diets and to the production of the diets themselves is necessary. As chronicled in this book, many of the most significant breakthroughs in mass-rearing technology came as results of (or in connection with) adoption of technology and equipment borrowed from the food industry. The use of flash sterilizer equipment for mass rearing of various moth larvae (Spark and Harrell 1976, Tillman et al. 1997) and the integration of this equipment with industrial equipment for tray forming and form-fill sealing prompted a huge increase in the quantity and quality of insects that could be produced. The adoption of this technology also had an impressive economic impact (Tillman et al. 1997). Recently, it was shown that adopting the food industry technology of twin screw extrusion was a breakthrough in the mass rearing of pink bollworms (Edwards et al. 1996).

As will be further discussed in Chapter 2 on the history of artificial diets and in Chapter 4 on why certain diets are more successful than others, there was a peak of productivity and interest and high degree of acceptance of research on rearing techniques that were based on artificial diets during the late 1960s and the 1970s. During this period, several of the most important advancements in insect food science were made, including the application of large-scale food processing equipment such as flash sterilizers and form-fill seal machines to produce and package insect diets and the use of highly nutritious foods such as wheat germ, bean meals and flours, and vitamin supplements. Clearly such advancements represent a hybridization of the two fields, entomology (insect-rearing aspects of insect science) and food science. Some of the most noteworthy advancements in insect-rearing systems came as a result of the adoption of information and techniques from food science.

1.3.4 Dietetics versus Nutrition

There has long been confusion in the rearing community between the disciplines of insect nutrition and insect dietetics. In general, nutrition has been a science aimed at understanding the requirements and function of food components, while dietetics has traditionally been a more practical application aimed at developing diets with less attention to *how* they work than *that* they work. It would seem that these disciplines should complement and support one another, but after studying the dynamics of these fields over their 100-year-plus history, I have become convinced that the misunderstanding has hampered progress in both disciplines (Cohen 2001).

The two disciplines are at ends of continuum. Insect nutrition is a very basic science and insect dietetics is very applied. The standards for the two disciplines (though not explicitly defined) have been very different from one another, and sometimes these differences have been a basis for frustration for practitioners of each approach. For the pure nutritionists, the adoption of diets with whole foods such as wheat germ, soy flour, chicken eggs, or beef liver is not of much value because these foods are so chemically complex that it is not possible to pinpoint why they are effective at fulfilling an insect's nutritional needs. Therefore, the nutritionists (or other scientists who subscribe to pure nutrition standards) are prone to reject or discount studies that report on the efficacy of components that are undefined chemically. The fact that several diets have been formulated using such undefined components and showing very good results in terms of producing vigorous colonies of insects at low costs does not change the opinions of the nutritional purists. They view such studies as being of no help in understanding a target insect's nutritional needs.

To the pure nutritionist, the salient questions are, for example, "Does the insect require tryptophan in its diet and if so how much?" "What purpose does the tryptophan serve?" If the insect in question does not have a requirement for tryptophan, this question would be apropos to a nutritionist: "Does the insect have its own metabolic pathway to produce tryptophan, or does it have a symbiont population that is producing this amino acid?" The same types of questions would apply to each and every nutrient that a given insect uses. In the end, when pure nutritionists have completed their mission, they will know each and every dietary component that the target insect uses and what role each plays in the insect's life.

In contrast, the pure dietetics expert would not focus on what the components do, but instead, on diets that work to support excellent profiles of growth, development, and reproduction—all at a cost that makes a rearing program economically feasible. As discussed elsewhere in this book, especially in Chapters 4 and 10, the more purified the ingredients, the more they cost and the lower the overall nutritional value. Therefore, dietetics experts have gravitated to the whole food-type ingredients such as wheat germ and other such whole foods. Dietetics specialists avoid purified ingredients because of their expense and difficulty in handling and add them as supplements, only if they are absolutely needed.

Singh (1977) distinguished between insect nutrition and insect dietetics as a matter of degree of practicality. The pedagogy of insect nutrition has been built upon the use of diets that were as carefully defined (and pure) as possible because any impurities would cast doubt on the exact nature of each and every nutritional component. If, for example, a source of protein such as casein was added to an otherwise chemically defined diet, the phosphate, minerals, and other impurities might be nutritionally useful to the target insect, and the casein may be providing cryptically required or useful ingredients that were not accounted for in the fastidious formulation of the diet.

In an insightful summary of the state of nutrition and dietetics, Beck (1972) credited Gottfried Fraenkel with having had the vision to "shift the emphasis from the purely biochemical determination of minimum requirements for various amino acids, vitamins, etc., to a broader consideration of what we might call 'insect dietetics.'" Beck went on to say, "We have now reached a point where we are beginning to appreciate realistically that the effects of an insect's dietary substrate are not simply nutritional in the strict sense. We must also deal with the influence of factors affecting digestion, utilization, and conversion as well as factors affecting metabolism, form determination, reproduction, longevity, and general behavior." This eloquently stated position is as timely today as it was three decades ago, and its call to regard insect dietetics and nutrition in an integrated, holistic manner should be the defining direction of current and future studies.

1.4 DIET IN THE CONTEXT OF A REARING FACILITY

The artificial diets that are the subject of this book do not occur in vacuums. They are used in a context of rearing facilities that vary from program to program in their purpose and scope, and they are used with a variety of insect species with dynamic populations. Whatever the purpose and scale of a given rearing program, there are several satellite concerns that are related to the usage of the diet. These matters must be considered separately as contexts of the diets and the insects' interactions with the diet. These satellite components include the genetics of the populations that are being reared, the complex of environmental factors, the microbial interactions, the rearing facility's characteristics, and the personnel that run the rearing program. If any of these components goes awry, no amount of diet quality will rescue the insect colony from the likelihood of failure in the overall rearing program.

1.4.1 Genetics of the Colony and Domestication

Much of what is understood of the genetics of colonized insects comes from the general field of population genetics. Fortunately, there is a strong body of information about the basic principles of population genetics, but relatively little information has been derived about the genetics of insects in captivity and the dynamics of genes in insectaries. The application of the principles that underlie the potential changes that take place in insectary gene pools was explained by Bartlett (1984, 1994). The size of populations and their gene pools are inherently very small in laboratory-reared insects, and only a few hundred to a few thousand insects are brought into the laboratory to begin a colony.

This number is a small fraction of the total field population in a given location, and that field population is generally a small fraction of the total number of individuals in the target species. Once the insects are brought into the laboratory, Bartlett points out, they are further reduced in number by inadvertent selection of the subset of the captive population that are able to survive under the greatly simplified (compared to nature) rearing conditions. In nature, the target insects had choices of gradients of moisture, temperature, light, nutrient density, and a great number of other parameters that are made homogeneous in the laboratory setting.

The cultivation of insects under laboratory conditions inherently imposes conditions that cause the colony populations of insects to change profoundly across the entire captive population's genetic structure. Such changes can occur with violations of the premises of the *Hardy–Weinberg equilibrium*. The Hardy–Weinberg equilibrium dictates that the gene pool of a population will remain in equilibrium if these conditions are met: Populations are large; mating is random (panmictic); there is no significant influx or efflux of gene flow (immigration or emigration); there is no selection of any of the genes or traits in the population. The rearing situation inherently violates all of these equilibrium criteria. Laboratory populations are small; they are manipulated in such a way that there is selection for *laboratory-fit* rather than *field-fit* characteristics such as tolerance to very simplified environments that lack the gradient-rich circumstances of the field (including thermal, light, humidity gradients, choices of foods, and numerous other factors). Furthermore, relative to the size of laboratory populations, removal of insects to serve their assigned purposes and influx of field insects to enrich the genetic diversity or to bolster sagging populations are violations of the emigration/immigration rule of the Hardy–Weinberg equilibrium. Also, most colonies are structured like *mini-islands*, rendering it impossible to have completely random mating. The mini-island-isolation phenomenon stems from keeping insects in cages, which effectively reduces the population size to the number of insects in each cage. The consequence of this departure from Hardy–Weinberg equilibrium conditions is an accelerated, intense departure from the field populations from which the populations of laboratory-reared insects were derived.

Preventing the undesirable departure from field equilibrium conditions under laboratory conditions is a difficult matter (Bartlett 1984, 1994). The deviation from equilibrium conditions probably cannot ever be completely averted, but the maintenance of populations that are as large as possible and deliberate, well-planned efforts to reduce forces of selection can be helpful toward the maintenance of near equilibrium. For example, in most rearing settings, adult insects are caged in moderate to large numbers to allow reproduction. Whenever possible, the largest possible population should be brought together to allow panmictic mating. The cages in Figure 1.3a and b demonstrate two possible mating situations used in a *Lygus hesperus* rearing program. In the smaller cage (a), about 800–1,000 *L. hesperus* adults are present; in the larger cage (b), between 8,000 and 10,000 adults are present, presenting a 10× greater opportunity for panmictic, large-scale mating. The system of harvesting newly laid eggs from gel packets on top of the cage (Figure 1.3) allows the choice of the large cage system with no increase in the labor intensity of egg collection as compared to the labor involved in the smaller cage system. However, it must be noted that in such a rearing system based on using large populations collected in one large cage, rather than a series of smaller cages, the communication of disease throughout the larger group is much more likely than what would take place in smaller rearing units. Therefore, it is a trade-off between maintaining genetic diversity and preventing communication of pathogens. The decision as to which strategy should be followed can result from empirical tests on a case-by-case basis.

Within the framework of population genetics is a subject that is at once neglected in literature on insect colonies and yet of keystone importance: *domestication*. It surprises many laypeople and even some entomologists that silkworms (*Bombyx mori* L.) have been domesticated for about 5000 years and honeybees (*Apis mellifera* L.) for nearly the same amount of time. While honeybees can become feral and live apart from human intervention, silkworms are so deeply entrenched in the domestication process that they are unable to maintain continuous generations without human care (husbandry).

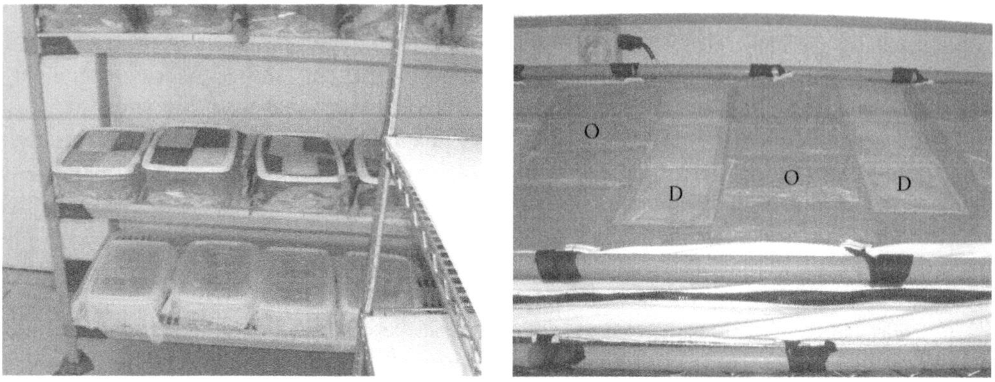

(a) (b)

Figure 1.3 Two possible mating situations used in a *L. hesperus* rearing program. In the smaller cage (a), about 800–1,000 *L. hesperus* adults are present; in the larger cage (b), between 8,000 and 10,000 adults are present, presenting a 10 times greater opportunity for panmictic, large-scale mating.

Over their domestication period, silkworms have been both deliberately and inadvertently selected for a variety of traits, including silk productivity, strength of silk, efficiency of food utilization, ease of handling, multivoltine reproduction (more than one generation per year), and many other clearly advantageous features for silk producers. However, over the five millennia that they have been cultivated, there are many other features or qualities for which *B. mori* have been selected, many of these selections being inadvertent or accidental. For example, it is evident from Figure 1.4 that about 20% of domesticated silkworm energy budgets are devoted to the production

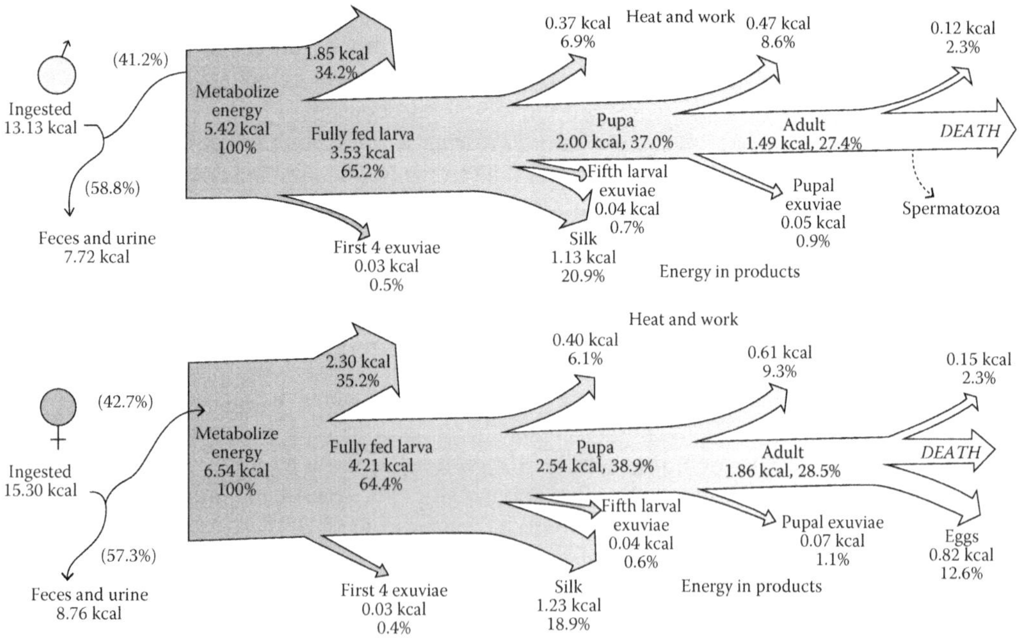

The energy economy of male and female *Bombyx mori* (based on data of Hiratsuka, 1920)

Figure 1.4 Adaptation from Waldbauer (1968) and Hiratsuka (1920) energy budget and utilization by silkworms.

of their silk cocoons. This is disproportionate to the allocation of materials and energy to pupation allocations by other insects that were not selected for silk production.

In another example, Rao et al. (2006) discussed the various characteristics that appeared in silkworms that produced two different shapes of cocoons: oval and peanut-shaped. Within these two broad groups, there were associated traits that included cocoon weight, proportion of larvae that produced cocoons, cocoon shell weight, silk filament length, and many other traits. Similarly, Ramesha et al. (2012) demonstrated the nutrigenic selection factors that had gone into increasing the nutritional efficiency of utilization of subspecies of mulberry that were matched to the genetically selected silkworms, all resulting in improvement of conversion of plant biomass to silk.

1.4.1.1 Domestication in Other Insects

Examples of domestication in rearing, especially with regard to diet-related factors, abound in the literature and in unpublished discussions of insects. However, it is often the case that the discussions are hidden as subtexts of other discussions, so it can be difficult and haphazard to find direct statements about how far domestication has progressed with a given insect. The best documented cases of genetic selection are found in drosophilid flies. In *Drosophila melanogaster*, there have been many well-documented instances of pleiotropy where a seemingly simple trait such as vestigial wings includes several other factors such as longevity, number of bristles on the scutellum, and number of ovarioles (Caspari 1952).

Some comments from R. T. Yamamoto (1969) on the laboratory strain of tobacco hornworm are very telling about the extent of the far-reaching effects of domestication in colonized tobacco hornworms:

> The culture was started from moths caught in light traps at Oxford, N. C., in the summer of 1965 and continuously inbred since then. Many aberrant morphological types were observed initially, and attempts to subculture them were unsuccessful. Larvae were selected for their ready establishment on the artificial diet, their growth performance on the diet, their size, and their lack of irritability on being handled. By selecting for these characteristics at each subsequent generation, we appear to have succeeded in developing a strain well-adapted to the laboratory. The larvae reared on the diet are bluish in contrast to the greenish hue of larvae reared on host plants. The greenish color can be partially restored by adding dried pulverized leaves of host plants, such as tobacco, to the diet. It has been reported that diet-reared moths respond to higher intensities of light than plant-reared moths, which result may indicate insufficient plant pigments in the diet (Sparks 1968). Reproduction is not affected (Yamamoto 1968), although subtle differences in host-finding behavior are manifested and are currently being studied.

The hornworm colony described by Yamamoto has several features that characterize the process of domestication. First, the colony is small, started from a small founder population, and in Yamamoto's own words *inbred*. There was deliberate or conscious selection for *tameness* (*lack of irritability on being handled*) and, especially pertinent to this discussion, ability of the insects to grow rapidly on the artificial diet. Yamamoto's comments about the blue color of the *Manduca sexta* and how the green color can be partially restored with addition of some ground natural host plant lead to a most interesting and well-documented relationship between diet and insect characteristics.

1.4.1.2 Environment: Physiological Ecology in the Rearing Facility

The field known as physiological ecology can provide valuable insights into the laboratory rearing situation. As the hybrid name implies, physiological ecology (or environmental physiology) is a formal discipline that deals with environmental factors and their implications in the physiology or functioning of target organisms. The major topics in this discipline are temperature relations, salt and water balance, gas exchange, and all other aspects of environmental/organism interplay. Rearing rooms are environments and that they have microclimates and contain microhabitats, just as outdoor ecosystems—but more simplified versions. Insects in nature are free to move through their environment where there are gradients of temperature, humidity, light, other electromagnetic energy, and chemical gradients (such as plant aromas, pheromones, allomones, kairomones). Insects in our rearing domains are captives that are confined in what can be inhospitable settings.

As a general rule (with some interesting exceptions described by Heinrich [1998]), insects are poikilothermic ectotherms, organisms whose body temperatures vary widely due to the heat exchange between their bodies and their environment. In nature, most insects reach their set point temperature by transferring heat from a radiant source (infrared or visible light), by direct contact with surfaces (conduction), or by transfer to surrounding fluid air or water (convection). This can include heat gain, heat loss, or both. The summation effect of the heat exchange is that the insect in nature witnesses a range of body temperatures that reflect the exchanges with the environment and sometimes behavioral activities such as basking in direct sunlight, ducking under a leaf surface, and other elective or voluntary measures. Such activity allows insects to attain body temperatures that are usually adaptive to the insect in question. For example, insects that undertake voluntary elevation of their body temperatures can raise their metabolic rates, speeding the processes of digestion, growth activities, and reproductive efforts, among other outcomes. A general rule is that for every 10° change in body temperature, all metabolic processes change proportionately by two- to threefold.

The thermal gradient in most insects' environments is so steep that in making spatial choices, an insect can select microenvironments that range by 20°C or greater (Edney 1977). This means that by some simple behavioral choices such as emerging from a burrow in the soil to basking on top of a plant or rock, an insect can vary its metabolic rate by as much as sixfold! In terms of elevation of digestive rates alone, such differences in metabolic processes can have profound impact on the well-being of an insect. Those who work with diurnal species should be especially aware of the deprivation to which they are subjecting their insects when they culture them in laboratory situations where there are no gradients or opportunities to undertake voluntary elevation of body temperature.

The next aspect of physiological ecology is the relationship of insects' size to their susceptibility to potential harmful heat and water exchange with their environment. There are a couple of physical rules of thumb that govern heat and water exchange: (1) The greater the surface area of an organism, the more susceptible that organism is to heat and water exchange with its environment, and (2) the greater the mass of an organism, the lower is the surface area to mass ratio. As a consequence of these two precepts, smaller insects have higher surface to mass ratios than do larger insects. For example, Table 1.1 shows an estimation of the relative surface to mass ratio of insects of two different weights, based on surface to mass calculations from the relationship provided by Edney (1977) that surface area in cm^2/mg of body weight is equal to 0.117 $mass^{0.67}$. Table 1.1 shows a comparison of the body mass of neonate hornworms (*M. sexta* L.) and fully grown larvae in terms of surface area and surface to mass ratio of the smaller and larger larvae.

The smallest insect has the highest surface to mass ratio and is the most susceptible to water loss or heat gain (or loss). This is a basic physical explanation for the high degree of vulnerability of newly enclosed larvae and nymphs. They start off with little water inertia or thermal inertia, and they can lose water through their relatively large surface area. The larger insects not only have a lower surface to mass ratio, but they also have an absolutely higher reserve of water and a great deal more protection from heat gains or losses that could be life threatening.

To compound this problem that the neonates and early developmental stages face with regard to their small size and high surface to mass ratios, there is another danger that comes from our rearing practices: creation of a *winter desert* or an *insectary desert*. This is especially a problem in the winter when outside temperatures are low, in many regions at or below freezing; and we draw in air from these cold conditions and warm that air to typical rearing room temperatures (very commonly 27°C or about 80°F). This greatly increases the drying power of the air, changing its humidity from near saturation for the outside air (80%–100% relative humidity) to less than 40%–50% in the warmed rearing domain. The reader is reminded of this phenomenon that should be familiar to anyone who has suffered from sinus trouble and dry skin during the winter—a direct result of

Table 1.1 Surface to Mass Relationships of Typical Insects Such as Lepidopteran Larvae That Are Neonates (10 mg) and Large, Near-Pupal-Stage Larvae (10,000 mg)

Weight of insect	10 mg	10,000 mg
Surface area (SA) in $cm^2 = 0.117$ (wt. in mg)$^{0.67}$	1.1 cm^2	115.6 cm^2
Number of cm^2 per mg insect = SA/mg body wt.	0.12	0.012

having created an *indoor winter desert* in our homes or workplaces. It is a simple result of taking air that has a given amount of water (absolute humidity) and greatly increasing that air mass's ability to hold water by elevating the temperature but not increasing the amount of water that is present. This increase in the drying power of the air can become a huge problem to a neonate insect, especially if it is having a hard time finding its diet, which for many insect-rearing domains is the only source of water that this insect is provided. From Table 1.2 it can be demonstrated that a cubic meter of saturated air at 0°C holds 4.85 g of water, and if that volume of air is brought into a rearing room held at 25°C at which temperature it requires 22.97 g of air for saturation, the relative humidity will be $4.85/22.97 \times 100 = 21.1\%$. Combining these issues—higher surface to mass ratios of smaller insects compared to larger ones (Table 1.1) and the increase in the air's drying power when outside air is warmed in the laboratory—there is a greatly enhanced desiccation threat that especially affects smaller insects (neonates). This accounts, at least in part, for the higher mortality of small larvae or nymphs in the rearing setting. Compounding this problem is the fact that the dry air described here also desiccates diet surfaces, adding further stress to developing insects. Readers are encouraged to use Table 1.2 to estimate consequences of any temperature changes that they are considering.

Table 1.2 Water Content in g/m³ of Air at Temperatures Ranging from 0°C to 35°C at Relative Humidities Ranging from 10% to 100% Saturation

	10%	20%	30%	40%	50%	60%	70%	80%	90%	100%
0°C	0.48	0.97	1.45	1.94	2.42	2.91	3.39	3.88	4.36	4.85
5°C	0.68	1.36	2.04	2.72	3.4	4.07	4.75	5.43	6.11	6.79
10°C	0.94	1.88	2.81	3.75	4.69	5.63	6.57	7.51	8.44	9.38
15°C	1.28	2.56	3.84	5.12	6.4	7.68	8.96	10.24	11.52	12.8
20°C	1.72	3.45	5.17	6.9	8.62	10.35	12.07	13.79	15.52	17.24
25°C	2.3	4.59	6.89	9.19	11.48	13.78	16.08	18.37	20.67	22.97
30°C	3.03	6.05	9.08	12.11	15.13	18.16	21.18	24.21	27.24	30.26
35°C	3.95	7.89	11.84	15.79	19.74	23.68	27.63	31.58	35.52	39.47

Source: Haynes, W.M., *CRC Handbook of Chemistry and Physics*, 91st edn., Taylor & Francis, Boca Raton, FL, 2010.

It is also relevant that the metabolic rate is affected by temperature in the Q_{10} relationship described so that for every 10°C change in temperature, the overall use of energy, oxygen, growth rate, development rate, and other metabolic parameters is either accelerated by two- to threefold or decreased by this amount, respectively, with increases or decreases in temperature.

1.4.1.3 Forcing Insects through the Bottleneck Stresses

What has been touched upon briefly here and explained in more detail in Chapter 8 is that in their natural environments, insects have a great deal of choice (environmental heterogeneity) and that such choices as temperature, humidity, and light gradients, variation in nutrients, and opportunities to avoid antinutrients all lend themselves to an insect's seeking and often finding a zone of optimal conditions. Although some people may consider nature as cruel, insects in their natural setting are well adapted to deal with nature's *harshness* including predators, diseases, temperature extremes, damaging radiation, wind, rain, water currents, and water deficits, and any number of other potentially threatening conditions are all challenges to which insects are prepared through evolutionary processes to face successfully. Inherently in the rearing process, insects are taken out of their natural settings and have imposed upon them stresses that they never face in nature.

Rearing is driven by convenience and economic needs and is hindered by a lack of understanding of most of the insects' natural needs. Although rearing experts try to present their insects with optimal conditions, inadvertently, they usually subject the insects to what indeed are hostile conditions by imposing stresses in terms of temperature, water balance, nutrition, crowding, and other stresses. Other stresses include antimicrobial agents, which are not only directly toxic but also harmful indirectly by killing the insect's natural flora. Diets that are structurally and nutritionally overly simplified and that do not offer feeding choices can be stressors. Forcing insects to use alien sites for oviposition can stress them. When I speak to groups on the subject of rearing, I try to make this point: "Insect rearing is not rocket science; rocket science pales by far in complexity next to insect rearing." The points about *bottlenecks* and *winnowing* (Bartlett 1984) are further explained and documented with case studies in the section, to follow, on domestication (Section 1.4.1).

1.5 INSECT DIETS AND REARING: WHY SCIENCE AND TECHNOLOGY RATHER THAN ART?

One of the issues that insect diet and rearing specialists have argued is whether diet and other rearing issues are science or art. A third term, technology, also has a place in this discussion. Cohen (2001) argued that if rearing specialists were to increase their own status in entomology and to enhance the contributions that they make to insect science, they must practice diet and other rearing studies as a science and a technology. To clarify what is meant here, the word *art* connotes skill in a practice that does not require understanding of cause-and-effect relationships, whereas the foundation of science is understanding and advancement of cause and effect. In art, mechanistic grasp of the world around us is not nearly as important as intuition or feelings (emotions). In art, predictions about outcomes of natural phenomena are not as important as conveying an inner sense of truth. The opposite can be said for science where the better the understanding of mechanisms and cause and effect, the more predictable and reliable are the outcomes of our understanding. Technology may have elements of both art and science, but it leans more toward science in its ability to use understanding to determine and control outcomes. Technology relies heavily upon predictable and practical outcomes. With art, practicality is of little or no consequence.

These descriptions convey what insect diet and rearing studies have been when they were functioning at their most useful and at their best in terms of solving problems in how to provide all the requirements for captive or colonized insects for the many purposes described earlier in this chapter. In several places in this book, questions about mechanisms or the underlying principles were raised continually to explain and to circumscribe the practices in diets and other rearing issues. Even when diets and rearing techniques were developed with a heuristic approach (discussed further in Chapter 10), it has been found frequently that understanding the scientific basis of rearing practices helped in both problem solving and improvements in these rearing systems. An excellent example of the importance and value of developing a scientific understanding of the diet and rearing systems that we use is to be found in the development of artificial diets for tarnished plant bugs and western tarnished plant bugs. The technical aspects of this work are discussed in Chapter 4 on what makes diets succeed or fail. But the story behind this advancement is not to be found in the literature.

When I arrived at the Biological Control of Insect Laboratory (USDA, ARS) in Tucson, Arizona in 1979, Jack Debolt had already succeeded in developing a diet for a *Lygus* species—a task that had not been accomplished in at least 20 years of research by an impressive range of researchers, no less than the great Erma Vanderzant. By the time that I got to Tucson (as a postdoctoral research associate with the University of Arizona and fresh from my doctoral work with analytical biochemistry and physiology at the University of California at Riverside), Debolt and Ray Patana had devised a feeding system for *L. hesperus* where Debolt's diet slurry was loaded into Parafilm packets and placed on top of the cages. Debolt and Patana had also developed an oviposition system where packets were filled with a carrageenan gel (Gelcarin® 812 from FMC), allowing *L. hesperus* to punch their eggs into the gel packets where the eggs could complete their development without recourse to plant materials. Prior to these inventions, *Lygus* spp. had to be fed plant materials such as green beans and broccoli and given green beans for oviposition, which in plant bugs (Miridae) were obligatorily within plant tissues. The system was a fantastic success and was used for producing parasitoids to manage *Lygus* populations, and the system was used worldwide for producing *L. hesperus* and its parasitoids for research and biological control. The program led by Harry Graham in Tucson became a model system for mass-rearing host–pest insects for the production of parasitoids, and other scientists at the Tucson lab, including Glenn Jackson and Walker Jones, developed many outgrowths of the system, allowing field-testing and industrial accomplishments that could not have been developed without the Debolt diet and the Patana/Debolt oviposition system.

That's the success story, but behind the story, there is another story—a very complex one about art and science. In testing many iterations of early versions of the incipient diet for *L. hesperus*, Jack Debolt had devised a bioassay system that was nothing short of remarkable for medium-scale testing of different diets. The diets were tested in shell vials or condiment cups, and Debolt had devised Plexiglas® *hotels* for the cups, allowing multiple individual bioassays to be conducted. These were the days before 96-well ELISA plates were in fashion, and Debolt had to tap into his considerable skills as a hobbyist who used Plexiglas for building accoutrements for his model trains and other related hobbies. He used these skills to construct an efficient bioassay system that would, importantly, allow him to test the diets individually, as these mirids are quite cannibalistic, and many bioassays in his works and those of other researchers were flawed by the confusion about whether the test diet was adequately nutritious or whether the extra nutrition acquired by the insect's cohorts was the factor that tipped a feeding test into success (or failure in case insects that died during the experiment were killed by another lygus bug or from malnutrition). Building bioassay *hotels* required a skill that must be called *art*. Designing the assays to meet the feeding characteristics and other biological aspects of the target insects was science. Jack Debolt had explained to us that he came up with the idea of using chicken eggs because he understood (from his research, observations, and controlled feeding experiments) that *L. hesperus* is what we call zoophytophagous and therefore required (or profited by) a supplementation of animal (insect) biomass. But he also said that he got the idea of using chicken eggs from his family culture where his mother put eggs in almost any food possible because she felt that eggs provided extra flavor and nutrition that made her cooking special. So using eggs was cultural and definitely art, except for the part that came from careful understanding of the insect in question: science!

When Jack Debolt approached his assignment of developing a diet that would support mass rearing of *L. hesperus*, he followed the tenets or wisdom of the entomological community, that as an insect with piercing and sucking mouthparts, *L. hesperus* must be finding liquids and sucking them through its mouthparts (see Figures 7.3 and 7.4). This being the assumption Debolt (and the rest of us) believed that if we could isolate the (hypothetical) liquid food and analyze it, we could duplicate this liquid with an artificial diet whose composition was modeled after the natural foods, which these insects were consuming. Even after the Debolt (slurry) diet became established as an excellent food for *L. hesperus*, it was still believed that having a more simple liquid would allow greater efficiency and productivity in *Lygus* rearing efforts. So we set out, spending much of the decade of the 1980s, trying to extract the liquid from green beans and from the diet. We would macerate green beans, press them in tissue grinders, centrifuge and filter the fluids, and analyze them for proximate nutrient profiles (protein content, amino acids, minerals, lipids, and carbohydrates). We did the same with the diet, and it must be added that we sometimes wore out suction pumps that strained to put the liquid portion of green beans (or other vegetables such as broccoli) and the slurry diet through various filter systems, but the small amount of liquid that we were able to extract failed to support even a single instar transition for *Lygus* bugs subjected to testing in the Debolt bioassay system.

This is where the scientific understanding of the lygus feeding system failed us until I revisited *L. hesperus* feeding biology in the mid-1990s and discovered through careful observations with a video microscopy system and through enzymatic analysis that these mirids, like the hemipteran predators that I had worked with throughout the 1980s, used extraoral digestion. We found that the centerpiece of lygus (mirid) feeding was the injection of salivary digestive enzymes that attacked structural proteins and stored carbohydrates and that the liquefied plant materials (and insect internal structures) were ingested as a slurry formed from the previously intact host tissues. This explained why the Debolt slurry diet worked and the liquid diets that we tried to make from fresh plant materials or Debolt diet did not work to support growth and development. The discovery of this paradigm had far-reaching consequences in improvements to diets (resulting in the development of the NI diet), which allow larger-scale mass rearing of *L. hesperus* and supported for the first time growth, complete development, and reproduction of *L. lineolaris*. The new scientific understanding of *Lygus* feeding prompted dozens of studies on the enzyme biochemistry and feeding physiology and influenced the concepts behind lygus control through systemic pesticides and especially through strategies for genetic modification of host plants.

The take-home messages from this anecdote are that intuition and heuristic approaches in a context of study of published studies on rearing *Lygus* spp. and related Hemiptera were useful. Art had some place in the process of diet and other rearing system development, but all of the accomplishments would have happened faster and probably more effectively if we had the scientific understanding of the principles of the Hemipteran feeding mechanism and related biology, including reproduction.

In the 30 or so years of work on development and improvements in the *Lygus* spp. rearing systems, there were millions of dollars of research efforts (probably tens of millions when all research operations are taken into consideration), and many (most) of these efforts led to blind alleys, which could have been averted with a better basic understanding of our research subjects. The art, such as techniques in mixing, pipetting, weighing, and other processing techniques, will always be a part of scientific studies, but it must always be remembered that art without science as the central direction will always be haphazard.

SELECTED BOOKS AND JOURNALS ON FOOD SCIENCE AND FOOD TECHNOLOGY

Books on Food Processing

Brennan, J. G., J. R. Butters, N. D. Cowell, and A. E. V. Lilley. 1990. *Food Engineering Operations*, 3rd edn. Elsevier Applied Science, London, U.K.
deMan, J. M., P. W. Voisey, V. F. Rasper, and D. W. Stanley. 1976. *Rheology and Texture in Food Quality*. AVI, Westport, CT.
Fellows, P. J. 2000. *Food Processing Technology*, 2nd edn. CRC Press, Boca Raton, FL.
Heldman, D. R. and R. W. Hartel. 1997. *Principles of Food Processing*. Aspen Publishers, Inc., Gaithersburg, MD.
Kent, N. L. 1983. *Technology of Cereals*, 3rd edn. Pergamon Press, Oxford, U.K.
Laurie, R. A. 1985. *Meat Science*, 4th edn. Pergamon Press, Oxford, U.K.
Lissant, K. J. 1984. *Emulsions and Emulsion Technology. Part III*. Marcel Dekker, New York.
Turner, A. 1988. *Food Technology International Europe*. Sterling Publications International, London, U.K.

Books on Food Chemistry

Damodaran, S., K. L. Parkin, and O. R. Fennema (Eds.). 2008. *Fennema's Food Chemistry*, 4th edn. CRC Press, Boca Raton, FL.
deMan, J. M. 1999. *Principles of Food Chemistry*, 3rd edn. Aspen Publishers, Inc., Gaithersburg, MD.
Fennema, O. R. 1996. *Food Chemistry*, 3rd edn. Marcel Dekker, Inc., New York.
Nielsen, S. 1998. *Food Analysis*, 2nd edn. Aspen Publishers, Inc., Gaithersburg, MD.

Books on Food Microbiology

Jay, J. M. 2000. *Modern Food Microbiology*, 6th edn. Aspen Publishers, Inc., Gaithersburg, MD.
Jay, J. M., M. J. Loessner, and D. A. Golden. 2005. *Modern Food Microbiology*, 7th edn. Springer, New York.
Marriott, N. G. 1997. *Essentials of Food Sanitation*. Aspen Publishers, Inc., Gaithersburg, MD.
Marriott, N. G. 1999. *Principles of Food Sanitation*. Aspen Publishers, Inc., Gaithersburg, MD.

Journals

Agricultural and Food Science in Finland
American Journal of Clinical Nutrition
Analyst
Animal Feed Science and Technology
Applied Environmental Microbiology
Aquaculture Research
Archivos Latino Americanos de Nutricion
Biochemistry and Biophysics Research Communication
Biological Trace Element Research
Biotechnology and Bioengineering
British Journal of Nutrition
British Poultry Science
Crop Science
Deutsche Lebensmittel-Rundschau
Ecology of Food and Nutrition
European Food Research and Technology
Fisheries Science
Food Biotechnology

Food Chemistry
Food Reviews International
Food Science and Technology International
Food Technology and Biotechnology
Grasas y Aceites
Indian Journal of Biochemistry and Physiology
Industrial Crops and Products
International Journal of Food Microbiology
International Journal of Food Science and Technology
International Journal of Food Sciences and Nutrition
Journal of Agricultural and Food Chemistry
Journal of Agricultural Science
Journal of Animal Physiology and Animal Nutrition
Journal of Animal Science
Journal of AOAC International
Journal of Applied Poultry Research
Journal of Biological Chemistry
Journal of Biotechnology
Journal of Cereal Science
Journal of Chemical Ecology
Journal of Food Composition Analysis
Journal of Food Science
Journal of Food Science and Technology-Mysore
Journal of Muscle Foods
Journal of Nutrition
Journal of Nutritional Biochemistry
Journal of Plant Nutrition
Journal of the American College of Nutrition
Journal of the Japanese Society of Food Science and Technology
Journal of the Science of Food and Agriculture
Journal of Trace Elements in Medicine and Biology
Lipids
Nahrung-Food
Nutrition Research Reviews
Nutritional Research
Plant Foods for Human Nutrition
Poultry Science
Proceedings of the Nutrition Society
Seed Science Research

CHAPTER 2

Diet Terminology and History of Insect Diet Science

2.1 INTRODUCTION TO DIET TERMINOLOGY

Singh (1977) pointed out that diet terminology has been used in very imprecise and ambiguous ways such as authors calling diets "containing starch, casein or wheat germ described as 'chemically defined.'" Singh goes on to say that "To some authors, a 'synthetic' diet is a mixture of nutritive substances, with perhaps a plant preparation with yeast, or vitamins or sugar added; to others it is a mixture of pure chemicals only." Dougherty (1959) provided a concise and logical set of definitions that have been used by many authors to give consistent meanings to diet formulation terminology. Dougherty described holidic diets as ones whose components are completely known and oligidic diets as ones whose components are not fully or even nearly well characterized. Meridic diets fall in between, with some components well characterized (or defined) and others poorly defined. Meridic diets can be considered intermediary between holidic diets and oligidic diets (Dougherty 1959). However, these terms retain ambiguity when it comes to nuances of purification. Soy protein is closer to being defined than soy meal; the extraction process involved in producing soy protein has removed many lipids, complex carbohydrates, sugars, and other components of soy meal (or soy flour), but the proteins are mixed, rather than a single protein species, and there are some lipids, sugars, and minerals that are present in soy protein preparations.

As neophyte insect diet professionals work their way through the literature on diets, they will be struck with all of the terminology used in this field. One reads about natural and artificial diets, defined, nondefined, chemically defined, holidic, meridic, oligidic, and aseptic diets. Furthermore, expressions abound describing target insects as monophagous, polyphagous, oligophagous, trophic generalists, and trophic specialists, as well as entomophages, carnivores, zoophages, detritivores, saprophages, herbivores, phytophages, xylophages, gramnivores, and many other terms that describe insects' feeding habits.

The terms *diet* and *medium* require special clarification. Diet is the most generic term indicating whatever the insect eats and medium (plural: media) generally indicates a diet that has been made artificially or synthetically. A monophage is an organism that eats a single kind of food (i.e., one species of host plant or one species of host or prey); an oligophage eats a few species; and a polyphage eats many species. Often, ecologists use the expression specialist for the monophage and generalist for the polyphage. An entomophage is an organism that eats insects either feeding as a predator (i.e., it kills the prey and then eats it, feeding on multiple prey in its lifetime) or as a parasitoid (i.e., living in or on an insect—a host—while the host is alive). The term *carnivore* connotes that the organism is eating other animal material (which could be a vertebrate or an invertebrate), though the term is sometimes used in a sense that is restricted to consumption of vertebrates. The potential ambiguity of the term *carnivore* and its possible implications of strict vertebrate application make the term *zoophage* more precise, meaning any organism that eats another animal. By contrast phytophage or herbivore connotes any animal that eats plants. Xylophages eat wood (termites and wood roaches being prime examples); gramnivores eat grain, and saprophages or detritivores eat dead and decaying materials. Animals with mixed feeding habits are known as omnivores, but the more specialized terms, zoophytophage and phytozoophage, have emerged recently to describe insects that eat both other animals and plants.

One of the most imprecise of diet-related terms is *natural* diet. The inherent imprecision in this term stems from the fact that insect feeding behavior is often cryptic and difficult to observe and also the fact that many insects are polyphagous (cosmopolitan) in their feeding habits. Even those insects that are thought to be monophagous (i.e., feed on one food source) or oligophagous (i.e., feed on few foods) may feed on a variety of different tissues of the same host. For example, whiteflies (such as members of the species *Bemisia tabaci*) can be notoriously cosmopolitan in their choices of host plants (having been reported to utilize more than 400 species of plants) and are said to be phloem sap feeders, but there are indications that they may feed facultatively on plant tissues other than phloem, including xylem and mesophyll (Cohen et al. 1998). Therefore, when we discuss the *natural food* of *B. tabaci*, what do we really mean? Are we talking about optimal host species? Is the phloem sap the natural diet, or must we include some xylem sap and even some mesophyll fluids, or organelles from mesophyll cells? If we are trying to model an artificial diet after the natural food, how do we apportion the diet? The same difficulties are found in tissue feeders such as *Lygus* bugs and even to specialists such as boll weevils. Despite the fact that feeding by *Lygus* bugs has been a topic of intense study, it remains obscure exactly which tissues or cells are targeted by *Lygus* bugs' pinpoint feeding mechanism (Cohen 2000b, Wheeler 2001). Likewise, the specific tissues or cells, which are selected by boll weevils or by pink bollworms, are clearly not simply indiscriminant consumption of entire bolls.

The antonym for *natural diet* is *artificial diet*. Another term used infrequently is *synthetic diet*, often used as an approximate synonym for *artificial diet*. Yet another term that suggests that a food is not part of the insect's habitual, *natural* feeding regimen is the phrase *factitious diet* or a *factitious host*. Understanding the nuances between the terms, natural, artificial, and factitious diet will be useful to those entering the literature of insect diets. As an example, when lacewing larvae (such as *Chrysoperla rufilabris* Burmeister Neuroptera: Chrysopidae) feed on a variety of soft-bodied insects and insect eggs (e.g., aphids, the eggs of noctuid moths, and various scale insects), they are consuming *natural* food. When we bring them into the laboratory and feed them, as it has traditionally been fed, the eggs of the Mediterranean flour moth *Ephestia kuehniella* Zeller (Lepidoptera: Phycitidae), we are using a *factitious host*. *Ephestia* eggs, though they are real insects, are considered factitious because the lacewings, in their natural environment, would not encounter flour moth eggs. We are using an *artificial diet* when we provide a diet, for example, of meat paste, cooked chicken eggs, yeast, sugar, water, and antimicrobial compounds (Cohen and Smith 1998, Cohen 1998). It should be noted that the latter diet is not a fully defined or *holidic diet*. The artificial diet described here is known as an *oligidic diet*, meaning that few, if any, ingredients are chemically defined and chemically pure. If some or several of the other components were chemically pure or defined, we would describe the intended food as a *meridic diet*.

Chemically defined ingredients are components that have been highly purified and subjected to tests of purity. Because it is impossible to produce material that is 100% pure, the standards and limits of purity of a chemical are generally stated by reputable suppliers. For some substances that may be used as diet ingredients, the purity of chemicals can be in excess of 99%. Obviously, there are some glaring ambiguities in these terms. If 9 out of 10 ingredients are not defined and 1 ingredient is, it would not be very meaningful to call the diet meridic.

Two other terms commonly encountered in diet and nutrition literature are *essential nutrient* and *nonessential nutrient*. A nutrient is essential if the target organism must use the substance in its metabolism, but it lacks the ability to synthesize the substance on its own. This means that the substance must be acquired from the organism's diet. Another way to express this is to call the substance in question a *dietary essential*. In contrast, a substance that has nutritional value but can be produced through metabolism of other substances is a nonessential nutrient. This does not mean that nonessentials are not important or valuable to the target organism. These concepts are further discussed in Chapter 3 on the nutrient functions, Chapter 5 on chemistry of nutrients, and Chapter 7 on the feeding biology and metabolism. Another related concept is that of a nutrient's ability to *spare* another substance, that is, to replace it. For example, if an insect could use the sulfur-containing amino acid methionine to replace another sulfur-containing amino acid cysteine, we would say that methionine can spare cysteine. Appendix A contains a glossary of important terms used in diet literature and throughout this book.

2.2 HISTORICAL ASPECTS OF INSECT DIET SCIENCE AND TECHNOLOGY

There are several places in this book where historical aspects of diet and rearing topics are discussed. This raises a question about why so much emphasis is placed on history in a book that is intended to help explain how and why insect diets work. Of course, there is the effort to give credit to authorities who have brought us to where we are in diet and rearing technology. Such credit helps to *empower* people who made useful contributions to help enable them to make further contributions. The other major reason for providing historical perspective is that engendered in the history is a *rationale* and an explanation of *mechanisms* and cause and effect relationship (the science of diet and rearing technology). There are so many examples that document this assertion, but a simple recourse to the use of wheat germ in diets makes the case. The first use of wheat germ, reported by Vanderzant et al. (1959), was not in itself a great success, but it led to the study reported by Adkisson et al. (1960a) where pink bollworms could be reared in continuous generations, indefinitely. In the diet described by Adkisson et al. (1960a), there was a major breakthrough in pink bollworm rearing for the first time after more than a decade of failures in terms of finding a diet that could completely replace cotton bolls as a rearing material. The logic and rationale of using wheat germ stemmed from the observations by Vanderzant et al. (1959) that another boll feeder, boll weevils, found wheat germ stimulating for oviposition, and the combination of works on boll weevils and bollworms led by Vanderzant proved that no cotton plant material whatsoever was needed to support complete growth, development, and continuous reproduction in bollworms. If the success of wheat germ use was limited to bollworms and boll weevils, the studies cited here would have been complete successes and of scientific and historical importance. But these studies proved to be the foundation of countless other successes with hundreds of species of insects in thousands of studies and rearing programs. The impact of wheat germ in insect rearing, clearly, is off the charts in its importance!

Human interactions with insects go back to the earliest phases of human evolution, where *Homo sapiens* had inadvertent associations with houseflies, cockroaches, biting flies, lice, and many other pests or disease-carrying associates, not the least of which is the bubonic plague/rat/flea/human interactions that wiped out large portions of the human population. Also, from the onset of agriculture, there were agricultural pests that negatively impacted humans' efforts to cultivate crops and livestock. There were also historical uses of insects as food, still practiced by many cultures today (see Chapter 16). Many of these associations can be considered a form of domestication, albeit inadvertent such as in the case of houseflies and household cockroaches. However, what can be considered classical examples of domestication—where humans deliberately cultivated a useful species—started with the cultivation of silkworms and honeybees, somewhere around 5000 years ago (Barber 1992, Dalley 2002). However, though these acts of deliberate domestication involved rearing efforts or husbandry, it was not until the turn of the twentieth century that efforts were begun to provide insects that we wished to cultivate with *artificial foods* (or artificial diets).

Although the pioneers in this discipline probably did not know to what extent that they were working in a new domain (insect dietetics), they were indeed blazing the trail to what is emerging as a separate discipline. The early uses of insects such as for silk or honey production must have suggested to those working in these areas that it would be convenient to have more control over the useful insects that produced these products than could be gained with reliance upon *natural* foods. Indeed, throughout the history of insect manipulations, for both research and practical applications, it has been a goal to have more convenient food sources than the foods available from nature in unprocessed forms. Early on, there was also a desire to determine the nutritional requirements of various organisms in what we now call nutritional science.

Several advancements that are highly instrumental in the advancement of insect diet science and technology are from other fields. These include the work with microbes and their role in food spoilage and contamination, including those of Kircher (demonstrating that milk contained bacteria, later to be shown by Pasteur to be causative agents of spoilage) and yeasts first visualized by von Leeuwenhoek in 1680. The various advancements in food preservation were highly significant in insect diet technology, including the advent of canning by Appert in 1810, food freezing in 1842, steam sterilization in 1843, autoclaving in 1853, and the works of Pasteur, beginning in 1854, which set the stage for a gentle, nondestructive heat treatment of many foods. The use of lowered water activity as a preservative in drying milk was basic to the means of preservation of many insect diet materials, and the use of chemical preservatives, beginning with sodium benzoate in human foods, was a major breakthrough in food preservation and processing technology.

2.2.1 Early Work with Artificial Diets

In 1908, Bogdanov was the first to rear an insect entirely on artificial diets, the subjects being blowflies (*Calliphora vomitoria*) fed a medium of peptone, meat extract, starch, and minerals (Singh 1977). Other pioneering efforts at rearing (reviewed by Singh 1977) were those of Loeb (1915) who reared *Drosophila* sp. for five generations on a simple medium (sugars, ammonium tartrate, dipotassium hydrogen phosphate, magnesium sulfate, and water), Guyenot (Guyenot 1913a–c, 1914a,b) who also reared *Drosophila*, and Zabinski (1929) who reared cockroaches (*Periplaneta orientalis* and *Blattella germanica*) on ovalbumin, starch, saccharose, and agar. In the decade following these works, Michelbacher et al. (1932) broke ground with a diet that contained agar, casein, and several other nutrients such as meat products. Later in the 1930s, Fraenkel and his associates published several diets based on casein, and other pioneers such as Beck continued using casein in dozens of diets developed for a wide range of insects.

The discussion of these early diets raises the point about the major breakthroughs in artificial diet technology. Besides the advent of casein in insect diets, the other major components that were revolutionary in the scope of insects that they supported were agar, yeast products (both of which were used in many of the earliest diets that succeeded in supporting insect rearing), and wheat germ, which was brought into diet technology in the late 1950s. The rationale behind the use of agar, casein, yeast products, and wheat germ is discussed in detail in Chapter 4, where the basis of successful diets is explained.

The derivation of knowledge and techniques from the field of microbiology has been a profound influence on insect artificial diet science and technology. Early in the history of diet development, diet pioneers recognized that microbial contamination could compromise the diets themselves and our understanding of the diets. They were also strongly influenced by the fundamental practices of microbiology, including the materials and techniques used for microbial media. An example of the adoption of microbiology into insect dietetics is found in a keystone 1917 paper Baumberger (1917b). He starts with a quotation from an earlier paper (Baumberger and Glaser 1917):

> Five or six bananas were mashed up in 500 cc of water. This was allowed to infuse on ice over night, after which the liquid was passed through cheese cloth. Powdered agar-agar was then added in the proportion of 1 and 1/2 gm to 100 c.c. of the banana infusion. This was then heated until the agar had dissolved. The liquid was then filtered through a thin layer of absorbent cotton into test tubes. The tubes were then plugged, sterilized and slanted in the customary manner.

This paper provides the first use of agar as a gelling agent to make a *solid* artificial diet for an insect. It should further be noted that the diet was presented as a slant, in the fashion common to what was then current practice in microbiology. Baumberger went on to write:

> As pointed out in the above article, one of us (Baumberger 1917a) had observed that the bacterial growths which always develop on this medium "do not seem to harm the larva" and the mold which sometimes appears "is usually destroyed by the larvae just as soon as they hatch." This question was further investigated and it was found (Baumberger 1917a) that the principal food of Drosophila is yeast and the flies cannot develop on banana which is kept free from micro-organisms. Delcourt and Guyenot (1919) had previously (unknown to the author) published similar conclusions and Loeb and Northrop (1917) had confirmed them shortly before the author's report (1917a) was sent to press.
>
> It is therefore very well known that the food of Drosophila is yeast and the prime necessity of any medium for rearing this fly must be either abundant food for yeasts to grow upon or the presence of large numbers of yeast cells. If a medium is made of sterile compressed yeast and agar-agar it serves as a perfect food for flies which have been freed from microorganisms; however, if living yeasts develop young larvae are usually killed. As adult flies usually carry living yeast cells upon theta this medium would be difficult to use for work in heredity. A nutrient medium for yeast would best suit the needs of geneticists. Into such a food the adults or pupae would carry living yeast cells which would ferment the sugars and produce odorous substances which cause oviposition by the female fly. Larvae on hatching would spread the yeasts throughout the medium, thus increasing growth and alcoholic fermentation which may prevent the development of injurious microorganisms.

In this paper, other papers from Baumberger's laboratory, and a series of papers from Guyenot's laboratory (Delcourt and Guyenot 1910, Guyenot 1913a–c, Guyenot 1914a,b), the feeding biology of *Drosophila* spp. was described in amazing detail and with impressive insights. These authors set the stage (along with Loeb 1915, Loeb and Northrup 1917) for inquiries into defined diets to establish the most irreducible nutrients required by insects. Collectively, these authors demonstrated the efficacy of using live yeast and yeast products in diets, which was the basis for the majority of insect diets that were developed over the century since the decade of 1910–1919. They (especially Baumberger) introduced agar as the fundamental solidifying agent in insect diets. Building from the agar and yeast-based diets and Michelbacher et al. (1932) including casein, the foundation was ready for the fourth component to be introduced: wheat germ. In the decades from the 1917, through 1959, hundreds of diets for scores of insects were developed. But it was not until 1959 when diets with the *big three* components would be completed with the addition of wheat germ. This point is more fully developed in Chapter 4.

It is noteworthy that in September 2014, a search of Web of Science under the keyword, *Drosophila* produced 127,739 papers. The first 100 papers were published in the 13 years between 1905 and 1918. The next 100 papers appeared in the next 6 years and so on with an exponential increase in reports on various *Drosophila* species. The earlier papers dealt with basic biology and rearing conditions, but once the rearing protocols became stabilized and reliable, the topics became more varied with genetics being the dominant topic, and ecology, evolution, physiology, morphology, and other aspects of biology also appearing frequently. More recent articles treat subjects such as nutrigenetics (Laturney and Billeter 2014) and gut homeostasis (Buchon et al. 2013), although the literature on diet improvements was fairly well stabilized (reduced in numbers of papers on the topic) by the time Bridges and Darby (1933) published their diet studies on optimization of pH in *Drosophila* media. But there remains a trickle of diet papers on *Drosophila* into the late 1970s (reviewed by Singh 1977). It is also interesting that despite many thing that were discovered early on in *Drosophila* research, several topics were not explored until relatively late in the history of research on this genus. Examples are studies of the gut flora and their relationship to homeostasis or overall health in *Drosophila* (Steinhaus 1940, Chu 1945, King 1954). There is a huge body of work on the nutritional and evolutionary ecology of *Drosophila* spp. and their interactions with yeast and host plants. This topic is covered in Chapter 13 on microbiological features of rearing where special attention is paid to *Drosophila* as a model system. Finally, it must be emphasized that without the pioneering accomplishments of *Drosophila* diet and rearing system researchers, the vast body of knowledge of almost every aspect of biological systems that has emerged from research with these tiny flies would not have been possible. It can be argued convincingly that the foundation of modern biological science was dependent upon the breakthroughs in *Drosophila* diet and rearing research.

2.3 OTHER HISTORICAL DIETS AND HISTORICALLY SIGNIFICANT CONCEPTS

A major breakthrough in insect nutrition by Hobson demonstrated that calliphorids (blowflies) required a nutritional factor shown to be cholesterol (Hobson 1935). One of the hallmark differences between the nutritional requirements of insects and most other species of animals is the absolute requirements of the sterol nucleus in insects, and Hobson's remarkable work set up many of the studies that were to follow on specific differences and similarities between insect nutritional requirements and those of other animals. On the heels of Hobson's work came the revolutionary studies of Fraenkel, who can be argued to be the father of modern insect nutrition and diet science. Fraenkel identified *factors* that were nutritional essentials in grain insects and set out much of the experimental protocol that has been used over the years to discover nutritional requirements (Beck 1972). Fraenkel is also responsible for the concepts of plant secondary compounds and token stimuli as key factors in insect feeding (Beck 1972). Fraenkel (including more than a dozen papers with Blewett over the decades of the 1940s and 1950s) established the foundation of insect nutritional science methodology with work on vitamins, plant secondary factors, the "hidden nutrient" concept (often referred to as "nutrient factors"). This influential body of work and approaches to insect dietetics and nutrition followed the pioneering studies of Chapman (1924), where flour beetles (*Tribolium confusum*) were the model insects. It is fascinating to consider how much our current day concepts of insect nutrition and dietetics was shaped by these studies and the nature of the target organisms. These works established the fundamental experimental design that not only allowed careful dissection of the nutritional components and dietary factors that influenced or determined fitness in the test insects, but they even became the foundation for a series of pioneering nutritional ecology experiments that helped establish the concept of ecological niche and other major concepts of general ecology provided by Thomas Park in his elegant experiments throughout the 1930s, 1940s, and 1950s (e.g., Park 1932).

Fraenkel's works in the early 1940s proceeded through the 1960s and can be considered as major stimuli in developing the *golden age* of insect nutrition and dietetic advancement led by such figures as Stanley Beck, Rex Dadd, G. R. F. Davis, Ken Hagen, H. L. House, Thomas Mittler, and Erma Vanderzant. Each of these authors, in his or her way, injected into insect nutrition and dietetics the kinds of mechanistic questioning or cause and effect isolation that steered the field into becoming a true science. The specific accomplishments of these pioneers are explained in several places elsewhere in this text.

In terms of economic and social impact, the diets and rearing systems for screwworms (Melvin and Bushland 1936, 1940) represent an unrivaled historical first. These systems led to the development of the sterile insect techniques that have had a huge impact on management of various fruit flies, pink bollworms, and several other taxa beyond the original projects aimed at screwworm eradication (Knipling 1966). These accomplishments advanced the diet and rearing fields by stimulating numerous other projects and productive lines of research in areas such as quality control, basic and applied nutrition, and genetic manipulation of target insects (Gingrich 1972). From the proof of principle of genetic manipulation as a major tool in screwworm management came the robust field of genetic pest management that is thriving today and making advancements in insect control by utilizing forces of transposable elements, meiotic drive, genetic basis of insect disease resistance, transgenic insects for disease control, engineered underdominance, and many other approaches (Gould).

In terms of economic and intellectual advancements of diet development and utilization, the development of pink bollworm diets deserves special explanation. Adkisson et al. (1960b) made one of the most significant breakthroughs in insect diet science by using wheat germ in their diet for pink bollworms (*Pectinophora gossypiella*). Subsequently, this nutrient was used to revolutionize diets of numerous other phytophagous insects (e.g., use of this nutrient in Heliothine diets by Berger 1963). Unfortunately, the rationale for using wheat germ was not fully explained in the original paper, but a careful examination of its nutritive properties reveals why this is such an excellent food for a broad spectrum of insects.

First, it has a high nutrient content as revealed in Table 3.4. This table shows that wheat germ has an impressive protein content of about 23%, a substantial mineral content, including an iron concentration that rivals beef liver (Table 3.4), and a high lipid content that is rich in polyunsaturated fatty acids and phytosterols. Except for ascorbic acid, wheat germ contains a sizeable amount of most vitamins known to be required by insects (Chapter 3). Wheat germ (Table 3.4) contains a complete complement of amino acids, essential and nonessential. The arrangement of nutrient components in wheat germ (the matrix structure) lends itself to stabilization of the dietary components. The fiber content of wheat germ is an excellent bulking agent that helps promote normal passage of foods through the target insects' alimentary canal (Chapter 7). Although these rationales for using wheat germ were not presented in the paper where its use in insect diets was first introduced, it is clear, in retrospect, that these qualities are at least part of the reasons why this material had prompted a revolution in artificial diets for dozens of insect species. Other noteworthy historical accomplishments are discussed in the context of specific topics in the chapters to follow, and others are listed in Appendix B.

CHAPTER 3

Function of Insect Diet Components

3.1 INTRODUCTION TO FUNCTIONAL ASPECTS OF DIET COMPONENTS

The classes of components that are commonly added to insect diets include carbohydrates, proteins, lipids, vitamins, and minerals. For clarity in discussion functions of diet components, it is useful to adopt a term that includes complex components—ones that are not chemically pure and that contribute major nutritional value and possibly other factors such as palatability and texture, possibly even stability and bioavailability. The terms *dietary staples* or *complex foods* seem to meet these qualifications. So in speaking about components such as soy flour, wheat germ, vertebrate liver, chicken eggs, and yeast, the terms *complex foods* or *staples* convey the multiple contributions made by such components. To amplify this concept, ground plant materials such as powdered cabbage leaves, dried and ground weed leaves or stems, and hemolymph from a host insect used in a semiartificial diet for parasitoids would all be complex foods or staples.

Other ingredients commonly added to diets are emulsifiers, stabilizers, gelling agents (or stabilizers), pH modifiers, and preservatives, which may include antimicrobial agents and antioxidants. Other functional components that are added, often incidentally, are phenolic compounds, flavonoids, terpenoids, and other factors that are only recently coming to the attention of the food science and entomology communities (Carroll et al. 1997, Johnson and Felton 2001). Interestingly, some factors that have antinutrient qualities also find their way into insect diets. This includes digestive inhibitors, lectins, agents of oxidative stress (reactive oxygen species [ROS]), and a variety of other potentially deleterious substances.

3.2 ESSENTIAL VERSUS NONESSENTIAL NUTRIENTS

Before the function of individual nutrients is surveyed, it will be useful to clarify what nutritionists and biochemists mean by the concept of *essentiality* of given nutrients. First, all nutrients utilized by insects are processed in *metabolic pathways*. After the nutrients are digested and absorbed (covered in Chapter 7), they are transported to the appropriate cells where they are used as appropriate components of metabolism. The metabolic pathways are covered in detail in various references such as general biology and biochemistry texts. Chapter 7 explains the path of nutrients from preingestion preparation through absorption and finally delivery to cells where the metabolic pathways are active in incorporation and utilization of nutrients. However, a simple example of how nutrients participate in metabolic pathways will help further the current discussion.

An example of metabolic pathways found in most insect cells includes the glucose metabolism pathway that is involved in the generation of the energy transduction molecule adenosine triphosphate (ATP), a substance often called the *energy currency* of cells. Cells need ATP to power mechanical and chemical activities, and they get it largely from anaerobic (without oxygen) and aerobic (with oxygen) breakdown of the widely used fuel, glucose, and, in other circumstances, other sugars, lipids, or proteins. Typical of metabolic pathways, the production of ATP is an indirect, multiple step process, which is controlled by a series of specific enzymes. The process is also highly organized in terms of spatial order. The anaerobic pathway of transduction of glucose energy to ATP energy takes place in the cytoplasm of most cells and is called glycolysis (literally the taking apart of sugar). It involves 11 steps that require enzymes and vitamin-based cofactors and results in the production of several products that are starters for the aerobic part of ATP production, which takes place in the cell's *powerhouses*, the mitochondria. The bottom line for this process is that over 20 enzymes, several vitamin-based cofactors (such as thiamine, riboflavin, and niacin), minerals (such as magnesium, zinc, and iron), and cytochromes (metal-bearing electron carriers) are involved in this well-regulated, complex process that results in the breakdown of a molecule of glucose into carbon dioxide and water and the transfer (conservation) of bond energy that results in the synthesis of 38 ATP from 38 adenosine diphosphates (ADPs) and 38 inorganic phosphates. It must be understood that virtually every energy-requiring reaction or process that is done by living things is accomplished via the utilization of the energy delivered by ATP.

Because this pathway of ATP synthesis at the expense of glucose results in the degradation of sugar, it is known as a catabolic (breakdown) pathway. The other type of metabolic pathway (the anabolic pathway) involves the building up or synthesis of molecular structures. These pathways take place within the cells of the insects that are our rearing subjects, and they are generally common to most other organisms, including other animals, bacteria, fungi, and plants. Of course, there are some characteristic differences in some metabolic pathways according to the taxonomic or phylogenetic status of the organisms in question. For example, some organisms (such as plants, algae, and some bacteria) possess light-driven pathways for the production of complex organic molecules in a metabolic process known as *photosynthesis*. However, many pathways of organisms, in general, are common to most living things, including the cellular respiration pathway for extraction or conversion of useful chemical energy from fuels such as sugars and fats.

Returning to the concept of essential nutrients, if an organism in question must have a given nutrient to carry out one of its defining pathways, but if it lacks the metabolic ability to produce that given nutrient, it must obtain that substance from its diet. Such a nutrient that can be obtained only from the diet is referred to as an *absolute essential* or simply as an *essential nutrient.*

For the purposes of this discussion, the amino acids valine and glutamic acid can serve, respectively, as examples of essential and nonessential amino acids. Both of these compounds are components of many insect proteins. This being the case, insects must have both of these amino acids present to build their body proteins. In the case of glutamic acid, insects can get this compound from their food or they can build it from raw materials such as sugars or lipids, as long as they have a source of nitrogen in the form of an amino group ($-NH_2$). So in general, as part of the normal metabolism known as protein synthesis, insects can build their own glutamic acid in whatever quantities they require, simply by using a carbon source such as a sugar or a lipid. The pathway for glutamic acid synthesis (as is the case for synthesis of other nonessential amino acids discussed in Chapter 4) is very versatile and can include many kinds of sugar (glucose, fructose, galactose, and others) or any of several lipids (including various kinds of fatty acids and sterols). In sharp contrast, the pathway for synthesis of valine is absent in insects (as it is in most animals); therefore, any valine that is needed for protein synthesis must be gotten from the diet. The only exception to this is in cases where insects have microorganisms known as symbionts living within them and where those symbionts have the metabolic pathways to produce an adequate amount of valine from raw materials provided by the insect host. This will be discussed further in the chapters on metabolism and microbe/insect interactions (Chapters 7 and 13).

3.3 PURPOSES OF INDIVIDUAL DIET INGREDIENTS AND NUTRIENT FUNCTIONS

Artificial diets for insect generally must contain the following components: a nitrogen source—usually proteins (but sometimes free amino acids), lipids, carbohydrates, vitamins, and minerals–and also stabilizers, preservatives, and often *fillers* or bulking agents. Most (if not all) successful diets contain special components that do not have direct nutritional function, but they stimulate normal feeding responses and are called *token stimuli*. As token stimuli, dietary substances do not serve any nutritional role: they do not serve as energy sources, building blocks for synthesis, cofactors for enzymatic pathways, or any other role served by true nutrients. The true nutrients serve a variety of functions, but essentially, they are the raw materials of the metabolic pathways (discussed in Chapter 7) and as structural components that give insects their physical organization, or they are the minerals that play various functional roles in insect metabolism and physiological activities such as nerve impulses and muscle contractions. The token stimuli (including many plant secondary compounds such as rutin, sinigrin, and gossypol) are components that evoke feeding but serve no (known) function in metabolism or structural organization of the insects that consume them.

In what is now a classic study, Ma (1972) showed that the mustard compound sinigrin is a stimulant that evokes a food biting response in the cabbage butterfly larva, *Pieris brassicae* L. Sinigrin has no known metabolic function in this (or any other insect), yet it is instrumental in the normal feeding process of this insect that specializes on host plants in the mustard family. By contrast, sucrose, a common plant sugar, stimulates the swallowing response by *P. brassicae* larvae; but sucrose does not evoke the biting response. Therefore, in order for *P. brassicae* larvae to perform the entire normal feeding response of biting and swallowing, both sinigrin—a feeding (or biting) incitant—and sucrose (a stimulus that triggers swallowing) must be present (Schoonhoven 1972). Because sucrose evokes a feeding response (or part of a feeding response), it must be regarded as a *feeding stimulant*, but because it is also nutritive, it cannot be called a token stimulant. In contrast, because sinigrin stimulates a feeding response but is not of any known nutritional value, it is considered a *token stimulant*. Several reviews of literature on feeding stimuli, feeding deterrence, and the various stereotypic eating responses are presented by Schoonhoven (1972) and by Chapman and deBoer (1995). There is also further discussion of these issues in Chapter 7 of this book.

3.3.1 Proteins (Nitrogen Source)

Most insects, except for true sap feeders such as aphids, whiteflies, cicadas, and many leafhoppers, use whole proteins as their principal source of nitrogen. The proteins (polypeptides) are broken down into their amino acid components, which are absorbed and circulated to cells where they are resynthesized into the proteins that make up the insect's body (muscles, parts of cell membranes, enzymes, certain hormones, etc.). As a rule, insects (like people) require 8–10 *essential* amino acids (methionine, threonine, tryptophan, valine, isoleucine, leucine, phenylalanine, lysine, arginine, and histidine). The structure of each of the protein amino acids is shown in Figure 3.1. As discussed earlier, these amino acids (sometimes known as the *rat essentials* because they were originally shown to be required per se in rats) must be present in the insect's food. Other protein amino acids (serine, asparagine, aspartic acid, glutamine, glutamic acid, alanine, cysteine, glycine, tyrosine, and proline) are used by insects in building their proteins, but they are not considered *essential* because they can be synthesized by the insects using their own metabolic pathways. It must be emphasized that in nature (i.e., in most foods), the amino acids that are present are mainly present as components of proteins (i.e., long chains of amino acids that are bonded together in stable *peptide* bonds, characterized in Chapter 5). When we use food substitutes that are *hydrolyzed* (such as soy or yeast hydrolyzate), we are forcing the insect to use an unnatural form of its nitrogen source, which is now

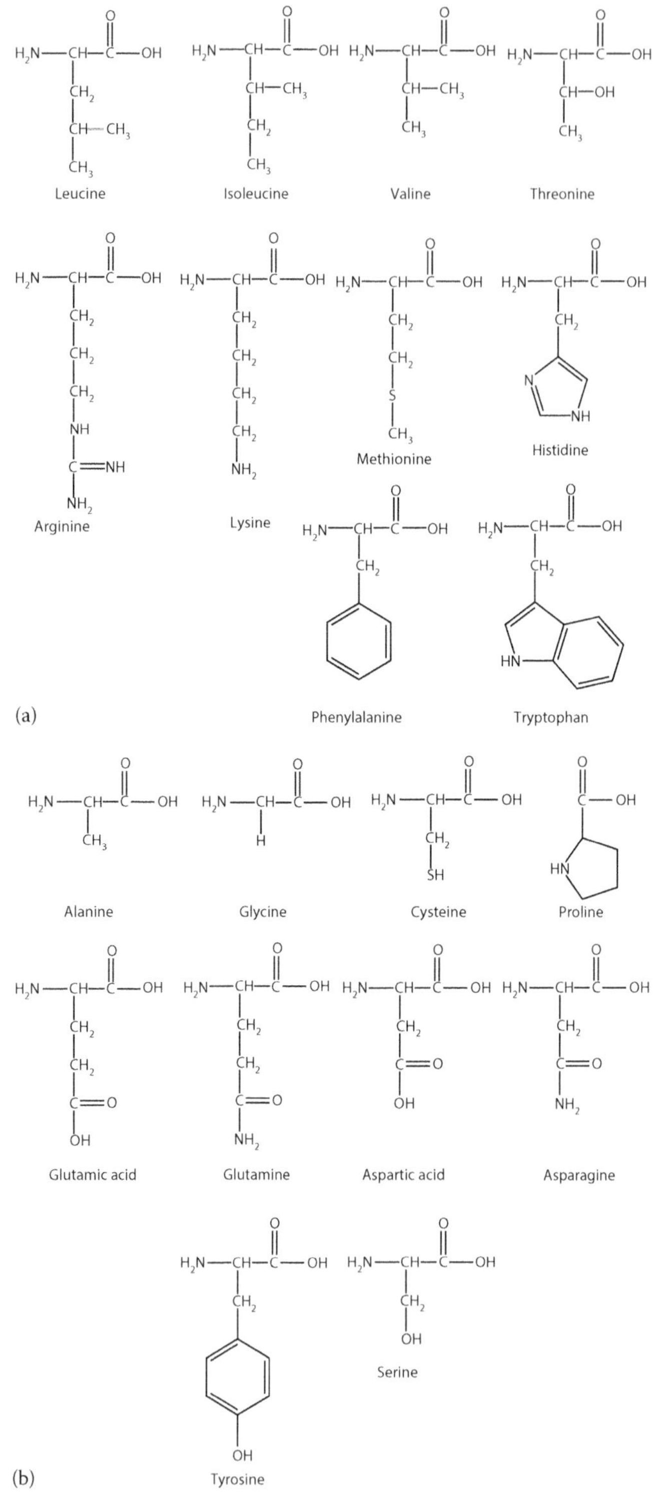

Figure 3.1 Structure of each of the 20 protein amino acids, including 10 insect-essential amino acids (a) and 10 nonessential amino acids (b).

loaded with free amino acids (some of which—especially tryptophan and threonine—have been destroyed by the hydrolysis process). Free amino acids may not be as palatable as the protein form of the nitrogen component, and they contribute to increases in the osmotic pressure (which may be desiccating to the insect's gut or to the insect, in general). Although we can know only indirectly and through complex and often ambiguous experiments that a diet component has an attractive, repellant, or indifferent taste to an insect, we do know that certain free amino acids impart an off-taste or repulsive quality to humans (Damodaran et al. 2008). Interactions between insects and their symbionts involve the supplementation of essential or other key amino acids by the microbial guests. This is discussed further in Chapters 7 and 13.

In hydrolyzed foods, proteins and polysaccharides that may be toxic or in some other way disagreeable to the insect are destroyed by the hydrolysis process. An example of this is in the fermentation process of soy products where various antinutrients are destroyed by the hydrolysis achieved by the microorganisms that carry out the process (Fukushima 1991). Many toxins (especially macromolecular ones) are destroyed by processing (most prominently heating) the dietary components. For example, raw soy flour, wheat germ, and meals made from various legumes have a large number of lectins and digestive enzyme inhibitors that are rendered edible by toasting and/or autoclaving. Changes in the nutrient and antinutrient characteristics of diet components are discussed in the chapter on food processing (Chapter 12) and the chapter on nutritional ecology (Chapter 11).

Protein digestion and absorption efficiencies and overall protein bioavailability are features of how proteins function in insect diets as they are in vertebrate foods. Damodaran (2008) discusses this issue in terms of *protein quality*, which is a composite of the amino acid profile of a given protein and its digestibility and absorption qualities. The presence of all essential amino acids in appropriate quantities confers upon a protein the potential of being a high-quality nutrient. For example, animal proteins such as egg yolk vitellin and milk proteins, especially caseins, contain all of the essential amino acids in high quantities and well-balanced proportions, including the amino acids, tryptophan, methionine, leucine, isoleucine, and lysine (Damodaran 2008). In insects, the balance of amino acids has been demonstrated to be important in only a few species such as honeybees (Standifer 1967).

3.3.2 Lipids (Including Sterols, Oils, Fats, Phospholipids)

The importance of lipids in insect nutrition has been underestimated. Probably many failings in insect dietetics stem from underprovision of the right amounts and types of lipids. For example, seed-feeding lepidopterans can readily digest oils and fats (triglycerides or triacylglycerols); however, leaf feeders digest oils and fats poorly, yet they require fatty acids (Turunen 1979). When fatty acids were presented as components of more polar compounds—phospholipids—they were easily digested, absorbed, and utilized by leaf feeders. Furthermore, all insects require a source of dietary sterols, yet because it is difficult to dissolve sterols, they are often omitted, lost, or provided in the wrong form. For example, a strict plant eater that may require plant sterols such as β-sitosterol or campesterol may be given cholesterol (Figure 3.2), which it cannot process (Svoboda et al. 1975, Svoboda 1984, Svoboda and Lusby 1986).

Lipids function as building blocks of cell membranes (especially sterols), hormones (e.g., sterols are converted into ecdysteroids or molting hormones, and fatty acids are converted to juvenile hormone), nutrient transporters, sources of energy, and structural material (carbon skeletons) for building other molecules. The pathways for producing sterols and unsaturated fatty acids are not reversible. This means that while an insect can use extra sterol for energy or for building carbohydrates, it cannot reverse the process and build sterols from carbohydrates. Because insects, unlike vertebrates, cannot make sterol to support their needs, they must get it from their diet, thus making sterols, by definition, essential nutrients. Sterols and other lipids, known as complex lipids, serve as membrane components, giving cellular membranes specialized characteristics, especially with

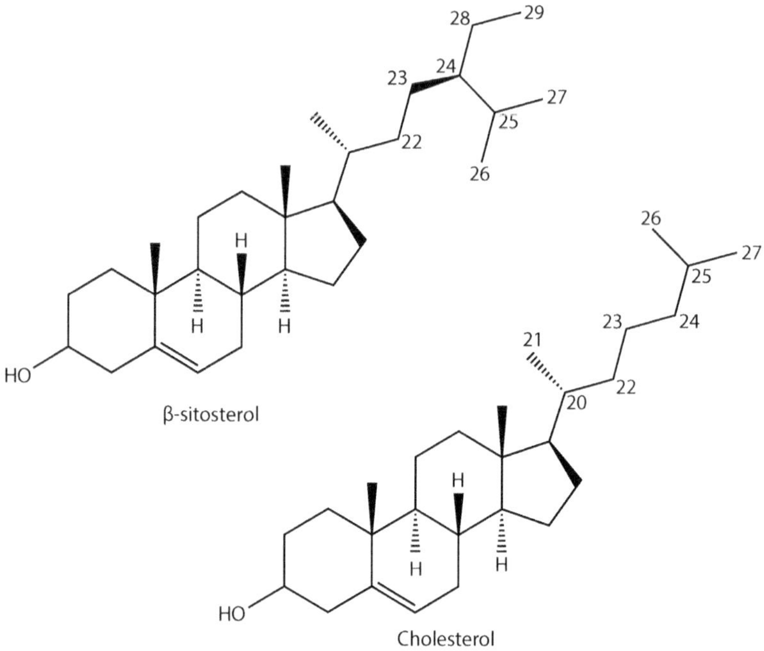

Figure 3.2 Structure of the sterols cholesterol and β-sitosterol.

regard to the import and export of materials into and out of cells. These lipids help modify the proteins that are components of receptor mechanisms that give the highly cell-specific functions that characterize special function tissues (Lehninger et al. 1993).

Lipids are uncharged or nonpolar, making them insoluble and immiscible in water. The noncharged, nonpolar nature of lipids results from the predominance of hydrocarbons as is seen in the drawing of a fatty acid in Figure 3.3. The repeating units of carbon and hydrogen are noncharged (nonpolar), and only the COOH end of the molecule (the acid part of the fatty acid) is charged (polar) and miscible with water. By contrast, the glucose molecule, an example of a carbohydrate, is shown with its numerous polar OH (hydroxyl) functional groups, making this molecule highly soluble in water. Because of their poor solubility and miscibility with water, lipids require special

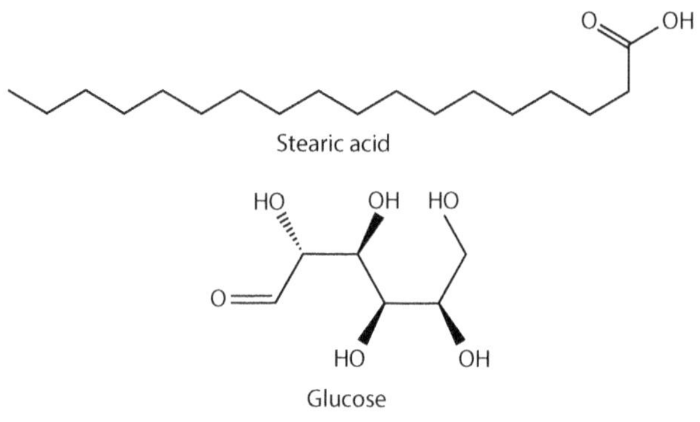

Figure 3.3 Structures of a nonpolar, lipid-soluble nutrient (stearic acid) and a polar, water-soluble nutrient (glucose). Note the large number of repeating carbon–hydrogen units in the lipid and the proportionally large number of charged OH groups in the water-soluble molecule.

transport mechanisms, which usually include lipoprotein carriers such as the molecule called lipophorin (lipid-bearing). The other side of the coin is that because of their nonpolar nature and the similarly nonpolar nature of cell membranes, lipids can easily cross cell membranes and become incorporated into cells without special entry mechanisms that are required by polar molecules, such as sugars, amino acids, and many minerals. This issue is treated in greater detail in Chapter 7, which deals with the logic and mechanics of digestion, transport, and metabolism.

3.3.3 Carbohydrates (Polysaccharides, Oligosaccharides, and Monosaccharides)

Insects use carbohydrates as building materials and as fuels. Also, the insect cuticle contains a polysaccharide (chitin) made of amino sugars. Some carbohydrates cannot be digested or utilized by most insects (e.g., cellulose), but they may be useful as fillers (bulk) in diets that help promote intestinal mobility (Chapter 7). Some insects, especially phytophagous ones, fail to thrive on diets that are low in (less than 50%) carbohydrates (House 1974a). The type of carbohydrate must be fitted to the specific insect. Certain sugars that are usable by some insects cannot be used by others. For example, the sugar melibiose, an α-galactoside, can be digested by several species of flies but not by honeybees (Gilmour 1961). Likewise, the sugars raffinose and stachyose, both α-galactosides, can be digested only by insects that possess the specialized enzymes α-galactosidases (Chippendale 1978). A sugar that is digested by a wide variety of insects, maltose (containing an α-glycosidic linkage), and a sugar that is not digested by most species of insects, lactose (containing a β-glycosidic linkage), are both shown in Figure 3.4. Both of these sugars are disaccharides.

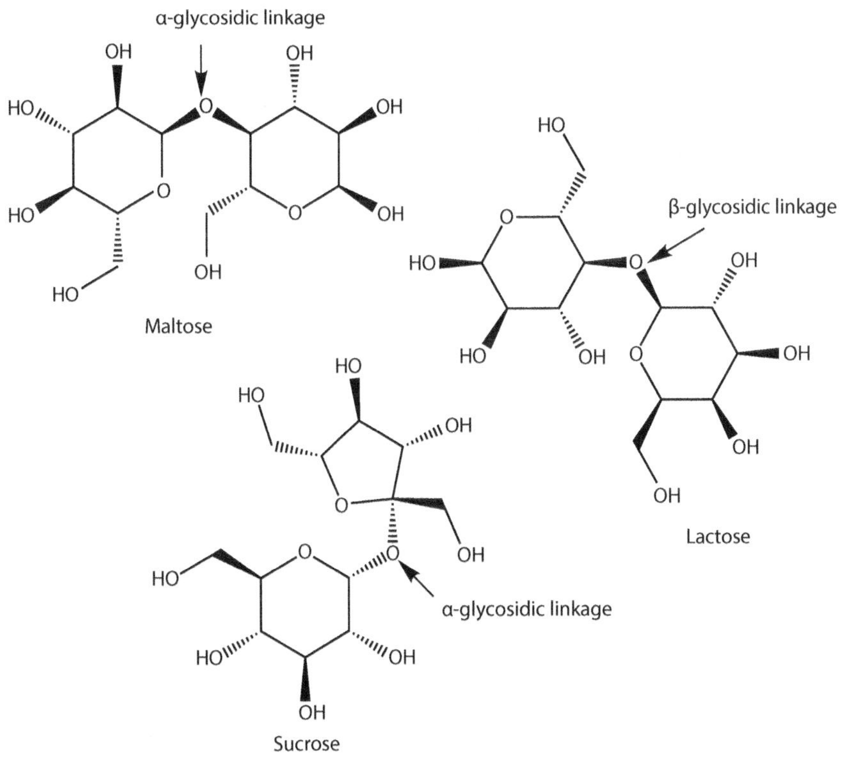

Figure 3.4 Structures of three disaccharides: maltose, an α-glucoside that is digestible by most insects; lactose, a β-glucoside that is not digestible by most insects; and sucrose, an α-glucoside that is digestible by many insects. Despite the superficial similarities in structure, the three sugars differ greatly in their nutritional values.

There are some interesting and unexpected responses to sugars, including the discoveries by Jules Silverman at North Carolina State University who showed that there are varieties of cockroaches that are averse to one of the most common nutritional sugars: glucose (Silverman and Bieman 1993, Wada-Katsumata et al. 2013). The authors (Wada-Katsumata et al. 2013) further demonstrated that the level of aversion response was in peripheral sensory neurons that generally respond to sugars as phagostimuli, but that in glucose-averse German cockroaches, a genetically determined response causes the gustatory response neurons to interpret glucose as a bitter stimulus, which German cockroaches reject. The authors pointed out that this mechanism evolved rapidly in response to evolutionary pressures from the use of glucose-containing baits used to control cockroaches.

Carbohydrates in insect diets and within the insect bodies are also components in glycoproteins. Recent developments in biochemistry of proteins have shed light on the very complex and intricate function of the carbohydrate portion of glycoproteins as sites of recognition for proteins that serve as channels and receptors for movements of materials in and out of cells. The lectins are a class of glycoproteins that have roles as agglutinins, antibiotics, and toxins. Some of the most toxic substances known belong to this interesting class of glycoproteins, including ricin from castor beans (Leathem and Brooks 1998).

3.3.4 Vitamins

Despite the fact that we have known for over a century that animals require trace amounts of certain organic structures called vitamins, our understanding of these factors in insects is frustratingly limited. The vitamins are conveniently divided according to their solubility, either water- or lipid-soluble vitamins. In general, the water-soluble vitamins have a relatively short half-life in insects, because they are readily excreted and lost from the insect's metabolic pool because of their solubility. In contrast, lipid- or fat-soluble vitamins tend to remain in the insect because they remain compartmentalized in lipid stores.

3.3.4.1 Water-Soluble Vitamins

This group (Figure 3.5) includes the B vitamins, vitamin C (ascorbic acid), and some miscellaneous compounds, such as choline and carnitine, a compound essential to meal worms. The B vitamins function as cofactors in many metabolic pathways (such as in energy utilization [thiamine, riboflavin, niacin] or as growth factors [biotin and folic acid]). Vitamin C is essential for many phytophagous insects, serving as a phagostimulant, antioxidant, and in other capacities—including cuticle sclerotization and possibly other defensive reactions. Vitamin C is very susceptible to degradation, especially when it is present in solution, exposed to heat, light, oxygen, or free radicals. A group of substances of emerging importance are the other antioxidants, including ascorbic acid, some phenolics, and flavonoid compounds. They may have key roles in the protection of insects from microbes, dietary toxicants, and other kinds of threats (such as attack by free radicals induced by environmental stresses). Some of the antioxidants also fall into the lipid-soluble category discussed in the following.

Most of what we know about the functions of vitamins in insects is derived from findings of vertebrate nutritional science. There are not, in insect studies, specific vitamin deficiency diseases ascribed to given vitamins such as the mammalian conditions of beriberi, rickets, or scurvy. Nutritional deficiencies in insects have been linked with such vague symptoms as poor growth rates, lowered fecundity or fertility, reduced body weight, or other conditions that do not help pinpoint a specific inadequacy. What is to follow is a specific vitamin-by-vitamin survey of the functions ascribed to each compound and a brief summary of the most abundant sources of each and information on the relative stability. Because no recommended minimum daily requirements have been

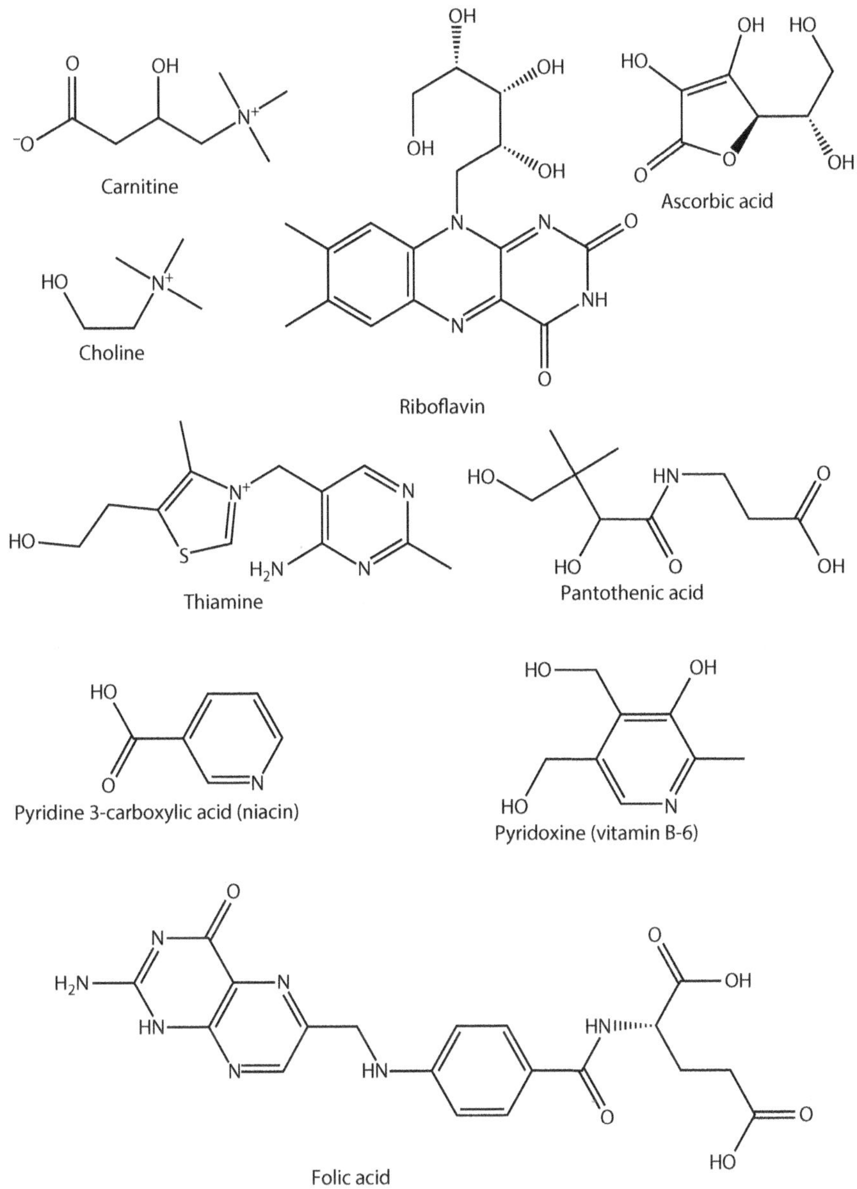

Figure 3.5 Various water-soluble vitamins.

established for any insect and because this discussion is meant to broadly cover insects, in general, no effort will be made to suggest a dosage range. However, later in the chapter, a table of the composition of vitamins suggested by several authors will be provided.

Ascorbic acid is most commonly present in its L-ascorbic acid form, a component most abundant in several kinds of fresh fruits and green tissues of plants. It occurs, for example, in amounts as high as 90 mg/100 g of fresh broccoli, 90 mg/100 g of fresh sweet green peppers, and 1900 mg/100 g freeze dried sweet green peppers (Anonymous 2013). It is present in much lower concentrations or absent in plant components that are not green or not fruits. This means that when grains and other seeds are used as main diet components, they must be supplemented with ascorbic acid in all cases where the target insects require this vitamin. As is the case with several other food components, ascorbic acid has functions both in and as a factor of the metabolic pathways of the organism that has ingested it.

Ascorbic acid has been shown to be essential to many species of insects, especially ones that are phytophagous. First, ascorbic acid is known to serve as a phagostimulant for phytophagous insects (Ave 1995). The first demonstration of essentiality of ascorbic acid in any insect's diet was that of Dadd (1957) who showed that this vitamin was required by the desert locust, *Schistocerca gregaria* Forsk. Gilmour (1961) in a comprehensive survey of the five decades of work that preceded his review mentioned no studies that showed an ascorbic acid requirement in insects. Although ascorbic acid can be synthesized by some species of insects, it must be present in the diet for many other species (House 1974a). Evidently, these species cannot produce this vitamin de novo. Lehninger et al. (1993) point out that ascorbic acid acts as a an antiscurvy factor by serving in the pathway for collagen synthesis in vertebrates, and although ascorbic acid had not been demonstrated directly to function in the collagen synthesis pathway in insects, it is possible that the vitamin acts in synthesizing the insect extracellular matrix, which is partially composed of collagen. Gregory (2008) explains the importance of ascorbic acid as an antioxidant both in the interaction with other food components and in the organisms that ingest this vitamin.

Thiamin (vitamin B_1; also spelled thiamine) is a cofactor in biochemical pathways of energy transduction from the chemical bonds of carbohydrates and lipids to those of high-energy phosphates, especially ATP. In these energy conversion pathways, thiamin is a cofactor in decarboxylations. Without thiamin, energy-processing reactions such as the citric acid cycle cannot take place. Also carboxylation and decarboxylation reactions involve thiamin. Deficiencies of thiamin have been shown to cause accumulation of pyruvic acid in insect tissues (Gilmour 1961). An important fact about thiamin is that it is next to ascorbic acid in its vulnerability to degradation from heat and oxidation. Edwards et al. (1996) made a special point of using thiamin as the indicator of possible destruction by heat, pressure, and mechanical action in their studies of pink bollworm diet processed by twin screw extruders. This is discussed further in Chapter 12.

Riboflavin (vitamin B_2) is probably essential to most insects, though in some species, the requirement is masked by production by microbial symbionts (Gilmour 1961). Metabolically, riboflavin functions as a cofactor for the flavoproteins. These complexes of riboflavin and protein act as carriers for electrons and hydrogens to the cytochrome system. In this capacity, riboflavin is crucial in the energy metabolism pathways involved in ATP production.

As with thiamin and riboflavin, niacin (and its derivative nicotinamide or various niacin esters) is involved in energy transduction pathways. As a part of the electron and hydrogen carrier nicotinamide adenine dinucleotide (NAD) and nicotinamide adenine dinucleotide phosphate (NADP), niacin is instrumental in serving the functions of the cytochrome system whose chief function is ATP synthesis. Also like thiamin and riboflavin, niacin's essentiality can be masked by the production of this vitamin by microbial symbionts (Gilmour 1961). As with the other energy pathway cofactors, a deficiency of niacin results in reduction in or loss of ability to use fuels as sources of ATP production, and such deficiency also manifests itself in retarded growth and development, as well as various structural deformities (Gilmour 1961). It also seems that the specific chemical form of the niacin complex may determine the usefulness of these vitamins to given insect species (Gilmour 1961). The niacin complex exists in a variety of forms in food matrices, and the type of processing, especially heating, influences the forms that predominate after processing is completed (Gregory 2008).

Pyridoxine and its phosphate derivatives (vitamin B_6) are involved in several pathways of amino acid metabolism. However, requirements for this vitamin seem to be very of species specific, and it cannot be said to be essential to all insects (Gilmour 1961). As part of its involvement in amino acid synthesis and degradation reactions, pyridoxine is involved in the processing tryptophan into various pigments, and a deficiency in this vitamin can manifest itself as an abnormality in pigmentation and in frass color (Gilmour 1961). As explained for the niacin complex, the various forms of pyridoxine have different degrees of biological activity, and the processing of foods containing this vitamin determines the predominance of the given forms (Gregory 2008). Pantothenic acid is essential to all insects, except for those that have this vitamin supplemented by microbial symbionts. It is a cofactor of coenzyme A, which is involved in transfer of acyl groups in metabolic pathways involving intermediate metabolism of carbohydrates, lipids, and amino acids (Gilmour 1961).

Biotin and folic acid are carriers for one carbon groups in intermediate metabolism pathways. Biotin is widely found in many foods, and deficiencies of this vitamin are rare, except where the egg white protein, avidin, is consumed in large amounts (Lehninger et al. 1993). Biotin deficiency in insects slows larval growth and decreases fertility of adults (Gilmour 1961). A variety of biotin precursors have been shown to be suitable to spare biotin as have several fatty acids or sources of fatty acids (products of some biotin-involved pathways) in some insect species (Gilmour 1961). A major portion of our knowledge of biotin, folic acid, and other B vitamins in insects was revealed by a series of studies by Fraenkel and Blewett through the 1940s and 1950s. These insect vitamin pioneers had to work with crude materials, having to extract their own nutritional supplements using techniques of water or chloroform solubility to prepare extracts from yeast, which they transferred to vitamin-free substrates such as pure glucose and vitamin-free casein. In an interesting passage, Fraenkel and Blewett (1942) made this comment: "These results make it very likely that the 'insoluble' factor in question is in fact biotin" in discussing restoration of viability from a diet with a yeast extract. The authors go on to write, "The final proof must, of course, wait until a test can be performed with pure biotin in place of insoluble yeast fraction. Although biotin has recently become available in the United States, we have so far not been able to procure a sample and have to postpone this *experimentum crucis* until it arrives." Fraenkel and Blewett added the observation that the deficiency of the *factor* was less severe when starch was used instead of glucose; and they hypothesized that the less severe deficiency was attributable to biotin being an impurity in starch. The point led to the painful realization that many experiments with vitamin (and mineral) deficiencies were compromised by cryptic, trace vitamins and minerals in ostensibly pure test materials.

Besides its role in metabolism of one carbon structures, folic acid is also an essential factor in nucleic acid synthesis and as a pigment precursor (Gilmour 1961, Chapman 1998). Although folic acid is essential, in some species it can be spared by chemically similar pteridines (Gilmour 1961). Other water-soluble factors include choline, carnitine, cyanocobalamin (vitamin B_{12}), and lipoic acid that are not universal requirements for insects, but have been implicated as improving growth or fertility in some species (Gilmour 1961, Chapman 1998). Choline is a component in polar lipid metabolism, including the production of cell membranes. Carnitine is also involved in lipid metabolism and serves a major role in movement of lipids in and out of mitochondria for lipid degradation pathways associated with energy metabolism (Lehninger et al. 1993).

3.3.4.2 Lipid-Soluble Vitamins

The vitamin A complex (β-carotenes and their carotenoid relatives) is essential for the formation of eye pigments and other pigments and for normal growth (Figure 3.6). The carotenoids are also among the most potent antioxidants, and their lipid solubility makes these compounds susceptible to placement in the lipid compartments of cells (membranes and vacuoles) where they can prevent damage to these delicate and important structures (Gregory 2008). Vitamin E (α-tocopherol) is known to be a fertility/fecundity factor, but it is also an antioxidant, and it probably has other functions (Gregory 2008). These (and probably other lipid-soluble factors) are very sensitive to oxidation by light, free radicals, excessive heat, or simple aging. Like many diet components (including unsaturated lipids), they are subject to becoming stale, rancid, or generally degraded by such abuses as long storage, being unrefrigerated, exposure to light, microbial contamination, or exposure to prooxidants.

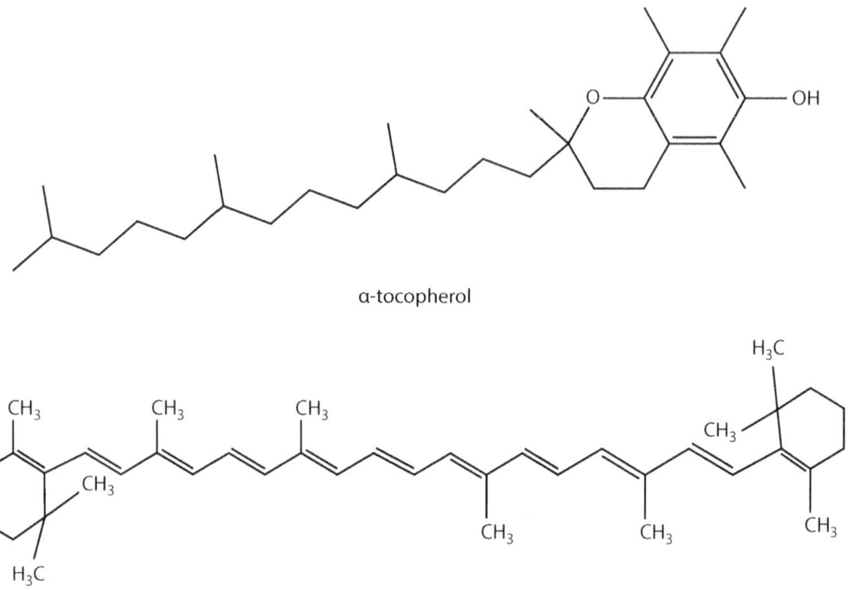

Figure 3.6 The lipid-soluble vitamins α-tocopherol and β-carotene.

3.3.4.3 Vitamin and Other Nutrient Deficiencies

One of the gaps in insect nutrition and dietetics is in knowledge of a specific set of symptoms that would be useful in diagnosis of vitamin deficiencies. In fact, this is the case for all nutrient classes. In mammalian nutrient deficiency syndromes, there are specifically characterized symptoms that can be used to diagnose the problem of nutrient inadequacy. For example, scurvy (a deficiency in vitamin C) manifests itself with symptoms such as gum degeneration (bleeding and swelling), tooth loss, bleeding under the skin surface, stiffness of joints, and slow wound healing (Lehninger et al. 1993). Similarly, beriberi, caused by thiamin deficiency, has specific symptoms that are manifested by loss of neural function and even more specifically by an elevated level of pyruvate in the blood (Lehninger et al. 1993). In insects, no such specific syndromes are known. Frequently, malnourishment can lead to wing deformities, lower body weights, and size reduction, but these effects have been observed in a wide variety of deficiencies, including various vitamins, amino acids, and lipids, as well as mineral malnutrition (Gilmour 1961, Cohen 1981). One of the few (and most) useful assessment tools for nutritional evaluation is the use of mandibular gland development used in honeybees (Standifer 1967).

3.4 MINERALS

Mineral mixtures are deliberately added to many diets as salt mixtures, but the majority of ingredients, unless they have been rigorously purified, contain some minerals. Therefore, the overall mineral composition of a diet is not identical to the salt mixtures that are added to the diet. The pink bollworm diet (Adkisson et al. 1960) discussed in depth in Chapter 4 is a good example of this point. This diet contains 8 g of Wesson salt mixture, which contains the minerals specified in Table 3.1 (and Table 3.2, containing the AIN salt mixture for comparison). Each mineral (i.e., each salt) contains a cation (positively charged ion) and an anion (negatively charged ion). Calcium carbonate,

Table 3.1 Wesson Salt Mixture

Ingredient	Amount (%)
Calcium carbonate	21
Copper sulfate $5H_2O$	0.039
Ferric phosphate	1.470
Manganese sulfate (anhydrous)	0.020
Magnesium sulfate (anhydrous)	9
Potassium aluminum sulfate	0.009
Potassium chloride	12
Potassium dihydrogen phosphate	31
Potassium iodide	0.005
Sodium chloride	10.5
Sodium fluoride	0.057
Tricalcium phosphate	14.9

Table 3.2 AIN Mineral Mixture 76

Ingredient	Amount (g/kg) {%}
Calcium phosphate (dibasic)	500 {50}
Cupric carbonate	0.30 {0.03}
Ferric citrate	6.0 {0.6}
Manganese carbonate	3.5 {0.35}
Magnesium oxide	24 {2.4}
Potassium citrate	220 {22.0}
Potassium sulfate	52 {5.2}
Zinc carbonate (70% ZnO)	1.60 {0.16}
Potassium iodate	0.01 {0.001}
Sodium chloride	74 {7.4}
Sodium selenite	0.01 {0.001}
Chromium potassium sulfate	0.55 {0.055}
Sucrose, finely powdered	118 {11.8}

for example, when dissolved in water, dissociates into Ca^{++} and CO_3^{--}. The calcium ion has a double positive charge, and the carbonate has a double negative charge. Likewise, all the other salts have cations (monovalent or divalent or trivalent) and anions (counterions) to balance the charge. Also, some salts are hydrated, for example, copper sulfate, which is listed with a dot and $5H_2O$, meaning that it is hydrated with five water molecules. The hydration state is considered when calculating the amount of a given mineral such as copper in a given weight of a hydrated salt. Failure to consider the hydration state can lead to overestimates of the other elements in a salt. The hydration state also influences the solubility of the salt. Also, some salts have three kinds of ions, such as potassium aluminum sulfate and potassium dihydrogen phosphate.

In Table 3.1, the compound listed as tricalcium phosphate is also known as tribasic calcium phosphate, tricalcium orthophosphate, tertiary calcium phosphate, or Calcigenol Simple (Windholz 1983). Confusingly, the same compound, $Ca_3O_8P_2$, with a molecular weight of 310.20 is indicated by all these terms. Two other calcium phosphate compounds exist and must be clearly distinguished from the tribasic or tricalcium phosphate, monobasic calcium phosphate ($CaH_4O_8P_2$) with a molecular weight of 234.6 and dibasic calcium phosphate ($CaHPO_4$) with a molecular weight of 136.06. These three compounds have distinctly different characteristics and substitution of these compounds is risky. Various phosphates compounds are widely used as buffers in their sodium or potassium forms.

They occur as KH_2PO_4, K_2HPO_4, and K_3PO_4 called monobasic, dibasic, and tribasic potassium phosphate, respectively. Each of these compounds has a monobasic, dibasic, or tribasic sodium form. The form in the Wesson salt mixture is the monobasic form, referred to in Table 3.1 as potassium dihydrogen phosphate. There is a great difference in the buffering capacity of the three forms of potassium phosphate as well as a substantial difference in the amount of potassium that is being added to the diet as a given amount of each of the three forms is used. The tribasic form will introduce three times the amount of potassium that the monobasic form will add to the diet, given the same weight of each is called for in the diet formula.

3.4.1 Required Minerals and What They Do in Insects

Nearly three decades ago, House (1974a) pointed out that mineral nutrition in insects was the most poorly understood aspect of insect nutrition as a whole. This situation has been virtually unchanged. The reasons for the paucity of information stem from the difficulty in performing definitive nutritional studies, such as the assurance that all ingredients in a diet are entirely free of a given mineral that is in question. It is nearly impossible, for example, to be sure that purified amino acids, the source of water, the gelling agents, or other additives are free of zinc, copper, or iron so that a diet void of each of these minerals can be devised. The problem is exacerbated by the difficulties of rearing insects on defined diets and getting robust growth, development, and reproduction that can be used as a basis for establishing experimental control groups. The difficulties described here (and emphasized by Fraenkel, Beck, Dadd, and House, to mention a few of the pioneers who provided guidelines) were roadblocks to nutritionists who might have attempted the daunting efforts of establishing mineral requirements but were discouraged by the difficulty of this task. For example, it is difficult to reconcile information about the functions of calcium, copper, iron, manganese, and zinc with the reports that boll weevils and fruit flies require none of these minerals (House 1974a). It is likely that traces of these minerals must have been present in the food or passed on by parental generations of the insects in these studies. It is against the current understanding of insect physiology, for example, that normal muscle function could occur in the absence of calcium and that normal energy utilization could occur without iron, which is essential to the cytochrome chains that are ubiquitous in aerobic biological systems.

3.4.1.1 Functions of Specific Minerals

All animals require minerals in their diets, including phosphorous, chloride, calcium, potassium, sodium, manganese, magnesium, iron, copper, and zinc. Potassium is involved in numerous chemical reactions and is a component in the structure of many substances, including lipids (phospholipids), some proteins, and nucleic acids. The energy-transferring compounds, including ATP, all rely on forming and breaking bonds with phosphate groups; therefore, it can be said that phosphate is absolutely essential to the entire process of bioenergetic activity. The various cellular control reactions that involve kinase-type enzymatic actions all rely on phosphorous transactions. Appropriate ratios of potassium to sodium or magnesium to sodium stimulate insect feeding responses (Cohen 1981).

Chloride (an ionized form of the element, chlorine) is also universally required by all organisms. Chloride is involved in the maintenance of membrane potential (i.e., electrical charge), which is a key part of the various actions of *excitable tissues and cells* such as muscle cells and neurons. Chloride also serves as a factor in several enzymatic reactions. For example, starch digestion by some amylases is chloride dependent or chloride enhanced. Potassium is an essential component in actions of excitable tissues, as is sodium. These two minerals are also involved intricately with regulation of pH in the cells and body fluids of insects and virtually all other organisms. All three of these minerals, chloride, potassium, and sodium, are involved in water regulation processes. Calcium is involved in muscular excitation and regulation of muscle responses to stimuli, and it also acts as a bridge between molecules, so it has a structural role in invertebrates, as well as a structural component of bone in vertebrates. Calcium is also a cofactor in several enzyme-driven reactions.

Magnesium functions in the glycolysis pathway involved in the conversion of carbohydrates to yield energy and in numerous enzyme actions in other pathways, including hexokinase, glucose-6-phosphatase, and pyruvate kinase (Lehninger 1993). Manganese is a cofactor in several enzyme actions, especially with metalloenzymes such as arginase and ribonucleotide reductase (Lehninger 1993). Zinc is a cofactor in many enzymes, including carboxypeptidase, carbonic anhydrase, and alcohol dehydrogenase (Lehninger 1993). Copper is a cofactor in several enzyme processes, including those involving cytochrome oxidase.

Iron is very important in several biological processes, including several enzyme reactions such as in pathways that biosynthesize DNA and RNA, in amelioration of oxidative stress (antioxidant activities), production of 20-dehydroxyecdysone (an ecdysis hormone), a part of the cuticle formation process, as part of the process of nitrogenous waste product synthesis, and as a component in the cytochrome system used in the conversion of stored chemical energy into useful ATP energy. Several of the various metabolic processes that require iron are listed in Table 3.3, and it is evident from this that many essential metabolic activities are dependent upon iron. For example, all growth and reproductive processes would come to a halt if iron were not present as a component of DNA and RNA synthesis reactions. All oxidative cellular respiration reactions would cease without iron, and free radical damage would occur on a wholesale level without this key mineral. Other processes such as ecdysone conversion to 20-hydroxyecdysone, toxin degradation by cytochrome P-450, and phenylalanine metabolism that is involved in neurotransmitter actions and cuticle formation would not be possible. Even the waste product (and nitrogen storage and antioxidant) uric acid would not be synthesized without iron.

Yet despite the fact that iron has such pivotal roles in the overall metabolic and physiological well-being of all insects, there are several ways that iron that is in the wrong place, at the wrong time, and in the wrong form can cause severe damage and in fact can act as a life-threatening toxin. Some of the eccentricities and paradoxes concerning iron in the diets and in the insects are discussed in Chapter 8 on complexity in diets.

Table 3.3 List of Metabolic Functions of Iron in Insects

Process	Enzyme(s)	Function
DNA and RNA synthesis	Ribonucleotide reductase, amidophosphoribosyl-transferase	Purine metabolism/synthesis
Cellular respiration	Cytochromes	Electron transfer for ATP production
Oxygen metabolism	Catalase, superoxide dismutase, oxygenases	Destruction of free radicals and hydrogen peroxide
Tricarboxylic cycle (Krebs cycle)	Aconitase	Perpetuation of precursor steps to energy metabolism
Steroid hormone production	Cytochrome P-450	Conversion of ecdysone to 20-hydroxyecdysone
Dealing with various toxins	Cytochrome P-450	Detoxification of various toxins
Phenylalanine metabolism	Phenylalanine hydroxylase and tyrosine hydroxylase	Neurotransmitter production, cuticle formation, and melanization
Purine metabolism (waste product and pigment compounds)	Xanthine oxidase	Uric acid production
Tryptophan metabolism	Tryptophan pyrolase	Step in pigment (ommochrome) metabolism
Oxygen transport in a few species of insects	Hemoglobin	Carrying oxygen to cells in some species of flies

Source: Locke, M. and Nichol, H., *Annu. Rev. Entomol.*, 37, 195, 1992.

Selenium is becoming increasingly well recognized as an antioxidant. Aluminum, nickel, and molybdenum are known to be cofactors in several enzyme reactions from plants or vertebrates, but they have not been shown to be used in insects. Fluoride and iodide have not been documented as having a role in insect nutrition (although they are present in the Wesson Salt Mixture, listed here).

Minerals cannot be biosynthesized; so if an insect requires a mineral, that mineral must be present in the diet in adequate amounts and appropriate form. It is possible for certain minerals to replace one another (i.e., have a *sparing* effect) such as zinc and manganese, which can replace each other in certain carboxypeptidases, amino peptidases, and other metalloenzymes. It is also possible for certain environmental minerals to displace essential minerals and thereby act as toxins. For example, rubidium and cesium can replace potassium, and at high enough concentrations of these trace minerals, they can become toxic. The phenomenon of the displacement by these minerals is used as a convenient marking tool for insects in field studies.

3.4.1.2 Immunological Roles of Minerals

One of the most important and until recently most poorly appreciated roles of minerals in insects is in the defensive functions in immunological/defensive responses and in protection against oxidative stress. The minerals listed in Figure 3.7 are well documented in various defensive roles. Zinc and manganese are involved in cuticle and peritrophic matrix formation and organization as cross-linkers for proteins, chitin, and glycosaminoglycans. Calcium can be thought of as the *master signaling material*. For example, Krautz et al. (2014) discuss the *damage signals* that calcium communicates in, also called *calcium flashes*. These flashes or bursts of calcium release stimulate the damaged cells to release hydrogen peroxide, which in turn summons hemocytes to move to the damage area. The other function of the hydrogen peroxide (whose production was stimulated by calcium) is the direct killing of invasive organisms such as bacteria, fungi, and parasitoids. A case of failure of defenses in honeybees is documented by Erickson et al. (1997) where the tyrosine nodules found in honeybees exposed to varroa mites were hypothesized to be an incomplete reaction of the bees' immune response.

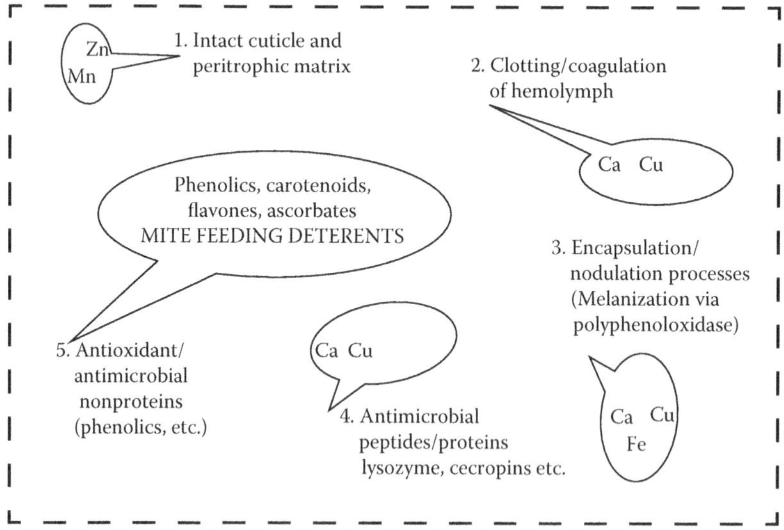

Figure 3.7 The minerals involved in *Drosophila* and honeybee and other insects' immune responses and antioxidant and antimicrobial defenses.

Copper is a moiety of polyphenol oxidase and as such, it is pivotal in the initiation of coagulation, encapsulation, nodulation, and melanization (Bidla et al. 2005). It is also part of the reaction sequence where cuticle is formed where it is involved in the polyphenol oxidase reactions in cross-linking proteins and chitin. This process is, of course, important in the first line of defense in insects—cuticular impenetrability. Other minerals of great importance in the protection of insects against stress include iron (Dunphy et al. 2002 and also discussed further in Chapter 8), which is important in antibacterial actions as well as its role in the cytochrome system and other defensive actions. Likewise, zinc, copper, iron, and manganese are important cofactors in actions of superoxide dismutase protection from oxidative damage almost universally (Corpas 2001).

3.4.2 Bioavailability of Minerals

It is not enough that minerals are present, but there must be a high enough degree of bioavailability in them to be useful to the species in question. Works on bioavailability of minerals (and other nutrients for that matter) in insects are lagging far behind those in the literature on vertebrates and even plants. The problem of bioavailability of minerals is essentially a digestive system issue. However, using models from vertebrates and other organisms, including plants and microbes, we can gain insights as to the importance of bioavailability of minerals, especially where the issue of mineral forms—especially counterions—is considered.

This complex topic starts with the solubility of different forms of minerals. If the mineral in question is present in an insoluble form, it will not be properly (uniformly) distributed in the diet. A focus on mineral mixtures added to diets is very revealing in terms of solubility. For example, Wesson salt mixture is composed of about 21% calcium carbonate and about 14.9% tricalcium phosphate or nearly 36% compounds that are listed in the Merck Index as *practically insoluble in water*. Their solubility is increased as pH is made more acidic, but at a common pH of insect diets of about 5.0, the calcium compounds in Wesson salts are going to dissolve only in trace amounts. Yet, it is common for diets that call for Wesson salts or similar mixtures (e.g. Yamamoto 1969 or Beck and Joachim 1976) to prescribe amounts of salt that are one or two orders of magnitude higher than would be soluble at the pH of these diet formulations. Therefore, if there is any hope of getting the diet components into the insect's system, it would have to be by ingestion of particulate portions of the Wesson salt mixture interspersed with other nonsoluble diet components. This raises another question about bioavailability of these minerals, how much of the mineral mixture's components can be absorbed under the gut lumen's physiological state? This topic will be discussed after mineral palatability is explained.

Clearly, if a mineral is going to be available to an insect (bioavailable), it must be palatable (or disguised in a palatable matrix). This is a prominent issue with people who use insects as feeders for reptiles, frogs, birds, fish, and other captive vertebrates that have high dietary calcium requirements. In tests with house crickets (*Acheta domestica* L.) consuming gels with various types of calcium compounds, the order of acceptance was as follows: calcium lactate >> calcium citrate > calcium phosphate monobasic = calcium phosphate dibasic = calcium phosphate tribasic > calcium carbonate (Cohen, unpublished data). The gels were made with 1% carrageenan, 0.05% calcium compound, and about 99% deionized water. The palatability of various minerals is another factor that constitutes part of bioavailability. In Figure 3.8, honeybees were provided with solutions of various mineral salts with citrate as the anion. Although all of the minerals tested suppressed acceptance of the syrup, iron had the strongest suppression effect. This limited information is strongly suggestive that further work is needed on mineral acceptability, interaction, and overall bioavailability in insect diets.

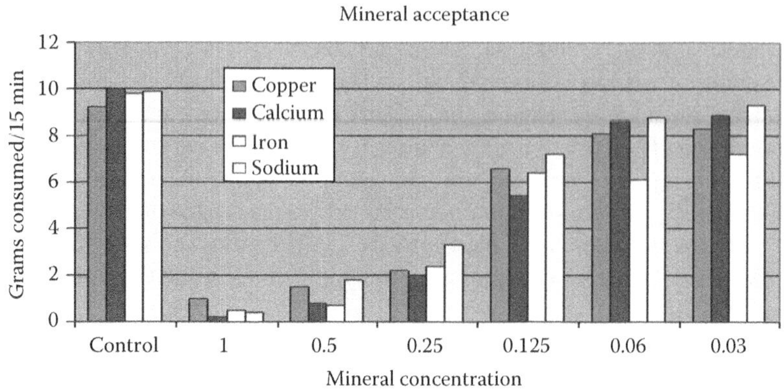

Figure 3.8 Honeybee acceptance of copper citrate, calcium citrate, iron citrate, and sodium citrate salts. Tests were independently conducted for each salt reported in percent of total diet weight, in 50% sucrose syrup (From Cohen, unpublished data).

3.5 FEEDING STIMULANTS

Many nutrients double as feeding stimuli (including sugars, some amino acids, lipids, ascorbic acid, and minerals). However, there are many cases where a substance is not used as a nutrient, but it does stimulate some part of the feeding process (biting, chewing, swallowing, etc.). Such substances are known as token stimuli (such as gamma amino butyric acid, sinigrin, a variety of waxes, and several plant secondary compounds). This topic is further discussed in Chapters 4 and 7.

3.6 PROTECTIVE INGREDIENTS

These are the substances that we add to diets to prevent microbial contamination, oxidation, or other means of destruction of nutrients. This includes (1) antibacterial agents such as streptomycin sulfate and chlortetracycline; (2) antifungal agents including sorbic acid, methyl paraben, propionic acid, and formalin; and (3) antioxidants such as ascorbic acid, tocopherol, and BHT. Many of these substances are toxic to insects in even fairly low concentrations. Also, many of these substances are very unstable under conditions of overheating, being kept in solution too long, exposure to light, or exposure to prooxidants. This will be further discussed under the diet treatment section.

The nutritional biochemistry of antioxidants is emerging as an important topic in human nutrition, and it is starting to gain interest in the insect biochemistry community. First, it must be made clear that within the great range of substances that have been identified as antioxidant, there is incomplete understanding of other functions (other than antioxidant properties) of many kinds of naturally occurring chemicals. For a long time, substances such as rutin, quercetin, and many other members of the phenolic family were thought to be present in plants for antiherbivore actions. More recently, however, it has become increasingly clear that many of these substances and myriad others (such as anthocyanins, lycopene, beta-carotene, and astaxanthin) are also key antioxidants that are very protective to the organisms that ingest them. Many foods, including cooked meats and certain cooked or raw plant materials, contain molecules that become (some by photoactivation) or generate *free radicals* and/or *oxygen singlets*. These free radicals attack various structures in the organisms' cells. For example, DNA and cell membrane lipids are common targets of free radical attack. These attacks are part of an aging process that manifests itself in a variety of unhealthy ways. Even insects are affected adversely by these attacks from free radicals, and we are only beginning to learn the

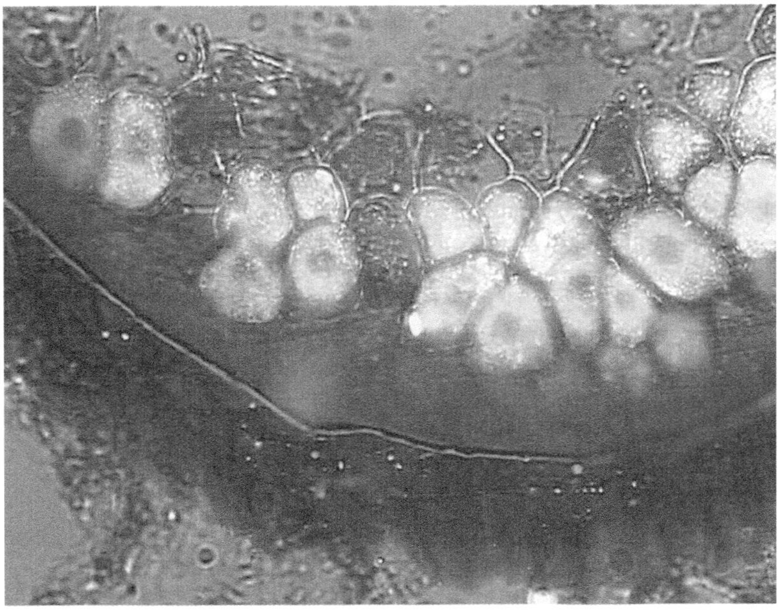

Figure 3.9 Cellular compartments containing membrane-bound lipids in wheat germ, stained with a fluorescent lipid marker.

liabilities of these rouge substances (see, e.g., Johnson and Felton 2001). What is becoming increasingly clear, however, is that certain kinds of antioxidants are useful (possibly absolutely essential) to many (possibly all) insects, and good insect husbandry demands that we respect these needs.

Diet stability is partly determined structure and ultrastructure of diet components. If, for example, the lipids and lipid-soluble vitamins are arranged in such a way that these *hydrophobic* (lipid-soluble or *lipophilic*) components are hidden deep within protein molecules, the protein coatings protect these components from oxidation (i.e., destruction). If, however, a processing technique is used that unravels the lipoprotein complex and exposes the lipophilic materials to more aerated parts of the diet, these components may now be subject to degradation. Heating, detergents, and harsh solvents are potential forces that can unravel protective proteins. Also, lipids are often present in biological materials in association with cell membranes and stored in vacuoles or other cellular compartments (Walstra 1996).

It is evident from Figure 3.9 that the lipids (which are stained with the fluorescent dye, Nile blue) are compartmentalized in only certain portions of the wheat germ. In this figure, the nonlipid material is stained a darker color that does not fluoresce, so a contrast can be seen with the lipids, and the nonhomogeneous distribution of lipids is evident as is the fact that some of the lipids have *escaped* from the wheat germ fragment and are marked by arrows labeled ELSs (escaped lipid spheres). These ELSs are more exposed to oxidative forces that can degrade them by oxidation of vulnerable molecular sites, especially where double bonds (known as points of unsaturation) exist.

3.7 *NUTRITIONALLY INERT* INGREDIENTS PROVIDE TEXTURE

Diet texture may be modified by the use of fillers (such as cellulose in various forms–powders, grits, flakes) and gelling agents (such as agar, gums, and carrageenan). Some nutritionally inert components are added to diets deliberately as bulking agents or carriers for other substances. For example, Thompson (1975) used Sephadex beads for carrying lipids into the diet matrix. Several authors have reported using various kinds of microparticulate cellulose such as Cellufil® or other cellulose-based fillers (Singh 1977). Other fillers are added unintentionally. For example, there are inert portions of wheat germ, soy flour, various bean meals, and other plant-derived materials. These components act as carriers for lipids and lipoproteins as well as bulking agents that may contribute to the stimulation of peristalsis (House 1974b) and other normal digestive processes (Chapters 4 and 7). Figure 3.9 (was 3.7) shows the distribution of lipids in a fragment of wheat germ that also contains cell walls that consist of cellulose, which is one of the components that are naturally occurring, inert, background material. The extent to which these bulking agents or naturally occurring, inert materials are conducive to normal gut function has not been demonstrated in insects as they have been in vertebrate models (Stevens and Hume 1995). However, as is apparent from the many successful diets that employ these bulking agents (Singh 1977), it seems likely that they have similar benefits to insects as they do in vertebrates. Regardless of their effects on gut motility and assimilation kinetics, bulking agents certainly contribute to desirable textures of solid and semisolid diets, and they reduce the dependence upon expensive gelling agents to improve texture.

3.8 IMPORTANCE OF pH AND ITS INFLUENCE ON DIETS

Acidity or alkalinity (pH) imparts several features to diets. A diet's pH influences its palatability, its stability, the activity of preservatives, the solubility of nutrients, and probably among other factors. Most antifungal agents work only at acidic pH. Even without antibiotics, bacterial growth is suppressed at a lower pH. The substances most often used to lower the pH of diets are hydrochloric acid, acetic acid, and phosphoric acid. Sorbic acid and propionic acid are often used as antifungal agents, but they also lower the pH of diets. Some acids are commonly used in human foods and have been used in insect diets including, benzoic acid, citric acid, lactic acid, formic acid, and tartaric acid (Singh 1977). Bases such as sodium hydroxide, potassium hydroxide, sodium carbonate, and sodium bicarbonate are used to raise pH. For reasons of palatability, microbial control, and suppression of oxidative and hydrolytic deterioration, insect diets are generally designed to remain in the slightly acidic range.

3.8.1 Buffers

Many ingredients that are added to diets for purposes other than pH stability also happen to act as buffers, that is, agents that resist changes in pH. For example, proteins are inherently very good buffers, and the proteins in soy flour are noteworthy for their ability to help hold slurries of soy flour in water at slightly acidic pH. However, many diets employ buffers to hold pH as constant as possible. Such buffers include the phosphates and sulfates of sodium, potassium, magnesium, and calcium. Buffers can also be of nutritional benefit. For example, the addition of potassium or magnesium compounds makes these minerals available to insects for nutritional needs, and they also serve as phagostimulants (Cohen 1981).

3.9 WATER CONTENT (PERCENTAGE) AND WATER ACTIVITY (a_W)

The concepts of water content and water activity (a_w) are helping explain how artificial diets work and why they sometimes fail. First, it is sometimes overlooked that water is the most fundamental nutrient. Without the appropriate amount of water, all life processes fail. While some organisms can use metabolic activities (oxidation of fuels) to manufacture enough water to sustain their life processes (e.g., some desert insects described by Edney 1977), most organisms need formed water in their foods or from a drinking source. Regardless of the accommodations made to support insects, inadvertent creation of water stress can be disastrous to a rearing program. Also, a_w is a key factor in the chemical reactions and physical characteristics of diets.

As a rule, the normal amount of water present in the insect's natural food is required in an artificial diet. For example, leaf feeders such as cabbage loopers or beet army worms are adapted to food that is about 90% water. Beet armyworms that thrive on fresh cotton leaves would be stressed on wilted leaves that were only 80% water or on an otherwise nutritious artificial diet with only 80% water (Cohen and Patana 1982). Even with the right percentage of water, an insect could be water-stressed by a diet whose nitrogen content was too high. Such a diet could cause the insect to excrete extra waste nitrogen forcing it to excrete an inordinate amount of water to rid itself of toxic nitrogenous wastes (Edney 1977). Conversely, nutritional stress is imposed by providing too much water to an insect that is adapted to feeding on foods that are concentrated in nutrients. Failure to recognize this fact and apply it to hemipterans (such as Lygus bugs and stink bugs) has been the basis for many shortcomings in rearing these insects (Cohen 2000a,b).

Even in situations where water content is adequate, water activity (a_w) can be inappropriate in a diet. The term a_w is a thermodynamic concept indicating the availability of water present in a given material. Water activity is a measure of the potential of water to move from one region to another. The a_w is expressed in terms of equivalent relative humidity; so an a_w of 1.00 (=100%) is the equivalent of 100% relative humidity, and 0.50 is equivalent to 50% R.H. A gel made of 5% of the gelling agent carrageenan and 95% water has a water activity of nearly 1.00, but 5% salts and 95% water many have a water activity of less than 0.80 (e.g., depending on which salt is used, NaCl contributing to a much lower water activity than an equal weight of KCl). This difference in water activity results from the fact that carrageenan does not bind the water to anywhere near the same extent as does the salts. Despite the fact that both solutions have equal amounts of solids and water, the water in the salt solution is much less mobile and less available for absorption and to support life processes. This unavailability of water (which is apparent to anyone who drinks sea water) affects the target insects and the microbial contaminants. The possibility of using this characteristic of a_w to optimize diets to reduce contamination is discussed in Chapter 13.

3.10 NUTRITIONAL PROFILE OF FIVE DIET PROMINENT COMPONENTS

Table 3.4 shows profiles of the nutrient composition of five materials (complex foods or staples) that have proven to be excellent bases for insect diets: wheat germ, soy flour, egg yolk, broccoli florets, and beef liver. First, it is clear that wheat germ has several features that make it an excellent source of nutrition: (1) a very high protein content, with a well-distributed profile of amino acids, including a good representation of all the *insect-essential* amino acids. (2) It has a high lipid content, which will be discussed in a later chapter in terms of the importance to insects, especially the polyunsaturated fatty acids. (3) It is abundant in trace minerals with what we now know is an *insect hospitable* high ratio of potassium and magnesium to sodium and ample amounts of iron, zinc, copper, manganese, and selenium—all of which are discussed in the chapter on nutrient function. (4) The vitamin content is fairly high with the exceptions of ascorbic acid (vitamin C), vitamin E, and vitamin A, or the precursors, which are members of the carotenoid family. Several decades after the publication of Adkisson et al. (1970) on pink bollworm diet, it was learned that wheat germ contains some antinutrients, including lectins (the major one known as wheat germ agglutinin [WGA]) and digestive enzyme inhibitors that impede that activities of proteolytic enzymes. Fortunately, most of these antinutrients are detoxified by various degrees of heating, especially with adequate amounts of water present—conditions satisfied by formulation conditions for most insect diets.

Soy flour has a very similar profile to wheat germ with the exceptions that soy has a higher protein and lipid content than wheat germ. This differential is reversed for carbohydrate content. Both wheat germ and soy have high vitamin contents, except for ascorbic acid. The amino acid contents

Table 3.4 Nutritional Components (Amount per 100 g) of Wheat Germ, Soy, Egg Yolk, Broccoli Florets, and Beef Liver

Component	Wheat Germ[b]	Soy Flour[c]	Egg yolk	Broccoli Florets	Beef Liver
Water	11	3.8	48.8	90.7	69
Energy (kcal)	(360)	441	358	28	143
Protein	23	35	17	3	20
Total lipid	10	22	31	0.4	3.9
Carbohydrate	51	34	2	5.4	5.8
Fiber	13	9.7	0		
Ash	4	6	7	0.9	1.3
Minerals					
Calcium	0.039	0.188	0.137	0.048	0.006
Iron	0.006	0.006	0.004	0.0009	0.007
Magnesium	0.239	0.369	0.009	0.025	0.019
Phosphorus	0.842	0.476	0.488	0.066	0.318
Potassium	0.892	2.041	0.094	0.325	0.323
Sodium	0.012	0.012	0.043	0.027	0.073
Zinc	0.012	0.004	0.003	0.4	0.004
Copper	0.0009	0.002	0.03	0.05	3.3
Manganese	0.013	0.002	0.07	0.023	0.026
Selenium	0.00008	0.00008	0.00001	0.000003	0.000041
Vitamins					
Ascorbic acid (vitamin C) (mg)	0.0	0.0	0.0	93.2	22.0
Thiamin (mg)	1.9	0.4	0.2	0.07	0.26
Riboflavin (mg)	0.5	0.9	0.6	0.12	2.8
Niacin (mg)	6.8	3.3	0.02	0.6	12.8
Pantothenic acid (mg)	2.3	1.2	3.8	0.5	7.6
Vitamin B-6 (mg)	1.3	0.4	0.4	0.16	0.9
Folate (mg)	0.28	0.23	0.15	0.07	248
Vitamin B-12 (mg)	0.0	0.0	0.003	0.0	0.07
Vitamin A (IU)	0.0	110	1945	3000	35346
Vitamin E	0.0	0.0	3.2	1.7	0.67
Lipids					
Saturated fatty acids	1.7	3.2	9.6	0.054	1.5
14:0	0.01	0.06	0.1	0.0	0.04
16:0	1.6	2.3	6.8	0.047	0.47
18:0	0.06	0.9	2.4	0.007	0.96
Monosaturated fatty acids	1.37	4.8	11.7	0.024	0.51
16:1	0.03	0.06	0.9	0.0	0.04
18:1	1.33	4.8	10.7	0.024	0.47
Polyunsaturated fatty acids	6.01	12.3	4.2	0.17	0.84
18:2	5.29	10.9	3.6	0.04	0.35
18:3	0.72	1.5	0.1	0.13	0.0
Cholesterol	0.0	0.0	1281	0.0	354
Amino acids					
Tryptophan[a]	0.3	0.5	0.2	0.029	0.29
Threonine[a]	1.0	1.5	0.9	0.091	0.92

(Continued)

Table 3.4 (*Continued*) Nutritional Components (Amount per 100 g) of Wheat Germ, Soy, Egg Yolk, Broccoli Florets, and Beef Liver

Component	Wheat Germ[b]	Soy Flour[c]	Egg yolk	Broccoli Florets	Beef Liver
Isoleucine[a]	0.9	1.7	0.9	0.109	0.92
Leucine[a]	1.6	2.8	1.5	0.131	1.9
Lysine[a]	1.5	2.3	1.4	0.141	1.4
Methionine[a]	0.5	0.5	0.4	0.034	0.51
Cystine	0.5	0.6	0.3	0.020	0.31
Phenylalanine[a]	0.9	1.8	0.7	0.084	1.1
Tyrosine	0.7	1.3	0.8	0.063	0.79
Valine[a]	1.2	1.7	0.9	0.128	1.24
Arginine[a]	1.9	2.7	1.2	0.145	1.26
Histidine[a]	0.6	0.9	0.4	0.050	0.55
Alanine	1.5	1.6	0.9	0.118	1.19
Aspartic acid	2.1	4.4	1.6	0.213	1.92
Glutamic acid	4.0	6.7	2.1	0.375	2.71
Glycine	1.4	1.6	0.5	0.095	1.15
Proline	1.2	2.0	0.7	0.114	1.06
Serine	1.1	2.0	1.4	0.100	1.0

Source: Anonymous, USDA Nutritional Database for Standard Reference, Release 27 (accessed February, 2015), 2013.
[a] Indicates that the amino acid is found to be essential for many species of insects.
[b] Wheat germ.
[c] Soy flour, full-fat, roasted.

of wheat germ and soy are both rich in all essential amino acids. Egg yolks, broccoli, and beef liver all have much higher water contents than do wheat germ and soy, but once the water contents are corrected for, the nutritional composition of all these foods is fairly similar. An exception to this is the high ascorbic acid content of broccoli and beef liver.

3.11 OVERVIEW OF DIET ADDITIVES

The purpose of most additives is to prevent the degradation of the foods by the general array of phenomena that are collectively categorized as equilibrium processes (Lindsay 1996). The same concept can be applied to insect diets. As discussed in Chapters 3, 5, and 8, insect diets and most foods are generally in a nonequilibrium state where components that we wish to have associated tend to dissociate. Chemicals deteriorate by oxidation or hydrolysis; flavors and aromas evaporate; regions of low water activity absorb moisture; and numerous other forces take place decreasing the palatability and nutritional quality of diets.

3.12 EMULSIFIERS

Emulsifying agents are stabilizers. They are chemicals that cause lipid-phase materials and aqueous-phase materials to mix and retain a long-term interaction or interfacing with one another. Broadly speaking, there are two classes of emulsifiers in insect diets, natural ones and artificial agents. The natural ones include the diet components that are otherwise nutritional but which can serve a dual role of nutrition and emulsification. Many proteins and phospholipids (polar lipids, in general) act as natural emulsifiers. Egg yolk proteins and phospholipids are among the most excellent natural emulsifiers. Milk proteins, soy proteins, and soy lecithin (phospholipids) are also excellent and widely used emulsifiers, both in insect diets and in human foods. The most commonly used group of artificial emulsifiers in insect diets is the polyoxyethylenesorbitans known as Tweens. Emulsifiers are simply a special form of stabilizer. The mechanism of emulsification is that these molecules have both a polar and nonpolar region within the same molecule. The polar end can form stable associations with water, while the nonpolar end can associate with lipids. The mixed character of emulsifiers encourages stable complexes of polar and nonpolar portions of diets that would otherwise dissociate as do typical oil–water interfaces.

3.13 GELLING AGENTS AND STABILIZERS

Gelling agents improve insect diets in four ways: (1) They render a high water content mixture into a solid (or gel) state so that solid-feeding insects are accommodated and that insects that tunnel will not have their food collapse on them. (2) They help preserve the mixed state of the diet components, preventing settling of more dense materials and floating of the less dense ones. (3) They help preserve the nonequilibrium conditions that help prevent the reactions that take place between ingredients. (4) Some gelling materials such as proteins, pectins, and starches act as nutrients. Forming gels and increasing viscosity of diets are ways of altering diet *textures*. In food science and technology, increasing viscosity is known as *thickening*, but this term has not yet gained recognition in insect diet literature. Gelling a group of ingredients that have been treated with emulsifiers provides an extra assurance that the lipid-compatible and water-compatible components will remain in place, contributing to the stability of the diets organization. This is further discussed in Chapter 4 on how diet organization into matrices or dispersions enhances diet quality. The specifics of texture analysis by rheological techniques are discussed extensively in Chapter 15; however, some of the functional aspects of texture in relation to other diet components are discussed here.

The mechanism of gelling is discussed in greater detail in Chapter 5. Briefly, gels form as a result of hydration of the macromolecules called gelling agents. The gel is an association of water molecules with the long, often branched gel formers. Once water is bound to the macromolecules, the freedom of movement that is seen in liquid water is no longer present, and the restriction of movement of the water is said to be a gel. This restriction of flow is a stabilizing feature of gels.

Carbohydrates are the most common gelling agents in foods. They include gum arabic, guar gum, locust bean gum, carboxymethylcellulose, carrageenan, agar, starch, and pectin (Lindsay 1996). These gelling agents are collectively known as *hydrocolloids*. The only protein commonly used strictly as a gelling agent is gelatin, a partially hydrolyzed form of collagen. Sometimes, the term *gelatin* is used to connote carrageenan, but it is most commonly used in reference to the collagen-derived protein. BeMiller and Huber (2008) characterized the predominant carbohydrates that are used to gel and thicken, pointing out that in human foods various forms of starch fill this role far beyond all other agents combined. Although starch is not commonly added to insect diets deliberately to gel or thicken the diet (i.e., as a texturizing agent), it serves both as a nutrient and as a texturing agent.

The texturizing characteristics of starch have not been evaluated in insect diets as they have been in human foods. For example, it would be useful to know the comparable rates of passage of materials through the intestinal tract with a digestible gelling agent such as starch versus a nondigestible gelling agent such as agar or carrageenan. Such a comparison would give diet development specialists a basis for substituting less expensive starch for more expensive agents such as agar or carrageenan. For example, a recent Sigma catalog lists unmodified wheat starch at about $5/kg and one of their lowest priced agars at about $100/kg. Obviously, the possibility of making such a substitution is an excellent topic for further investigation, given the fact that gelling agents are often one of the most expensive ingredients in insect diets. It would also be useful to have a precise understanding of the role played by cryptic starch or starch that was not deliberately added to insect diets in contributing to desirable texture and overall nutrition. Such cryptic starch would include that which is present in wheat germ, soy flour, rice or corn meal, and that which is present in a large variety of plant materials that are commonly used as diet components.

Carbohydrates are the most common thickening and gelling agents (both in insect diets and in human foods). They include starches (amylase and amylopectin), agar, guar, locust bean gum, xanthan, carrageenan, aligns, pectins, and gum arabic. The principal differences in the chemistry of these macromolecules are in the types of sugars present, the presence or absence of side groups such as sulfates, the type of linkages between the sugars (α- or β-linkages), and whether or not the structures are linear or branched. For example, starches and celluloses differ tremendously in their digestibility by insects and most other animals, yet it is the simple difference between the α-linkages of starches and the β-linkages of celluloses that make the starches highly susceptible to digestion by most species of insects and the celluloses indigestible to most insects (except for termites and wood roaches).

3.13.1 Polysaccharides as Gelling and Texturizing Agents

The term *polysaccharide* refers to polymers of monosaccharides (single sugars) that are linked by glycosyl bonds in long chains (more than 20 monosaccharides long, sometimes in the tens of thousands of monosaccharides) that are linear, branched, or a combination of linear and branched. Polysaccharides may consist of only one type of sugar and is thus called a homoglycan, or they may be composed of two or more kinds of sugar and are called heteroglycans. Examples of these are starch, glycogen, and cellulose—all homoglycans—agar and carrageenan are heteroglycans.

3.14 ANTIOXIDANTS

Recent literature on the roles and nature of antioxidants is emerging as one of the most dynamic areas of current research into mammalian, plant, and insect homeostasis. The well-known, well-established, and natural antioxidants, ascorbic acid, tocopherols, and carotenes, are only a few of the many compounds that are related to these and many others from different chemical families. For example, proteins, nucleic acids, purines, and a large variety of lipids have antioxidant potential as do anthocyanins, isoflavonoids, and a variety of other naturally occurring compounds (Damodaran et al. 2008, Lindsay 1996, Anonymous 1999). In human foods and in pharmaceuticals, several artificial antioxidants are used to protect delicate components, especially lipids whose double bonds are susceptible to oxidation. These compounds include butylated hydroxyanisole (BHA) and butylated hydroxytoluene (BHT). Neither of these nor any other food antioxidants have been widely used in insect diets. The potential toxicity of BHT to insects is discussed in Chapter 8. The methods of analysis of antioxidants and free radical scavengers are discussed in Appendix A, where a detailed protocol is presented for using a standard assessment technique, the 1,1-diphenyl-2-picryl-hydrazyl (DPPH) antioxidant or free radical scavenger assay (Sharma and Bhat 2009).

3.15 ANTIMICROBIAL AGENTS

As discussed in several places in this book, insect diets are ideal targets for microbial contaminants to utilize, especially under culture conditions where the insects are held at higher than standard room temperatures. Because of the threat of microbial contamination, antimicrobial agents, especially antiprotozoan, antifungal, and antibacterial agents that are chemically based, have been used as additives to insect diets for more than a half century. A tremendous variety of these agents have been used, and these chemicals are discussed in detail in Chapter 13.

3.16 FLAVORING AGENTS

Although there is considerable data regarding various chemicals that serve as recognition stimuli for insects, there has been less attention than expected in the use of such stimuli in improving consumption of artificial diets. On the other hand, the use of flavoring agents has been widely exploited in the development of baits and attractants for a large number of insects (Bernays and Chapman 1994, Chapman and deBoer 1995). The term *sign stimuli* encompasses a variety of sensory signals, including chemicals, which animals use to recognize key features of their environment, including specific hosts (Bernays and Chapman 1994).

3.17 COLORIZING ADDITIVES

It is well known that color plays a part in recognition and acceptance of foods by insects (Bernays and Chapman 1994). However, there has been little attention to the enhancement of artificial diet acceptance by use of colorizing additives, which are commonly used in human foods to add to the attractiveness or *eye appeal* of foods. In the human food industry, it is a common practice to use food coloring in such foods as soft drinks, candy, and pastries to enhance the attractiveness of these commodities (Lindsay 1996). Foods, including some insect diet components, often have natural colors, which are associated with various nutrients, including plant pigments such as chlorophyll, carotenes, xanthophylls, quinones, anthocyanins (and other flavones), and betalaines (von Elbe and Schwartz 1996). Most of these compounds or chemical families not only add color to foods, but they also serve as antioxidants. Although bright color is not associated with every good antioxidant compound (e.g., ascorbic acid is not brightly colored), it is a good rule of thumb that if a food has a bright color, it will contain high concentrations of some type of antioxidant. For example, red grapes contain higher concentrations of antioxidants than do white grapes, and red peppers contain higher concentrations of antioxidants than do yellow peppers (Anonymous 1999a, 2013).

3.18 BULKING AND TEXTURIZING AGENTS

Bulking and texturizing agents are considered separately from gelling agents, but gelling is certainly an aspect of modifying texture. As is more thoroughly discussed in Chapter 7, the role of dietary fiber materials (e.g., cellulose, pectin, and starches) in the motility of the digestive tract and in the availability of nutrients or in the potential impact of antinutrients is well documented in vertebrates (Stevens and Hume 1995). However, such documentation is much less abundant in insects. However, the generalization that bulk materials have an impact upon gut residence time of foods has been empirically recognized for several decades (Waldbauer 1968). The components that act as bulking and texturizing agents tend to be macromolecular aggregates and as such have properties (shape and charge) that make them bind minerals (i.e., they may act as chelating agents), vitamins, lipids, and other key dietary components. These effects can be most valuable in carrying otherwise insoluble or intractable materials into the diet in a stable and biologically available form. Also, bulking agents can serve as insulators that prevent undesirable reactions from taking place between components such as the interactions that take place between iron or copper with ascorbic acid, increasing the tendency for lipid peroxidation to take place (discussed further in Chapters 4, 5, and 8).

3.19 CHELATING AGENTS

As is discussed in the Chapter 5, on the chemistry of food components, many of the metals in insect diets are reactive, and they can form complexes or catalyze reactions that destabilize diets. Many naturally occurring substances are or contain chelating agents. For example, the organic acids (citric, malic, tartaric, oxalic, and succinic), polyphosphoric acids (ATP and pyrophosphate), and proteins are all excellent chelating agents that occur commonly in diet ingredients. Chelating agents suspend metal ions in solution and prevent precipitation of such metal ions in solutions whose pH is suitable for retaining the metal in a soluble form. So, for example, if a calcium (Ca^{++}) ion is not chelated, it can form an insoluble carbonate compound, especially if the pH of the solution is elevated above neutral to a mildly or strongly basic level. Once the insoluble calcium carbonate is formed, this compound is apt to precipitate and become nonhomogeneously distributed and subsequently unavailable for ingestion. The same type of outcome can result from complexes of magnesium, iron, manganese, and other metallic minerals; but the presence of a chelating agent can prevent the loss of these minerals from the diet's available nutrient pool. If the chelating agent is not present as a natural component, it may be useful to add deliberately as artificial chelators. Figure 3.10 shows a model of a molecule of ethylenediaminetetraacetic acid (EDTA) chelating a divalent cation, calcium. Note that the calcium (Ca^{++}) ion has been captured and sequestered within the grasp (i.e., chelation) of the two acid groups (COOHs) that are part of the acetate complex ($-C_2COOH$). The EDTA has the capacity to hold another divalent ion in the upper portion of the model molecule, or it could hold four monovalent cations (such as Na^+ or K^+). It is the very strong tendency that such substances as EDTA, EGTA, citric acid, and phytic acid have for sequestering metal ions that gives these molecules the property that we call chelation ability. Organic complexes such as heme groups (Figure 3.11) can also deliver certain minerals in a very efficient manner. This delivery is discussed further in Chapters 4, 5, and 8.

In insect diets, the deliberate use of chelating agents has been practiced infrequently, or at least, it has been seldom expressed that a component was added deliberately for the purpose of chelation. This is despite the fact that the value of sequestration of insect diet components by chelators (sometimes referred to as sequestrenes) has been known since the mid-1960s (Mittler 1972). Mittler (1972) provided a useful explanation of his use of EDTA and various other chelating agents to get minerals to remain in solution. In this account of his odyssey of diet development for aphids, Mittler explained that he had added cholesterol to the liquid diet in the form of a dispersion by mixing a cholesterol solution of acetone with the aqueous phase and then boiling the diet to remove the acetone. A most fascinating aspect of this account is that early on in this research, Mittler found that during the overall processing of the diet, a cloudy mixture resulted from efforts to include all the components, and the researchers attributed the cloudiness to the cholesterol, only to later learn that

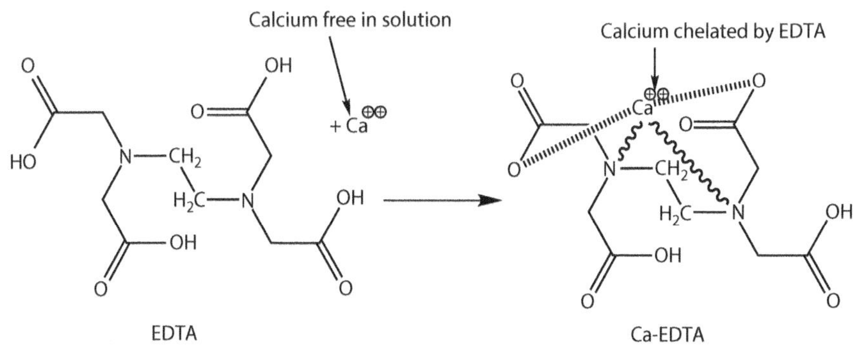

Figure 3.10 Model of a molecule of EDTA chelating a divalent cation.

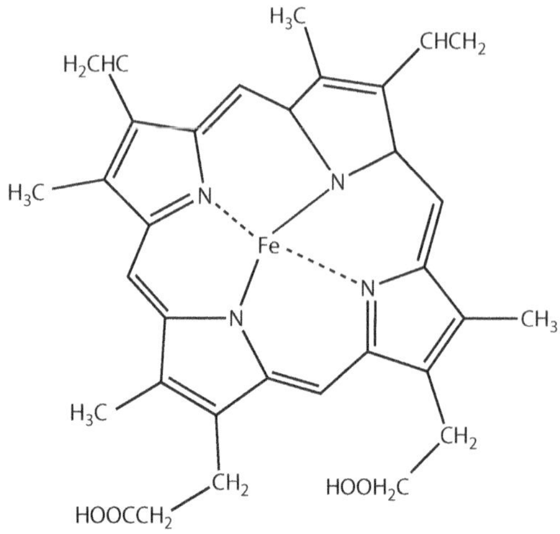

Heme

Figure 3.11 Structure of the heme molecule with iron at its center.

the precipitate was a magnesium/phosphate interaction that was taking place as the pH of the solution was raised to more strongly basic levels. Mittler's research team eventually concluded that the failure of the aphids to thrive on the diet was not a result of the loss of cholesterol but rather from the loss of essential minerals, a loss that resulted from attempts to provide minerals that had not been chelated and that had been subjected to unfavorable pH. Mittler also commented on the irony of his later discovering that the cholesterol was not even necessary as a dietary additive but that rather the symbionts in the aphids provided the required sterols. This view was later questioned by Campbell and Nes (1983) who argued that the symbionts in aphids are not metabolically capable of providing sterols. This complex issue is further visited in Chapter 13. The discussion by Mittler (1972) is further treated under the topic of the importance of using a proper order of mixing components in the processing of diets. The importance of chelating agents in forming stable dietary matrices cannot be overemphasized (Mittler 1972).

CHAPTER 4

What Makes a Diet Successful or Unsuccessful?

4.1 OVERVIEW

The subject of this chapter is difficult because of its complexity and intricacy, but most especially because it is inherently reliant on circumstantial evidence to explain why certain diets succeed while others fail. Even the concepts of success and failure are complex, because they can be subjective and dependent upon the specific situation and application. These complexities have motivated a careful search of historical progression of some of the most well-established diets. The effort to understand the reasons why certain diet components seem to work well in our goal to rear healthy and reliable populations of insects has led inexorably to several main ingredients that crop up frequently and in a large variety of diets, with respect to the diversity of species that these diets support. In this chapter, what was sought whenever possible was a rationale for the use of a component and the mechanism of its role in diets.

Over the past century of research, only a few diets have been developed that can be clearly considered as fully successful. While *success* is a subjective term, there are several qualities that can be taken as key requisites for a highly useful diet. Such a diet will support robust feeding, development, growth, reproduction by unlimited numbers of continuous generations, and populations that rival those taken from the field in healthy behavior and physiology (such as the last possible outcome listed in Table 4.1). These exceptional diets and their successful derivatives share certain features that explain their high quality.

The differences between these successful diets and the many less successful ones (those that give outcomes described in the first few rows of Table 4.1) are a basis for understanding how diets should work: (1) The top-notch diets contain appropriate feeding stimuli to elicit complete and hearty feeding responses. (2) Top diets contain all essential nutrients in appropriate amounts and adequate amounts of beneficial components. (3) These diets, as defined here, are all dispersions of a complex of ingredients that constitute a well-organized and compartmentalized matrix. An important part of the matrix concept is the organization of the diet components in physical relationship to one another. Are the components dissolved, or suspended in water or lipid portions of the diet? Are the components such as proteins and carbohydrates in molecular entanglements with one another? What are the size relationships of the diet components (Table 4.2) in the context of their distribution in the diet, and how does the distribution mesh with the feeding apparatus of the insects (i.e., are the components capable of being sensed and ingested?) The specific organization of these matrices assures that each component that is required or at least useful is present in a context that makes the said component biologically available. (4) Key components meet the

Table 4.1 Possible Outcomes of Feeding Experiments with Artificial Diets, Based on Hypothetical Tests Carried Out with Insects That Can Be Laboratory Reared on Natural Diets

Profile of Experiment	Possible Explanation of Results
1. No feeding—no growth	Missing phagostimulant(s) or faulty texture
2. Skimpy feeding—no growth or poor growth, no or poor reproduction	Missing phagostimulant(s), missing nutrient(s)? Toxin
3. Robust feeding—no growth or poor growth, no or poor reproduction	Missing nutrient(s)[a], nutrient imbalance, wrong matrix, trace of toxin
4. Robust feeding—growth (and development) but no or poor reproduction	Missing or low levels of nutrient(s)[a], nutrient imbalance, trace of toxin
5. Robust feeding—growth (and development) with limited reproduction	Missing or low levels of nutrient(s)[a], nutrient imbalance, trace of toxin
6. Robust feeding—growth (and development), good reproduction over many generations	No problems, good nutrient balance, good matrix, no toxins, or toxins at concentrations or in forms that insect can handle

[a] Nutrients may be either missing or not biologically available.

Table 4.2 Perspectives on Size of Diet Components

Length	Example of Object in This Size Range	Detectable by	Diet Components in This Range	Specific Stokes Radii of Certain Molecules[a]
0.1–1.0 nm	Small to medium molecules	Below microscopic threshold	Molecules of water, simple sugars, medium-sized molecules	Water = 0.15 nm Sucrose = 0.47 nm
1.0–10.0 nm	Depth of cell membrane	Electron microscope	Macromolecules (proteins, starches), high-density lipoproteins (~10 nm)	Dextran = 2.2 nm Bovine serum albumin = 3.6 nm
10.0–100 nm	Ribosomes (smaller organelles of cells)		Macromolecules, low-density lipoproteins (~20 nm)	
100–1000 nm	Cell nuclei		Lipid micelles, chylomicrons (~50–200 nm)	
1.0–10 μm	Bacteria, small cells	Light microscope	Finest grains of flour, small egg yolk particles	
10–100 μm	Amoeba, large cells, span or gape of neonate beet armyworm		Various plant and animal cells, small grains of flour	
100–1000 μm	Larger insect cells, length of neonate beet armyworm		Small particle size in plant meals and flours	
1.0–10 mm	Small- to medium-sized insects	Human vision	Larger particles in coarse meal, aggregates	
10–100 mm	Chicken egg yolk, larger insects		Whole diet aliquots	

[a] Partially derived from Buchanan et al. (2000).

requirements of *palatability*, *nutritional completeness*, *bioavailability*, and *stability*. (5) As organized matrices, these diets offer appropriate components with a chemical stability and spatial order that suits the needs of the target insect's feeding apparatus. (6) These diets are designed to be properly preserved so that they maintain a freshness and wholesomeness that is in accord with the needs of the insects. The concept of freshness may be modified in considering diets for insects that are adapted to feed on rotting or decaying materials, but even with these insects, the diet must retain a certain nutritional inertia or integrity. (7) All diets that have succeeded in supporting continuous generations of relatively large, fecund progeny contain at least some amount of undefined nutrients. The undefined nutrients are often described as the diet's *base* or the *staple* portion of the diet or *complex food* (terms such as *wheat germ-based, soy flour-based, pinto bean-based,* or *meat-based* diet are frequently used). The other ingredients in such diets are considered supplements to the staples. (8) In successful diets, the antinutrients are eliminated from the diet or rendered completely nontoxic. (9) The proportions of the macronutrient classes are in accord with the target insect's feeding adaptations. (10) Other than chemically defined (usually liquid) diets, components in successful diets exist as aggregates of diet staples (complex foods), water, and other functional components, which are present as solutes, hydrocolloids, or complex entanglements of proteins, carbohydrates, and complex lipids.

The last point means that if an insect is adapted to feeding on a food such as broccoli florets, which contain ~91% water, 3% protein, 0.4% lipid, and 5.2% carbohydrate, a high protein diet (e.g., >10%) with less than 80% water and a high fat, high carbohydrate content would be unlikely to succeed. For insects with very specialized diets, which we may consider as *exotic* diets (such as xylem sap-feeders, wool-eating, wax-eating, wood-eating insects, blood eaters, endoparasitoids, dung-eaters, or insects that feed on plants with a very unique secondary chemistry), special accommodations must be made.

Once the minimal requirements are met, the proportions of gross nutrients are correct, and the insects feed heartily on the diet, it would seem that everything should be in place for successful rearing. However, even after all these basics have been met, many diets still fall short. After more than three decades of studies of natural and artificial diets, this author is convinced that once the minimal, balanced requirements are provided, the deciding factor that determines a diet's success, the most important feature of a successful diet, is in the nature of its organizational matrix. Evidently, it is not enough to have an essential component such as cholesterol, iron, or an essential amino acid present but equally important is the presence of a suitable arrangement of the diet components, an appropriate diet *matrix*. The components must be present in a matrix that protects the component and also makes it available.

The availability can be subdivided into the categories *accessibility* and *bioavailability*. Accessibility pertains to the characteristics of the nutrients and feeding stimuli as they meet requirements of the specialized feeding structures and sensory apparatus. Simply, the diet must be in the appropriate form, both chemically and physically, and the components must be arranged in a way that they are within reach of the mouthparts so that once the diet materials are detected, they can be ingested. The bioavailability of all diet components depends, for example, on whether the lipids are dissolved in lipoproteins, suspended lipid micelles, suspended chylomicrons (lipoprotein aggregates), or embedded in aggregates of insoluble carbohydrates. Examples of such aggregates and their matrix are presented in Figures 4.1 through 4.4, and they are discussed later in this chapter. As a comparison with these artificial diets, a cotton leaf with a whitefly is shown in Figure 4.5 to demonstrate the relatively high degree of variety of potential feeding targets that are present in insects' natural foods.

The structure and interaction of diet subunits at various scales (molecular, macromolecular, or huge aggregates) are what characterize the dynamics of diet function. Using lipids as an example, the arrangement of the lipid-containing subunits is what determines if each component is suitable for ingestion, digestion, and absorption than would be the free lipid form. Likewise, minerals often require a physical context of chelating agents and absorption facilitators such as ascorbic acid for iron and manganese. Amino acids that are present in proteins that are not digestible may as well be absent from the diet, because their absorption will not be possible if they are not first digested down to their free amino acid components. Even if given nutrients are ingested, the presence of absorption competitors or inhibitors will cause the nutrients to be passed unused (egested) from the

Figure 4.1 A beet armyworm (BAW) head superimposed on a wheat germ diet. This figure shows the diet at 40 ¥ magnification (original) with the relatively small scale of the BAW head in relationship to the relatively sparse food components. These components are visualized in the upper left insect at 100 ¥ magnification (original) and illuminated with fluorescence so that the lipid micelles can be seen as bright spots confined to the wheat germ matrix. The other inset shows a close-up of the BAW head, with the mandibular span apparent.

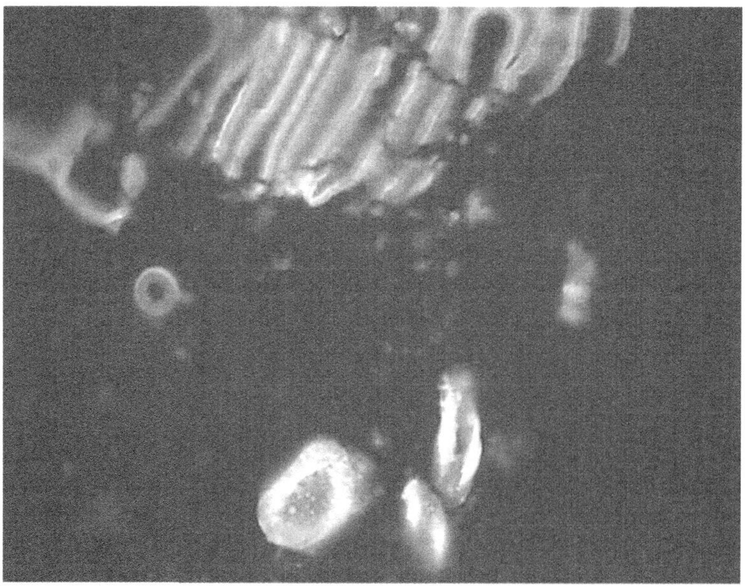

Figure 4.2 The matrix of a lepidopteran diet (Adkisson et al. 1960b) at 400 ¥ (original magnification) and visualized with fluorescence showing the lipids and their carbohydrate matrix. The large expanses of dark area in this and the next figure are nutritionally inert gel. (From Adkisson, P.L. et al., *J. Econ. Entomol.*, 53, 759, 1960b.)

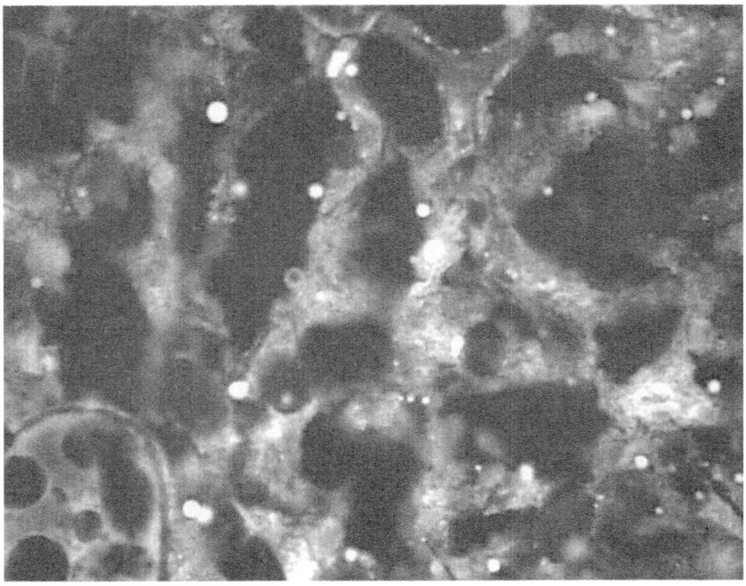

Figure 4.3 Plant bug diet (Cohen 2000b) at 400 ¥ (original magnification) and illuminated by fluorescent microscopy. (From Cohen, A.C., *Southwestern Entomol.*, Suppl. 23, 111, 2000b.)

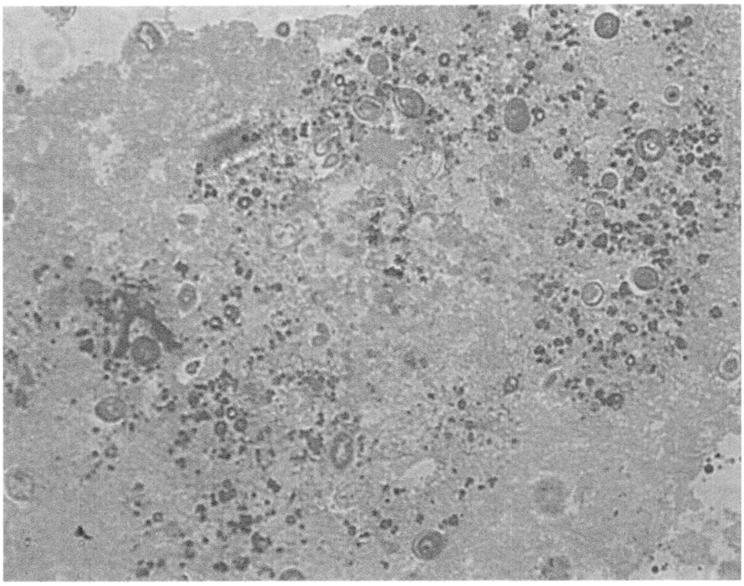

Figure 4.4 Entomophage diet (Cohen and Smith 1998) at 400 ¥ (original magnification), showing the distribution and intricate relationship between the egg and meat components of this matrix. (From Cohen, A.C. and Smith, L., *Biol. Control*, 13, 49, 1998.)

digestive system. The sum total of all these interactions is what defines the character and dynamics of the structural matrix of diets. The size relationship of these components is presented in Table 4.2.

Protection from the degradation of labile diet components is the other aspect of matrix function in good diets. Lipids that are surrounded by layers of proteins or insoluble carbohydrate macromolecules are more likely to be protected from the various forces of deterioration such as

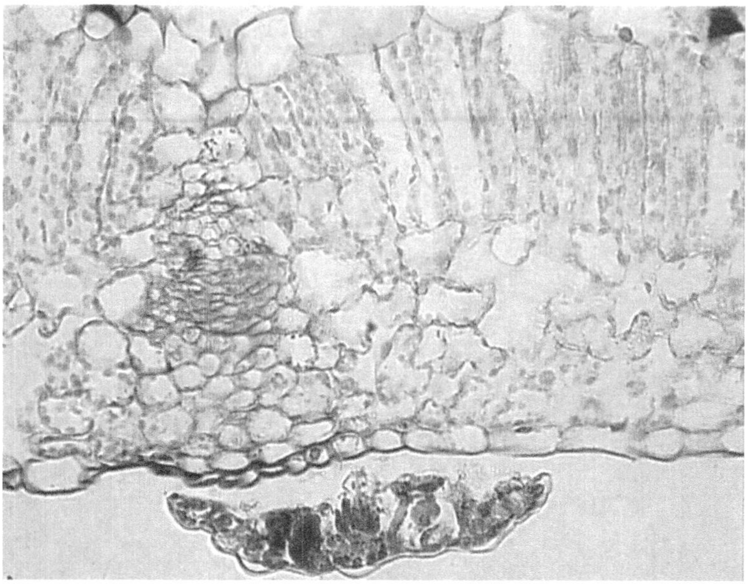

Figure 4.5 A whitefly nymph on the underside of a cotton leaf, feeding on the vascular bundle, which is evident as a dense mass of hollow-looking cells. Also evident in this figure are the various plant tissues, including upper and lower epidermis, palisade layer, and spongy mesophyll, and the chloroplasts within the latter two cell types.

ascorbate-/iron-induced lipid peroxidation (discussed in detail in the following and in Chapters 5 and 8). Likewise, copper, iron, zinc, or any other potential pro-oxidant species are immobilized by gelling, adsorption to macromolecules, or by chelation—all aspects of the matrix characteristics of a given diet. This potential for stability that results from the dispersion character or matrix qualities has been well explored in human foods (Walstra and von Vliet 2008), but it is neglected in literature on artificial diets for insects.

In re-examining diets from a perspective of organization (rather than simple nutrient composition), successful diets do retain at least some semblance of compartmentalization. Except for diets that are true solutions, most insect diets exist as dispersions. The majority of insect diets that are successful resemble the foods described as *manufactured foods* (Walstra and von Vliet 2008). Such foods are structurally complicated, because they "contain several different structural elements" that vary widely in size and state of aggregation (Walstra and von Vliet 2008). These foods are characterized as "filled gels, gelled foams, materials obtained by extrusion or spinning, powders, dough, and so forth" (Walstra and von Vliet 2008). The parallels between the dispersive qualities of foods and insect diets are compelling.

Insect diets share with foods six important consequences that result from their being in a dispersed state. (1) Because they are in different compartments, they are not in thermodynamic equilibrium, and therefore, they are continually subject to change: movement toward equilibrium. (2) Flavor components are in separate compartments, leading to the probability that the sensory responses of the insects rely on recognition of the separate components to have their phagostimulation mechanisms fully stimulated. Depending on the spatial arrangement of the components and the characteristics of the target insects' mouthparts, the ingestion of all essential components will be influenced by these sensory attributes. (3) The dispersive quality of the diet also relates to the ability of insects to bite or probe the diet and to apply their extraoral and postoral digestive processes (discussed in detail in Chapter 7). (4) The solvents in the diet (mainly water but possibly some lipids) are resisted in their tendencies toward bulk flow. This impedes the transfer of heat throughout the diet during processing, but it also protects the diet from interactions between components that would have destructive consequences (such as enzymatic degradation or metal-catalyzed degradation of lipids). (5) The dispersive state influences the visual qualities of the diet, including light reflection, transparency, color, and so on. (6) Because the dispersed state of the diet makes it inhomogeneous, at either a microscopic scale or a macroscopic level of organization, the diet materials are inherently unstable and will tend to degrade to a more homogeneous and random (disordered) state. This can also lead to a separation of components that were held in place by weak forces that can be overcome with time.

An example of how the matrix (or dispersive) character can be protective is to be found in egg yolks where lipids and nucleic acids are protected from stored iron by a special matrix arrangement. Although egg yolks contain substantial concentrations of iron, this potential reactive oxygen species (ROS) is held safely by chelation and sequestration within the phosvitin matrix, which insulates iron from the sensitive components such as lipids and nucleic acids, which are oxidatively degraded by iron (Jacobsen et al. 2001, Lee et al. 2002).

In contrast to the protective potential of matrices, there are cases where the wrong components are in the wrong place at the wrong time, and the matrix is responsible for a destructive outcome. One of the best documented examples of matrix-based oxidative destruction is associated with eggs: an iron-based, ascorbic acid–induced lipid oxidation (peroxidation) that takes place when the iron is freed from its egg yolk phosvitin sequestration and acts as an ROS (Thomsen et al. 2000, Jacobsen et al. 2001). Iron, which ordinarily is bound by phosvitin, emerges from or reaches the surface of the phosvitin molecule where it can come in contact with lipid micelles that contain vulnerable triglycerides that can be peroxidized by the iron, which is further activated by the reducing ability of ascorbic acid (Thomsen et al. 2000). Interestingly, another complicating twist to this iron, phosvitin, ascorbic acid, lipid interaction within given matrices is that upon being heated adequately, the egg yolk proteins physically stabilize their matrix (Anton et al. 2000), a condition that would reduce the access of the iron to the lipids, thus slowing the ROS degradation reactions.

4.2 TERMINOLOGY REGARDING SUCCESS AND FAILURE OF DIETS

Although hundreds of formulations for artificial diets have been published over the past 50 years (e.g., Singh 1977, Singh and Moore 1985), only a handful of these can be considered as fully successful in terms of truly replacing the target insects' natural foods. However, the concept of diet success must be qualified.

For a program in basic nutrition, a completely defined diet that might cost over $100/kg and that supports development from first to second instar larval form might be considered successful. Such a diet could be used in component deletion tests to help ascertain the essentiality of each nutrient. However, if what is required is a diet that will support mass rearing, such a diet would be useless.

In contrast, even if a diet made of unpurified materials or *whole* foods were satisfactory for mass rearing at a cost of pennies per thousand insects, it would not be considered successful to the nutritionist who is trying to understand the specific requirements for nutrients in their most chemically simple form. This difference in perspective has caused many problems in our efforts to advance our basic understanding and practical applications of that understanding to development and advancement of insect diets. The field of insect feeding biology has become divided into two camps: the *basic science group*, which includes *pure nutritionists*, nutritional ecologists, and neurobiologists, and the *applied group*, which includes those who study what has come to be known as *insect dietetics* (Singh 1977, Beck 1992). The underlying assumptions of the two groups differ sharply from one another. The basic science researchers subscribe to the tenet that diets must be composed of purified ingredients so that their nutritional value can be objectively assessed. The applied researchers reject defined diets as being of little practical value. The differences in the approaches and objectives of these two camps have led to unfortunate discounting of one another's work by two groups that really need one another's support (Cohen 2001).

As discussed in Chapter 3, the fundamental assumption of researchers of basic feeding science is that only through the use of highly purified diet components can one discern that a given nutrient is indeed required by a target organism. If one is testing the essentiality of a given amino acid, vitamin, lipid, or mineral, the diet complex must contain an exactly controlled amount of the substance in question. If one hypothesizes, for example, that selenium is an essential mineral, a test diet must be formulated that completely lacks selenium or contains suboptimal levels of this factor. Then, if the target insect performs less successfully on the selenium negative or low selenium diet than it does on a diet with optimal selenium, and if the two diets (the variable and the control) are otherwise completely identical, it can be concluded that this mineral is essential to the target insect. Once the essentiality of the factor in question is established, the optimal and tolerable ranges can be established by the same types of controlled experiments. It is also possible to use these types of tests to determine if any other nutritional factor can substitute for (spare) the selenium. Besides serving to establish the essential nutrients, another advantage of this line of research is that it can provide information about the function of the nutrient in question (Cohen 1992).

However, there are drawbacks of this line of inquiry. First, the rigorous standards of purity are so difficult to achieve that there is often a doubt that there were hidden sources of the nutrient in question. The purest nutrient chemicals (minerals and amino acids) listed in the Sigma, Aldrich, or ICN catalogs are said to be more than 98% or 99% pure. This means that almost 1% or 2% of the material is something other than the substance in question. For example, the Sigma listing for sodium chloride states that the purest form of potassium chloride (SigmaUltra) contains less than 0.01% sodium and lesser amounts of sulfate, aluminum, calcium, copper, iron, magnesium, phosphorus, lead, and zinc. Indeed, less than 0.0005% zinc is a small amount, but if potassium chloride is to be used in a defined diet that is intended for testing the zinc requirement of a target insect, these trace amounts of zinc cannot be completely ruled out as cryptically supplying at least part of the zinc requirement. This becomes especially problematic when one considers that all the other nutrients in the *pure* diet also contain contaminants. The problem of determining essentiality of nutrients is especially difficult for trace minerals because of their ubiquitous nature (Dadd 1968, Mittler 1972), but there are similar purity problems in organic components. Any researcher who has studied amino acids using a highly sensitive chromatographic technique, for example, can attest to the fact that the *pure* methionine purchased from a chemical supply company or purified in the researcher's laboratory will have at least several impurities that appear on chromatograms to haunt the worker who is trying to establish completely defined diets. Likewise, *pure* sugars such as sucrose contain impurities, including other kinds of sugar such as glucose and fructose; *pure* fatty acids contain their impurities. These problems are further discussed by Mittler (1972) and in a more recent review by Thompson (1999). In one of the earliest papers on insect nutrition, Fraenkel and Blewett (1942) discuss the concept of insects having their biotin requirements met by impurities in highly purified starch, where insects cannot develop on a diet where glucose is the sole carbohydrate source, but they can thrive on diets where starch is used to replace glucose. This point steered the field of insect nutrition for the past 70 years, and it remains a major source of doubt in interpreting classic addition/deletion studies.

Besides the purity issues, the pursuit of determining nutritional essentiality is complicated by the huge number of possible nutrients in a diet system. There are 20 protein amino acids, at least 10 minerals, 5–10 lipids, and about 10 vitamins. Besides these 50 odd individual nutrients (Singh 1977), there are also feeding stimuli. The researcher trying to determine the complete nutritional requirements of a given species must first develop a nutritionally adequate diet that is completely defined and then prepare and test diets that are complete except for the single nutrient in question. This means that at least 50 diets must be tested to gain a comprehensive grasp of the target insect's nutritional requirements.

A factor that raises the problem to nightmarish complexity is that of relative proportions of each nutrient in relation to all the other nutrients. House (1974a) has summarized the problem of nutrient profiles, concluding that the proportions of given nutrients is as important as the presence of any given component. So, for example, sodium can compete with potassium, calcium with magnesium, and arginine with lysine. The excess of one of these pairs of nutrients can cause absorption or metabolic problems that can effectively create a deficiency of a nutrient that is otherwise present in adequate amounts. Also the presence of certain other factors can cause problems in the bioavailability of nutrients.

One of the most notable and recently recognized examples of this phenomenon is the plant compound known as phytic acid. Phytic acid is present in many fruits, vegetables, and seeds. This compound chelates or binds the iron, manganese, and calcium that is present in foods or food additives. The bound minerals are withheld from the absorption process (discussed in Chapter 7) and are passed from the digestive system along with waste products. Thus, a nutrient deficiency is set up, even when the desired nutrient is present in what would otherwise be adequate amounts. Other components can act as chelating agents that impact, either negatively or positively, the absorption or bioavailability of a nutrient (Miller 2008).

4.3 MINIMAL NUTRIENTS (THE *SIMPLE NUTRIENT* MODEL)

An underlying paradigm of insect dietetics and nutrition has been the concept that there is a set of simple, irreducible nutrients for every species of insect. Simple nutrients, by definition, are the required compounds that cannot be reduced to a smaller or more chemically simple form without losing their nutritional value. For example, a protein such as casein, which contains all of the amino acids generally found in proteins as a whole, can be reduced (digested or hydrolyzed) into its component 20 kinds of amino acids, yielding a pool of all the amino acids used by a given insect in its life processes. According to the *simple nutrient* model, the whole protein molecule or the pool of amino acids can be used interchangeably as either a whole protein or the 20 kinds of amino acids in their amounts that were originally present in the protein. Each amino acid, in turn, can be further broken down into its component elements, carbon, hydrogen, oxygen, and nitrogen (and for methionine and cysteine/cystine—sulfur).

The essential amino acid L-methionine constitutes about 2.4% of casein by weight, and a methionine molecule consists of 5 atoms of carbon, 11 hydrogens, 2 oxygen atoms, 1 nitrogen, and 1 sulfur. A nutritional *law* is that arrangements of these elements other than the special and very exact arrangement that we call L-methionine will not fill the nutritional role of this amino acid, and if there is not an adequate amount of this amino acid in the target insect's diet, that insect will fail to thrive simply based on the L-methionine deficiency. In fact, the requirements for the L-methionine are so specific and particular that if efforts were made to substitute D-methionine (see Figure 4.6 for a comparison of the two forms of methionine), the insect would still fail to thrive. The aforementioned nutritional *law* points out that the L-methionine is a nonreducible nutrient and that no combination of carbons, hydrogens, oxygens, nitrogens, and sulfurs would substitute for the L-methionine form. The same logic applies to all the other essential amino acids from the casein molecule. In fact, even another sulfur-containing amino acid cysteine will not serve as a substitute for the essential L-methionine, nor will valine substitute for leucine or isoleucine for leucine, and so on. This is what is meant by an essential nutrient. However, with some nutrients that are used in the insect's metabolic pool, there are some degrees of freedom of substitution. For example, through a process known as transamination, pyruvate (a metabolite in a sugar degradation pathway) can be converted into alanine at the expense of glutamate, which is converted into α-ketoglutarate with the movement of the amino group. This reaction is illustrated in Figure 4.7. Through transamination reactions and other metabolic pathways, all of the nonessential amino acids can be synthesized

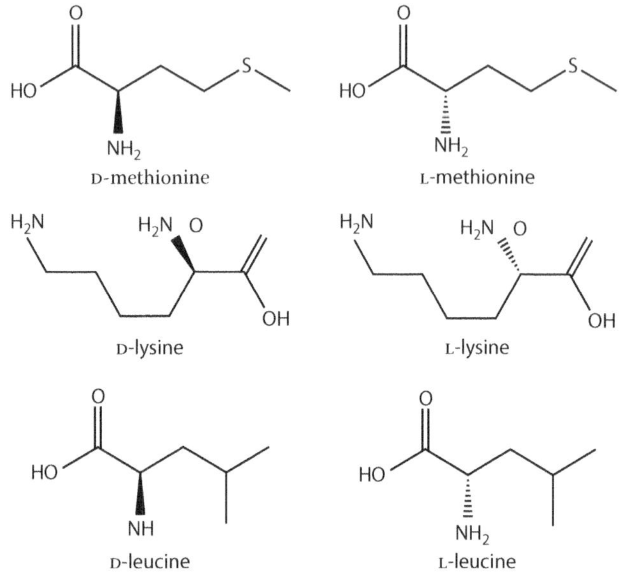

Figure 4.6 The two forms of the essential amino acids methionine, lysine, and leucine (the D and L forms). Despite the similarity of these paired structures, only the L form of each can be used by insects.

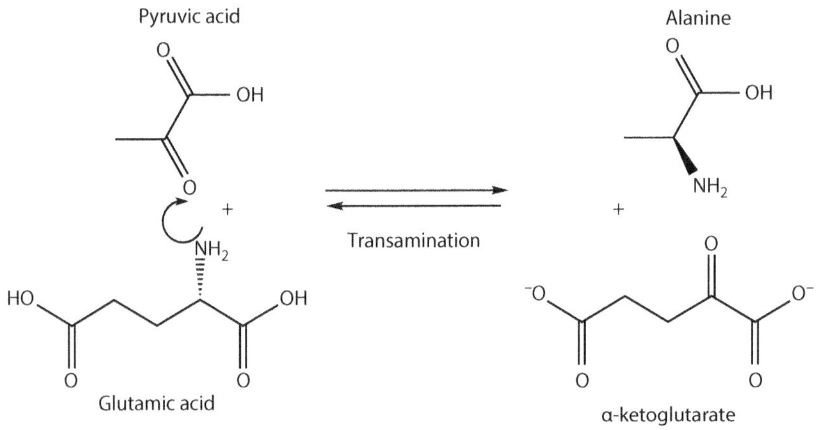

Figure 4.7 An example of the process of transamination, where nonessential amino acids can be synthesized. In this figure, the amino group from glutamic acid is transferred to pyruvic acid, converting the former to α-ketoglutarate and the latter to alanine.

or degraded to be used as fuels in case there is an excess of amino acids in an insect's diet or a deficit of more suitable energy sources. The concept of metabolic pathways and the types of logic and synthesis and degradation processes that take place in these pathways is further discussed in Chapter 7. However, it should be clear from these examples that with regard to various nutrients and biochemical components, there are limitations to the freedom of metabolic function that an insect (or any organism) can perform.

Returning to the casein molecule, it is another aspect of nutrient law that regardless of the source (i.e., from casein, soy protein, *Helicoverpa zea* vitellin, or wherever it originates), the L-methionine that goes into the nutrient (metabolic) pool can be used wherever that specific amino acid is required. Casein-derived L-methionine is *exactly and completely* equivalent to L-methionine from

H. zea vitellin, as it is to every other L-methionine. This is not to say that L-methionine does not display differential bioavailability when derived from different proteins or is occurring in different food or diet matrices. But the difference is in the environment of the methionine molecules, not from the methionine molecules themselves.

4.4 *MINIMAL NUTRIENT* CONCEPT

Insect biochemists have tried for the past century to reduce the complexities of insect feeding requirements to the identification of the simplest components (or irreducible components) that meet the nutritional needs of target species, and to determination of the function of each of those components. The composite of all those substances can be regarded as the *minimal nutrients* required by a target insect. However, the development of a robust base of knowledge of the minimal nutrient requirements of insects has been elusive.

Most often in insect nutrition studies, the recognition of an essential nutrient was the best that could be achieved, with the function of each nutrient being left ambiguous or vague. This leaves much of our current understanding of insect nutrition dependent upon studies of vertebrate nutrition, much of the basic information on insect nutrition being deduced from rat, mouse, and guinea pig studies. Only a few of these studies have been repeated with insect subjects. This is especially problematic because of vast diversity of insect species and myriad feeding habits of those species. Also, because of the long evolutionary history of insects in diverse feeding niches, it is of limited value to try to transfer what was learned about one species to another species. What has also made it difficult to attain the desired profile of nutritional requirements is that each factor is influenced by a large variety of internal and external conditions, including the interaction of nutrients with one another and with other factors. This interfacing (interaction) of the components or the absence of interaction is the result of the nature of the *matrix* of the diet, another name for the *physical and chemical organization of the diet components.*

4.5 RULES OF NUTRIENT SAMENESS, NUTRIENT PROPORTIONS, AND COOPERATING SUPPLEMENTS

House (1974a) articulated three principles of nutrition:

1. The *rule of sameness*, which states that all insects have more or less the same nutritional requirements. This is to say that most species of insects require the same 10 *essential amino acids*, and most species require a sterol nucleus such as cholesterol or a phytosterol.
2. The *principle of nutrient proportionality*, which House describes as an amendment to the rule of sameness in its recognition of the fact that different proportions of certain nutrients are characteristic of the needs of different species.
3. The *principle of cooperating supplements*, which states that some nutrients can substitute for one another or that certain nutrients that are either stored or originate from symbionts can act together to satisfy an insect's nutritional needs.

These principles are still largely valid, and they can help add insight as to how insect diets work or fail. As a very general rule, insects prosper from most of the materials listed in Table 4.3, and for those listed as absolute essentials, most insects studied to date share the requirement for them, especially when symbionts are excluded. However, it is also clear that the inclusion of these *essential* nutrients generally does not constitute a diet upon which most insects can thrive in continuous generations. As for the nutritional principle of proportions, the relative amounts of each nutrient vary from species to species and even from one life stage to another or from one gender to another.

Table 4.3 Minimal (Irreducible) Nutrients Shown to Be Useful or Essential to Insects

Amino Acids (All in the L-Form of Stereoisomer)	Lipids	Carbohydrates	Water-Soluble Vitamins	Lipid-Soluble Vitamins	Minerals
Arginine[a]	Cholesterol[a]	Starch	Ascorbic acid[a]	Tocopherol[a]	Calcium[a]
Histidine[a]	β-sitosterol[a]	Glycogen	Thiamine[a]	Vitamin A (various carotene derivatives)	Chlorine[a]
Isoleucine[a]	Stigmasterol[a]	Pectin	Riboflavin[a]		Copper[a]
Leucine[a]	Campesterol[a]	Maltose	Pyridoxine[a]		Iron[a]
Lysine[a]	24-methyl-cholesterol[a]	Sucrose	Nicotinic acid[a]		Magnesium[a]
Methionine[a]	Palmitic acid	Raffinose	Pantothenic acid[a]		Manganese[a]
Phenylalanine[a]	Palmitoleic acid	Stachyose	Biotin[a]		Phosphorous[a]
Threonine[a]	Stearic acid	Mellizitose	Folic acid[a]		Potassium[a]
Tryptophan[a]	Oleic acid	Glucose	Choline[a]		Sodium[a]
Valine[a]	Linoleic acid[a]	Fructose	Carnitine[b]		Sulfur[a]
Alanine	Linolenic acid[a]	Galactose	Cyanocobalamin (B_{12})[a]		Zinc[a]
Aspartic acid	Arachidonic acid	Mannose	Inositol[a]		Selenium
Asparagine		Ribose			
Cystine/cysteine					
Glycine					
Glutamic acid					
Glutamine					
Proline					
Serine					
Tyrosine					

[a] The nutrient has been shown to be absolutely essential in more than one species of insect.
[b] The nutrient has been shown to have growth-promoting activity but is not essential.

There is strong evidence that entomophages have higher requirements for nitrogenous compounds than do phytophages (Thompson 1999), and that among phytophagous insects, guilds that feed upon leaf materials better utilize phospholipids and more poorly utilize fats (triacylglycerols) than do insects, which feed on seeds, the former being low in fats and richer in phospholipids than seeds, which are rich in fats and fairly poor in phospholipids (Turunen 1979). Yet, even when the proportions of the specific nutrients or the major nutrient classes are close to the values reported for their natural foods, many insects fail to thrive on certain diets.

As indicated previously and throughout this book, the early views and what have become the prevailing views of insect nutrition are that there is a very mechanistic explanation for diet successes or failures, and the most likely cause of failures is the *missing nutrient hypothesis*. This appealing concept is very logical, but it may be overly simplistic. A more complex explanation is that the nutritional composition in a context of organizational structure or appropriate compartmentalization may explain much further the bases of success and failure. What is presented in the following section is an effort to examine a few of the diets that are highly successful by the criteria described earlier to rationalize how the interplay between the essential components and the organizational matrix works together to make these diets excellent media. The rationale is necessarily based on circumstantial evidence, as well as from direct studies.

4.6 EXAMPLES OF EXCELLENT DIETS AND WHY THEY ARE SUCCESSFUL

In this section, examples of diets that were landmark advancements in diet science and technology are discussed. What is attempted here is to give perspective about how innovations regarding diets had profound influence on the field. This seemingly simple concept is difficult to document and it required extensive analysis—almost a detective's analytical approach to tracing back a diet innovation to an original source. While one can argue that the original use of a substance such as *wheat germ*, *casein*, *agar*, or a *yeast* product in insect diets is solely of historic value, it becomes evident with the *forensic trail* seeking that the rationale and basis for success or failure can be found in this type of investigation. Understanding the rationale is a powerful foundation for the development of new diets, improvements and maintenance of quality in existing, successful diets. It is an especially powerful basis for analysis of why failures might be taking place in the rearing setting. The type of understanding discussed here is especially useful in the context of a well-developed process control program. It is in this context that the history and rationale for several diet components is examined—especially casein, wheat germ, agar and yeast products.

4.6.1 Major Influences on the Road to the Most Successful Diets: Casein, Wheat Germ, Beans, Agar, and Yeast Products

Along with discussing *successful diets* on a case-by-case basis, it will be helpful to trace the history of four of the major components that have been the most significant contributors to the success of the vast majority of insect diets and are the basis for thousands of publications and hundreds of rearing programs: *casein*, *wheat germ*, *beans (several types of legumes)*, *agar*, and *yeast products*. These *game changing*, keystone dietary components will be discussed in terms of their historical importance and the stated or implied rationale for their inclusion in what can be considered the most influential diets in the development of diet science and technology.

Although *casein* had been isolated in the late nineteenth century and used in microbial media since 1900, there was a lag in casein utilization in insect diets. Casein was first included by Michelbacher et al. (1932) as an insect diet component for flesh fly larvae (*Lucilia sericata* Meig.). The paper (published in the *Journal of Experimental Biology*) includes a rich framework of discussion of other macronutrient components (or *complex foods* or *staples*) that the authors tried (egg yolk, egg white, various meat components and extracts, *fish juice*, and several other substances that had been used in microbiological media) before they arrived at using casein. The authors explained that they used casein because of its demonstrated nutrient character and also because it was chemically well defined. They also used extra purification efforts to remove extraneous, undefined components.

The paper by Michelbacher et al. (1932) is remarkable in its thoroughness, including extensive demonstration that the diets and the insects used were axenic and the details in explaining the importance of diet consistency, order of mixing, and other features that should be the envy of modern diet literature. However, despite the high quality of this paper and the profound importance of casein in insect diets for the past eight decades, the paper was cited only 22 times, according to the Web of Science as of July 2014, and only one of those citations (Vanderzant and Reiser 1956) credits Michelbacher et al. as the source of casein as a diet component. This gap in acknowledgement of foundational contributions has hindered advancements in the development of the field of insect and rearing science and technology (Cohen 2001). It is an unfortunate reality that anyone wishing to thoroughly understand the nature of a diet staple such as casein or wheat germ must perform a sort of a close detection analysis of the literature.

Because of the gaps in literature citations, tracking casein (or other key diet components) is nearly impossible. A search of key words such as *casein* and *insect* in the Web of Science produces 197 references where these two words are present in titles, abstracts, or key words. To illustrate this point, although Michelbacher et al. published the first paper on use of casein in insect diets in 1932, there was no direct or explicit citation of casein in insect diets until 1987. This "citation gap" is especially unfortunate in light of the fact that many diets used (and published) during this "gap" utilized casein as an important diet component. However, in many cases, it is not even mentioned anywhere in the paper, or it gets brief mention in a list of ingredients. Furthermore, this search also failed to reveal several pivotal papers on casein-containing diets by the Texas A&M group (led by E. Vanderzant and P. Adkisson), the University of Illinois group (G. Fraenkel's group of collaborators, which later included the team of G. Waldbauer's colleagues), or S. Beck's group (from the University of Wisconsin). Together, these three groups produced well over 50 papers on rearing insects with diets containing casein as a major nutritional source.

Interestingly, though casein was clearly a keystone component of all these diets and nutritional experiments, there is very little attention to the nature of the casein products used in these diets. Michelbacher et al. (1932) tested several casein products, including some purified forms of this milk protein, to help them optimize their diets through heuristic techniques and rationale/incremental testing practices, discussed in the following text and especially in Chapter 10. In the other diets, with the exception of stating that the casein was vitamin free or soluble, the nature of the casein used in the formulations was not addressed, creating or reinforcing a sense that casein is a simple, uniform material.

On the contrary, casein is actually an incredibly complex material, as food chemists such as Swaisgood (2008) emphasize. Bovine casein occurs in at least four major forms (α-casein, β-casein, γ-casein, and κ-casein, and these forms can be further subdivided). Also a soluble form of casein known as *caseinate* present as a sodium salt or other cation salt is sometimes used under the generic name *casein*. The nutritional value, digestibility, interaction with other diet components, bioavailability, and palatability are all related to the nature of casein used in the diet. Failure to understand and heed the implications of using various forms of caseins can lead to unexpected and unnoticed precipitation, removal of sterols, vitamins, or minerals from the nutrient pool; failures of gelling agents can be caused by some casein products but not by others. Among many serious problems that can arise from misunderstandings about casein are formation of pockets of casein-rich and casein-poor zones, leading to a widely observed, but poorly documented lack of developmental homogeneity of developing insects.

Another important misunderstanding about casein is where casein hydrolyzates are used instead of complete polymer forms of casein. Often, researchers who wish to use a highly soluble source of small peptides and free amino acids will use casein hydrolyzate (as they will sometimes use soy, yeast, or other hydrolyzed materials). This raises the issue of nomenclature involved in casein-based products, many of which are used as microbial media or cell culture media. The complexity and variety of these products becomes apparent from visiting a catalog or website from a supplier such as Sigma Chemical Company or BD Technical Manual. The latter document describes the nature (and complexities) of several tryptone, peptone, and several products with trade names. An important aspect of the complexity of the various hydrolyzate products, including casein hydrolyzates, is that different processes are used to produce the materials: tryptic hydrolysis and pancreatic enzyme hydrolysis, for example. These different processes and differences in the casein quality used in the starting material result in sweeping differences in the nature of the hydrolyzate, which may contain longer peptide chains, a combination of chain lengths, or free amino acids (in products referred to as *casamino acids*).

Yet another ambiguity in dealing with casein has been in failure to specify the purity of the casein product being used. Besides the complexities of casein products discussed earlier, there are also very important differences in such products in relation to the total mineral content (ash content), phosphorus and calcium content, lipid or sugar content. The more of these impurities that may be present, the lower is the actual protein content; so if a casein with 54% protein is used in a diet that was designed to provide target insects 87% protein (as discussed by Swaisgood (2008) and posted in an online catalog from United States Biochemical Corporation, Cleveland, OH), there is a high likelihood that protein malnutrition will result from the inadvertent substitution.

Having all these complexities of casein discussed, it would seem that the nature of this nutrient in insect diets should be complete. However, because casein is widely used in diets in a supplementary or redundant nutrient function along with other macronutrients, it will be discussed further in a later part of this chapter.

4.6.2 Wheat Germ Diets and Their Influences on Diet Science and Technology

One of the most important contributions to the advancement of artificial diets for insects was and continues to be the publication of a paper describing the inclusion of wheat germ in a diet for the pink bollworm *Pectinophora gossypiella* (Adkisson et al. 1960a). This work came out of the laboratory of the renowned insect biochemist, Erma Vanderzant, whose contributions in several areas of insect nutrition are noteworthy and historical. The importance of this diet cannot be overemphasized in the history of insect diets; though the paper has been cited directly only 186 times as of June 2014, there are hundreds of diets that are derivatives of this formulation based on wheat germ. The use of wheat germ in the diet opened the door to rearing countless insects from numerous species and has impacted billions of dollars in programs of research and insect control. Although no economic assessments are available, when one considers the numerous large-scale programs, all based on wheat germ diets and all having been part of mass-rearing efforts that totaled billions of insects per year for about 40 years, it becomes feasible the vast economic and environmental efforts have been leveraged by this seemingly simple breakthrough.

This landmark paper reported that in diet-fed insects, pupal weights were slightly lower than those from pupae derived from cotton bolls. Development period was equal to that from cotton. Oviposition yield was superior to cotton-derived insects, and there was 81.5% larva to adult survival. These excellent results and the remarkable success of diets based on this wheat germ formulation recommend the advantage of examining this diet carefully to develop a rationale that explains why it was such a successful diet. However, there are some complications in tracing the history of wheat germ in this diet and what must be considered a predecessor, the wheat germ diet developed by the Vanderzant group in the USDA, ARS laboratory at Texas A&M, Vanderzant et al. (1959). This wheat germ diet was incorporated in a series of tests of diets for boll weevils (*Anthonomus grandis* Boh.), where the Vanderzant group was trying to identify and isolate feeding stimuli and nutritional factors that would be optimal for rearing both larval and adult boll weevils, and they were trying to determine whether or not they could depart completely from diets that contained no cotton fractions (cotton bolls being the natural host for *A. grandis*). The abstract for the paper by Vanderzant et al. (1959) is presented here:

> Artificial diets are described for the adult boll weevil (*Anthonomus grandis* Boh.). Basal diets contained an enzymatic hydrolyzate of casein, glucose, corn oil, cholesterol, Wesson's salts, choline, ascorbic acid, B vitamins, water, and agar. Mixed pollen in amounts of 0.1 to 5 grams per 100 grams of the basal diet stimulated feeding and oviposition. Five generations of weevils were obtained on artificial diets without cotton-plant parts by rearing from egg to adult on a soybean protein diet and allowing the adults to feed and oviposit on a diet containing mixed pollen. Oviposition occurred when pollens from different families of plants were used. The only substance other than pollen that appreciably stimulated oviposition was *wheat germ*. A basal diet containing homogenates of heated cotton plant parts, squares, flowers, germinated cottonseed, and young green seedlings also caused weevils to feed and oviposit. Tests with modifications of the basal diets are described. Substances that repelled the boll weevil when included in the basal diets were alfalfa meal, forage juice, distiller's dried solubles, yeast extract, and beef extract.

The abstract is a historical first as it mentions wheat germ (italics added) for the first time in literature on insect diets.

Although the wheat germ did not offer spectacular results in this early diet for boll weevils (certainly not dramatic enough outcomes that the power of this material to revolutionize insect diet technology was evident), there was an important breakthrough, where wheat germ was shown to induce strong oviposition responses, and evidently, it was evident that it stimulated feeding in boll weevils. There is no record of the rationale that induced the Texas A&M team to use a wheat germ diet for pink bollworms, which had been, prior to 1960, elusive targets for diet development. However, Vanderzant (1967) did explain some of the features of wheat germ, which helped explain why this material proved to be a major factor in diets for many species of insects (possibly in retrospect or possibly in accord with her knowledge of wheat germ in 1959). Whatever the motivation, Adkisson et al. (1960a) and the Adkisson et al. (1960b) used wheat germ in several formulations of diet for pink bollworms, and the success of this innovation clearly opened the doors to testing wheat germ in many other diets.

This raises some important issues about the pathways of diet development. First, how much does each component of a successful diet (or series of diets) contribute to the overall success of the diet? This is, in a sense, a Pareto type of question that asks for assignment of proportions of credit for a component's contribution to the total outcome. In this model of hypothetical contributions, it is recognized that base materials, called *macronutrients* in this book, such as wheat germ, yeast, soy flour, cottonseed meal, and pinto bean meal, contribute major portions of the nutritional profile (proteins, carbohydrates, lipids, most vitamins, and minerals), but certain factors (called *micronutrients* in this book) may be missing such as sterols, ascorbic acid, or vitamin B_{12}. How diet components interact is discussed later in this chapter, but for now, it is valuable to examine a formulation such as the diet of Adkisson et al. (1960a), and ask to what extent can each component be deleted or replaced with another material. In this diet, bioassays showed that wheat germ as the base nutritional material was superior to the casein-based diet. Later tests showed that casein was more of a supplement than an essential part of the diet. By using wheat germ, the Texas A&M rearing team was able to eliminate cellulose, glycine, cysteine, α-tocopherol, and cholesterol. They were even able to reduce the amount of agar in the diet because of texture improvements that came from wheat germ.

In light of the tremendous influence that this diet has had on the progress of insect diet science and technology, it is very impressive to find the extent to which the authors credited other works with having led to the current paper. Much to the authors' credit, they presented the background work that led to their own work, citing Beckman et al. (1953) as having established the possibility of rearing pink bollworms on an artificial medium. This quote reflects the tone of this work, "Research by Beck and Stauffer (1950) which led to a purified casein medium for the European corn borer, *Ostrinia* (formerly *Pyrausta*) *nubilalis* (Hbn.) provided the basis for the development by Vanderzant and Reiser (1956b) of a similar type purified casein medium on which the pink bollworm could be successfully reared." The authors went on to write, "The casein medium not only provided a method for future work pertaining to the dietary requirements of the pink bollworm, but it also proved valuable in the development of a rearing medium for laboratory cultures of the boll weevil, *Anthonomus grandis* Boh. (Vanderzant and Davich 1958)." The authors further explained the connections between the several diets, which resulted in the use of corn oil to meet the pink bollworm's requirements for linoleic acid. The clarity of explanation of methods, the presentation of data on bioassays and comparisons with field-derived insects, and the honest recognition of prior art make this paper stand as a model work for diet studies in all these regards, except for the lack of rationale for the use of wheat germ, which is now recognized as the most noteworthy and remarkable contribution made by this publication.

The nutritional profile of the Adkisson et al. diet (Wheat Germ 2 Diet from Tables 4.4 and 4.5, for the commercial version) contains about 4% protein, about 0.4% lipid (including the intrinsic plant sterols that exceed 1% of the lipid content from the wheat germ), about 5% carbohydrates, and various vitamins and minerals that were added as vitamin and mineral mixtures, as well as those present in the wheat germ. Table 4.6 shows the nutritional profile of wheat germ, revealing a combination of virtually all nutrients presented in Table 4.3 as essential and beneficial to insects. Both casein and wheat germ offer a complete complement of amino acids. All essential lipids are present, evidently in the wheat germ. The form of the diet, a stable gel, supports the feeding mechanism of the insects. It would appear that these features suffice to explain that the diet is successful, because it meets the standards of providing all essential nutrients in suitable proportions, all

Table 4.4 Wheat Germ Diets Developed by Adkisson et al. (1960)

Ingredient	Casein Diet (g)	Wheat Germ 1 Diet (g)	Wheat Germ 2
Casein, vitamin free	5.0	3.0	3.5
Cysteine hydrochloride	0.1		
Glycine	0.15		
Wheat germ		3.0	3.0
Sucrose	5.0	5.0	3.5
Wesson's salts	1.2	1.0	1.0
Cholesterol	0.05		
Corn oil	0.25		
α-Tocopherol	0.01		
Choline chloride	0.1	0.1	0.1
Cellulose	4.0		
Agar	3.0	2.0	2.5
Sodium alginate	0.5		0.5
Vitamin mixture[a]	1.0 mL	1.0 mL	1.0 mL
Water	80.0 mL	80.0 mL	85.0 mL

[a] The vitamin mixture used for the casein and wheat germ 1 media did not contain inositol.

Table 4.5 Modification of the Vanderzant–Adkisson *Special Wheat Germ Diet* as Offered by ICN

Ingredient	Amount (g/kg before Water and Vitamins are Added)
Vitamin free casein	28
Sucrose	27.5
Wheat germ	24
Alphacel, nonnutritive bulk	12
Cholesterol U.S.P.	0.05
Linseed oil	0.2
Wesson salt mixture	8.0

nonessential but beneficial nutrients, appropriate feeding stimulants, and suitable texture to allow feeding (including tunneling) activities that are normal to the pink bollworms that are targets and to alternative species that are also nutritionally supported by this diet. There are about 2 mg of phytosterols in the 3 g of wheat germ (based on observations by Toivo et al. 2001). There is a trace of cholesterol in the casein; so between the casein and wheat germ, the sterol requirements of the boll worms are evidently satisfied without recourse to inclusion of additional sterols. Another point that deserves further attention is the use of casein in this diet and in countless other insect diets. The rationale for using casein has been that this milk protein supplies a good balance of essential amino acids. However, a comparison of the profile of essential amino acids in casein and in wheat germ (Figure 4.8) reveals that these two sources have very similar patterns, raising the question about whether or not casein is redundant with respect to amino acid profiles when other rich protein sources are available such as with wheat germ, soy, or other high-quality proteins.

However, there is much more to the success of this diet than is indicated in this first analysis. The nutrients in the diet are such that they meet the bioavailability requirements. The form of the diet, a gel that retards mass movement of the water and its dissolved constituents and that contains compartmentalized nutrient particles, adds both to the accessibility of the diet and its stability. The stabilized compartments contain lipids that are sequestered from lipo-oxygenases, iron, and other components known to attack lipids by peroxidation reactions or free radical–instigated chain reactions. Such reactions remove nutritious lipids from nutrient pool, replacing them with toxic and unpalatable short chained fatty acids and other rancidization products. Other factors are the compartments, which

Table 4.6 Composition of Yamamoto (1969) Diet for *M. sexta* Larvae

Components of Yamamoto Diet for *M. sexta* (1969)		
Wheat germ	100	g
Casein, purified	45	g
Sucrose	40	g
Torula yeast	30	g
Salt mixture	15	g
Ascorbic acid	4	g
Sorbic acid	1.5	g
Methyl-*p*-hydroxy benzoate	1	g
Cholesterol	0.5	g
Vitamins		
Nicotinic acid	10	mg
Calcium pantothenate	10	mg
Riboflavin	5	mg
Thiamine hydrochloride	2.3	mg
Pyridoxine hydrochloride	2.3	mg
Folic acid	0.2	mg
Biotin	0.2	mg
Others		
Agar	20	g
Formaldehyde (10% solution)	20	mL
Water	980	mL

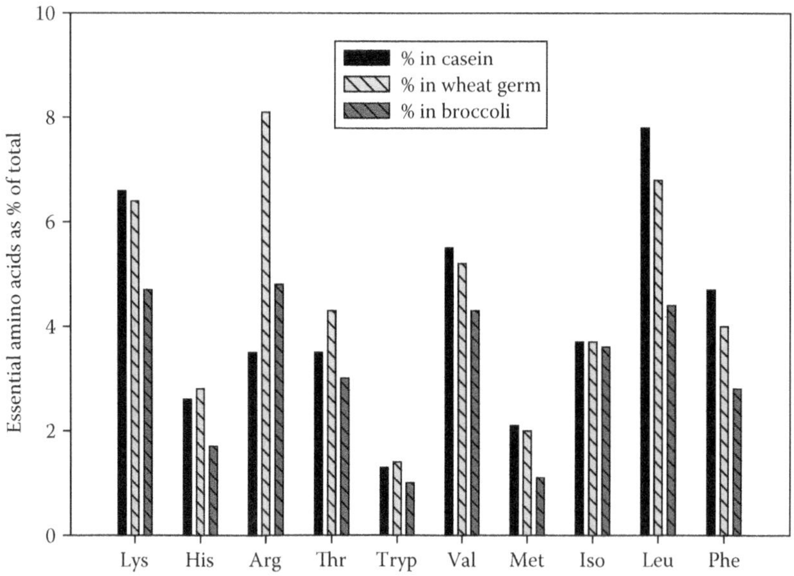

Figure 4.8 Profiles of the insect essential amino acids present in broccoli, casein, and wheat germ. The profiles are represented as percents of the total amino acids.

also serve to protect the nutrients from microbes that cannot reach the full complement of growth-promoting substances. The gel-stabilized compartments are antimicrobial by virtue of their structure. It is noteworthy that the Adkisson diet as originally described does not contain additives that are deliberately added to prevent microbial growth. When strictly sanitary conditions are maintained, there are minimal problems with microbial contamination. However, Adkisson et al. reported (1960a) that they later began using a 0.2% mixture each of methyl paraben, butyl paraben, and sorbic acid. The preparation of the diet involves a blanching but not a sterilization of the diet ingredients, and nonsterile utensils or insects would introduce a considerable number of environmentally abundant microbes. However, the early results where the insects could be successfully reared without microbial inhibitors would testify to the potential of this diet to resist microbial attack.

Examination of Tables 4.4 and 4.5 reveals that the wheat germ diets contain fewer ingredients than their predecessor diets; yet, they were shown to be superior, proving that the extra additives, such as cysteine and glycine, were not necessary. The lipids in wheat germ replaced the corn oil, α-tocopherol, and cholesterol that were added to the so-called casein diet. Also, glycine and cysteine were deleted from the wheat germ formulations as was the added bulking agent, cellulose. Interestingly, the commercial formulation (Table 4.5) produced by ICN Pharmaceuticals Inc. (Costa Mesa, CA) includes a bulking agent in addition to that which is provided by the wheat germ. In retrospect, it seems likely that the cellulose was redundant with the bulking or textural properties conferred by the wheat germ. The issue of bulking agents is discussed in several places in this book, especially in Chapter 15, dealing with diet texture issues. However, while the importance of bulking agents in some diets cannot be disputed, in other cases (such as the pink bollworm diet), the added bulking agent is clearly redundant.

The inclusion of the casein raises some interesting questions. First, could the casein be replaced by additional wheat germ or some other relatively complete food such as soy flour, another legume meal (other than soy), or yet another nutritionally complete plant material. The use of casein in hundreds of insect diets raises questions about whether or not it is included just for tradition or if it is the best protein source. As a milk protein, it is complete in terms of the amino acid composition and compared with many other proteins, it is relatively inexpensive. Also, casein is often thought of as a pure protein, but this is not accurate.

Various listings of the composition of casein list this product as vitamin free, soluble (casein salts), α-casein, β-casein, γ-casein, κ-casein, and various casein hydrolyzates. Among these listings, two casein products from USB (United States Biochemicals) are listed as containing as little as 54% protein and as much as 87% protein, the remaining materials being water, ash (including a wide range of trace minerals), and often a considerable amount of phosphate (more than 6%) and various lipid and carbohydrate groups. Swaisgood (1996) explained that there are at least four major kinds of casein from bovine milk and that these proteins form micelles that give them unique properties in terms of interaction with their matrix components (water, sugars, lipids, and minerals). Casein potentially imposes many unintended consequences such as binding calcium, which may have positive or negative effects in relationship to diet functions such as gelling properties. In fact, the observation (Cohen, unpublished) that addition of milk products to agar-based diets reduces the gel strength is possibly related to the calcium/agar interactions. Also Serafini et al. (2009) showed that casein interacts strongly with polyphenols, reducing the antioxidant potential of these compounds in mammalian blood.

A little background on the processing that is involved in casein production can help to clarify the nature of this important diet ingredient. This information is derived from Swaisgood (1996) and an Internet site GEA Niro. Fresh milk (whole or skim) is treated with an acid, usually lactic acid, to render a pH of 4.6, which causes the caseins to precipitate, and they can be recovered and separated from whey fluids by centrifuging, filtering, and/or decanting. This process is a common part of cheese-making, and in fact most of the protein in cheeses is of a casein type. A form of casein that is water soluble is known as caseinate, which is often used in insect diets and added as a sodium, potassium, or calcium salt and is produced by reacting casein solids with NaOH or CaOH. The steps in the process of casein production include spray drying or fluid bed drying, processes where the casein is heated. Any part of this complete process may have influence on the nature of casein components in terms of diet interactions, nutritional value, palatability, bioavailability, and stability. The fact that the term *casein* can include so many hidden variables makes the use of the word almost meaningless in terms of repeatability of diet research or utilization. Therefore, researchers and diet technologists who use a form of casein should always specify the source and whatever other identifying information they can obtain on the product in question. In keeping with this statement, editors and reviewers should not only allow authors to be specific, but they should demand it.

As far as the value of casein as its functionality in insect diets is concerned, its long history of use does not unequivocally affirm its status as an optimal dietary component for all insects. Among the earliest reports of casein as the major or sole source of protein or amino acids, Fraenkel and Blewett (1942) used casein-containing diets to try to replace whole wheat in three species of stored grain beetles. They found that a diet of casein (50 parts), glucose (50 parts), cholesterol (1 part), McCollum's salt mixture (1 part), and yeast extract (15 parts) could nearly equal whole wheat as a rearing medium. Despite the lesser level of fitness attained with the wheat replacement diet, this study and many more since 1942 confirm that casein is a substantial contributor to many insect diets' usefulness. However, there is a tremendous need to be cautious in using casein without a careful vetting process, and there is further need to explore the relationships between the various casein products and their efficacy, their bioavailability, their interactions with other components in diets, and their nutritional usefulness relative to other protein sources.

Interestingly, the economics of providing the protein (and the component amino acid) requirements via casein, wheat germ, or purified free amino acids is dramatically altered according to the source. Taking the essential amino acid methionine, at a cost of about \$35 per kg of protein from casein and a cost of about \$2 per kg of protein from wheat germ, the cost per gram of methionine is \$0.04 in casein, \$0.002 in wheat germ, but for pure methionine, even when purchased in kilogram quantities, the cost is \$0.17 per g. Besides the quality of the insects, the other important considerations are degree of difficulty in producing the diet and the expense of the ingredients. Obviously, on a per ingredient basis, wheat germ is a superior component in terms of labor and ingredient costs.

4.6.3 Overview of the Yamamoto Diet for *Manduca sexta*: A Model for Many Lepidoptera Diets

The diet discussed here was cited 295 times as of the June, 2014, and scores of insect diets were based on the formulation described by R. T. Yamamoto (1969), which is wheat germ based and casein based with the addition of torula yeast as an additional base material and agar to give the diet a firm gelled texture or consistency (Table 4.6). Besides the macronutrients (wheat germ, casein, and torula yeast) and a texturizing component (agar), Yamamoto also incorporates antimicrobial agents and micronutrients (vitamin mixture, minerals, and sterol in the form of cholesterol). This paper has some valuable (and unusual) explanation of rationale for components such as antimicrobial agents and a thoughtful discussion of the selection process (which is actually *domestication*); however, it unfortunately does not include a rationale for use of torula yeast. This paper and the other Yamamoto paper (1968) on rearing *M. sexta* are also important historically and scientifically because of the huge body of science that was based on use of the sphinx moths derived from this colony. The extensive treatment given to the Yamamoto diet is especially merited when it is considered how much studies on *M. sexta* have provided so much of what we understand in neurobiology, toxicology, genetics, ecology, plant/insect interactions, and countless other subjects where the basic nature of the insect reflects the basic relationship between the diet and the insect produced on that diet. Unfortunately, many publications of the studies that actually depended on the Yamamoto diet failed to cite this important reference. This point is discussed in the following text where the diet of Bell and Joachim (1976) is discussed.

Yamamoto (1969) explains the special presentation features where the diet is fed to the insects on a wire platform that allows the frass to drop to the floor of the rearing container to prevent contact between the frass and the diet. Yamamoto further explains that the *M. sexta* frass readily builds microbial growth and therefore must be separated from the diet by provision of a screen with holes that are 0.635 × 0.635 cm, a dimension determined to allow the large (more than 10 g) larvae to feed while even their largest frass pellets could drop through the screen away from diet slabs, which were provided on top of the screen. The feeding design of the containers also included a mesh circular ladder that larvae could use to gain access to the diet in case they had dropped to the floor of the container. This special feeding and frass management design is important in relationship to the nearly impossible task of preventing *M. sexta* colonies from becoming overridden with mold. This raises questions of finding a balance of the appropriate antimicrobial agents discussed in the following text.

Other important features of the *M. sexta* rearing system include the use of what Yamamoto refers to "20 ml of a 10% formaldehyde solution." It is an unfortunate ambiguity that makes it impossible to know the exact concentration of HCHO that was intended. The gas called formaldehyde dissolves in water to a point of saturation that is 37% by mass, giving a 100% solution called *formalin* (Budavari 1996). It is likely (based on common usage in the vernacular of histologists at the time of the Yamamoto publication) that the solution to which Yamamoto was referring was 10 parts of formalin and 90 parts water, which is actually a 10% *formalin* solution (rather than 10 parts of formalin and 30 parts water, which would result in a 10% HCHO in 90% H_2O). The ramifications of the formalin/formaldehyde ambiguity are not trivial. Yamamoto states that he tried many different kinds of antimicrobial agents in the *M. sexta* diet and found that none gave the control that he was able to get from the HCHO solution that he specified. This ambiguity is common in descriptions of formalin utilization in insect diets as is evident from examining dozens of diets listed in Singh (1977). It is noteworthy that formaldehyde is widely known to be acutely toxic and a significant carcinogen (Budavari 1996) and since the dangers of this chemical had become widely established in the 1980s, many rearing groups tried to eliminate it or greatly reduce its use in routine rearing operations. However, it is an unfortunate reality that replacement of formalin solutions has proved difficult as has been the case with *M. sexta* rearing at North Carolina State University for more than four decades after Robert Yamamoto published the diet discussed here. This difficulty remains a problem, though several further improvements were made by Bell and Joachim (1976) for their colony at the Metabolic Research Laboratory run by the USDA, ARS in Fargo, ND.

The colony of *M. sexta* that Yamamoto (1968, 1969) established had been maintained at North Carolina State University until 2011, and it served as a starter colony for many research programs and educational programs in North America. Progeny from the colony originally established in 1966 were propagated in the USDA, ARS facility in Fargo, North Dakota, and they were adopted by several commercial companies, where the *M. sexta* and diet based on the Yamamoto formulation remain important components of educational and research programs.

4.6.3.1 Adoption of the Yamamoto Diet by Bell and Joachim

The Yamamoto diet was adopted by Bell and Joachim (1976), and these authors included many useful details of the rearing process for hornworms and pink bollworms, including two modifications of the Yamamoto diet: the addition of raw linseed oil at a rate of 0.4% of the diet and a reduction of the water content by 25%. The Bell and Joachim paper is without question an important contribution, but it is interesting that this paper was cited 917 times according to the Web of Science as of June, 2014 compared to the 295 citations of the Yamamoto diet, which was the base diet which was only slightly modified. It is especially interesting that though citations of rearing procedures (including diet formulations) point to Bell and Joachim (1976), there is no place in this paper where the entire diet formulation is specified as is presented in Table 4.6. This raises some important questions such as: how did the researchers formulate the diets if the amounts of each component were not specified in the Bell and Joachim (1976) paper? Can a researcher wishing to use similar techniques to produce insects as research subjects duplicate the diet with the information provided by a reference that does not contain the complete formula? And finally, what is the meaning and value of citations if they do not inform readers sufficiently to allow them to judge the scientific merit and the reasonableness of the interpretation?

It is also an unfortunate reality that citations of a researcher's publications are a major means for the scientist's promotion, funding, and retention in the field, not to neglect the empowerment for the researcher to help develop the fields of rearing science and entomology (and other biological disciplines such as genetics and ecology, which rely strongly on the availability of reliable, high-quality insects). Because this is such an important issue, it is suggested here that an example of how diet specialists and other researchers can treat references. This excerpt from a paper by Ely and Jungreis (1977a) is an excellent example of how a diet or rearing technique derived from a secondary source should be cited:

> Experimental animals: Tobacco hornworms used in these experiments were derived from an inbred colony maintained by Dr. Lynn M. Riddiford, University of Washington, Seattle. Larvae were individually reared in plastic containers on the diet of Yamamoto (1969), modified after Bell and Joachim (1976) and Riddiford (personal communication), under an 18L: 6D photoperiod regimen at 23–25°C, according to the methods outlined in Ely and Jungreis (1977b).

4.6.3.2 Diets Containing Yeast and Agar: Rationale and Historical Perspectives

As discussed in several places in this book (Chapters 2 and 10 and elsewhere in this chapter), a very important root of insect diets is their early relationship to microbiology, as the revolution in microbial technology and culture emerged in the early twentieth century. Besides the materials and sterile techniques being developed in microbiology, there was also an intellectual quest to find the most irreducibly simple nutrients that could support organisms of interest. Just as in the nineteenth century biologists were actively dissecting organisms to see how their underlying structures made them work, microbiologists (and then entomologists, in hot pursuit) were using components such as tryptone, peptone, yeast extracts, sugars, tartaric acid, and of course agar to try to cultivate insects. Therefore, various forms of yeast (live, heat-killed, and extracts) and agar were used early on for cultivating insects, especially *Drosophila* species, which were of huge importance in genetic studies that were flourishing in the late 1800s and early 1900s. Studies of mutations, gene linkage, crossing over, and many other discoveries that came from *Drosophila* genetics would have been much harder to achieve without the early culture techniques that served as the roots of rearing other kinds of insects such as flour beetles and blow flies. This takes us back to the four main components of the Yamamoto diet and derivatives of that diet.

Besides wheat germ and casein, the Yamamoto diet contains yeast and agar. Both of these materials have been used in hundreds, possibly thousands, of insect diets dating back to the early 1900s when diets for *Drosophila* species were being developed to support the exploding field of insect genetics. Agar seems to have been used first in an insect diet by Baumberger (1917a,b). The inclusion of yeast was a reasonable extension of the observations such as those of Guyenot (1913b) and again shortly thereafter by Baumberger (1917a), where it was noted that the principal food of *Drosophila* was the yeast *on* the rotting fruit, rather than the fruit itself. During this early period, the basics of insect diets was aligned with the rapidly developing field of microbiology and microbial culture techniques. The understanding and application of the principles of sterile media and the concept of nutritionally complete media were emerging from the early 1900s through the 1940s and 1950s. Therefore, use of agar as a solidifying (gelling) agent was a logical outgrowth of familiarity with microbiology by people such as Baumberger and later by Michelbacher et al. (1932). This is also in concert with many other instances where insect dietetics drew intellectual assets from microbiology, including the use of casein, hydrolyzates, or extracts from meat and plant products as well as peptone, tryptone, yeast extract, and yeast autolysate.

In tracking the progression of insect diets, agar and agar substitutes (various hydrocolloids) are found in countless diets. The use of a gelled diet, that is, hydrocolloid-containing diets, is one of the most important foundations of insect dietetics, and this topic is discussed in several places in this book, including discussion of the function of diet components (Chapter 3), diet chemistry and physics (Chapter 5), diet/feeding system interactions (Chapter 7), diet processing and quality control (Chapters 11 and 12), and the chapter devoted to diet texture and rheology (Chapter 16).

The subject of yeast products in insect diets is one of the most complex topics in this book. Because yeast products are crucial in terms of nutritional value, palatability, and functional interactions with other diet components, it is of utmost importance that rearing and diet professionals understand the basics of yeast components in their rearing systems. Starting with the yeast used in the Yamamoto diet, torula yeast, the question of how certain yeast products are either *deal makers* or *deal breakers* in hundreds of different diets for more than 1000 insect species must be appreciated. The torula yeast in the Yamamoto diet is not described in terms of product source or how the yeast was processed. This lack of specification is a most unfortunate problem in hundreds of papers that include formulations for artificial diets. A cursory review of the diet formulations described by Singh (1977) include *yeast, baker's yeast, brewer's yeast, powdered yeast, yeast cake, dry yeast, active yeast, yeast extract*, and *yeast hydrolyzate*. Sometimes authors specify whether the yeast was active versus inactive, and sometimes they specify the supplier. So where does the torula yeast suggested by Yamamoto fit into this murky question?

In a quest to understand the yeast product that insect diet professionals may be using, they find a formidable and baffling literature in peer-reviewed journals and an even worse conglomeration of popular articles and Internet postings about characteristics of the many yeast products on the market. Fortunately, there are a few peer-reviewed articles that help sort out the picture. For example, Bekatorou et al. (2006) provide a comprehensive review of different kinds of yeast products used as human foods, but this review also offers an excellent overview of what insect diet specialists use in preparing media for target insects.

Summary of yeasts and yeast products: The three questions that must be answered are as follows: (1) Is the yeast live or killed? (2) What species of yeast is being used? (3) How was the product processed (Reed and Nagodawithana 1991)? The questions about live yeasts and species (as well as other trophically based insect/microbe interactions) will be treated further in Chapter 13. The second question about species of yeast is very important in light of the fact that several studies have shown in ecology studies (e.g., Dobzhansky et al. 1956, Starmer et al. 1976, Anagnostou et al. 2010) that several species of insects are differentially responsive to different species of yeast (and other fungi for that matter). In accord with the derivation of commercially available yeasts from the human food and the livestock/pet food industries, the genera of yeast in insect diets are from the genera *Saccharomyces*, *Candida*, *Kluyveromyces*, *Pichia*, and *Torulopsis*. By far, the most common species used in insect diets are *Saccharomyces cerevisiae* and *Candida utilis* (formerly *Torulopsis utilis*); *S. cerevisiae* is the major yeast utilized in insect diets.

A great deal of confusion exists with the names and actual products derived from *S. cerevisiae*, which is called live, active yeast, baker's yeast, brewer's yeast, and nutritional yeast, to mention the most common terms. Much of this discussion is derived from Bekatorou et al. (2006) and Stone (2006). The live or active yeast is discussed in Chapter 13, but sometimes authors mention baker's yeast: the yeast is live, ready to be used to leaven, flavor, texturize, and add nutrition to bakery products. However, what is supposed to be an inactive yeast product, incapable of fermentation or other metabolism, is used in diets, but not a product of brewing. According to Bekatorou et al. (2006), this inactive yeast (usually *S. cerevisiae*) is grown on molasses or some other nutritive media materials (soy products, wheat products, etc.), and this material is sometimes called *nutritional yeast* (Stone 2006). Another term that helps explain the nutritional characteristics of *nutritional yeast* is *primarily grown yeast*, indicating that the yeast is not a by-product of some other process such as the brewing industry fermentation.

The major name of the yeast that was derived as a secondary product of the brewing industry is brewer's yeast. This is probably (based on an informal survey of the literature) the most commonly used yeast product in insect diets. Brewer's yeast comes mainly from the spent fermentation material from the beer industry. It is from the dregs of the beer-making process, and as such, it contains materials such as hops chemicals and a complex mixture of postfermentation products. While it retains much of its nutritional and palatability characteristics, it must be washed and debittered. The debittering process is accomplished generally by treating the spent yeast with alkaline reactions and washing, then filtering or centrifuging. Debittering is needed for brewer's yeast that is to be used as a food because of the bitter, off flavor that is imparted by the hops used in brewing beer. Primarily grown yeast does not require debittering, and some manufacturers of so-called nutritional yeast products take extra steps to enrich the yeast by adding materials to the media such as selenium salts, methionine, or other factors that make yeast even more nutritious.

This leads to a very important aspect of the yeast products used in insect diets: there can be huge differences between products, depending not only on the species or strain of yeast but also upon the nature of the media on which the yeast was grown. Another factor that must be considered by diet personnel is that the way the yeast was processed after cultivation can also have strong bearing on the nutritional, bioavailability, and palatability qualities of the yeast component. For example, Stone (2006) points out that the extent of heating of the postcultured yeast can greatly impact the bioavailability of the proteins: the question of whether or not the yeast was autolyzed or hydrolyzed and whether the hydrolysis was done enzymatically or with acids; further, it is relevant to yeast quality to consider whether the yeast was treated with a high salt osmotic shock in the lysis process. To be explicit, the kinds of yeast products that are sharply different from one another may be from different species (*C. utilis*, *S. cerevisiae*, or *Kluyveromyces marxianus*, to name the three most common species), and the yeasts may be by-products of the brewing industry, wood processing waste treatments, or dairy waste treatment, or they may be noncellular materials such as extracts or lysates (including autolysates vs. hydrolyzates from acid or enzymatic treatment). The nonliving, cellular forms of yeast are generally killed with heat, such as pasteurization, baking, toasting, pressure/steam treatment, and the cells may be washed to remove the media in which the yeast colonies were grown, or they may be produced with their growth media as part of the product matrix. For example, sweet dairy whey is treated with *K. marxianus* or one of the other dairy yeast species that is capable of metabolizing lactose (such as *Saccharomyces fragilis*). After fermentation of dairy whey (which is about 70% lactose on a dry weight basis), the yeast and the remaining medium still contains a high content of proteins, lipids (including cholesterol), vitamins, and minerals.

This mixture of *spent* medium and yeast cells can be dried, powdered, and packaged as a product called Wheast™ (Hagen and Tassan 1970). The derivation of the product is from the cheese industry where milk was converted into curds and whey. The whey, having a high lactose content (and lactose being a digestive problem for some humans and many nonhuman species, including many insects), was a waste material that caused disposal problems in sewage systems because of the effects of lactose on the biological oxygen demand (BOD) in waste treatment systems. As a solution to this problem, engineers and microbiologists applied a simple technological solution: use dairy microorganisms such as specialized dairy yeasts to break down the lactose. The nutritional profile of brewer's yeast is presented in Table 4.7, and the specialization of different yeasts to different sugars is captured in Table 4.8. Not only did the lactose amelioration work, but the yeast and whey fermentation product, once dried, proved to be a highly nutritious product that was no longer considered a waste material but rather a primary food substance. The trade mark *Wheast*™ was used, and the product was sold as a health food, which was tested in several insect diets, including one published by Hagen and Tassan (1970) for lacewings. Although what the product which came from fermented dairy whey is no longer available because of other uses that were found for sweet dairy whey, a few suppliers still sell a product called Wheast, though this may be a mixture of brewer's yeast and dairy whey. While it may be attractive to insects and nutritious, if it is not derived from

Table 4.7 Gross Composition of Brewer's Yeast, a Heat-Killed Yeast

Gross Composition of Yeast Biomass	
Moisture	2%–5%
Protein	42%–46%
Nucleic acids	6%–8%
Minerals	7%–8%
Lipids	4%–7%
Carbohydrates	30%–37%

Source: Reed, G. and Nagodawithana, T.W., *Yeast Technology*, 2nd edn., Van Nostrand Reinhold, New York, 1991.

Table 4.8 Utilization of Four Sugars by Different Yeast Species

	Yeast Species		
	Saccharomyces cerevisiae	***Kluyveromyces marxianus***	***Candida utilis***
Glucose	+	+	+
Sucrose	+	–	–
Lactose	–	+	–
Xylose	–	–	+

Sources: Adapted from Reed, G. and Nagodawithana, T.W., *Yeast Technology*, 2nd edn., Van Nostrand Reinhold, New York, 1991; Stone, C.W., Yeast products in the feed industry yeast products in the feed industry: A practical guide for feed professionals, http://en.engormix.com/MA-feed-machinery/formulation/articles/yeast-products-feed-industry-t243/800-p0.htm, 2006.

fermentation of sweet dairy whey as was the original product, it is likely to be a very different material than the Wheast that was sold in the 1960s and 1970s.

The soluble forms of yeast inner contents are different from whole yeast cells where the nutrients are completely or partially contained within cell walls (active yeasts, inactive brewer's yeast, or inactive primary grown yeast = *nutritional yeast*). All of this information pertains mainly to the common yeast products made from *S. cerevisiae*. What about the products made from yeast species other than *S. cerevisiae*?

The most important example of non-*S. cerevisiae* yeast: torula yeast such as the product used in the diet for *M. sexta*. Torula yeast was originally a product of the wood/paper industry where processors needed to dispose of wastes in an economical and environmentally friendly way. They found that the yeast *C. utilis* (formerly *T. utilis*) could metabolize the wood pulp waste, including xylose sugar, which predominates wood slurry liquid. Most species of yeast (Table 4.8) and other microbes cannot metabolize wood pulp sugars (Bekatorou et al. 2006, Stone 2006). It was found that from an unpalatable wood product waste, a nutritional and very palatable product could be produced with considerable commercial value. Torula yeast that has been separated from the parent medium, heat-killed and spray-dried, is used not only as a human food additive (sometimes with controversy) and a material useful in insect diets (Vargas et al. 1994) and for trapping insects such as olive fruit flies (McPhail 1937). Unfortunately, the rationale for using torula yeast in diets has been stated explicitly neither in the Yamamoto paper (1969) nor in other reports, except for the recent paper by Chang (2009) on diets where torula is used instead of other kinds of yeast. Chang (2009) reported experiments where she compared torula yeast with several other yeast products from *S. cerevisiae* in diets for medfly, *Ceratitis capitata*, and the melon fly, *Bactrocera curcurbitae*, with little difference between the torula yeast product and other products.

Case I in Figure 4.9 shows two instances of antagonism between components. Dairy products, including casein, reduce the capability of agar to bind water to form gels, so the casein is *antagonistic* to gel formation. This point is amplified in detail in Chapter 15 where texture and

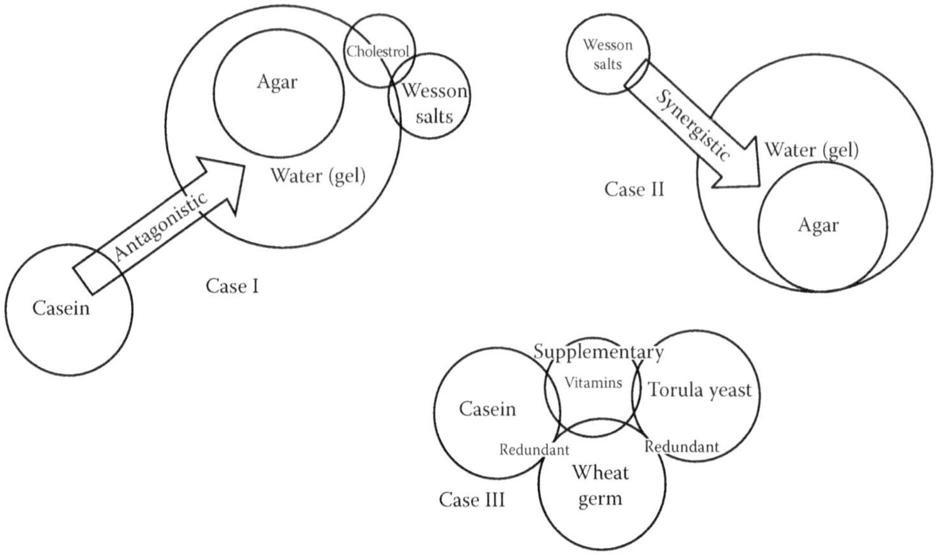

Figure 4.9 Three cases that illustrate possible interactions between components in the Yamamoto (1969) diet: antagonism between components (Case I), synergism (Case II), and redundancy and/or supplementation (Case III).

rheology are discussed. In Case II (Figure 4.9), the interaction known as *synergism* is depicted where some of the minerals from the Wesson salt mixture (notably calcium ions and potassium) enhance (synergize) the cross-linking of gels (Chapters 3 and 5), making stronger gels that are better able to hold water. In Case III (Figure 4.9), the illustrated interactions are *redundancy* and *supplementation*. Because the casein, torula yeast, and wheat germ all provide the essential amino acids, it is possible that there is a surplus of each amino acid. This surplus would render as redundant the presence of torula yeast as a source of essential amino acids in addition to those in wheat germ. However, should some kind of damage or interference take place where digestion or absorption of the proteins in the wheat germ was hindered, then the proteins (and their component amino acids in the casein and torula yeast) would serve as a *safety factor*. So redundancy is not necessarily a harmful interaction. However, if the redundancy took the form of a dilution of other nutrients, or if it became a metabolic burden, such as in the presence of extra amino groups from the protein causing an excess of the nitrogenous materials, which could result in a toxic amount of ammonia or uric acid, the redundancy could become a problem. Finally, where a nutrient material, including a staple or complex food, is lacking an adequate amount of vitamin or mineral, then another diet component such as the vitamin or mineral mixture can function as *supplements* to the main diet components.

4.6.3.3 Bean (Legume) Diets

The diet for noctuid Lepidoptera developed by Shorey and Hale (1965) has been cited 887 times as of October 2014, and has been the basis for hundreds of other papers on bean-based diets. This diet (derived from the 1963 cabbage looper diet by Shorey) is one of the simplest artificial diets and is one of the most economical to produce. It is composed of either lima beans (2133 g) or pinto beans soaked for 24 h, and blended with brewer's yeast (320 g), ascorbic acid (32 g), methyl paraben (20 g), sorbic acid (10 g), formalin (20 mL), agar (128 g), and water (6400 g). The diet is remarkable because of the breadth of species that it supports, the simplicity in synthesizing it, and the low cost of the raw materials (no casein and no vitamins except for ascorbic acid are important reasons why the diet is so economical). The base material (dry beans) can be stored safely for years, and only the yeast is somewhat perishable. Ascorbic acid is stable, provided it is kept dry and refrigerated. Many diets for insects other than those described by Shorey and Hale (1965) have been reared on derivatives of the bean diet (e.g., a cactus moth diet described by Carpenter and Hight (2012) is essentially the same as the Shorey and Hale diet with white kidney beans replacing the lima or pinto beans).

The impressive performance of legume-based (bean) diets begs the question: why are these diets so nutritious, palatable, accessible to many species, and so stable? At least part of the answer to these questions is the high protein content of legume seeds, the storage component of other nutrients (starches and lipids), the complete mineral content, and the stability that comes from the evolutionary strategy of legumes to keep their reproductive structures from deteriorating before they have had a chance to germinate. Of course, this factor can introduce antinutrients such as enzyme inhibitors that must be deactivated for some insect species before legumes can be used as a completely effective food. And this aspect of legumes is treated at length in the chapters on soy bean chemistry (Chapter 5) and diet processing (Chapter 12).

One last point about the use of beans in the Shorey (1963) and the Shorey and Hale (1965) diets: what was the basis for using beans in this diet? This is the same question that was raised in the context of wheat germ in the Texas A&M diets. There are no comments in either of the papers on bean diets as to what the original impetus was to use beans in the diets of insects that were not bean feeders (originally cabbage loopers, which feed more on vegetative parts of host plants than on seeds). Interestingly, Shorey and Hale (and earlier Shorey paper) were not the first to report research with legumes in insect diets. Fukuda et al. (1960) reported success with using soy flour and soybean curd in diets for silkworms (diets supplemented with ground mulberry leaves). So the question remains unanswered as to what inspired the use of legume beans in insect diets, but it is certainly one of the major breakthroughs in insect rearing and the history of insect diet science and technology.

4.6.4 How Organizational Matrices Contribute to Successful Diets

If the ingredients of the Adkisson et al. (1960) diet (Tables 4.4 and 4.5) were broken down to their irreducible form such as the breakdown listed in Table 4.3 (free amino acids, simple sugars, minerals that were not bound to macromolecules, etc.), would the diet be as good as the diet in its existing complex macromolecular form? Two alternative hypotheses regarding this question are fundamental to insect nutrition and dietetics: (1) The whole diet is no greater than the sum of its (irreducible) parts. (2) The whole diet is greater than the sum of its parts. Hypothesis 1 has been the working concept behind most experiments conducted in insect nutrition and dietetics throughout the history of these fields. It is seemingly a more mechanistic hypothesis than the second one, but it also may suffer from being an oversimplification of the potentially complex mechanisms that govern the character of food dispersions (discussed earlier).

The question of the efficacy of whole macromolecular structures was raised early on by Naylor (1964), but the suggestion that the whole was greater than the sum of its parts was not accepted, probably because it went against the grain of the insect nutrition community at that time. At the time that Naylor presented data on his intriguing experiments, the prevailing idea in the insect nutrition community was that the most irreducible level of nutritional organization was at the size and relatively low complexity of simple organic molecules such as amino acids, simple lipids, simple carbohydrates, vitamins, and minerals. If all of these components were presented in a diet, which stimulated feeding but failed to support robust growth, the results were attributed to either of two major causes: (1) Some cryptic nutrient (factor) must be missing or (2) the nutrients present must not be present in suitable proportions.

The concept of the nutritional *factor*, an undiscovered but key nutrient, became a household word among insect nutritionists after several notable discoveries such as the demonstration that carnitine was an essential nutrient for some insects (Fraenkel 1958) and the demonstration of the efficacy of ascorbic acid in several species of phytophagous insects (Chippendale and Beck 1964). In fact, in their quest for unraveling what was then unknown about vitamins, Fraenkel and Blewett published a series of studies (e.g., Fraenkel and Blewett 1947a–c, 1948) that established the foundation of vitamins, especially the water-soluble ones. However, it is possible that the concept of hidden nutritional factors has been overused as a reason why certain diets fail to work, especially when defined diets are used in a context that ignores the complex nutritional matrix that insect feeding systems are adapted to confront. This is a topic that deserves further attention by researchers that will apply creative mechanistic investigations to the hypothesis that the matrix of the food is a major determinant of the food's value to a given species.

4.6.5 Screwworm Diets: A Landmark Success Story in Insect Diets and Mass Rearing

One of the most remarkable success stories in the annals of entomology is the use of sterile insect techniques for the eradication of the screwworm *Cochliomyia hominivorax* (Coquerel) (Diptera: Calliphoridae) from most of North America (Scruggs 1978, Taylor 1992). Taylor (1992) pointed out that the eradication program for this insect employs the largest insect mass production system in the world as indicated by the fact that between the mid-1950s and the late 1980s, more than 400 billion of these insects had been reared, sterilized, and released. The program that was spearheaded by E. F. Knipling and R. C. Bushland in the late 1950s employed billions of laboratory-grown, sterilized male screwworm flies to confound the mating system of the wild insects (Scruggs 1978, Taylor 1992). This system required a rearing system based on an artificial diet that was both economically and biologically feasible. In a 2-year period alone (1958–1959), Scruggs (1978) noted that 3.7 billion sterilized pupae were released in the southeastern United States. By the late 1980s, over 400 billion insects had been reared, sterilized, and released (Taylor 1992). The roots of this work are traced back to the diets devised by Melvin and Bushland (1936, 1940; see Table 4.9).

The diet of Melvin and Bushland (1936) was historically important, because it was the first diet that was shown to completely replace live hosts for a parasitic insect and was further capable of supporting mass production. That diet contained three parts whole milk, one part citrated calf blood, two parts ground lean beef, and 0.5% formalin and yielded pupae that weighed 40–60 mg—more than the weights of screwworms reared in guinea pigs (45 mg) but less than those reared in calves (75 mg) (Gingrich 1972). The cost was \$0.30/1000 pupae, an economically acceptable amount, but the lower than desirable pupal weights prompted Melvin and Bushland (1940) to develop a new formulation (two parts water, two parts beef, one part blood, and 0.24% formalin), which produced larger larvae. With this formulation, the screwworm unit was able to produce millions of pupae per day, but as economic issues arose (e.g., the growing pet food industry, which became competitors for the meat products in the screwworm diet), the quest began again for cheaper but equally nutritious materials.

The recent and current generations of screwworm rearing have been and are still aimed at the reduction of costs of rearing this insect while maintaining or improving its quality so that it is highly competitive in the field, thus making it even more economically and biologically feasible as a component for this environmentally friendly means of pest control. As Taylor (1992) described, the efforts

Table 4.9 Melvin and Bushland Diet

Component	Amount	Unit
Water	1000	mL
Ground beef (finely ground, lean)	1000	g
Beef blood (citrated with 5 g of sodium citrate/500 mL)	500	mL
Formalin	5	mL

Sources: Melvin, R. and Bushland, R.C., A method of rearing *Chochliomyia americana* C. & P. on artificial media, U.S. Bureau of Entomology and Plant Quarantine, Report ET-88, Washington, DC. 2pp, 1936; Melvin, R. and Bushland, R.C., *J. Econ. Entomol.*, 33, 850, 1940.

Ingredients are thoroughly blended in order specified. Larvae are fed for 3 days, then transferred to a tub of fresh diet for 3 days, and then transferred to a tub of sand for pupation.

to improve the diet moved from the original Melvin and Bushland formulations that had several components replaced, including horse meat as a less expensive substitute for beef, through a *hydroponic* diet, and then back to a gelled diet. The hydroponic diet consisted of dried whole chicken egg, dried whole bovine blood, a milk substitute called calf suckle, sucrose, dried cottage cheese, and formalin, all suspended in water. This diet, which was developed by Gingrich et al. (1971), was intended as an inexpensive substitute for meat products, which have increased in cost over the past several years. The chapter by Gingrich (1972) is an excellent resource on the application of basic nutritional science to expand our grasp of feeding and nutritional requirements of screwworm larvae and how that knowledge was applied to the development practical improvements in diets for these insects that have had millions of dollars and environmental benefits that cannot be assessed. For example, Gingrich (1972) pointed out how the basic research that determined that choline was essential to the screwworm larvae and that this information was used as an impetus to find high choline-containing materials (such as egg yolk) for replacement of the meat components. Likewise the basic research on the intolerance of screwworms for certain sugars and for high carbohydrate concentrations in general led to formulations that contained appropriate amounts of suitable carbohydrates that supported healthy growth of larvae.

Initially, the hydroponic diet was presented on cotton or acetate, but it was found that the use of a gelled form of the diet reduced labor and handling costs incurred with the strictly liquid/suspension form of the diet materials (Taylor 1992). It should be noted that the terminology used in reports about this diet can be misleading in the implication that the diet is a simple, conventional liquid. As is the case with literature on diets for predators, plant bugs, egg parasitoids, and several other insects, the so-called liquid diets are actually slurry diets that consist of particulate materials that are suspended—not dissolved—in their aqueous medium. Suspensions can have much higher nutrient concentrations than can solutions. They also have numerous other properties that give them a special place in insect diet considerations (high viscosity, impeded flow, inclination to separate, among many other features). This point is covered in more detail in another part of this chapter.

The tremendous importance of the mass-rearing system for screwworms has encouraged continuous efforts at improving the diets and other rearing components for these insects. For example, a series of excellent studies of replacement materials for blood and meat components of the adults' diet (Chaudhury et al. 1998, 2000, 2011) and inexpensive gel replacements (Chaudhury and Alvarez 1999, Chaudhury et al. 2011) have been reported. Considering the fact that ~140 million flies are produced per week at the USDA facility at Chiapa de Corzo, Chiapas, Mexico, that this demands ~27,000 kg of dry food, which costs U.S. $42,000 (or $2,184,000 per year), it becomes clear why it can be of tremendous economic importance that replacement materials be tested and incorporated (Chaudhury and Alvarez 1999). Furthermore, considering that much of the spent material from the diet, frass, and the insects themselves must be disposed of as a waste product, there are large-scale environmental concerns incumbent in this program. For example, the disposal of the waste materials is made more environmentally friendly when a gelling material such as a starch-grafted sodium polyacrylate gel (Hampton Roads Repackaging, Chesapeake, VA, USA) can be substituted for more expensive materials such as Water-Lock G-400 (Grain Processing Corporation, Muscatine, IA, USA); it is important that the new material is recognized as nonhazardous and suitable for disposal in approved landfills Chaudhury and Alvarez. And it is, of course, essential that the use of this material does not compromise the quality of the insects that are the products of this program. It should be noted that recently, Chaudhury et al. (2011) reported on an addition of cellulose as a solidifying and texturizing agent for screwworm diets, again in the interest of saving money while producing the highest quality of insect possible.

4.6.6 Diets for Tarnished Plant Bugs

Prior to the invention of the Debolt (1982) diet, more than 20 publications appeared reporting efforts to develop diets for *Lygus hesperus*, *L. lineolaris*, and other closely related plant bugs

(Miridae) or to provide feeding information on these species that would be helpful toward diet development. As pointed out by Cohen (2000b), despite the fact that several studies implicated *Lygus* spp. as targeting solid materials in their host plants (and prey), all of the earlier diet development efforts for these species were based on strictly liquid diets that were solutions or very dilute suspensions of lipids combined with aqueous solutions of hydrolyzates (yeast, soy, and casein) or defined ingredients such as free amino acids, sugars, and other simple molecules. Unlike the previous diets for *Lygus* spp., Debolt diet was a complex slurry, which contained particles of wheat germ and lima bean meal mixed with several defined components (i.e., a meridic diet). This diet succeeded in supporting more than several hundred continuous generations of *L. hesperus*, and has served as a basis for several mass-rearing programs dedicated to the production of parasites and for various biological investigations of these pests. The complexity of the diet and the considerable expense of several ingredients prompted investigations into simplification and cost reduction of this excellent diet, and the replacement diet, designated by Cohen (2000b) as the NI diet, proved to be not only simpler to produce and about 1/10 of the cost of the Debolt diet, but it also proved to yield a greater biomass, more eggs per female, shorter development times, and a higher survival rate than the earlier diet (Tables 4.10 and 4.11).

The reasons why both of these diets succeed in supporting robust, evidently unlimited production of the tarnished plant bug and the western tarnished plant bug are evidently that they satisfy all the requirements mentioned earlier in this chapter: (1) They induce robust feeding (indicating that they include appropriate feeding stimuli and/or token stimuli), allowing the insects to use their extraoral digestive process to select and process key nutrients prior to ingestion. (2) They contain all the essential nutrients (all the amino acids, lipids, vitamins, minerals, and any other cryptic factors, not yet recognized—should such factors exist). (3) They exist in the form of a complex matrix of supermacromolecular structure (lipoprotein/glycoprotein complexes with cross-linkages to polysaccharides—as depicted in Figures 4.1 through 4.4). (4) They contain organization that allows both the preservation of components such as unsaturated lipids being protected from peroxidation by iron, copper, and zinc, thus preserving the diet throughout storage and cage life. (5) They contain organization that leads to the components' bioavailability. Each of the figures shows the complex and heterogeneous arrangements of the lipid or lipoprotein components (red-staining spheres) in relationship to the other components. Figures 4.1 and 4.2 are sections of the Adkisson et al. (1960) diet with the rigid cell walls of the wheat germ component surrounding several diet components. Likewise, a histological section of the arthropod diet (Figure 4.3) (Cohen 2000b) shows the protective matrix of plant

Table 4.10 Major Diet Components (≥1 g per kg) in the NI Diet and the Debolt Diet in Grams of Material per Kilogram (Cost in U.S. Dollars)

Component	NI Diet	Debolt Diet for *L. hesperus*
Chicken egg yolks	120 (0.12)	0
Whole chicken eggs	55 (0.04)	144 (0.10)
Wheat germ (toasted)	80 (0.22)	36 (0.10)
Lima bean meal	120 (0.17)	36 (0.05)
Soy flour (toasted)	20 (0.06)	0
Sucrose	10 (0.12)	22 (0.22)
Lecithin	4 (0.27)	0.04 (0.03)
Vitamins (Vanderzant)	3.2 (0.38)	7.2 (0.86)
Brewer's yeast	1.8 (0.05)	0
Honey solution (50%)	8.9 (0.03)	0
Salt mixture	0	2.3 (0.50)
Casein hydrolyzate	0	14.3 (0.96)
Gelcarin	0	2.2 (0.22)
Water	573	730
Acetic acid (10%:90% water)	3 (0.01)	0

Table 4.11 Other Diet Components (<1 g per kg) in the NI Diet and the Debolt Diet in Grams of Material per Kilogram (Cost in U.S. Dollars)

Component	NI Diet	Debolt Diet for *L. hesperus*
Chlortetracycline	0.01 (0.04)	0.4 (1.84)
Streptomycin sulfate	0.01 (0.05)	0.1 (0.50)
Formalin	0.4 (0.01)	0.4 (0.01)
Niacin	0	0.8 (0.91)
p-Aminobenzoic acid	0	0.8 (0.90)
Cholesterol	0	0.4 (0.25)
RNA	0	0.4 (0.31)
Tween 80®	0	0.7 (0.05)
Linoleic acid	0	0.14 (1.23)
Propionic acid	0.3 (0.01)	0
Total cost of materials	$1.46	$9.04

Prices based on local (Starkville, MS) supermarket prices of liver, high fat ground beef, fresh eggs, baby lima beans, wheat germ, and soy flour. The other prices are based on a 1998 Sigma catalog.

materials arranged in conjunction of other diet components. The Cohen and Smith (1998) entomophage diet is characterized in Figure 4.4 by the matrix relations of the egg yolk and meat components.

One of the most conspicuous differences between the two diets is in the water content, Debolt diet's being ~84% water, the NI ~70%. Despite this fact, both diets have a consistency of slurry, the Debolt diet being slightly runnier. The NI diet has ~1.7× the protein content of the Debolt diet, and the abundance of essential amino acids is higher in the NI diet, implying a higher overall protein quality in the NI diet. Importantly, the lipid content is more than three times as high in the NI diet as it is in the Debolt diet. There are key differences in the amounts of key lipid nutrients: for example, cholesterol is about 2.5 times that of the Debolt diet. The polyunsaturated fatty acids in the NI diet are four times those found in the Debolt diet. The vitamin content of both diets is very similar, except for ascorbic acid, which is more than twice as much in the Debolt diet as it is in the NI diet. Also, vitamin B-12, inositol, and choline are all higher in the Debolt diet than they are in the NI diet, largely because of the differences in the amount of vitamins added. It is noteworthy that a considerable amount of the vitamins present in the NI diet are naturally occurring in the food components such as the wheat germ, soy, lima bean meal, and especially the egg yolks. It is also noteworthy that by far the greatest contribution to the cholesterol content is that which is derived from eggs (specifically the egg yolks). Although they are not listed, plant sterols are more abundant in the NI diet than they are in the Debolt because of the greater amount of plant material present in the NI diet. According to Toivo et al. (2001), the dried plant materials in these diets (wheat germ, soy flour, and lima bean meal) contain roughly 0.05 g of phytosterols (campesterol, β-sitosterol, stigmasterol, and stanols) per 100 g of material. Therefore, the NI diet, which contains 220 g of plant materials, contains about 10 mg of *cryptic* plant sterols, and the Debolt diet contains about 3.5 mg of phytosterols per kg of diet. It is possible that these phytosterols are especially well suited as nutrients for dedicated plant-feeding insects. It should further be noted that the Debolt diet contains added cholesterol, raising the total sterol content of this diet; however, it is not documented whether or not the cholesterol added to the diet is as suitable in bioavailability as the NI diet.

Some other noteworthy points are that the mineral content of the NI diet is higher than that of the Debolt with respect to each mineral, despite the fact that there is a special mineral mixture added to the Debolt diet. This is a result of the fact that all the *complete* ingredients have an ash or mineral content that is naturally present, often in a profile that reflects generalized mineral requirements. It is evident from Tables 4.12 and 4.13 that the sources of these minerals are the whole food components (e.g., eggs, wheat germ, lima beans). The same point applies to the vitamins, with the most notable

Table 4.12 Contributions of Various Components to the Nutritional Contents of Debolt Diet for *L. hesperus*

1.438 pt	Egg	Wheat Germ	Lima Beans	Misc. Additives	Salts	Vitamins	Casein + Additives	Σ
Water	100	4	3.7	730				838
Protein	18	8.3	7.7	12.5				49
Lipid	14	3.5	0.3	0.54				18
Carbo	1.8	18.6	22.8	22				65
Ca	70.6	14	29.2					113.8
Fe	2.1	2.3	2.7					7.1
Mg	14.4	86	80.6		86			267
P	256	303	138.6		550			1248
K	174	321	620.6		399			1515
Na	181	4.3	6.5		320			512
Zn	1.6	4.4	1		7.6			14.6
Cu	0.02	0.29	0.3		1.7			2.3
Mn	0.03	4.8	0.6		7.7			13.1
Se	0.04	0.003	0.003					0.047
Ascorb.	0	2	0			1944		1946
Thi.	0.09	0.68	0.18			1.8		2.8
Rib.	0.73	0.18	0.07			3.6		4.58
Ni.	0.1	2.45	0.55			7.2		10.3
Pant.	18.1	0.81	0.49			7.2		26.6
B-6	0.2	0.48	0.18			1.8		2.7
Folic	0.07	0.1	0.14			1.8		2.1
B-12 mg	0.001	0	0			21.6		21.6
Biotin	a	a	a			0.135		0.135[a]
Inositol	a	a	a			144		144[a]
Choline Chloride	a	a	a		a	360		360[a]
Vit. A IU	914	0	0					914
Vit. E	1.51	7	0.26			56		65
FA Sat.	4.46	0.59	0.06					5.11
FA Mono	5.47	0.49	0.02					6.0
FA poly	1.96	2.2	0.11	0.18				4.5
Chol mg	612	0	0	0.4				612.4
Tryp.	0.219	0.11	0.09	0.22				0.64
Thre.	0.86	0.35	0.33	0.58				2.12
Iso	0.98	0.31	0.41	0.64				2.34
Leu	1.54	0.57	0.67	1.36				3.95
Lys	1.29	0.53	0.52	1.2				3.54
Meth	0.56	0.16	0.1	0.36				1.18
Phe	0.95	0.33	0.44	0.8				2.52
Val	1.1	0.43	0.16	0.95				2.84
Arg	1.08	0.67	0.47	0.6				2.82
His	0.43	0.23	0.24	0.44				1.34

[a] Means that material is probably present but was not measured or reported.

Table 4.13 Contributions of Various Components to the Nutritional Content of the Cohen (2000a) Diet for *Lygus* spp. and Other Species

Ingredient	Yeast (1.8)	Yolk (120)	Egg (55)	Wheat Germ (80)	Lima Beans (120)	Soy (20)	Additives	Vitamins (3.2)	Σ (1000)
Water	0.13	59	41.2	4.5	12.2	1	573		698
Protein	0.7	20.2	6.9	23.3	25.8	6.9			83.8
Lipid	0.08	37.1	5.5	8.6	0.8	4.1	4		60.2
Carbo	0.7	2.1	0.7	39.7	76.1	7	20		128.8
Ca (mg)	1.2	164	27	36	97.2	41			366
Fe (mg)	0.3	4.2	0.79	7.3	9	1.3			23
Mg (mg)	1.7	10.8	5.5	256	269	85.8			376
P (mg)	23.2	585.6	97.9	917	462	98.8			2185
K (mg)	36	112.8	66.6	758	2069	503			3546
Na (mg)	0.9	51.6	69.3	3.2	21.6	2.6			149
Zn (mg)	0.1	3.7	0.61	13.3	3.4	0.78			22
Cu (mg)	0.009	0.03	0.008	0.5	0.9	0.58			2
Mn (mg)	0.01	0.08	0.01	16	2.0	0.46			19
Se (mg)	T	0.05	0.002	0.05	0.009	0.002			0.1
Ascorb. (mg)	0.005	0	0	4.8	0	0		864	869
Thi. (mg)	0.04	0.2	0.03	1.3	0.6	0.12		0.8	3.1
Rib. (mg)	0.1	0.77	0.28	0.66	0.2	0.23		1.6	3.8
Ni. (mg)	0.7	0.018	0.04	4.5	1.8	0.86		3.2	11.1
Pant. (mg)	0.2	4.57	0.69	1.1	1.6	0.32		3.2	11.7
B-6 (mg)	0.03	0.47	0.08	0.78	0.6	0.09		0.8	2.9
Folic (mg)	0.04	0.175	0.003	0.28	0.5	0.07		0.8	1.9
B-12 (mg)	0.0004	0.004	0.0006	0	0	0		9.6	9.7
Biotin	[a]	[a]	[a]	[a]	[a]	[a]		0.06	0.06[a]
Inositol	[a]	[a]	[a]	[a]	[a]	[a]		64	64[a]
Choline chloride	[a]	[a]	[a]	[a]	[a]	[a]		160	160[a]
Vit. A IU		2334	349.3	0	0	24			2708
Vit. E (mg)	0.001	3.8	0.58	14.5	0.9	0.39		25.6	45.8
FA Sat.	0.01	11.5	1.7	1.46	0.2	0.59			15.5
FA Mono	0.05	14.0	2.1	1.2	0.07	0.91			18.3
FA poly	0.0002	5.0	0.75	5.3	0.37	2.3	4		17.8
Chol (mg)	0	1281	233.8	0	0	0			1515
Tryp.	0.009	0.24	0.08	0.3	0.31	0.1			1.039
Thre.	0.04	1.1	0.33	0.98	1.11	0.3			3.86
Iso	0.04	1.0	0.38	0.86	1.36	0.34			3.98
Leu	0.06	1.77	0.59	1.58	2.22	0.56			6.78
Lys	0.06	1.6	0.49	1.48	1.73	0.46			5.82
Meth	0.01	0.5	0.22	0.46	0.33	0.09			1.61
Phe	0.03	0.86	0.37	0.93	1.48	0.36			4.03
Val	0.04	1.1	0.42	1.21	1.5	0.35			4.62
Arg	0.04	1.5	0.41	1.88	1.58	0.54			5.95
His	0.02	0.52	0.16	0.65	0.79	0.19			2.33

[a] Material is probably present but was not measured or reported.

exception being ascorbic acid, which is absent in most of the whole foods here. Most animal-derived materials, seed products, and other components, other than fresh fruits and vegetables, lack ascorbic acid. Therefore, for those insects that require this vitamin, special precautions must be taken to include it and to guard it so that it does not deteriorate during storage or during the active feeding period.

Another noteworthy point is that the materials in the diets are both protected from autodestruction while they retain a high level of bioavailability. It is well known, for example, that egg yolk proteins are among the most digestible and well-balanced proteins and that their being cooked increases their digestibility (Damodaran 2008). Not as well known, however, is the fact that the iron that is relatively abundant in the egg yolk and plant products in these diets can induce lipid destruction through peroxidation (Nawar 2008), and other components such as zinc and copper can further promote the destruction of nutritionally valuable lipids, especially in the presence of ascorbic acid (in a process known as ascorbate-induced lipid peroxidation, Chapter 5). However, the matrix in which the iron is held in its natural state in the foods used in these diets protects the lipids from destruction. Heat treatments are one of the most pervasive and effective diet treatments. Heating (Chapters 5 and 12) denatures destructive proteins, and it forms gelled complexes that immobilize the components that would otherwise diffuse into diet regions where destructive process could occur. This is a major factor in the strategy to improve the nutritional quality and bioavailability of components, especially eggs, in the NI diet. It is also the basis of the patented processes involved in the other diets described by Cohen (1998c, 2000b). One final point: a personal anecdote about the development of the Debolt diet and the Cohen diet is provided in Chapter 1 in a discussion of the benefits of using scientific inquiry as a basis for developing and improving insect rearing systems.

4.6.7 Diets for Drosophilids and Other Yeast-Containing Insect Media

As discussed in several places in this book, the importance of rearing saprophagous drosophilid flies (especially *Drosophila melanogaster*) and tephridid flies cannot be overemphasized in the advancement of science and society. Considering the scientific and economic importance of rearing drosophilid and tephridid flies, it is amazing that scientists who work with these flies have fairly well ignored the basics of these insects' feeding biology, mainly that their major food is the yeasts that grow on rotting fruits and other detritus-type materials, rather than the detritus itself. The documentation for yeasts as the feeding targets of these saprophagous flies comes from several excellent ecological studies summarized by Anagnostou et al. (2010). These authors made the following comment: "Unfortunately, the influence of dietary yeast species on the life-history of *Drosophila* has not been addressed any further since then, and most studies on ecological aspects in *Drosophila* are carried out using merely Baker's yeast, *Saccharomyces cerevisiae* Meyen ex E. C. Hansen (Saccharomycetaceae), as food source. Especially for *D. melanogaster* Meigen, being one of the most studied *Drosophila* species as well as a popular model organism, next to nothing is known about the influence of dietary yeast species on the insect's life-history." Anagnostou et al. (2010) further point out that not only are the flies selective about the yeast species upon which they depend for optimal nutrition, but they are adept at carrying the yeasts to their new feeding targets, inoculating the rotting materials, then feeding on them as the cultures accumulate significant biomass. It is important to recognize that saprophagous drosophilids and tephridids depend upon live yeast of the appropriate species, and the natural medium (rotting cactus, leaf materials, and fruits) can be a major determinant of the ability of these flies to thrive and meet their optimal fitness. In one of the most remarkable instances of specificity of a dietary requirement, Heed and Kircher (1965) showed that *Drosophila pachea*, a species restricted to a single host (senita cactus, *Lophocereus schottii*), requires the sterol Δ--^{7}stigmaster-3β-ol, which is present in the cactus tissues in considerable abundance. Heed and Kircher commented, "Our early tests showed that *D. pachea* could breed successfully on a standard medium composed of bananas, malt, corn syrup, and yeast only when the medium is supplemented with a cube of fresh or autoclaved senita cactus." This analysis of the remarkably fine-tuned and specific tritrophic relationship between *Drosophila pachea* and its host plant and yeast is an excellent representative model of the many kinds of specialization that is found in drosophilids and their yeasts of choice. This is discussed further in the chapter on the microbial relationships of various insects and the fungi on which they specialize.

4.7 HOW ORGANIZATIONAL MATRICES CONTRIBUTE TO SUCCESSFUL DIETS

If the ingredients of the Adkisson et al. (1960b) diet were broken down to their irreducible form such as the wheat germ breakdown listed in Table 4.3, would the diet be as good the diet in its existing form? Two alternative hypotheses regarding this question are fundamental to insect nutrition and dietetics: (1) The whole diet is no greater than the sum of its (irreducible) parts. (2) The whole diet is greater than the sum of its parts. Hypothesis 1 has been the working concept behind most experiments conducted in insect nutrition and dietetics throughout the history of these fields. It is seemingly a more mechanistic hypothesis than the second one, but it also may suffer from being an oversimplification of the potentially complex mechanisms that govern the character of food dispersions (discussed earlier).

The question of the efficacy of whole macromolecular structures was raised early on by Naylor (1964), but the suggestion that the whole was greater than the sum of its parts was not accepted, probably because it went against the grain of the insect nutrition community at that time. At the time that Naylor presented data on his intriguing experiments, the prevailing idea in the insect nutrition community was that the most irreducible level of nutritional organization was at the size and relatively low complexity of simple organic molecules such as amino acids, simple lipids, simple carbohydrates, vitamins, and minerals. If all of these components were presented in a diet that stimulated feeding but failed to support robust growth, the results were attributed to either of two major causes: (1) Some cryptic nutrient (factor) must be missing or (2) the nutrients present must not be present in suitable proportions.

The concept of the nutritional *factor*, an undiscovered but key nutrient, became a household word among insect nutritionists after several notable discoveries such as the demonstration that carnitine was an essential nutrient for some insects (Fraenkel 1958) and the demonstration of the efficacy of ascorbic acid in several species of phytophagous insects (Chippendale and Beck 1964). In fact, in their quest for unraveling what was then unknown about vitamins, Fraenkel and Blewett published a series of studies (e.g., Fraenkel and Blewett 1947a–c, 1948) that established the foundation of vitamins, especially the water-soluble ones. However, it is possible that the concept of hidden nutritional factors has been overused as a reason why certain diets fail to work, especially when defined diets are used in a context that ignores the complex nutritional matrix that insect feeding systems are adapted to confront. This is a topic that deserves further attention by researchers that will apply creative mechanistic investigations to the hypothesis that the matrix of the food is a major determinant of the food's value to a given species.

4.7.1 Interactions between Components in Diets

The components shown in Figure 4.9 represent the ingredients in the *M. sexta* diet developed by Yamamoto (1969), with wheat germ, casein, and torula yeast being the major sources of nutrition and the remaining ingredients being functional components that either supplement the main nutrients, provide texture improvements, reduce microbial growth, or increase palatability (note that these functions support the major factors in diets that are successful in supporting insect production (palatability, nutritional suitability, bioavailability, and stability). The arrows in this figure represent the kinds of interactions that are possible in the functionality of the diet components. The figure is intended to provide a model of component/component interactions, and Figure 4.9 gives three *cases* of interactions that are well documented.

4.8 VITAMIN AND MINERAL SOURCES IN SUCCESSFUL DIETS

Although many successful diets have no mineral or salt mixtures added, about half of all diets do have salt mixtures, and in some cases, they do contribute to the nutritional value. However, there are some very important differences in various salt or mineral mixtures. Comparisons of the two vitamin mixtures (the AIN and Vanderzant vitamins) that are commonly used in insect diets show some very important differences. The composition of these mixtures is listed in Appendix C. The most major difference between the AIN mixture and the Vanderzant mixture is in the absence of ascorbic acid in the AIN mixture and its abundance in the latter mixture. A second important difference is seen in the amount of sugar present in the two mixtures, with about 97% of the AIN being sucrose and about 65% of the Vanderzant mixture being the simple sugar, glucose. Unlike the Vanderzant mixture, the AIN formulation contains vitamin A, vitamin D_3, and vitamin K_2, all known to be required in most vertebrates whose vitamin requirements have been evaluated. However, with the possible exception of vitamin A, whose function is ambiguously interpreted in some insects (Gilmour 1961), these vitamins are not required by insects, in general. Other important differences include the lack of inositol in the AIN formulation and the fact that the α-tocopherol is in its more stable and slightly more water-soluble acetate form in the AIN mixture, rather than the less soluble simple α-tocopherol in the Vanderzant formulation. Inositol has been shown to be an essential or at least a beneficial nutrient in several species of insects (Vanderzant 1959). The functions of these vitamins are discussed in Chapter 3, and their chemical interactions in diets are described in Chapter 5.

4.9 ISSUE OF BIOAVAILABILITY

When accounting for why diets succeed or fail, the issue of bioavailability must be considered prominently. Bioavailability questions pertain to all nutrients, but they are most crucial in the domain of essential nutrients. Unfortunately, almost nothing is known of the bioavailability of nutrients in insects, and therefore, most of this discussion is based on what is known of bioavailability in vertebrates—more specifically, mammals. The questions of bioavailability are these: (1) Is a given nutrient required? (2) If it is required, is it present in the diet in amounts that are adequate (i.e., are minimum daily requirements met)? (3) If the amounts of the nutrient are adequate, are they in a chemical form that makes them useful to the target organism? (4) If the nutrient is present in a chemical form that is adequate, are there other factors present in the diet matrix that interfere with or enhance the absorption of the nutrient? The bioavailability questions must be considered in reference to each of the classes of nutrients, because each has its own peculiarities.

4.9.1 Bioavailability of Proteins and Their Amino Acids

As discussed further in Chapter 5, on the chemical nature of the nutrients, and Chapter 7 on digestion, amino acids, which are the constituents of proteins that must be freed from their parent proteins to be of nutritional value. It must also be recalled that except for the limited range of insects, such as most homopterans, that truly feed on plant saps (phloem sap and xylem sap), most insects gain the vast majority of their amino acids from proteins, rather than from the fairly limited pool of free amino acids. Cohen (2000c) pointed out that in predaceous insects, for example, more than 90% of the amino acids were derived from proteins. This is certainly the case in the vast majority of insect artificial diets. This means that the proteins in the diets must be digestible, and the digested products must be in a suitable matrix for absorption. It is intuitively apparent that proteins that defy digestion or that are, at best, poorly digested will not provide the amino acid nutrition that is expected of them. Differences in the digestibility of proteins vary not only from protein to protein but also from species to species.

4.9.2 Bioavailability of Minerals

Issues of bioavailability are potentially even more problematic for minerals than for proteins. First, it is likely that insects, like vertebrates, depend upon specific mineral receptor sites on gut cells for the absorption of essential elements. The factors that influence mineral bioavailability are the interplay between the absorption sites, the minerals, the competitors for absorption sites, pH of the local cell surface/lumen region, and the presence of mineral sequestering agents. For example, iron and zinc compete with one another for absorption sites, and an excess of one of these essential minerals could create a de facto deficiency of the other. A good diet will be balanced in the potential competitors so that such a de facto deficiency is not created. Such a balance is found in cottonseed meal, for example, which contains about 13 mg each of iron and zinc per 100 g of product. However, the presence and even balance of minerals does not guarantee that they will be in a suitable form to assure absorption. Chapters 5 and 8 contain detailed discussions about idiosyncrasies of iron availability in conjunction with the form of this mineral in diets (the crystalline phosphate form being less available than the amorphous phosphate form). A growing base of information is being accumulated about the facilitation of mineral absorption by various factors such as ascorbic and citric acid and acid pH at absorption sites. Finally, the prevention of absorption by chelating and sequestering agents such as phytic acid is becoming increasingly well documented.

4.9.3 Bioavailability of Vitamins

As is the case with the other nutrients, most of what we know about bioavailability of vitamins is derived from the research on vertebrates, mainly from studies on rats and domestic animals. It is dramatically evident from Table 4.14 that the bioavailability of biotin is greatly dependent upon the species of animal in question and the type of food (the matrix) where the biotin is present. Generally from considering the first three foods in this table, it appears that turkeys are generally more adept at absorbing biotin from all foods. However, the reversal of percentages absorbed with wheat as the biotin source indicates that bioavailability is not simply a consumer species phenomenon, but rather it is a complex combination of factors including characteristics of the consumer species and the food type (or food matrix). Also, the differences evident from this table indicate the danger of broad generalizations about herbivores (turkeys) being more efficient at utilization of plant materials than omnivores (pigs) or carnivores.

What further complicates the issue of bioavailability of vitamins in various foods is that the utilization efficiencies may flip-flop from one type of vitamin to another and are modified by other factors such as contents of minerals, lipids, proteins, and food processing (Gregory 2008). A most dramatic and well-documented example of the effect of processing is to be found in the interaction between the egg white protein, avidin, and biotin. It has long been realized that a factor in uncooked egg whites (avidin) binds irreversibly the biotin from egg yolks and other diet components and that the avidin is scarcely digestible, so it carries the biotin with it so that both are eliminated with undigested wastes. The simple cooking of the egg whites so that avidin is denatured prevents this

Table 4.14 Bioavailability (% Absorbed) of Biotin in Feedstuffs in Two Species of Vertebrates

Source of Biotin	Pigs	Turkeys
Soybean meal	55	77
Corn	4	95
Canola meal	4	65
Wheat	22	17

Source: Gregory, J.F., Vitamins, in: S. Damodaran, K.L. Parkin, and O.R. Fennema (Eds.), *Fennema's Food Chemistry*, 4th edn., CRC Press, Boca Raton, FL, 2008.

process and allows biotin to be available, providing that other factors that may influence bioavailability are met. These complexities are further complicated by evidence that the physiological state and even the genetic characteristics of the consumer also exert an unpredictable influence on the bioavailability characteristics. This issue is discussed further in Chapter 8 on chaos in insect diets.

Likewise, two commonly used mineral (salt) mixtures have strikingly different compositions. Most dramatically, the Wesson mixture has considerably less calcium than does the AIN mixture. It is interesting to note that insects' requirement for calcium is much lower than that of the *higher* vertebrates for which these mineral mixtures were originally designed. It is therefore likely that both mixtures have much more calcium than is required by most insects. Because calcium is strongly inclined to form insoluble compounds (Dadd 1968) and large, cumbersome matrices, it can have detrimental effects on the diet dispersion as a whole, especially if the AIN mixture is used. Other differences include the presence of fluoride in the Wesson mixture but not in the AIN mixture and, conversely, the presence of zinc, chromium, and selenium in the AIN but not in the Wesson mixture. Possibly one of the most important differences and one that might appear innocuous is the counterions (anions) for the iron. In the AIN mixture, the iron is in the form of ferric citrate, but in the Wesson mixture, it is ferric phosphate. Insects are known to respond differently to these two forms in terms of efficiency of iron absorption (Keena et al. 1998). These types of issues underlie the question of why certain diets are successful while others are not, and more research should be encouraged into determining how issues such as forms of minerals or matrices of nutrients have profound nutritional consequences. This topic is further discussed in Chapters 5 and 8.

CHAPTER 5

Chemistry and Physics of Insect Diets

5.1 INTRODUCTION TO DIET CHEMISTRY AND PHYSICS

Insect diets are chemically and physically complex and dynamic. Even before the insect ingests the completed diet, many interactions take place within and between components, both in processing and storage. Diet components as raw materials and within the completed diet are subject to chemical and physical laws that predispose their sensory and nutritional qualities as well as their stability. The chemical and physical characteristics of diets are products of the interactions of components in the context of the diet's organizational matrix or structure.

All insect diets, for example, contain water, often in substantial amounts. How that water behaves due to its own properties and to the interaction with other diet components can have a profound effect on the target insect. Likewise, each diet has a characteristic pH, which has a profound influence on many factors, including palatability, stability, and microbial interactions. The presence of oxygen and other oxidizing agents can compromise the quality of a diet by destruction of key nutrients. Also, these agents can also be destructive by degrading the chemical protection from antioxidants. The texture of the diet has far-reaching implications in terms of palatability, digestibility, and even stability. Several factors influence texture, including the choice of ingredients and the type of processing used in diet synthesis. Preservation of diets also depends on processing strategies.

5.2 BIOENERGETICS AND THE NATURE OF ENERGY IN INSECT DIETS

As was discussed in Chapters 3 and 7, one of the driving forces of biology of organisms in general is the need to acquire energy and transduce (transform) it into biologically useful form. Except for plants and the limited array of microbes that use photosynthesis (conversion of absorbed light energy into chemical energy), most organisms rely upon chemical energy to meet all of their biological needs for doing biological work (movement, information processing, growth, and reproduction). Heat gain from radiation, convection, or conduction may warm an organism to temperatures that enhance chemical reactions, but that heat does not do biological work. It is, again, only the harnessing of chemical energy from organic (carbon-based) materials that permits most organisms functions. Table 5.1 shows the energy values for the three major classes of macronutrients.

A primary rule of bioenergetics is that the energy that organisms use is present in *chemical bonds*. The bonds are connections between various atoms within a molecule via electrons shared by those atoms. A convenient (though somewhat oversimplified) view of bioenergetics in living systems is that organisms do their biological work by transferring high energy electrons from molecule to molecule. Biochemists call the energy that can be made available for such work as *free energy*

Table 5.1 Energy Values for Macronutrients in kcal and kJ

Chemical Class	kcal/g	kJ/g
Protein	4.0	16.7
Carbohydrate	4.1	17.2
Lipid	9.1	38.0

(as opposed to heat energy or energy of disorder = entropy). Biochemists measure and discuss chemical energy in terms of calories or joules (the International System of Units term for a unit work). A calorie is the unit of heat energy that is required to raise a gram of water 1°C (from 14.5°C to 15.5°C). A joule is 0.239 cal (or a calorie is 4.184 J). Because calories are more familiar, this term is followed in this book.

A confusing convention is that of using the terms, calorie, Calorie, and kilocalorie. Because the calorie is a relatively small value for describing common biological processes, most biochemists have adopted the use of the kilocalorie (kcal), which equals 1000 calories. The confusion arises from the convention of using the abbreviation Cal for kcal. Therefore, a Cal equals 1000 cal. The energy values of foods are almost invariably described in terms of Cal (which can be distinguished in writing but not in speaking).

A commonly used breakdown of foods is provided here. A popular chocolate and peanut candy bar (Snickers®) weighs 58.7 g (2.07 ounces), and has a total of 280 Cal (280,000 Cal) or about 1,172 kJ. Its fat content is about 14 g (or about 130 Cal from fat). Its sugar content is 30 g (or about 123 Cal from sugar), and its protein content is 4 g (or about 17 Cal from protein). The fat, carbohydrate, and protein contents add up to about 270 Cal; the remaining 10 Cal are attributable to minor nutrients that contribute energy (vitamins and nucleic acids) and some error from rounding off.

Frequently, in biological literature, when energy values are presented, they are arrived at proximately from calculations based on published values for the components (such as those values for proteins, fats, and carbohydrates presented in Table 5.1). However, some biological materials do not have known compositions, or sometimes more precise and direct values are desirable for energy contents, and these parameters must be measured. The most direct and common way that energy contents are measured is through the use of *bomb calorimetry*. A bomb calorimeter is a chamber made of heavy duty metal that has been machined to seal tightly after a sample is placed in it with an attached electric fuse. The sample is dried, turned into a pellet by pressure, weighed, and placed in the chamber. The chamber, after being sealed, has the air removed, and the air is replaced by pure oxygen to allow complete combustion of the sample. Next, the bomb is placed in a preweighed water bath, and the temperature is monitored before, during, and after the fuse and the sample are ignited and fully combusted. The heat transfers through the bomb and is absorbed by the water. After corrections for heat transfer efficiency and fuse combustion, the sample's energy content is calculated. So roughly speaking, if a 1 g sample were combusted in a bomb surrounded by 1 kg of water, and if the water temperature rose by 4°C, the energy content of the sample was 4 Cal/g.

5.3 NATURE OF WATER AND WHAT IT MEANS TO INSECT DIETS

> Water, water, everywhere, and all the boards did shrink.
> Water, water, everywhere, nor any drop to drink.
>
> *The Rime of the Ancient Mariner* by Samuel Taylor Coleridge

Water is one of the great wonders of the universe, and because it is so abundant, it and its miraculous properties are sometimes taken for granted. Yet to understand living systems, such as the insects in rearing systems, it would be useful to understand the nature and role of water.

Water covers the majority of the earth's surface. It is generally abundant in the atmosphere. It is a major determinant of weather and climate. It is *the* solvent of living systems. All life processes take place in a matrix of water. All metabolic processes that define the character of life depend on their watery basis. Water's heat capacity, degree of conductivity (to heat and electricity), is all important in how living systems function. The fact that water is a liquid at biological temperatures is essential to life as we know it. The ability of water to form lattice networks contributes to its relative stability (nonvolatility at biological temperatures) and its structure, which allows solutes to enter into solutions and macromolecules into gels—states of the matter of living systems that make life possible. The behavior of water due to its polarity (i.e., its having negatively and positively charged ends) contributes to the determination of protein structures and the behavior of membranes as semipermeable organizational devices. The ability of water (H_2O) to dissociate into a positively charged proton or hydrogen ion (H^+) and a negatively charged hydroxyl ion (OH^-) gives water the pH properties that are so important to all enzymatic reactions that make life possible at the kinetic rates required to allow the timing of living systems' processes. These hydrogen ions and hydroxyl ions are also instrumental components in the regulation of the pH that is necessary for living systems to function normally.

Water's ability to carry away heat as it evaporates makes possible the *air conditioning* that allows organisms to thermoregulate by the process of evaporative cooling. The tendency of water to condense out of the atmosphere under appropriate temperature conditions provides a source of available moisture for many organisms. The behavior of water in capillary systems has multiple ramifications, not the least important of which is the ability of water to move up through the vascular bundles of plants (in some cases moving against gravity up to the tops of trees a hundred meters above the roots). The turgor pressure of water gives living things a great degree of their form and structural integrity (consider, e.g., limp, water-poor plants versus well-watered, crisp, turgid leaves and stems). The fact that water is (anomalously) denser in its liquid state than in its solid state means that ice will float and that large bodies of water will not become great ice sinks that eventually bind most of the free water on earth into deserts of ice at the bottoms of lakes and oceans.

All of these properties and characteristics are also relevant to the way water functions in diets. These characteristics are discussed in terms of diet palatability, texture, nutritional value, stability, and the various other features of water.

5.3.1 Water Activity (a_w), Water Content, and Diet Quality

Water activity (a_w) is not the same thing as water content. Water content is the relative amount (percentage or ratio) of water to all other substances in a material. Water activity is the ratio of vapor pressure in a material to the vapor pressure of pure water at a given temperature. What this amounts to is the relative ability of water to be mobilized from one region or compartment to another (the activity). Water contents may range from 0% to 100%. Water activities range from 0.000 to 1.000. While it is true that water content and water activities tend to be positively correlated, there are important exceptions. The availability of water for chemical reactions and biological use is more reliably predicted by a_w than by water content. The differences between water content and a_w arise from the interaction of water and the other components in the matrix of materials (including diet ingredients).

For example, dissolving 350 g of sodium chloride in 650 g of water yields a solution whose water activity is ~0.75. Dissolving 350 g of lithium chloride in 650 g of water yields a solution whose water activity is ~0.50. In both cases, there are solutions that are 35% solute and 65% water, so the water content is the same but the water activities are very different. The reasons that explain the differences are complex thermodynamic concepts, but roughly, they pertain to the number of solute molecules (or ions) dissolved in and interacting with the water. Because of the differences in the atomic weight of lithium and sodium, the molar concentration of 350 g of lithium gives the solution an osmotic pressure of ~8.5 M, while the same mass of sodium gives a ~6.0 M osmotic pressure (or osmolality). Generally, higher osmolality is correlated with lower the water activity.

The key feature of water activity is the fugacity or the tendency for the water to leave its current location to move to another location (i.e., escaping tendency or fugacity). The water activity of foods is most commonly described by the term a_w; however, Damodaran et al. (2008) argue that the best expression of water activity is the ratio of the tendency of the water to leave the matrix in question (p) to the tendency of pure water to leave (p_0) or p/p_0. The factors that govern the values for p are the number of dissolved particles, the charge or other binding characteristics of the particles as well as temperature and pressure. Compared to simple or small solutes such as simple sugars and salts, macromolecules such as starches and proteins have, per unit of weight, very little effect on the water activity. This is despite the fact that they may be present in diets in considerable amounts. For example, agar and carrageenan may be present as 1%–5% of a water suspension (or dispersion), yet they lower water activity by less than 0.01 units.

Water activity is expressed as a decimal, and the highest possible value is 1.000, the activity of pure water. Recalling the term for calculating water activity, p/p_0, the logic of the value of 1.000 can be rationalized as follows: if we are dealing with pure water, its fugacity or escaping tendency (p) with no solute (pure water, by definition has no solute) and the tendency of pure water to leave (p_0) are equal, so dividing a number by its identity will equal 1.000. The decimal here is carried three places to the right to reflect the convention of measuring water activity to three places. From this, it can be further reasoned that by the inherent colligative properties of water (osmolarity, freezing point depression, boiling point elevation, depression of water vapor pressure), the more dissolved particles per unit of volume, the less water is inclined to move away from a given site (e.g., solutions or diets). This is expressed by lower water activity values approaching 0.000 where water is completely unavailable.

The net movement of water follows a water activity gradient. So if two compartments are separated by a membrane or other medium that is permeable to water, and if one compartment has water activity of 0.50 and the other 0.75, the net movement of water will be from the 0.75 compartment to the 0.50 compartment. This explains one of the main reasons why people cannot drink sea water to maintain a positive water balance. The sea water (because of the abundance of dissolved salts) is at a lower water activity than are the cells of the person's body. Therefore, the net movement of water (not salts) is from the person to the sea water in the gut. The same phenomenon applies to microbes in our insect diets. If the microbes are at a water activity of ~0.98 and the diet is at 0.99, water will go from the diet into the microbes, allowing them to maintain a suitable water balance. However, if we could safely bring the water activity of the diet down to 0.90, microbes with a water activity of 0.98 would lose water to the diet and constantly be fighting a water balance or water shortage problem. Table 5.2 shows the water activity of some typical foods and diet components. Table 5.3 shows the water activities that reduce or prevent growth of various kinds of microbes.

Table 5.2 Water Activity (a_w) of Various Foods, Insect Diets, and Diet Components

Water	1.0
1% agar gel	0.999
5% agar gel	0.999
Honey	0.500
Grape jelly	0.720
Soy flour	0.150
Lygus hesperus diet (Cohen 2000c)	0.998
Lepidopteran diet	0.999

Source: Cohen, unpublished data.

Table 5.3 Minimal Water Activity (a_w) for Growth of Specific Microbes

Microbe Genera	a_w
E. coli, Pseudomonas	0.95–1.00
Salmonella, Vibrio, Serratia	0.91–0.95
Many yeasts	0.87–0.91
Most molds	0.80–0.87
Most halophilic bacteria	0.75–0.80
Xerophilic molds (*Aspergillus chevalieri*)	0.65–0.75
Osmophilic yeasts and molds	0.60–0.65
No microbial growth	0.50 and below

Sources: Beuchat, L.R., *Cereal Foods World*, 26(7), 345, 1981.

5.3.2 Gradient-Based Water Contamination

A further consequence of differences in water activity between two (or more) compartments separated by a water-permeable medium (air, packaging material, surface film) is that the water will always tend to follow the gradient from higher to lower water activity. An important ramification of this in insect diet regimens is that dry materials are constantly in danger of becoming wetter. If flours or other powders with a typical water activity of less than 0.300 (such as yeast hydrolyzate or other similar dry materials such as soy flour, wheat flour, vitamin mixtures, salt mixtures) are being stored in a room with an atmosphere that is more than 0.300 (i.e., more than 30% relative humidity), the dry materials will always be in a nonequilibrium state with the surrounding air and will always be inclined to become wet. After a long enough period of exposure to the humid air (i.e., air with a higher water activity than the stored material), the material that was being stored has changed in several important ways.

First, material that has gained moisture from its surroundings is less concentrated than it was prior to gaining water weight. Second, the stored material has a higher water activity that is now more likely in the growth range of microbial contaminants (which were present as contaminants in the virgin material or were introduced when the package was opened), so the material is now subject to microbial deterioration and can serve as a contaminant to the diet in which it was going to be used (discussed further in Chapter 13). Third, at a higher water activity, chemical reactions are accelerated, including oxidation, which is destructive to most nutrients. Fourth, the activity of enzymes is higher at higher water activities, so enzymes that had not been totally deactivated can destroy materials that were expected to be present and intact in the diet.

What can be done to reduce the consequences of *water gradient contamination*? First, using suppliers of diet ingredients that are reputable and stand behind the quality of their products is the first line of defense. Second, establishing a utilization schedule, which includes recording of the date of receipt and date of opening, will assure that components have been recently obtained. This means that a 5-year-old vitamin mixture is not the most wholesome as a freshly purchased mixture, especially if the older mixture was opened years ago. It is very common in large-scale rearing facilities to buy bulk quantities of diet components and sometimes keep them around for years (decades in some cases). The economic value of this practice is offset by the loss of the colony or by the compromise of the insects' quality. Third, use of a container that can be resealed tightly reduces exposure to warm, wet air. If the product does not come in such a container, one can be provided by the insectary staff. Fourth, storage in a cool, dry place and limiting access to the storage facility help to reduce contamination. Following the instructions on each product regarding proper storage is prudent. For example, most antibiotics require storage at 2°C–8°C or below freezing. Fifth, for very moisture sensitive materials, a packet or canister of silica gel might be sealed in the

Table 5.4 Consumption by Colonies of *Lygus hesperus* of Diets with Various Humectants

Diet	NI Diet (Control)	NI + 10% Glycerin	NI + 10% High Fructose Corn Syrup	NI + 10% Proline
Amount consumed	1.5 g	0.1 g	0.4 g	0.3 g
Measured a_w	0.997	0.903	0.921	0.928

Source: Cohen, unpublished data.
Values are weights of diet consumed from 50 g packets presented for 2 h on rearing cages containing 8,000 to 10,000 adults.

container with the component. Sixth, having a quality assessment plan that is compatible with the facility's resources is a prudent practice. In larger facilities, it may be advisable to have a routine monitoring of water activity of the material. Water activity meters are moderately expensive (about $3000–$5000) but very easy to use. They can be used to process a sample in less than 5 min and give accurate readings of water activity of both diet ingredients and complete diets. More of this technology is discussed in the chapter on quality control.

Given the fact that there is no microbial proliferation at water activities below 0.50, it may be questioned why such activities are not routinely used to help preserve diets. This strategy is practiced in the food and pharmaceutical industries with the use of *humectants*. Commonly used humectants are glycerin, sugars (sucrose, glucose, and fructose—especially high fructose corn syrup), and the amino acid proline. First, it is difficult to provide an adequate percentage of water to meet most insects' water requirements and textural demands (except for stored products insects). Second, use of large amounts humectants to lower water activity to less than 0.60, for example, could impose upon the insects the same physiological stresses as those discussed for people trying to maintain water balance by drinking sea water. Third, the addition of whatever solute that is intended to lower water activity dilutes the concentration of the other nutrients. Even compounds that have a strong effect on lowering (a_w) such as NaCl, glucose, or proline would have to be added in amounts exceeding 30% of the diet's composition, thus displacing the other nutrients and lowering the overall nutritional value of the diet in question. Finally, the target insects must accept the diet as readily with the humectant as they do control diets that lack these agents. Table 5.4 shows that such acceptance may not be easy to achieve.

Some insects such as honeybees have, in effect, anticipated this strategy by producing honey, a food that has such low water activity (<0.60) that naturally inhibits microbial growth and is ideal for long-term storage of the sugars so important to bees. However, honey is not the exclusive food bees but is a supplement used in conjunction with pollen and water to provide a nutrient balance. The nurse bees that feed the brood make a mixture of pollen components and honey that they provide for brood rearing with an ideal balance of proteins, lipids, carbohydrates, vitamins, minerals, and, of course, water. It is also interesting that the pollen stored by bees is another ideal storage form for foods, nutrients that are preserved by a nearly intractable outer covering and with a very low water activity.

5.3.3 Moisture Sorption Isotherms

In relation to the earlier discussion of water content and water activity, the dynamics of water association with other diet components is extremely complex and therefore somewhat unpredictable. For example, there is a disparity between what is called the isotherms of desorption and resorption (i.e., the dynamics of water removal and water addition to materials). This is illustrated in Figure 5.1, which shows that over most of the hydration range of materials, a given moisture content will be characterized by different a_w, a phenomenon known as hysteresis. What is to be noted about the hysteretic effect is that when materials are desorbing water (i.e., losing water or being dried out), they have lower water activities at a given water content than they do at the same water content reached through resorption. As this phenomenon applies to diets, the water activity that is achieved in the

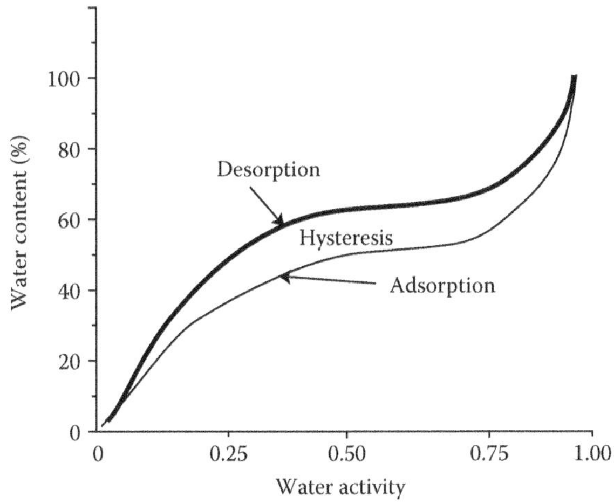

Figure 5.1 Graphic illustration of *hysteresis*: showing that water activity is greater for a given water content during *adsorption* than it is for the same water content during *desorption*.

synthesis of the diet after mixing and heat processing will be lower than what is achieved through the rehydration of that diet that had been dried (by freeze-drying, spray drying, or other means of water removal). Therefore, if a fully synthesized diet has a water activity of 0.950 and a content of 70 g of water per 100 g of diet, and if this diet is freeze-dried to remove the 70 g of water, when it is rehydrated by the readdition of 70 g of water to 30 g of dry material, it can be expected to have a higher than 0.950 a_w. This means that simple rehydration of diets that have been preserved by drying must require special attention to restore water activity to its original value. This may mean addition of less water, a humectant to compensate for the hysteretic effect, or reprocessing such as heating. The effects of these treatments on the target insects may be unacceptable, and this may mean that the desirable outcomes from having a lower water activity (e.g., lower microbial growth rates, reduced rates of oxidation, or other destructive chemical reactions) may have to be forfeited for the convenience of using freeze-dried or otherwise desiccated diets. Clearly, this matter requires more attention by researchers and rearing program managers. This is further discussed in Section 5.3.5.

5.3.4 Molecular Entanglements, Molecular Mobility, and Diet Stability

A major purpose of processing ingredients into diets is to provide nutrients in an accessible, a palatable, and stable form. A major reason why artificial diets are used is that natural foods quickly lose their freshness. Chemically, this means that these foods leave nonequilibrium conditions to go to their most probable state—equilibrium. Equilibrium conditions are tantamount to decay and disorder, breakdown of cells and their organelles, destruction of macromolecules, and oxidation of essential nutrients. This force that seems to be at war with the orderliness of living systems is entropy. What Damodaran et al. (2008) describe as the major task of food scientists and technologists applies to insect diet workers and can be paraphrased as follows: our major duty is to treat insect diets in such a way as to reduce the possibilities of the diet components reaching their equilibrium state. Damodaran et al. (2008) introduce several concepts of food science that are directly applicable to insect diet chemistry; these concepts deal with the stability of complex foods, which for our purposes include insect diets.

Similar to what Damodaran et al. (2008) said about food scientists and technologists, insect diet scientists and technologists must perform a balancing act. They are charged with production of diets that are economical, convenient and useful under conditions that are conducive to entropy or disorder (relatively high temperatures and humidity along with bright light). The conditions that are optimal for insect growth are also optimal for common environmental microbial contaminants. Insect diet specialists are charged with designing diets that at once compartmentalize potentially interactive components, detoxify or deactivate antinutrients, and kill microbes. The diets must serve these criteria while not destroying labile nutrients or palatability.

5.4 NATURE OF pH AND HOW IT AFFECTS DIET

In light of the fact that insect diets are aqueous-based formulations, it is inherent that they each have a characteristic pH. The definition of pH is the negative log of the hydrogen ion (H^+) concentration. The pH scale spans a range of 0–14, with 7 being neutral and the lower numbers being more acidic as 0 is approached and more basic as 14 is approached. The pH scale is logarithmic (to the base 10), so a solution with a pH of 5, for example, has 10× the hydrogen ion concentration than does a solution of pH 6 and 100× the hydrogen ion concentration of a solution of 7.

Virtually, all chemical reactions are strongly pH-influenced, and most organisms have very narrow pH ranges that are tolerated internally. Insects are no exception to these generalizations, and the range pH of foods that are acceptable or preferred by insects is generally within a fairly narrow scale (about two pH points or less being a reasonable *rule of thumb*).

5.4.1 Multiple Effects of pH

An important component of the palatability of foods is pH, and insects are no exception to this generalization. Diverse mechanisms that are pH related include texture effects and modification of flavors of various nutrient components. Also influenced by pH is the degree of stability of insect diets as in relation to microbial contaminants and chemical reactions that are both enzymatic and nonenzymatic. Figures 5.2 and 5.3 illustrate the actions of digestive enzymes of the predator *Zelus renardii* and the plant bug *Lygus hesperus*, showing the strong dependence of pH on the activities of these digestive enzymes. The peaks of activity for these enzymes, known as their optima,

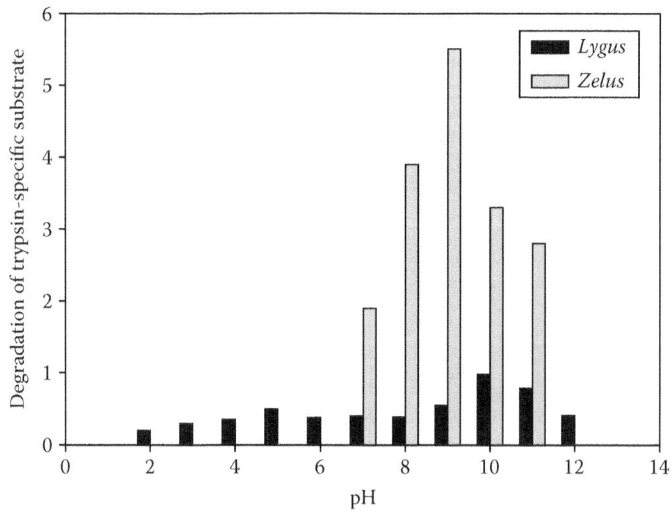

Figure 5.2 Actions of digestive enzyme (protease known as trypsin) of the predator *Zelus renardii* and the plant bug *Lygus hesperus*, showing the strong dependence on pH. (Data from Cohen, unpublished; Zeng, F. et al., *Insect Biochem. Mol. Biol.*, 32, 455, 2002a, Zeng, F. et al., *Compar. Biochem. Physiol Part B*, 131, 453, 2002b.)

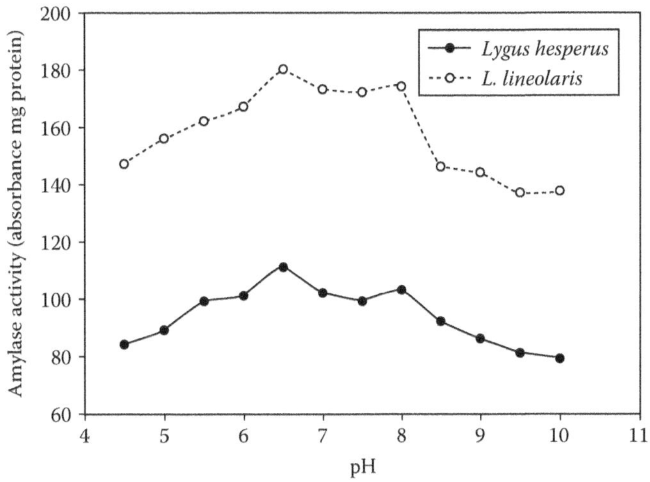

Figure 5.3 Actions of the starch-digesting enzyme (known as amylase) of two species of plant bugs *Lygus lineolaris* and *L. hesperus* showing the strong dependence on pH. (Adapted from Zeng, F. and Cohen, A.C., *Comp. Biochem. Physiol. B: Biochem. Mol. Biol.*, 126, 9, 2000.)

are the pH domains where the maximum digestive activity will take place. It is important to remember that *every enzyme in every* insect, plant, microbial contaminant, or insect diet has a similar pH dependency or optimum of activity. This is not to say that all enzymes have the same pH optima, but rather that they all have roughly a bell-shaped curve that describes lesser activities at lower or higher than optimal pHs and maximal activity at the pH optimum (or close to the optimum).

It is reasonable to project how these pH-related bell shaped curves can help determine not only palatability profiles for diets but also preservation profiles. For example, if a pesky microbial contaminant is dependent on near-neutral pHs, its growth can be severely discouraged by lowering the diet's pH below its optimum (a strategy discussed further in the chapter on microbes in the diet). Similarly, enzymes that have a potential for destroying nutrients (e.g., lipases, lipo-oxygenases, proteases, carbohydrases, polyphenol oxidase, and enzymes that destroy vitamins such as ascorbate oxidase) all have pH optima, which can be used as the basis for diet preservation.

5.4.2 Use of Buffers in Insect Diets

Buffers are key factors in the pH and other qualities of insect diets. Buffers are chemicals that resist changes in pH, so they exert a stabilizing effect on the chemistry of the diets. The meaning of resistance to change in pH is simply this: if some factor or force causes a hydrogen ion to be removed from a water matrix in a diet, the buffer replaces that hydrogen ion, and conversely, when a hydrogen ion is added to the aqueous matrix, the buffer removes that hydrogen ion. What is said about buffers' actions in regulating the hydrogen ion concentration also applies to their ability to regulate hydroxyl (OH^-) ions. For the sake of simplicity, this discussion will be confined to the regulation of hydrogen ions only.

The common compounds that are used directly and deliberately as buffers are various phosphate compounds such as mono- and disodium phosphate, mono- and dipotassium phosphate, acetates, citrates, and sulfates. Amino acids are good buffers, and some proteins have excellent buffering capacity. When amino acids and proteins are added to diets, it is usually for other reasons rather than their buffering capacity (i.e., for nutrition enhancement or for texture), but they do simultaneously contribute to the pH stability of our diets. This point emphasizes the complexity of insect diets and how the various components often have multidimensional, multifunctional roles. It is further emphasized that the choice of a buffer (or any other functional component of a diet) often has spin-off in other facets of diet function. For example, using potassium phosphate rather than sodium phosphate as a buffer will also increase the phagostimulatory quality of the diet for the many species of insects that show feeding preferences for high potassium diets over high sodium diets. Conversely, the selection of potassium-based buffers over sodium-base buffers may decrease the diet's palatability for hematophagous insects.

5.5 OXYGEN AND REACTIVE OXIDATIVE SPECIES (ROS) PRESENT IN DIETS

There are several components that can interact with diet ingredients in a type of destructive reaction known as oxidation. On a strict chemical basis, oxidation means the removal of electrons from a substance, and reduction is the addition of electrons to a substance. In the simple reaction known as burning hydrogen in the presence of oxygen to yield the product, water, two hydrogen atoms give up their electrons to an oxygen atom (which receives electrons). In this reaction, the oxygen is *reduced* and the hydrogen is *oxidized*. Similarly, when metallic iron interacts with atmospheric oxygen to form iron oxide, the iron is said to be oxidized, and the oxygen is reduced; that is, the iron loses electrons, and the oxygen gains them. Elements and compounds such as oxygen and several other substances that are avid electron acceptors, which aggressively remove electrons from substances in diets (and in other biological domains), are sometimes called reactive oxygen species or sometimes, reactive oxidative species (ROS). The ROS substances include pesticides; heavy metals such as cadmium, arsenic, lead, and mercury; free radicals; and a large number of biologically active substances that can damage various molecules in organisms, especially macromolecules such as DNA, RNA, and proteins.

Searches of recent literature reveal thousands of articles on various food components' ability to detoxify or ameliorate the ROS present in animals' bodies. One of the most exciting subjects that have emerged from all of this attention to antioxidants is the demonstration of various previously unrecognized substances that are now being shown to have various degrees of potency against ROS. For example, it has been long known that ascorbic acid, β-carotene, and α-tocopherol serve as antioxidants as part of their function as vitamins. However, recent attention and technological improvements in antioxidant assessment (e.g., Cao and Prior 1999, Gao et al. 2000) have demonstrated a wide variety of substances from various chemical classes that act as natural scavengers of ROS. These classes of compounds include arytenoids, flavonoids, proteins, lipids, and even some elements such as selenium.

5.5.1 Antioxidants

There are many kinds of antioxidants—substances that protect diets and insects from the attack by and subsequent damage from ROS. Broadly speaking, antioxidants occur in two major classes: those that are water soluble and those that are lipid or fat soluble.

5.5.2 Role of Antioxidants in the Insects' Metabolism

Evidence for the health benefits of various antioxidants has encouraged efforts to account for compounds that protect people against oxidative stress (Haslam 2001). Like mammals and other organisms, insects are known to be adversely affected specifically by free radicals and generally by oxidative stress (e.g., Timmermann et al. 1999), but there has been less attention to the role and nature of antioxidants in insect foods than there has to human diets and food supplements. In fact, it has pointed out that activity against reactive oxygen species and the oxidative stress that these substances induce includes both ingested substances such as ascorbic acid and insect-produced substances such as the purines, uric acid, xanthine, and hypoxanthine (Timmermann et al. 1999).

The essentiality of vitamins A, C, and E (including precursors and derivatives of vitamins A and E) among insects has been known for decades (e.g., Gilmour 1961), but the potential function of these vitamins as antioxidants in insects and in the diets themselves has been only recently recognized. In fact, there is often ambiguity in discussions of these vitamins, largely because of their multiple functions, including their antioxidant activity. Furthermore, the recognition of other compounds as antioxidants has emerged as a challenging component of human food science and nutrition, and the potential is great for application of these compounds in insect diets.

5.5.3 Role of Antioxidants and Their Function in the Diet

Many artificial diets for insects contain compounds that are deliberately added to serve as antioxidants. However, the persistence of these components in artificial diets for insects after the diets have been synthesized has not been reported (Cohen and Crittenden 2004). Also, although it has been well documented that a wide variety of naturally occurring substances possess antioxidant qualities (Stukey 1972, Lindsay 2008, Gao et al. 2000), the characterization of such antioxidants in insect diets had been neglected. This is unfortunate because an understanding of the contribution of such *cryptic* antioxidants to the diet would help us better understand why some diets fail and others succeed for certain insects.

There are special problems inherent in insect diets and the functions that they must serve that make them extremely susceptible to ROS-induced degradation, compared with human foods. For example, insect diets are often held at rearing temperatures that are optimal for microbial growth and acceleration of chemical reactions that are destructive to antioxidants and molecules that are sensitive to oxidation. Also, the preservatives that are added to human foods that are stored at temperatures above those of conventional refrigeration are not generally added to insect diets. Furthermore, the practice of using diets that are infested with insects (the fundamental purpose of the diet) inherently threatens the biochemical integrity of the diet.

In this vein, three widely used insect diets were examined to determine the effects of various storage practices on the antioxidant capacity (Cohen and Crittenden 2004). This showed that there was loss of the overall antioxidant capacity resulting from storage, even with refrigeration and freezing. It was also evident that some diets were more unstable than others and required greater care in preservation of nutrient quality.

5.5.4 Negative Effects of Excess of Certain Antioxidants

As the importance of antioxidants becomes increasingly clear, the question is raised about the safe levels of these chemically and biologically active substances. Unfortunately, there is little research on amounts of the various antioxidants that are optimal in diets, and clearly, there must be optimal amounts for each insect species being targeted. In one group of experiments intended to develop a bracket of optimum, suboptimum, and superoptimum amounts of various additives, it was demonstrated that ascorbic acid and α-tocopherol could not only be provided in less than optimal amounts, but that even slight excesses could be very destructive to the quality of *Lygus hesperus* (Cohen, unpublished data, 2003). Figures 5.4 and 5.5 show the effects of overabundance of ascorbic acid in relationship to mean fresh weights of 2-week postemergent females and egg production per day (Cohen, unpublished data, 2003). Although the mechanism of toxicity or deterioration of diet quality from use of these substances is not evident, clearly, they can be destructive if present in too great an abundance. Also evident in Figures 5.4 and 5.5 is the fact that similar concentrations of soy isoflavones induce a dose–response improvement in both adult weight and egg production.

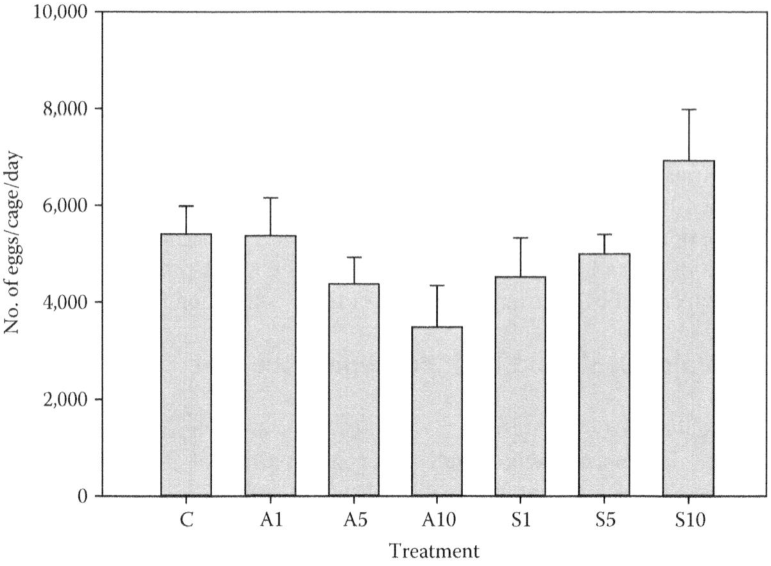

Figure 5.4 Effects of overabundance of ascorbic acid in relationship to egg production per day. C, control diet; A1, 0.1% ascorbic acid; A5, 0.5% ascorbic acid; A10, 1.0% ascorbic acid; S1, 0.1% soy isoflavone; S5, 0.5% soy isoflavone; S10, 1.0% isoflavone.

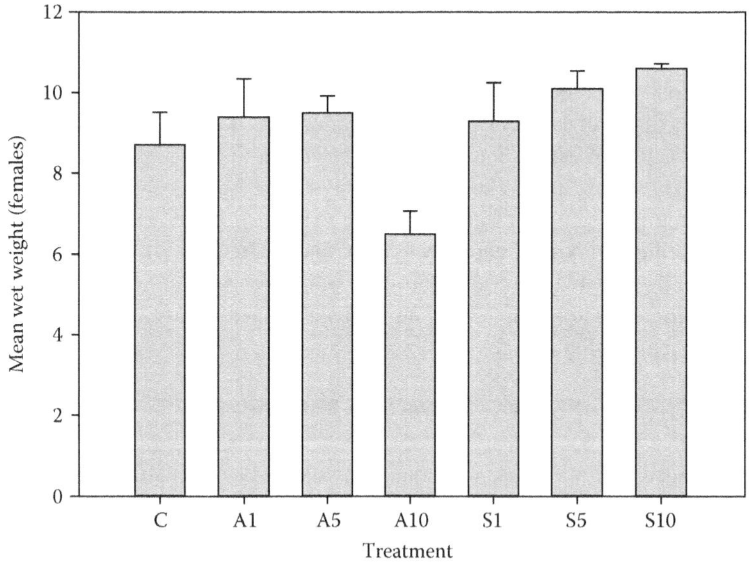

Figure 5.5 Effects of overabundance of ascorbic acid in relationship to mean fresh weights of 2-week postemergent females. C, control diet; A1, 0.1% ascorbic acid; A5, 0.5% ascorbic acid; A10, 1.0% ascorbic acid; S1, 0.1% soy isoflavone; S5, 0.5% soy isoflavone; S10, 1.0% isoflavone.

Interestingly, the antioxidant butylated hydroxytoluene (BHT), which is commonly added as a preservative to human foods to protect them against lipid oxidation, was toxic to the *L. hesperus* whose diet was modified with this substance (Anonymous 1999a, Enigl 1999). Figure 5.6 shows a simple dose–response curve for the increases in BHT resulting in proportional increases in toxicity as indicated by decrease in yield of colony biomass.

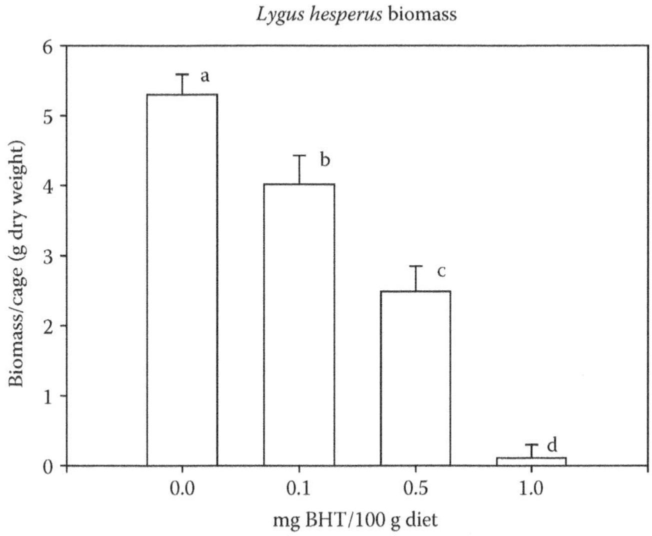

Figure 5.6 Simple dose–response curve for the increases in BHT resulting in proportional increases in toxicity as indicated by a decrease in yield of *L. hesperus* colony biomass. (Cohen unpublished data: Mean ± SE followed by the same letter are not significantly different, $P < 0.05$, PROC ANOVA, LSD test. SAS Institute, Cary, NC, 1999.)

5.5.5 Measurement of Antioxidants in Insect Diets

Appendix V contains methods of measuring antioxidants in diets, one that measures the total antioxidant capacity of the diet and another that specifically measures the ascorbic acid content. Both procedures allow the detection of possible degradation that has taken place and is affecting the overall diet quality. Both methods are based on alcoholic (ethanolic) extracts made from freeze-dried diet. These methods can be altered to meet specific needs of the quality control facility, and appropriate references should be consulted such as Nielsen (1998), Miller (1998), Woods and Aurand (1977), or Meloan and Pomeranz (1980). The details of several strategies of measuring the antioxidants in diets and diet ingredients are presented in Appendix IV. The rationale behind each of these measurements is presented here.

5.6 FACTORS THAT AFFECT DIET TEXTURE

The subject of diet texture is among the most neglected topics in insect diet science, yet it is a very important aspect of both palatability and stability of diets. Texture is a complex of several factors including viscosity, shear strength, plasticity, and penetration resistance. Diets may be liquids (including slurries that readily flow or take the shape of their container), semisolids, gels, and solids. The most important textural characteristic of liquid diets is viscosity, which can be conveniently thought of as internal resistance to flow. Familiar examples are water and pancake syrup. The syrup pours more slowly and moves with less ease through a fine tube than water. Warming the syrup makes it flow more readily. There is an increase in viscosity when water has sugar dissolved in it; the more the sugar, the higher is the viscosity. In solutions and suspensions (such as whole blood), the more dissolved substances or suspended particles, the higher is the viscosity. As viscosity increases, the ingestion of the liquid diet becomes increasingly more difficult and requires greater amounts of insect-generated suction to move the fluid from its source, through the mouthparts, and into the insect's gut. This is especially the case for the Homoptera and Heteroptera, because both classes use mouthparts that are long tubes with small openings. The movement of the fluid is described by Poiseuille's law, which explains that fluid flow through a tube (such as stylet-type mouthparts, the esophagus, or other feeding structures) is proportional to the pressure differential and the radius of the tube divided by the length of the tube times the viscosity. Given this simple relationship, insects (whose mouthpart length and diameter are essentially of fixed dimensions) must generate higher negative pressures (suctions) as viscosity increases.

5.7 PROCESSING HISTORY OF DIETS: PHYSICAL QUALITIES OF DIETS

Most insect diets, other than those that are simple solutions, contain extensive amorphous regions. These regions are not crystalline solids, but instead are more potentially mobile than are crystals. This opens the question of the molecular mobility in relation to diet stability. As discussed in several places in this chapter, the molecular mobility of diet components is an important factor in the determination of diet stability and quality. It is worthwhile to think of molecular mobility in terms of access of potentially reactive molecules to one another. If compounds that are pro-oxidants are sequestered into stabile, immobile compartments, they cannot gain access to other molecules such as lipids that are potential targets for oxidation reactions, and these lipids are therefore safe from potentially destructive forces that lurk within the diet itself. The molecular mobility is related to a_w, temperature, and the physical characteristics of the diet (is it a gel, a liquid, or a solid?). This topic is discussed further in Chapters 4 and 8.

5.7.1 Physical and Chemical Consequences of Processing

Both physical and chemical components are affected by the processing history of the diet. The water activity, molecular mobility, enzymes, enzyme cofactors, enzyme inhibitors, pH, amount and kinds of reactive oxygen species, nutrient quality, and numerous other factors are affected profoundly by how the diet was treated, especially in terms of heat and cold treatments.

5.7.2 Heating

The profound effects of heating on virtually all aspects of insect diets are sometimes overlooked. Heating processes are often thought of as being pertinent only to destruction of microbial contaminants and *activation* of the gelling material. However, there are more implications to heating than merely disinfection and activation of diet gelling agents. Table 5.5 summarizes the effects of heating diets.

The temperatures at which these processes occur vary from component to component and from diet to diet. In heating procedures that do not involve increases in pressure (i.e., where superheating does not take place), the highest temperatures reached are at the boiling point of water, ~100°C. This means that some heat-induced processes cannot take place under these limited heating conditions. Such outcomes as the killing of most spore-forming microbes, extensive reorganization of proteins, and extensive activation of macromolecule/water interactions will not be realized. Under the higher temperatures that can be reached with pressures, these outcomes can be achieved, but care must be exercised to make sure that overcooking does not negatively impact the diet.

5.7.3 Benefits of Heat Processing

As will be discussed at length in the chapter on microbial contamination, heating can be an extremely effective means of destroying microbial contaminants. And as will be discussed in the section on macromolecule/water interactions in this chapter, heat is required to fully activate most

Table 5.5 Effects of Heat Processing on Diet Components and Whole Diets

Benefits	Liabilities	Mixed Effects or Possible Negatives or Positives
Killing microbial contaminants	Killing beneficial microbes	
Activation of gelling agents	Creating overly firm gels	
Increasing protein digestibility	Decreasing protein digestibility	Must find balance between undercooking and overcooking
Denaturation (and destruction) of digestive inhibitors	Maillard reactions (irreversible binding of sugars and amino acids)	
Denaturation of harmful enzymes such as lipo-oxygenases and phenol oxidases	Nutrient destruction (ascorbic acid, unsaturated lipids)	
Increase flavor	Decrease flavor	Must find balance between undercooking and overcooking
Acceleration of desirable chemical reactions	Acceleration of undesirable chemical reactions	
Increases economic value of diet	Adds expensive step to diet production	

of the gelling agents that are used to stabilize diets and to give them a texture that is suitable for insects with certain feeding and/or tunneling requirements. However, even if microbes and gel activation were not an issue, heating would remain one of the most important parts of diet processing.

As indicated from Table 5.5, heating deactivates enzymes, many of which would otherwise remain active in many diet components, even after several other processing/purification steps have been performed. During steps such as size reduction of meats, eggs, and various plant-derived components, the natural, protective compartments are disrupted, and enzymes such as lipases, lipo-oxygenases, polyphenol oxidases, ascorbate oxidase, pectinase, amylase, catalase, and peroxidase are freed and potentially allowed to come into contact with other nutrients, which they can disrupt or deteriorate. Heating processes that range from mild blanching (a light, quick cooking of materials) or pasteurization to more heat intense processes such as flash sterilization or high temperature cooking in extruders can denature most or all of these enzymes, depending on the severity (the temperature and duration) of the process. These enzymes break down lipids, causing rancidity—production of toxic short chain fatty acids and removal of the polyunsaturated fatty acids. Also, they destroy vitamins, causing enzymatic browning, which reduces protein and carbohydrate availability. Finally, they can cause hydrogen peroxide and other pro-oxidant species of compounds to be produced—all to the detriment of diet quality.

5.7.4 Liabilities of Heat Processing

However, heating can have destructive effects along with the possible benefits. Just as heating changes the chemistry of components by driving off volatile substances, speeding chemical reactions, and develops flavors, it can also drive off some of the volatiles that are attractants or phagostimulants. In the sections on the chemistry of individual components, it is discussed how heating can destroy vitamins and form complexes such as sugar amino acid products (called Maillard products) that are nonnutritional (Damodaran et al. 2008). Also, the heat-induced improvement in protein digestibility can be offset by such severe overheating that digestibility is reduced (Damodaran 2008).

5.7.5 Chemical and Physical Effects of Cold Storage

Cold storage is widely used for reducing rates of degradation of diets and diet components. The rates of chemical reactions decrease by two to three times per 10°C drop in temperature. Also, cold storage—especially freezing—immobilizes components in diets, so potentially interactive components do not get access to one another. Slowing of the rate of diffusion takes place above freezing temperatures, but freezing dramatically decreases diffusion. Microbial activity also slows as temperatures decrease, but cold does not kill most microbes. In fact, the most common ways of preserving microbes include freezing, freeze-drying, and cryogenic treatments with liquid nitrogen or dry ice.

An adverse effect of freezing, especially if the process is done slowly, is that ice crystals form and disrupt the organization and compartmentalization of diets and diet components. Freezing is especially harsh to plant materials. Also, a great deal of deterioration takes place at the diet surface under refrigerator or freezer conditions. Desiccation, oxidation, and enzymatic reactions take place at below freezing temperatures and the stored materials undergo freezer burn and other degradations of their nutritional and palatability qualities. This is further discussed in Chapter 8.

5.7.6 Desiccation Processes

Freeze-drying is generally the least destructive process of diet and ingredient preservation (Damodaran et al. 2008, Cohen 1999). However, it is not effective in eliminating or ameliorating microbes (Jay et al. 2005) and toxins (Damodaran et al. 2008). The degree of change to different food components must be determined on a case-by-case basis, because different foods or food components can give dramatically different results, especially for most labile nutrients. For example, there were from 26% to 60% losses in vitamin C in freeze-dried green beans, but only 8% to 30% in peas and 3% in orange juice (Fellows 2000). At the same time, losses in vitamin A were much lower than those reported for vitamin C, and riboflavin and thiamin remained completely intact after freeze-drying. This point is especially pertinent to the development of quality control procedures and how important it is to select the appropriate components to test.

There are several other methods of water removal, mainly by application of heat. The various heat-based methods include hot-air driers, bin driers, cabinet (or tray) driers, solar driers, heated surface driers, drum driers, vacuum driers, and spray driers. Few of these have been applied to insect diet processing, except for the method used by Patana and McAda (1973) where diet flakes were produced from gelled diet that had been spread on waxed paper and subjected to hot air forced to flow over the diet surface. The flakes then were added to the top of the gelled diet as a means of increased surface area for the target insect larvae (*Heliothis virescens*) to hide, feed, molt, and simultaneously reduce contact with other larvae, reducing microbial contamination and cannibalism.

Generally, the processing is done by the manufacturers of diet components, but because the processing can greatly affect the insect-specific quality of the components, the nature of this process is discussed here. For example, it would seem that spray-dried eggs could be benignly substituted for fresh eggs in insect diets. However, there are evidently differences in the nutritional and/or sensory qualities of spray-dried eggs to some insects as was discovered by Debolt (1982) in his efforts to use spray-dried eggs to substitute for fresh eggs in the diet for *Lygus hesperus*. Review of the spray-drying process does not offer clear reasons for the loss of value that results from this process. Spray-drying takes advantage of the rapid rates of water loss that result from the enormous increase in surface area that is exposed to hot air in the process of atomizing the eggs (or other spray-dried products such as milk). The product is dried in 1–10 s and retains much of the sensory qualities when used as human foods (Fellows 2000). As anyone who has eaten reconstituted, powdered eggs have probably witnessed, there are differences in the sensory qualities that can be detected.

5.7.7 Purification of Diet Components

The concentration of nutritionally rich materials is a type of purification process. Some purification is done by insect diet workers, especially in using special additives such as plant components or extracts (e.g., Chang and Kurashima 1999, Blossey et al. 2000). Besides the application of mechanical separation techniques, including sifting or other particle size and gravity-driven separations, many purification steps are liquid based and involve extraction, filtration, membrane-mediated separation (such as dialysis), precipitation (or differential solubilization), or phase separations (mainly involved in lipid extractions). For the purifications that are done outside of the rearing facility, generally by commercial producers of the ingredients, it would be helpful for the rearing professional to know a considerable amount about the process whereby the ingredients were produced as this could have great bearing on their quality with respect to the target insects. For example, if a carrageenan had been produced by a given extraction process from a certain alga, a change in the process or the algal species can greatly change the acceptability of the product. Examples of this kind of change are given in greater detail in Chapter 8 on complexity in the insectary.

Examples of the differential solubilization process are given in this chapter where methods of purification of soy proteins are discussed. From this section, it can be seen that an extraction process that is not precise in the separation and selection of storage proteins from the biologically active whey proteins can have harmful results. The storage proteins are nutritious and void of antinutrients, while the whey proteins include toxic lectins, protease inhibitors, and other enzymes that are destructive to target insects.

5.7.8 Effects of Storage of Ingredients and Finished Diets

The purpose of storage is the practical convenience of having materials to be used in diets (or the diets themselves) on hand when they are required. The tacit assumption is that storage will not significantly contribute to the degradation of diet materials. The key word here is *significantly*. One of the most difficult parts of planning diet production systems is deciding how much the process in question affects the quality of the diet. The difficulty is that every processing step can have some effect on the diet's characteristics, but the long-range effects must be predicted based generally on incomplete or short-range information.

Certainly, ingredients and diets change after various periods of storage. With age, ascorbic acid turns from a clean-looking white to a faded yellow. Powders or fine crystals cake up into clumps that must be broken up before they could be used. Vitamin mixtures darken after their packaging is opened, and discoloration and off-smells characterize yeast that had been stored for months or even weeks. Upon observing these changes, insect diet professionals might rightfully wonder if these observed changes in substances matter to the quality of the diet. The decision as to when to discard and replace materials in which changes have been observed remains one of the most haunting questions that diet professionals must deal with.

A detailed answer to this question is to be found in the explanation of the chemistry of characteristics of each diet component. However, a generalization of the potential trouble spots should be useful. The major types of instability of ingredient and diet are moisture gains or losses, oxidation, enzymatic degradation, loss or gain of volatile substances, and microbial contamination/deterioration. The forces that contribute to these unwanted changes are the presence of water vapor, oxygen, active enzymes, heat, and light.

The effects of heat are discussed in Section 5.7.9, but it is noteworthy that every degree of temperature elevation to which the ingredients or diets are subjected greatly increases the probability of degradative reactions. Also, higher water content and the probability of degradation compared to lower water contents. Light induces many reactions directly (such as in the photodestruction of the B vitamin, riboflavin) and it also raises temperature. Oxygen is ubiquitous rearing and diet environments. It gains easy access to ingredients and diets and is an omnipresent force against diet stability. Like oxygen, microbial contaminants are ever present and ever vigilant to ensure their own survival at the expense of the ingredients or diets. Finally, these factors in concert amplify the risk that ingredients will not survive prolonged storage. Therefore, measures to protect diet ingredients and diets are recommended as the first step in diet protection (Chapters 6 and 11).

5.7.9 Effects of Heat on Diet Chemistry

The concepts of temperature and heat are sometimes confused. The *heat is the kinetic activity* of the molecules in a substance, and *temperature is the measure* of that kinetic activity. So in a substance of some given molecular composition, the higher the molecular activity (or molecular motion), the higher is the temperature. It should be noted that another quality of matter is *heat capacity*, which is roughly the relationship between uptake of heat and increase in temperature. Heat capacity is specific to different substances. Water, for example, has a higher heat capacity than iron, so a given amount of heat absorbed by a gram of iron will raise the temperature higher than the same amount of heat added to a gram of water. Because insect diets are largely composed of water, they have high heat capacity, and in line with this, diets have a high thermal inertia. What this means on a practical basis is that it takes a large amount of energy transfer to change the temperature of diets.

A related concept is *thermal conductivity*. Different materials have different thermal conductivity characteristics. For example, wood, styrofoam, or cardboard are better insulators or materials that retard heat transfer than copper, stainless steel, or cast iron. Thermal conductivity is important in thermal processing of strategies for diets. Insulation/conductivity characteristics are determinants of the heating systems that can be most efficiently and safely applied to diet processing. For example, the use of a steam kettle or autoclave to heat a large mass of diet introduces the problem of even and timely heating. While the core of the diet is awaiting heat transfer from the heat source (a steam jacket or hot air surrounding the diet), the sides of the diet mass are being heated and remain at the high temperature in danger of being overcooked while the core remains undercooked. Even continual stirring of the diet may not compensate for the time delay involved in attempting to heat a massive amount of diet evenly from the outside in. This problem of low thermal conductivity of diets (which are aqueous based and therefore characterized by high thermal inertia) is circumvented by using heating strategies that take advantage of diet geometry changes that lend themselves to faster and more uniform heating, such as flash heating using a heating coil. This issue and the strategies to deal with problems in thermal conductivity are discussed in greater detail in Chapter 12.

Chemical reactions, including ones in diets, are accelerated as temperatures increase and are decelerated with decreases in temperature. Rates of reactions increase or decrease by two- to threefold with every 10°C of increase in temperature. This includes all types of chemical reactions and several physical processes such as evaporation. This reaction rate/temperature interaction is described by the Arrhenius equation as discussed by Damodaran et al. (2008). Rates of reactions catalyzed by enzymes behave according to more complex kinetics described in a later section.

5.8 CHEMISTRY OF PROTEINS AND AMINO ACIDS IN DIETS

Except for diets for insects specialized in feeding on plant saps and except for certain defined diets, most insect diets contain proteins. As was discussed in Chapter 3, dietary proteins and their component amino acids function as building blocks for insect proteins, also as sources of energy, as components of hormonal peptides, and numerous other metabolic functions (also covered in Chapter 7). The purpose of this section is to clarify the chemical structure and nature of amino acids, peptides, and proteins as components of diets.

Amino acids, peptides, and proteins are nitrogenous compounds, the peptides and proteins being composed of specific sequences of amino acids. There are 20 protein different amino acids (whose structures are shown in Figure 3.5). Each amino acid has an amino group ($-NH_2$), an acid group (–COOH), and a variable group called an R group. Except for proline (which is technically classified as an *imino* acid), the only difference in structure between one amino acid and another is the structure of the R group. When amino acids are present individually, not bonded in a sequence that we call peptides, we refer to them as *free amino acids*. Some diets include free amino acids. However, most diets contain their amino acids in the combined peptide/protein state. Each amino acid can connect with every other amino acid in the manner described here in Figure 5.7 where they are covalently bonded to one another in what is called a peptide bond (Figure 5.7). When two amino acids are connected by a peptide bond, the structure is a dipeptide; three amino acids form a tripeptide, and so on. The peptide in Figure 5.7 consists of eight amino acids connected by bonds between the carboxy carbon (the acid or COOH carbon), and each peptide bond is illustrated by a dotted line. Peptides that are dozens of amino acids in length, at some arbitrary point, are called proteins. Peptides and proteins are linear, rather than branched as is the sometimes the case with carbohydrates, which can be linear, branched, or a combination of both linear and branched. However, proteins can have other kinds of molecules such as carbohydrates bonding with the R groups, especially the amino acids threonine and serine or with asparagine to form the very important class of proteins known as glycoproteins, which function as cell surface recognition sites, signaling molecules, lectins, and numerous other biologically important functions that depend not only on the protein core structure but also on the type and number of sugars at the glycosylation sites. Several proteins that serve as lipid storage or transport molecules are known as lipoproteins (including vitellin and lipophorin, respectively, found in insect eggs and insect circulation systems) and various plant proteins used for storage (such as glycinin and conglycinin from soy beans). Still other proteins form complexes with other groups such as the iron- or magnesium-bearing proteins that complex with heme or other porphyrins such as the active component in the chlorophyll molecule.

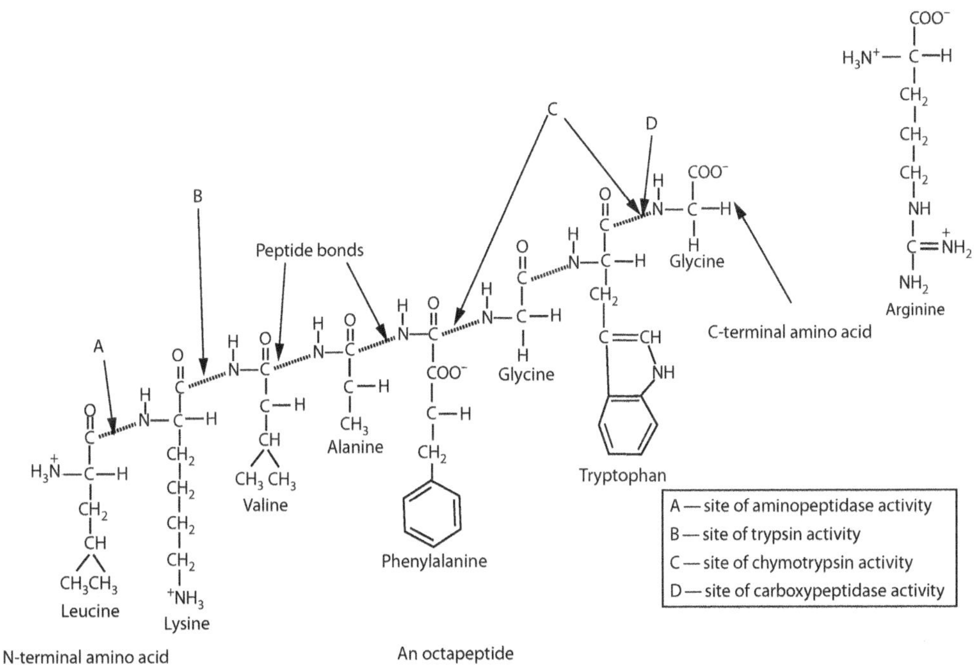

Figure 5.7 Each amino acid can connect with every other amino acid in the manner shown, where they are covalently bonded to one another in what is called a peptide bond. When two amino acids are connected by a peptide bond, the structure is a dipeptide; three amino acids form a tripeptide, and so on. The peptide in this figure consists of eight amino acids connected by bonds between the carboxy carbon (the acid or COOH carbon), and each peptide bond is illustrated by a dotted line.

The diverse nature of proteins makes them suitable for the many roles that they play as enzymes, which carry out thousands of different catalytic functions in virtually every life process and as the governors of nearly every step in all metabolic pathways. Their diverse structural character also disposes them to serve as storage materials, diverse structures, toxins, signaling systems, and countless other functions. This diversity of structure also is characterized by a somewhat frail nature, being influenced by temperature, pH, salt concentration, and various cofactors and inhibitors. Therefore, the all important structure of proteins can be distorted by even slight changes in temperature, pH, or other conditions to influence how well the proteins may function in their systems.

5.8.1 Functional Roles of Proteins in Diets

The functional roles of proteins in foods include binding fats, binding flavors, and storage, especially exemplified by the storage proteins from egg yolks, meats, and seed storage proteins such as glycinins and conglycinins (Damodaran 2008). These roles also apply to the proteins in insect diets. Proteins act as emulsifiers and film formers at interfaces between diet components. They may give the diets greater elasticity or other texture features that may be either desirable or detrimental, depending on the circumstances. Proteins bind ions, giving a buffering and stabilizing capacity to the diet, and they bind water to form gels or stabilized dispersions. A generally undesirable aspect of proteins is that with mechanical energy and the presence of air, they form foams or froths that may hinder the function of diets. They are important in increasing diet viscosity and in helping diets to retain a given shape, a feature that is especially important in the diets that also serve as living sites for insects that bore or burrow into their diets (such as boll weevils and pink bollworms). Proteins include enzymes that may destroy the nutritional value of diets (oxidases, lipases, proteases, and carbohydrases, among many other enzymes). Most important, proteins are the major source of amino acids that are to be used by the target insect as building blocks for its own proteins.

5.8.2 Character and Roles of Amino Acids in Diets

Amino acids, having both an acid group (COOH) and an amino group (HNH), can act both as acids (lowering the pH of a solution) or bases (raising the pH of a solution) and are therefore said to be ampholytes. This characteristic gives the amino acids the ability to act as buffers (substances that resist changes in the pH of their solution), both as free amino acids and within proteins. The solubility of free amino acids varies greatly, according to their R groups, ranging from cysteine and histidine, which are virtually insoluble and tyrosine whose solubility is 0.4 g/L to arginine with its solubility of 855 g/L and proline with a solubility of 1620 g/L (Damodaran 2008). Even if free amino acids were inexpensive and nutritionally valuable to insects, the low solubility of many of the amino acids, especially essential amino acids such as leucine (21.7 g/L) or isoleucine (34.5 g/L), would make the provision of an adequate amount of these nutrients difficult.

5.8.3 How Enzymes in Diet Ingredients Affect the Diet

Many, if not most, components of diets contain enzymes. Examples of typical enzyme-containing components are various meals and flours from seeds, including wheat germ, corn meal, and soy flour, as well as meat products, dairy products, and eggs. Certain specialty diets such as screw worm media contain blood, which is a source of several enzymes. In any such diet, if specific measures are not taken to inactivate the enzymes, they can continue to function in the completed diet. Such activity can be damaging to the diet's quality.

Examples of the types of enzymes are lipases, lipo-oxygenases, amylase, proteases, catalase, peroxidase, and polyphenol oxidase. The result of allowing these enzymes to function is the deterioration of diet quality, including the destruction of nutrients such as lipids, proteins, and starches (lipase, protease, and amylase, respectively), creation of off-tastes from rancidization that results from lipid deterioration (lipo-oxygenases), or formation of insoluble or indigestible complexes (polyphenol oxidase). The most simple, effective, and economical means of destroying these enzymes or denaturing them to cease their functioning is by heat. The amount of time required to denature these enzymes (called D, discussed in Chapters 11 and 13) varies according to the diet matrix and the temperature used to circumvent these enzymes' actions.

An excellent example of the protocols for dealing with viable enzymes is the methodology of soy processing so that soy can be used in insect diets.

5.8.4 Chemistry and Processing of Soy: A Case Study

Soy is one of the most highly nutritious, economical, and widely used insect diet components. Soy products are among the most widely used nutrients in human foods for many of the same reasons that they are useful in insect diets: high protein content, well-balanced amino acid profile, high vitamin and mineral contents, beneficial lipid profile, and relatively low cost. Soy products are also widely used because of their pleasing or acceptable taste qualities and versatility in an incredible range of foods including oils, lecithin, sauces, flour, tofu, miso, texturized meat imitations, soy milk, puddings, and numerous other uses (Fukushima 1991). The use of soy in insect diets has not gained as much versatile attention as it has in human and domestic animal diets, but it has gained growing acceptance (Cohen 2000c) and promises to become even more useful as processing technology becomes better understood in the insect diet community.

The main part of the soy plant used in human and insect foods is the bean seed. Mature beans are harvested and dried, and the seeds may either have their lipids extracted and separated into an oil fraction and a lecithin (phosphatidyl choline) fraction. The oil is nutritious and a valuable source of polyunsaturated fatty acids. The lecithin portion is also a good source of unsaturated fatty acids. These lipid structures and their chemistry are discussed in the next section on lipids. The methods of extracting the lipids are through mechanical pressure, steam extraction, or a combination of these. Some lipid extraction is performed with organic solvents that must later be removed before the oils and lecithin can be used in foods. Many insect diets contain soy oils and/or soy lecithin as nutrient sources and as emulsifiers (discussed in detail in Chapter 3).

The protein concentration of the soy seed before processing is about 40% and after lipid and hulls are removed, the remaining flakes have a protein concentration in excess of 50% (Endres 2001). After removal of the lipid components, soy solids remain and are available for further processing. These solid remains contain all of the protein, carbohydrates, and water-soluble vitamins and small traces of lipids. At this point, the remaining material may be toasted, roasted, or otherwise subjected to heat treatment and then ground into flour. Prior milling (flour making), the soy may have the seed coat removed. The flour or meal may next be used as a base for the extraction of proteins. The typical way of extracting proteins from soy is to perform a water extraction of the dry material; separate the suspended solids by gravity, centrifugation, or filtration; and removal of the aqueous phase for further processing. The aqueous phase contains most of the water-soluble proteins, some carbohydrates, and water-soluble vitamins and is further processed either with pH changes or heat to precipitate the most nutritious proteins, allowing the removal of the biologically active proteins (protease inhibitor, lectins, lipo-oxygenases, β-glucosidases, urease), which may be discarded or used in other processes. The proteins that precipitate are generally the storage proteins (known as glycinins and conglycinins), and the proteins that remain in solution are generally the whey proteins, which confer most of

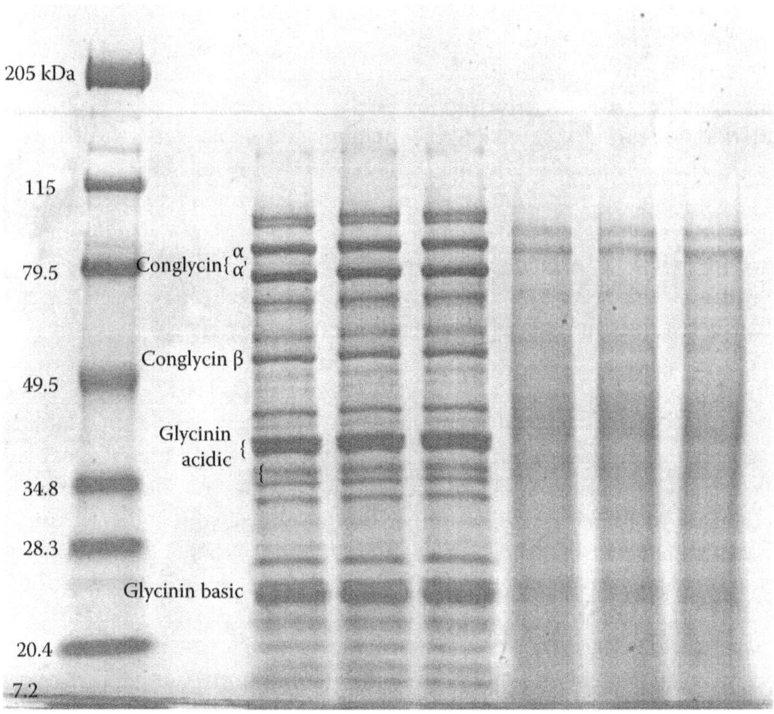

Figure 5.8 Complex of soy proteins separated by electrophoresis in SDS–PAGE. The gels show a series of molecular weight standards and the samples of proteins that had been extracted and then either not heated or subjected to autoclave treatment prior to electrophoresis.

the enzyme activity and other biological activities to soy. Figure 5.8 shows the complex of soy proteins separated by electrophoresis in sodium dodecyl sulfate (SDS)–polyacrylamide gel electrophoresis (PAGE) gel separation. The gels show a series of molecular weight standards and the samples of proteins that had been extracted and then either not heated or subjected to autoclave treatment prior to electrophoresis. What is evident from these gels is that there are numerous proteins that make up the total composition of soy flour as well as the other complex nutrients. It is further evident that the heat treatment removes a considerable amount of each protein from the soluble protein pool. However, what is not evident from these figures is that the apparently removed proteins have been redistributed into an insoluble aggregate of proteins and the lipids, carbohydrates, and minerals that form a highly complex matrix that is of higher bioavailability than the free proteins.

This example of processing soy proteins is potentially applicable to better understanding insect diets. First, it is evident that the soy plant produces the seeds for its own purposes of reproduction. The high nutrient content of the soy seed makes it a target for insects and other phytophagous organisms that make a living by consuming plant materials. Therefore, it is to be expected that the seed will have protection against the potential predators, including the insects that are rearing targets. These protections take the form of *antinutrients* such as trypsin and chymotrypsin inhibitors (which thwart the digestive process of would-be seed predators); lectins (which cause multilateral damage to digestive systems and other targets); and other biologically active components that are toxic, noxious tasting, or otherwise meant to discourage consumption of soy seeds. There are other potentially harmful biologically active components sometimes known as *housekeeping enzymes*. For example, the relatively high levels of urease found in soy and legume relatives are probably connected with utilization of urea in soils where these legumes are starting to grow. However, these housekeeping enzymes can have adverse effects on the target insects in whose diets raw or poorly cooked soy is incorporated.

The example of the characteristics of proteins as exemplified by the soy processing has broader implications beyond nutrient storage, enzymatic, and defensive functions of proteins. The solubility of certain proteins is another aspect of soy processing that is relevant to insect diet chemistry. The protein solubility was affected by temperature and pH, underscoring the fact that precipitation of the soluble proteins came as a result of the *denaturation* of the proteins from their *native conformation* to a *deformed* or denatured state. The denaturation process may be short term or permanent. Heating to extreme temperatures (above 60°C–70°C), pH extremes, mechanical agitation, and exposure to organic solvents are all means of denaturing proteins. It is also worthwhile to note that the proteins described in the soy processing example are all *globular proteins*, which have enzyme, storage, and defensive functions. However, proteins may also be *structural* as exemplified by proteins in muscle (actin and myosin), connective tissues (collagen), or cell membranes.

Processing and purification are involved in the detoxification of those soy proteins that would otherwise be toxic. One of the most important detoxification steps in the processing of soy proteins and a huge variety of other proteins is the heating step. The denaturation of the proteins and their subsequent aggregation into harmless masses resulted from the application of heat, and this process is enhanced by using pressure and suitable pH. It is noted here and in the chapters on nutrition and diet processing that the protein aggregates that have been denatured are also more suitable as foods, because the protein digestibility index increases dramatically as a function of denaturation.

In passing, it is noteworthy that proteins serve multiple roles as food or insect diet components. Besides being primary nutrients (as aggregations of amino acids) in themselves, they contribute flavor, texture, emulsification capacity, and buffering capacity to their matrix. The role of proteins and protein processing in insect diets is a poorly understood subject. In human foods, this is a subject of extensive information, and it is intriguing to realize that nearly the entire array of taste preferences, palatability, and stability of food components rest on the proteins in our foods and how these proteins were processed. For example, toasting soy proteins confers a desirable nutty taste to soy milk, and it averts the development of a bitter taste that nontoasted soy protein would otherwise confer to this product. Unfortunately, it is not clear how similar processing steps might affect the palatability of components in insect diets.

Although it has long been known that diet texture is a key factor in the attractiveness of a diet to its target insect, there has been virtually no research on the contributions of texture to phagostimulation properties of insect diets. This is especially unfortunate in light of the remarkably great texture diversity that can be conferred by proteins such as those derived from soy. Such proteins can be made to be solutes that yield low-viscosity solutions, or they can be processed to increase the viscosity of liquids. They can form gels without the aid of other gelling agents such as agar or carrageenan. Soy puddings and gelatin are widely used protein-based gels. Finally, proteins can be used to give diets very firm, meat-like texture, as is commonly done in soy processing. Even as the proteins lend texture to their matrix, they can be used as binders, carriers, or emulsifiers, adding and holding lipids in the diet. Again, the technology for all of this versatility is in place in the soy food industry, but it has been largely neglected in insect diet science.

5.8.5 Protein Complexes with Lipids and Carbohydrates

Protein structures are diverse in themselves, but they can become even more multifunctional when they form complexes with lipids (lipoproteins), with carbohydrates (glycoproteins), or with both (lipoglycoproteins). First, let us consider the proteins themselves and their possible structural diversity. If we consider a tripeptide (a three amino acid sequence), we can begin to see the potential diversity of proteins, especially larger ones. For simplicity, we might pretend that there are only three amino acids: alanine, leucine, and valine. The rules will be followed that amino acids will be bonded in peptide bonds, that the amino acid in the first position will have its amino group

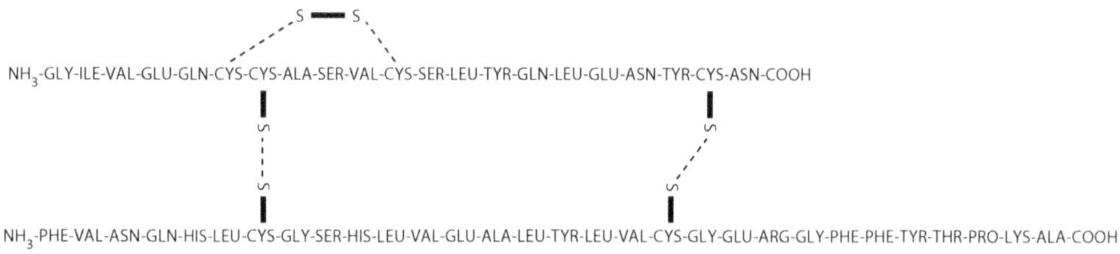

Figure 5.9 Amino acid sequence of the primary sequence of the simple protein insulin.

unbonded; the middle amino acid will have both its amino group and acid group bonded, and the amino acid in the third position will have its acid group unbonded; and finally that any of the three amino acids named can be in any of the three positions.

It becomes apparent that in our *3 slot* peptide, there are 27 possible unique combinations of the three amino acids (Table 5.6). It must be emphasized that each of these is a unique and biochemically different peptide from each of the other peptides. If we were to now extend this logic more realistically to the 20 protein amino acids that can occur in each of the positions, and if we further extend these *rules* to real biological systems with peptides and proteins have lengths from dipeptides to macromolecules with hundreds to even thousands of amino acids, it becomes obvious what an incredible diversity is possible (and exists) in the domain of protein chemistry. The complexity of protein chemistry is further expanded when we realize that the chains of peptide or protein structures may interact with one another, forming even more complex structures.

For example, even a relatively simple and short polypeptide (or protein), insulin has two polypeptide chains, an A chain of 21 amino acids and a B chain of 30 amino acids, and the two chains are connected by disulfide bonds (bridges) that occur in two places in the two chains. The protein insulin, which consists of 51 amino acids, was the first protein whose sequence was determined (Lehninger et al. 1993). As depicted in Figure 5.9, the simple protein, bovine insulin, has two peptide chains formed by specific sequences of amino acids bonded to one another by peptide bonds. There are amino terminal ends and carboxy terminal ends of these chains, and the chains are connected to one another by disulfide bonds. Close accounting of the amino acids in the insulin molecule reveals that most of the 20 typical protein amino acids are represented by at least one molecule of each, with the notable exceptions of aspartic acid, tryptophan, and methionine. Because both tryptophan and methionine are essential amino acids to most insects, bovine insulin would be a nutritionally inadequate protein source. The concept of essentiality of nutrients is discussed in detail in Chapter 3.

Table 5.6 Possible Combinations with Alanine Occupying Position One

Position One (Amino Terminal End)	Position Two	Position Three (Carboxy Terminal End)
Alanine	Alanine	Alanine
Alanine	Alanine	Leucine
Alanine	Alanine	Valine
Alanine	Leucine	Alanine
Alanine	Leucine	Leucine
Alanine	Leucine	Valine
Alanine	Valine	Alanine
Alanine	Valine	Leucine
Alanine	Valine	Valine

5.8.6 Undesirable Reactions of Proteins and Amino Acids

Some proteins that are glycosylated (i.e., have sugars associated with them) act as agglutinins (or agents of agglutination or *clumping*) of cells or macromolecular components. These proteins (some known as lectins) can also react with cell surface molecules that are key factors in the receptors for various molecules. For example, glucose and amino acids are absorbed into cells in the digestive system by glucose or amino acid transport systems that are regulated by specific receptors that can be blocked by several kinds of lectins (Hopkins and Harper 2001, Gatehouse et al. 1984). In a related phenomenon, lectins may also bind the peritrophic matrix of the insects, reducing the efficiency of this structure to act as a selective filter for nutrients and antinutrients. Other undesirable reactions of whole proteins are their actions as inhibitors of desired enzymatic reactions such as the digestion of nutrients by digestive system proteases, lipases, amylases, or nucleases. The proper heating of the diet components containing these antinutrient proteins can denature them, making them completely safe for inclusion in diets (Pusztai and Grant 1998). Heating proteins and amino acids with sugars such as glucose and fructose can cause a nonenzymatic browning reaction known as the Maillard reaction (Damodaran 2008). This reaction is discussed further in Chapters 5 and 12, but it must be emphasized that the overcooking that causes Maillard reactions can greatly reduce the nutritional value of proteins and their component amino acids. Finally, excessive heat (i.e., >200°C) such as that encountered on surfaces next to hot plates or at surfaces of steam kettles can cause a rearrangement of amino acids, a process called racemization. This rearrangement can destroy more sensitive amino acids such as aspartic acid, serine, cysteine, phenylalanine, or threonine, and can form toxic products such as D-amino acids (Damodaran 2008).

5.8.7 Chemistry of Lipids in Diets

The lipids have several important characteristics or properties that pose special problems in their inclusion in insect diets. The most immediately important characteristic is that most lipids that are of nutritional consequence are insoluble or only sparingly soluble in water. This means that special pains must be taken to include and retain lipids in insect diets, which are inherently aqueous media. The issue of compatibility of lipids and aqueous components of diets is best described by the statement, "Like dissolves like." Because lipids are predominantly nonpolar (or noncharged), they dissolve or readily associate with other noncharged, nonpolar, or neutral materials, including solvents. Organic solvents that are predominantly hydrocarbons (hexane, pentane, and benzene) are better solvents for lipids than are more predominantly polar solvents (methyl alcohol, ethyl alcohol, and ethyl acetate). All of these solvents are better at dissolving lipids than is water; this results in a partitioning of lipids and lipid-soluble molecules—a further aspect of compartmentalization in diets.

An integral part of lipid chemistry and the nutritional and biological qualities of lipids is that they are (compared to proteins and carbohydrates) very low in oxygen contents. For example, note that the ratio of oxygen to carbon in a typical trisaccharide is 18 oxygen atoms to 18 carbons; but it is 2 oxygen atoms to 18 carbon atoms in a fatty acid such as stearic acid or oleic acid. Besides making them very nonpolar and therefore insoluble in water, this low oxygen content (i.e., very *reduced* state) confers upon lipids a very high energy content. It is clear from 5 g that lipids contain more than twice as much energy per unit weight than do proteins or carbohydrates, both of which are far more oxidized than are lipids. As was noted in the chapter on the logic of metabolism, the high energy density of lipids makes them suitable for storage of energy in the most efficient and compact way that is compatible with biological systems. Besides giving lipids their high energy content, the structure of lipids is characteristically nonpolar and therefore strongly immiscible and insoluble in water. This point becomes clear when Figure 3.2 is examined. Illustrated in Figure 3.2 are the sterols, cholesterol, and sitosterol, a triglyceride (more properly known as triacylglycerol, but the more familiar convention is used here), and an example of a *polar* lipid, lecithin (more properly known as phosphatidyl choline). Less common dietary lipids not shown in Figure 3.2 are diglycerides (diacylglycerols), monoglycerides (monoacylglycerols), and free fatty acids. The complete digestion of triglycerides by digestive enzymes known as lipases results in the production of three free fatty acids and a molecule of glycerol (Figures 5.10 and 5.11).

Cholesterol is typically an animal sterol (in contrast with the common plant sterols, campesterol, sitosterol, and stigmasterol or the fungal sterol, ergosterol). Insects, unlike vertebrates, are unable to synthesize the sterol nucleus. Therefore, as discussed in Chapter 3, sterols are essential to insects. The structure of cholesterol is centered on the multiple-ringed sterol nucleus (Figure 3.2). The sterol

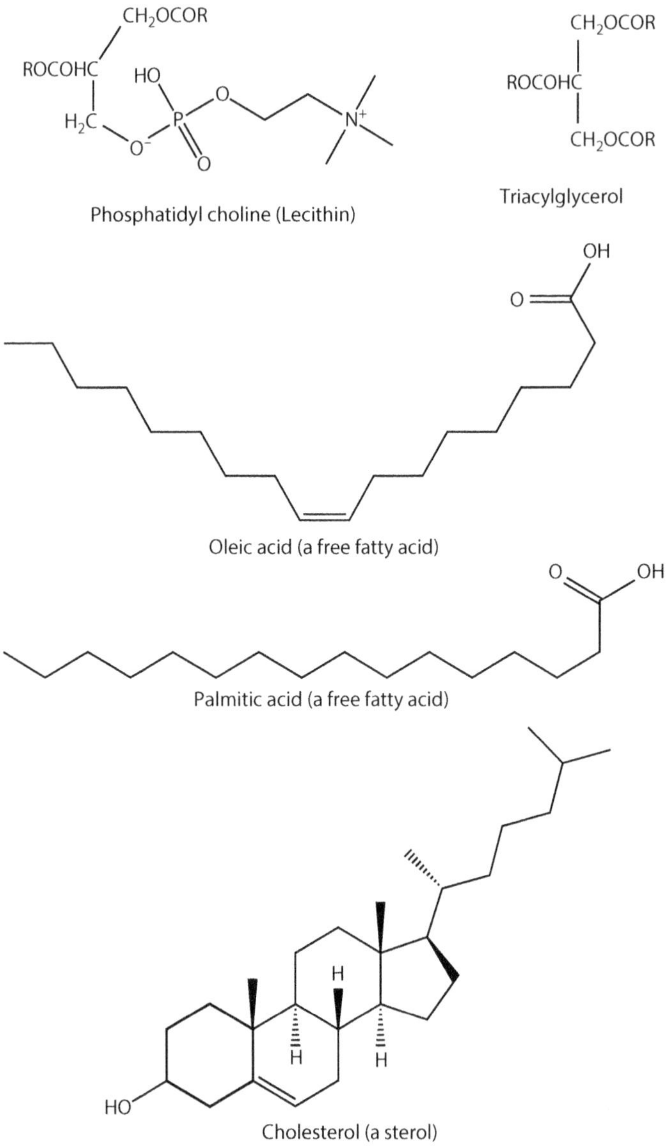

Figure 5.10 Various dietary lipids, including the sterol cholesterol, a triglyceride (more properly known as triacylglycerol, but the more familiar convention is used here), and an example of a *polar* lipid, lecithin (more properly known as phosphatidyl choline). Less common dietary lipids not shown in the figure are diglycerides (diacylglycerols) and monoglycerides (monoacylglycerols). This figure illustrates the structure of a polar-ended phospholipid shown in comparison with the nonpolar triacylglycerol (triglyceride); both structures are built on a glycerol molecule. The phospholipids are of key importance in the all important cell membranes and organelle membranes.

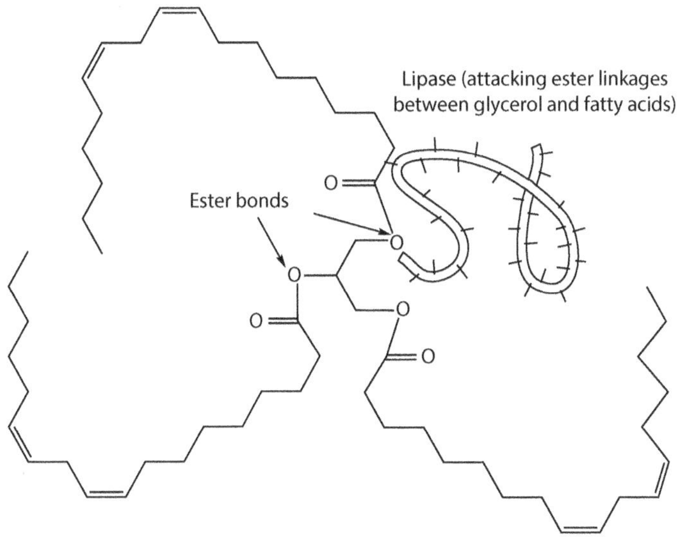

Triacyclglcerol (tri-linoleate)

Figure 5.11 Onset of digestion of a triacylglycerol by the enzyme called lipase. It is evident from this figure that either fatty acids or polar groups are substituted for the OH groups, via an ester bond (–O–).

nucleus is indicated by carbon numbers 1 through 20. The single oxygen in cholesterol and plant sterols is in an OH (hydroxyl) group bonded to the number 3 carbon. It is noteworthy that if the OH group or a fatty acid sterol ester involving this group (Figure 3.2) is not present, the sterol is useless to insects (Gilmour 1961). The number 24 carbon in plant sterols is bonded to carbons 23, 25 (as it is in cholesterol) and also number 28 carbon (a major departure from the structure of cholesterol). The ability of certain insects to break the bond between the number 24 and number 28 carbons is a key feature of ability of certain insects to use plant sterols for normal metabolic functions such as ecdysteroid synthesis and normal tissue growth. To get around the problem of insolubility of cholesterol, some tissue culture formulations call for *soluble cholesterol* (polyoxyethanyl-cholesteryl sebacate); however, use of this substance must be proven in case-by-case studies of the target insects. Also, because it is about three times as expensive as cholesterol, the soluble form of this sterol is limited in its scope of uses. Finally, as Mittler (1972) discusses, the poor solubility of cholesterol makes its use in artificial diets problematic and requires special measures to assure its presence in the diet.

The glycerides are built around the three carbon glycerol unit (Figure 5.10), with fatty acids (Figure 5.10) being substituted for the hydrogens from the OH groups. The site of substitution, where the fatty acid joins one of the three carbons in the glycerol molecule via an atom of oxygen, is called an *ester* (C–O–C), and that is the naming logic behind calling the various glycerides glycerol esters. Each of these linkages is formed via the removal of a water molecule as an alcohol group (OH) joins an organic acid (COOH). When triglycerides are digested, the ester linkage is broken as an enzyme adds a water, so the process is known as *hydrolysis* (= taking apart by water).

The glycerol molecule itself is fairly polar (charged) and therefore soluble in water. However, as the carbons in glycerol form esters with fatty acids, the resulting glycerides become progressively less polar and less soluble in water (and more soluble in organic solvents). The most abundant lipid class in most insects is the triglyceride group (Turunen 1979, Thompson and Barlow 1983). However, before dietary triglycerides can be absorbed by the insect gut and transported through the hemolymph by the carrier protein lipophorin, they must be broken down to diglycerides (Canavoso et al. 2001). Before being utilized for storage, the diglycerides in the target cells must be converted back into their original triglyceride form.

While the compact, hydrophobic, and rather stable structure of the triglycerides makes them ideal as storage components, it is the phospholipids that are very important components in membranes and membrane-mediated systems. The phospholipids (more generally, the polar lipids) are present in a variety of forms that include phosphatidyl choline, phosphatidyl ethanolamine, phosphatidyl serine, cardenolides, and numerous other compounds. These polar lipids are not only capable of forming complexes with proteins and covalent compounds with amino acids, but they can also become complexed with sugars, forming glycolipids, known to be important membrane components. Although some of these lipid groups can reach molecular weights approaching 1000, they do not tend to form the huge macromolecular complexes that proteins, carbohydrates, and nucleic acids form.

An important aspect of lipid (glyceride and fatty acid) chemistry is the degree of unsaturation of the fatty acids. It is especially pertinent to the chemistry of diets that the fatty acids that are unsaturated (Figure 5.10) are less stable than those that are saturated. It is evident from Table 5.7 that different foods have different proportions of saturated and unsaturated fatty acids in their lipids and that the more saturated lipids tend to be solid at room temperature. Unsaturation is a function of the number of carbon–carbon double bonds in a fatty acid. Fatty acids such as palmitic acid and stearic acid are completely saturated. That is to say that all the aryl carbons (those not part of the acid group) are bonded with single bonds to other carbons or with hydrogens (Figure 5.10).

Monounsaturated fatty acids such as palmitoleic acid and oleic acid have one double bonded carbon, meaning that each of those carbons could potentially gain a hydrogen atom, each, which would saturate them. Polyunsaturated fatty acids such as linoleic and linolenic acids have (respectively) two or three carbon–carbon double bonds.

The fats listed in Table 5.7 are mainly triglycerides. It is evident from this table that the fats with the proportionately lower amounts of saturated fatty acids are prone to be liquids at room temperature. Furthermore, the higher the degree and percentage of unsaturation and unsaturated fatty acids, the lower is the melting point characteristic of any given lipid. This is clear from Table 5.8, which shows that the greater the number of unsaturation points in a fatty acid, the lower is the melting point. It is further evident from this table that given a fixed number of unsaturation points, the determinant of melting point is the carbon number of the fatty acid. As we move from lower carbon numbers in the unsaturated fatty acid series to higher molecular weights, the melting point increases. The higher the melting point, the more stable the fatty acid can be expected to be. Therefore, generally speaking, those lipids that are solids at room temperature will tend to become rancid (oxidatively degraded to toxic, foul smelling, bad tasting compounds) less readily than those that are polyunsaturated and liquid at room or lower than room temperatures. This added stability is one of the reasons that food processors hydrogenate oils to form margarine. Of course, the process of hydrogenation is a conversion of some of the nutritionally valuable unsaturated fatty acids into the more stable but less nutritious saturated or partially saturated fatty acids.

Interactions of components within the diets can lead to degradation of the fatty acids, especially ones that are unsaturated. For example, the iron from eggs can attack the fatty acids in the yolk if

Table 5.7 Proportions of Saturated and Unsaturated Fats in Common Fatty Foods

	Fatty Acid%				
		Unsaturated			
Food	Saturated	Mono	Poly	% Fat	State at Room Temperature (~25°C)
Olive oil	~13	~74	~9		Liquid
Corn oil	~13	~24	~58	~99	Liquid
Margarine from hydrogenated corn oil	~15	~37	~23	~79	Soft solid
Butter	~51	~24	~3	81	Soft solid
Lard (animal fat)	~40	~45	~11	100	Hard solid

Source: USDA Nutrient Data Base.

Table 5.8 Common and Important Fatty Acids in Insect Diets

Lipid Name	Shorthand Designation Giving Unsaturation Numbers	Melting Point (°C)
Myristic acid	14:0	53.9
Palmitic acid	16:0	63.1
Stearic acid	18:0	69.6
Palmitoleic acid	16:1(Δ^{9})	−0.5
Oleic acid	18:1(Δ^{9})	13.4
Linoleic acid	18:2($\Delta^{9,12}$)	−5
Linolenic acid	18:3($\Delta^{9,12,15}$)	−11
Arachidonic acid	20:4($\Delta^{5,8,11,14}$)	−49.5

Source: Adapted from Lehninger, A.L. et al., *Principles of Biochemistry*, Worth Publishers, New York, 1993.

ascorbic acid is present (Nielsen et al. 2000, Jacobsen et al. 2001). Ascorbic acid, when added to foods containing egg yolk, frees and activates the iron that is otherwise associated with the yolk protein, phosvitin. The potential for this phenomenon is generally averted in the natural structure of egg yolk by the fact that iron is sequestered. This issue is further discussed in the chapter on matrix effects in insect diets and again in Chapter 8 on complexity in diet settings.

5.8.8 Getting Lipids into Diets

There are several strategies that have been used to include lipids into insect diets despite the fact that these diets are predominantly aqueous and repellant to lipids. The pre-eminent strategy is use of emulsifying agents, that is, substances that induce lipids to go into a micelle state. Micelles are spherical aggregations of lipids that have minimized their contact with their aqueous surroundings. They are a more stable and probable state than would be a solution of lipids in an aqueous continuous phase. Micelle-like chylomicrons and their component parts are illustrated in Figure 5.12. Figure 5.10 illustrates the structure of a polar-ended phospholipid shown

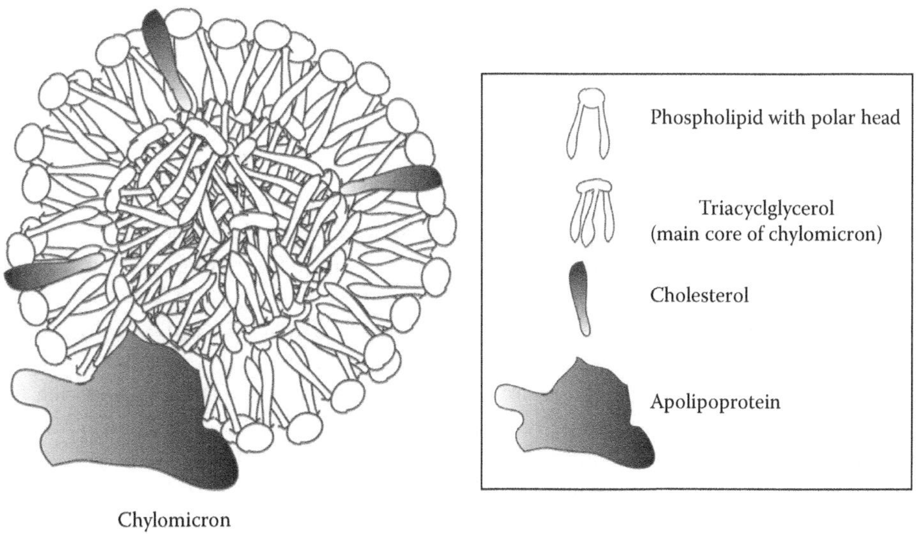

Figure 5.12 Model of a chylomicron. This apoprotein-containing structure, along with lipid micelles, is a principal carrier of lipids in insect circulation systems. These structures are a more stable and more probable state than are lipids in an aqueous continuous phase.

in comparison with the nonpolar triacylglycerol (triglyceride), both structures being built on a glycerol molecule. It is evident from 5 m that either fatty acids or polar groups are substituted for the OH groups, via an ester bond (–O–). The phospholipids are of key importance in the all important cell membranes and organelle membranes.

The dispersion of fat droplets in milk is a familiar example of lipid micelles. Milk that is homogenized has had the lipids reduced in size to decrease the likelihood of micelles accumulating into larger aggregates that are more likely to separate to form a butterfat layer. Milk is also a good example of a second way of stabilizing the lipids in an aqueous medium—association of lipids with proteins in what is known as *lipoprotein complexes*.

As was discussed in the aforementioned section on proteins, lipoprotein complexes are very important in the transport and storage of lipids. For example, in insects, the lipoprotein lipophorin is the agent of transfer of much of the lipid material transferred from the gut to various tissues where the lipids will be used or stored (Shapiro et al. 1988). These specialized lipid transport and lipid storage proteins have structures that suit them to function as carriers of hydrophobic (*water-fearing*) lipids in a highly aqueous environment (Levenbook 1985). Such proteins contain strategically located lipophilic zones that surround the lipids and hydrophilic zones that give these macromolecules stable interaction with the aqueous body fluids (Lehninger et al. 1993). Other notable lipoproteins involved in transport and storage of lipids are the yolk proteins, including vitellogenin (or vitellin).

With this understanding of how the hydrophobic nature of lipids is dealt with in living systems, there emerges a set of strategies to coax the lipids into diets and to keep them in some kind of stable interface with our predominantly aqueous diets. Probably the least satisfactory method that has been used, especially to produce chemically defined diets, is the introduction of lipids (most commonly cholesterol) into diets with organic solvents. Once efforts are made to distribute the solvents with their lipid solutes homogeneously throughout the diet, the solvent is evaporated with a stream of nitrogen bubbled through the diet. An improvement on this technique is to use emulsifiers such as the Tween compounds (Singh 1977). Because of their emulsification ability and because Tween compounds (polyoxyethylenesorbitans) are chemically defined, they are widely used in research on chemically defined diets. Lecithins from various natural sources (purified from egg yolks or soy beans) are also used as emulsifiers, but are not deemed as acceptable to those demanding stringent purity of diet ingredients. Finally, lipoproteins from various sources, including eggs, soy beans, and wheat germ, are used when purity is less of a concern than efficient lipid delivery.

5.8.9 Undesirable Reactions of Lipids in Diets

The undesirable events that can happen to lipids include their becoming separated from other diet components. Just as vinegar and oil dissociate in salad dressing and butterfat tends to separate from the aqueous phase of milk products, lipids can become separated from the aqueous diet components. Such separations lead to loss of availability of the lipids to the insects. The other common problem is the oxidation of fatty acids, especially at sites of unsaturation (double bonding between two carbons). The results of such oxidations are (1) the loss of the nutrient (e.g., linoleic acid) and (2) the gain of a toxic product such as butyric acid or valeric acid, both very unsavory tasting and foul smelling short chain fatty acids. Such oxidation reactions can be induced by heat, presence of enzymes such as peroxidases and lipo-oxygenases, or attack by free radicals. Such oxidations can be prevented by the use of antioxidants, the destruction of the enzymes by heating under proper conditions, and by care in removing oxygen or reactive oxygen species.

5.9 CHEMISTRY OF CARBOHYDRATES IN DIETS

Next to the proteins, the carbohydrates are the most chemically diverse group of molecules. The term carbohydrate means that water has been added to carbon, so it can be expected from this name that these molecules will contain carbon, hydrogen, and oxygen, often in an approximately 1:2:1 ratio. The carbohydrates can exist as small units of only a few carbons (e.g., *monosaccharides, disaccharides, trisaccharides*), intermediate units (*oligosaccharides*), and huge, macromolecular structures (*polysaccharides*) consisting of hundreds to thousands of single sugar molecules (Figure 5.13). The units can be repetitions of the same subunit (sugar), or they can be combinations of two or more kinds of sugars. They can be linear forming short or incredibly long chains, or they can be highly branched.

Compared with proteins, carbohydrates have more possible diversity as to the branching patterns. Proteins (polypeptides) are made up of carbon–nitrogen bonds called peptide bonds that can take place in only one part of the amino acid component (residue). In contrast, two glucose molecules, for example, can be joined by their number 1 carbons, their number 2 carbons, number 3, and so on. So it is possible to get 1–6, 1–4, 1–1 bonds between the pair of single sugars (monosaccharides) in question. The fundamental carbohydrate structure has another avenue of potential diversity, the addition (or substitution) of a functional group other than carbon, hydrogen, or oxygen. For example, we commonly find amino groups added to a sugar forming an amino sugar. Amino sugars have a great importance in the biochemistry of organisms, including insects, but none more important to insects than the use of glucosamine in the formation of the larger unit n-acetylglucosamine in the formation of the cuticular polysaccharide, chitin. Figures 5.14 and 5.15 are diagrams of the effects of heat activation of macromolecular carbohydrates, showing how heat unravels the starch and carrageenan molecules, allowing water and calcium ions to associate and link to the giant molecules. These associations, which are also influenced by pH, are the basis of gel formation.

Figure 5.16 shows examples of various small carbohydrates important in insect diets and insect metabolism, including a three-carbon carbohydrate, glyceraldehyde, a five-carbon sugar, ribose,

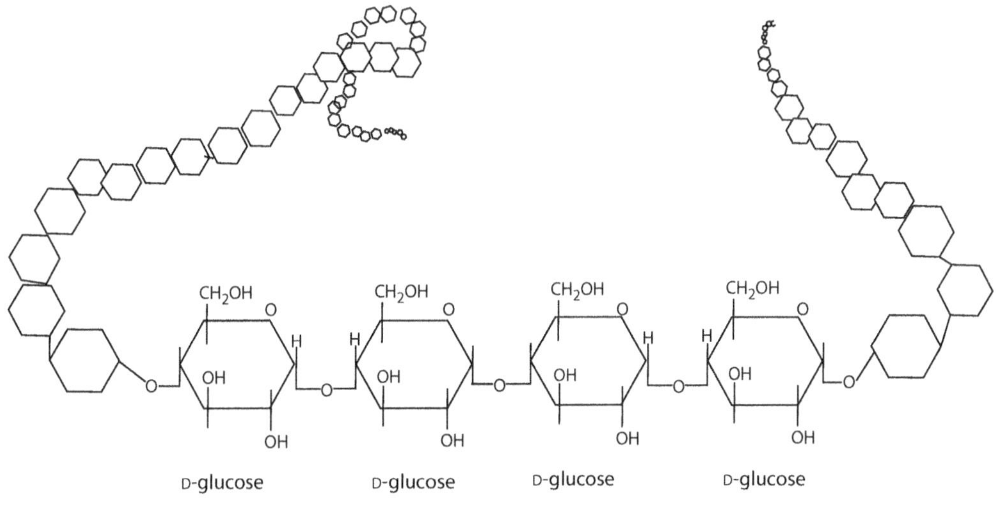

Figure 5.13 Structure of starch molecules, showing the 1–4 glycosidic linkages between the glucose subunits that characterize the bonding of these huge molecules. Not shown are the branching patterns that result from occasional 1–6 glycosidic linkages.

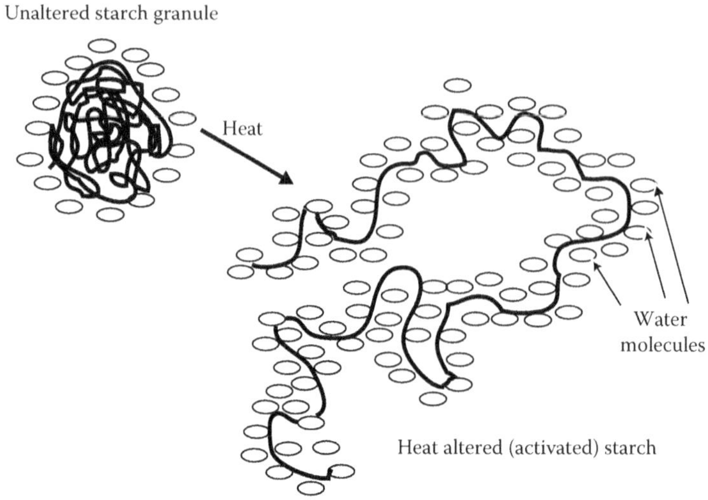

Figure 5.14 Representation of the heat activation of starch showing the unraveling of the polysaccharide chains and the association of water with the elongated strands.

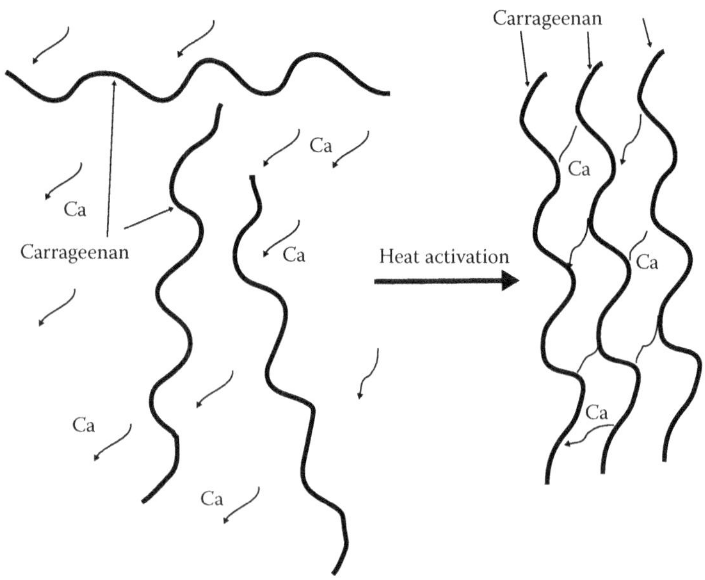

Figure 5.15 Model of the association of calcium ions with carrageenan. The cross-linking of the mineral ions with chains of many gel-forming agents is an essential part of the gel stabilization and strengthening process, which is also influenced by pH.

and several six-carbon sugars, glucose (= dextrose), fructose, and mannose. Not shown in Figure 5.16 is the structure of amino sugars such as glucosamine, which is simply a glucose molecule with an amino group substituted for one of the OH groups. Amino sugars are extremely important to insects as their polymerization results in the formation of chitin, the fundamental building structure in cuticle formation. Figure 3.4 shows some common disaccharides, including lactose, sucrose, and maltose. The bonds between these disaccharides, as well as those between various polysaccharides, contain an oxygen atom, which is a chief component of the glycosidic linkage. This glycosidic linkage can be oriented in on direction called an alpha linkage (α-linkage) or in

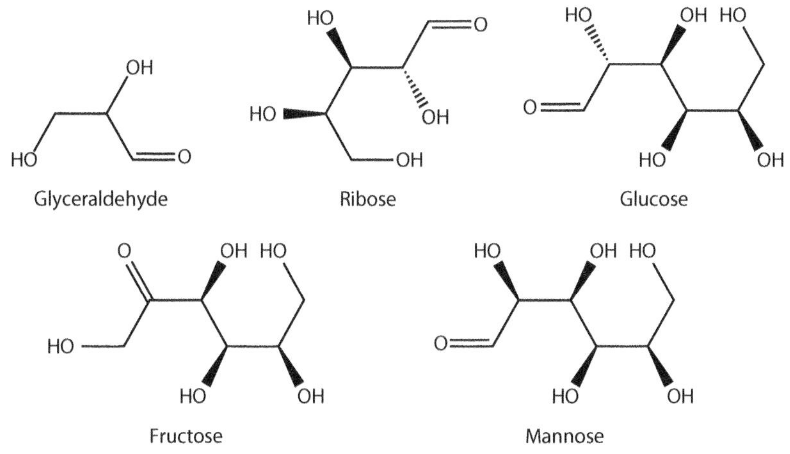

Figure 5.16 Examples of various small carbohydrates important in insect diets and insect metabolism, including a three-carbon carbohydrate, glyceraldehyde, a five-carbon sugar, ribose, and several six-carbon sugars, glucose (=dextrose), fructose, and mannose.

the opposite direction called a beta linkage (β-linkage). Specialized enzymes are required to break these α- or β-glycosidic linkages. These enzymes are generally classified as carbohydrases, and they have names that reflect their specialty, including α-glycosidase, β-glycosidase, α-glucosidase, as a few examples. The disaccharide, lactose (milk sugar) is an interesting example of a sugar that is a β-galactoside, and therefore requires a β-galactosidase to digest it into its monosaccharides, galactose and glucose. Many insects lack this enzyme and therefore cannot use lactose sugar as a nutrient. It is well known that many adult humans lack this enzyme or produce lesser amounts than is required to digest the lactose present in many dairy products. These people are said to have lactose intolerance. The syndrome of lactose intolerance in insects is clearly not as well documented as it is in humans, but it may be an important factor in the cases where milk and milk products were noted to have failed in insect diets.

Another example of digestive restrictions in use of a carbohydrate is in cellulose dynamics in some insects' natural foods and in artificial diets. Cellulose is a polysaccharide formed from connections of glucose molecules in a (β1–4) bonding sequence. Compared to plant starches and animal glycogen, which are formed from (α1–4) glucose to glucose bonds, cellulose is very linear (versus coiled as is starch) and is very strong in terms of high fiber strength. The strength of cellulose is evident from the fact that it is the fundamental material making up cotton fabric, paper, cardboard, wood, and a variety of other substances used for their structural strength. The cell walls of plants are composed of cellulose, giving this polysaccharide the distinction of being the world's most abundant organic compound. From the perspective of insect diets, cellulose is, because of the inaccessibility of the (β1–4) bond, indigestible by most insects. Those insects such as termites and wood roaches, which do utilize cellulose, do so thanks to the microbial symbionts that they support in their digestive systems. Also, cellulose is a bulking and binding agent.

Like cellulose, chitin is a structural homopolysaccharide, and like cellulose, chitin is not digested by the digestive systems of insects. A further resemblance to cellulose is that chitin is composed of individual (*N*-acetyl-D-glucosamine) joined by (β1–4) glycosidic linkages. Interestingly, the failure of insects to utilize chitin in their diets is not a result of the genetically based inability to produce chitinase. This enzyme is made by insects in their cuticles as they molt. However, the inability of insects to digest chitin is probably related to the possibility that digestion of ingested chitin would also be involved in the digestion of the chitinous intima in the foregut and hindgut of the insect (Cohen 1999). This could be subversive to the digestive system's organizational integrity.

Returning to the comparison of starches and cellulose, it is significant that the same kind of sugar, glucose, can be a highly useful nutrient if it is bonded in an (α1–4) linkage or a totally non-nutritional polysaccharide if bonded in a (β1–4) linkage. However, even where it is not utilized directly as a nutrient, cellulose may be used in diets to deliver lipids and other nutrients that may be resistant to mixing with water or that may be unstable. Also, it is known that bulking agents function in vertebrates to induce peristalsis and movement of materials through the gut (Stevens and Hume 1995). Although this function is not well documented in insects, bulking agents may have the same type of functional importance in this group.

5.10 CHEMISTRY OF NUCLEIC ACIDS IN DIETS

The nucleic acids, deoxyribonucleic acid (DNA), and ribonucleic acid (RNA) are present in small amounts in many insect diets, usually as co-occurring components with the desired nutrients (i.e., as a component of flour, wheat germ, or animal products). Because of the relatively small role that the nucleic acids play as nutrients, their chemistry will not be covered as extensively as the major nutrient groups. However, the importance of the nucleic acids cannot be overlooked in the biological and molecular status of this group of macromolecules. The structure of DNA is a long chain of the bases (adenine, thymine, guanine, and cytosine) connected covalently with the five-carbon sugar deoxyribose and elongated by connections between sugars via a phosphate group. The long chains are in the form of a twisted structure known as a double helix, and the bonding between chains is stabilized by hydrogen bonds (polar attractions) between the base pairs adenine and thymine and cytosine and guanine. RNA has a similar structure except that the sugars present are ribose units rather than deoxyribose, and the base uracil replaces thymine. Of course, this structure is at the heart of the genetic role played by DNA as the carrier of all the organism's information about all proteins that are going to be expressed, giving that organism its individual character and integrity. It is this code that dictates how foods (raw materials) will be converted into the materials that make up the organism with all its functions, peculiarities, and potentials. "You are what you eat" only goes as far as you are what your genes tell you what to do with what you eat.

5.11 CHEMISTRY OF VITAMINS IN DIETS

In an important sense, discussion of the chemistry of vitamins in insect diets is more complex than that of the other nutrient classes, because vitamins belong to a broad range of chemical *families*, which defy simple generalization. First, the vitamins fit into two very broad categories, the water-soluble and fat-soluble factors. Even within these classes, there are many very different subcategories and related to their structural differences, the vitamins are very heterogeneous in their behavior in diets under the different processing and matrix conditions. The subject of vitamin chemistry in diet matrices is further complicated by the various interactions with other food components, the interplay of pH and vitamin interactions and stability, and the effects of temperature in these potentially destructive reactions. Adding to the complexity of vitamin discussion is the fact that the same vitamin can have multiple chemical and nutritional effects. Such multiple effects are best represented by ascorbic acid and its derivatives.

5.11.1 Multifaceted Nature of Ascorbic Acid

A most dramatic example of the potentially ambivalent nature of some vitamins is to be seen in ascorbic acid, which has biochemical activity that qualifies it to be regarded as vitamin C. In a sense, vitamin C is the best known and best documented vitamin with a remarkable history of having been demonstrated to prevent the serious disease scurvy. It is well known that plants contain considerable amounts of ascorbic acid, especially in certain tissues where concentrations may exceed several hundred mg/100 g of fresh tissue (USDA Website Bulletin 2002). The function of the ascorbic acid is to serve as an antioxidant against the many reactive oxygen species (ROS) that plants are confronted with, including molecular oxygen (in its ground triplet state), singlet oxygen, superoxide anions, hydrogen peroxide, hydroxyl radicals, perhydroxyl radicals, and ozone (Buchanan et al. 2000). The ROS destroy cell membranes, nucleic acids, and other plant cell components, and this being the case, plants are in a constant struggle to quench or scavenge these ROS to eliminate their threat. The forces that generate ROS are light, wounding, herbicides, drought, pathogens, extreme heat or cold, and heavy metals (Buchanan et al. 2000).

Like plants, various environmental and internal processes in insects and other animals generate ROS that are subversive to cellular well-being and molecular integrity (i.e., homeostasis). The section on antioxidants covers this subject further, but it is important to recognize that the normal and the stressed metabolism of insects constantly generate ROS, and there is a constant need for insects, including those that we are trying to rear, to fight off these damaging species of chemicals. Ascorbic acid is a key component in this struggle against cellular disorder via oxidative stress.

Besides this function, and returning to the scurvy issue, ascorbic acid is an important component of the synthesis of collagens (a special protein associated with connective tissues). Because insects do produce collagens too (albeit in proportionately smaller amounts than do vertebrates), it may be a function of ascorbic acid in insects to serve in the collagen synthesis pathway. It may be that the insects that do not seem to require ascorbic acid in their diet are able to synthesize it on their own by the L-gulonate pathway (as can most vertebrates), a metabolic system, which is missing in humans, monkeys, guinea pigs, and some fish. The documented effects of ascorbic acid deprivation in insects are discussed in the chapter on nutrient functions.

Several forms or derivatives of ascorbic acid have been shown to have vitamin C activity in vertebrates, and these variants may also satisfy insect needs. Because L-ascorbic acid and its nonlipid-linked derivatives are highly soluble in water, they are readily excreted and therefore lost from the metabolic pool. Also, the high degree of polarity makes L-ascorbic acid and its polar derivatives, ascorbic acid sulfate and ascorbic acid phosphate, insoluble in lipids. Therefore, these forms do not afford dietary lipids protection against oxidation as does the lipid-soluble ascorbate palmitate (or other fatty acid derivatives of ascorbic acid) (Gregory 2008).

Of all the vitamins, ascorbic acid is probably the most labile to heat, light, and a variety of other harsh conditions. For example, the ascorbic acid content in broccoli goes from 56 mg/100 g fresh weight to 40 mg/100 g in raw versus cooked material (USDA). However, it is interesting to note that, at least for mammals, the bioavailability of broccoli increases by 20% as a result of cooking (Gregory 2008). Therefore, the destruction of this vitamin by cooking is almost completely offset by the increased accessibility resulting from improvement in digestion/absorption dynamics.

Because of the instability of ascorbic acid under aqueous conditions, some researchers have used the ascorbic acid-2-phosphate form (Koizumi et al. 1990, Chang and Kurashima 1999). Ascorbic acid is sometimes used in its lipid-soluble forms ascorbyl palmitate and ascorbic acid acetal (Gregory 2008). In its lipid-soluble forms, ascorbic acid is highly regarded as a protection against lipid oxidation (peroxidation losses), and in the lipid soluble forms, ascorbic acid activity is less likely to be lost to leaching (Gregory 2008). However, the cost and inconvenience of these alternative forms must be weighed against the benefits.

Also, although this aspect of ascorbic acid is discussed more thoroughly in other places in this book (the section of this chapter on minerals and the chapter on complexity), it is important to raise the issue of interaction of ascorbic acid and metal ions. This is especially important with respect to the catalytic interactions. First, the interaction between ascorbic acid and metal ions such as iron and copper has become better understood recently (Buettner 1988). This increased understanding has led to an explanation for the deterioration of ascorbic acid at neutral pHs, even in air saturated solutions. This loss has been documented as metal ion–catalyzed destruction. Even very small amounts (micromolar concentrations) of metal ions in an aqueous solution at neutral pHs cause pronounced losses of this vitamin (Buettner 1988). Although anaerobic losses of ascorbic acid in foods (and most likely in insect diets) are greatly reduced compared to losses where oxygen is abundant, metal ions do catalyze some ascorbic acid degradation.

Another complex and largely unexpected interaction of ascorbic acid and iron, copper, and other metal ions is the ascorbate-induced degradation of lipids via a peroxidation type loss. Although the physical chemistry of these reactions is very complex and not completely understood, it is becoming increasingly well documented that ascorbate can induce iron from complexes such as with the egg protein, phosvitin, to cause accelerated peroxidation of lipids in a variety of foods. This phenomenon leads to the unexpected (and therefore chaotic) role of ascorbic acid as a pro-oxidant in many foods, rather than an antioxidant, which is the conventionally thought of role of this vitamin and food additive.

Not to be confused with the role of ascorbic acid as a vitamin and phagostimulant in insect diets is its role as an additive that not only serves as an antioxidant that protects other nutrients (α-tocopherol, β-carotenes, lipids, folic acid, and proteins), but it also reduces the degree of non-enzymatic browning, which can cause loses of nutrients due to Maillard reactions (Gregory 2008).

5.11.2 Chemistry of the Other Water-Soluble Vitamins

The discussion of the chemistry of the other water-soluble vitamins is not as detailed as it was for ascorbic acid for several reasons. First, its importance as a vitamin and dietary additive has spurred an extremely robust body of research on this nutrient in human foods. Next, where it is present as a naturally occurring factor, it is generally present in higher concentrations than those of any of the other vitamins. For example, it is present in excess of two percent of the dry weight of green peppers (Koizumi et al. 1990). Also, it is probably the least stable of all the vitamins. Last (and related to the previous statement), it is probably the most chemically dynamic vitamin causing or inhibiting numerous chemical reactions in foods.

Thiamin (vitamin B1) occurs in a wide variety of foods that cross lines of plants, animals, and microorganisms. It is most common, in nature, in the form of thiamin pyrophosphate, and as an additive to diets and foods, the most common form is thiamin hydrochloride (Gregory 2008). It is clear from Table 5.9 that the stability of thiamin is greatly dependent on the food matrix in which it occurs. For example, the blanching and canning process with spinach causes losses of up to 80% of the original thiamin, while the losses in tomatoes amount to less than 20% of this vitamin. This difference is related to a complex set of conditions, but pH is very important, thiamin being much more stable at low pHs than it is at neutral or basic pHs (Gregory 2008). For practical purposes, it can be summarized that thiamin compounds are very stable under conditions of low water activity and at temperatures lower than 37°C, but at higher temperatures and especially at higher water activities (>0.4), thiamin becomes subject to rapid degradation (Gregory 2008). It is noteworthy that conditions that typify completed diets being held at rearing temperatures are conducive to substantial thiamin losses.

Table 5.9 Depiction of Stability of Various Vitamins by % Losses during Canning

Product	Biotin	Folate	B6	Pantothenic Acid	A	Thiamin	Riboflavin	Niacin	C
Green beans	—	57	50	60	52	62	64	40	79
Carrots	40	59	80	54	9	67	60	33	75
Spinach	67	35	75	78	32	80	50	50	72
Tomatoes	55	54	—	30	0	17	25	0	26

Source: Modified from Lund, D., Effects of heat processing on nutrients, in: *Nutritional Evaluation of Food Processing*, 3rd edn., Karmas, E. and Harris, R.S., eds., Van Nostrand Reinhold, New York, 1988, pp. 319–354.

Riboflavin (vitamin B2) and its derivatives called flavins is another vitamin that exhibits complex behavior in the matrix of foods and insect diets. The flavins are often associated with proteins. Riboflavin can become highly reactive in foods and insect diets and is thought to be responsible for substantial amounts of oxidative damage in foods (Gregory 2008). The flavins are notably photoreactive and are rapidly degraded by exposure to light, even in diet matrices. However, at lower pHs (<5.0) and at lower temperatures (below 30°C), flavins are less photoreactive than they are under higher temperatures and pHs (Gregory 2008). This pH sensitivity is evident in Table 5.9 where it is reported that 64% of the riboflavin was lost in the canning of green beans compared with only 25% that was lost in canning of the tomatoes (a more acid food than green beans).

The niacin family includes nicotinic acid and its derivatives, nicotinamide and nicotinamide adenine dinucleotide, as well as a wide variety of derivatives. Both nicotinic acid and nicotinamide are considered to be the most bioavailable forms of this vitamin. Heat processing tends to increase bioavailability of the niacin family. Also, the niacin family is stable at all pHs, making it one of the few vitamins that do not present a concern about adverse effects from processing.

5.11.3 Chemistry of Minerals in Diets

The activities of minerals (or inorganic components) in diets are complex; what Dadd said back in 1968 remains true today: this aspect of insect nutrition is "probably the most neglected area of insect nutrition." This statement pertains not only to the function of minerals (discussed in Chapters 3 and 4) but also to the chemistry of interactions of minerals with one another and with other dietary components. One of the most important and complex aspects of mineral chemistry in foods (and therefore in diets) is the acid/base interactions of minerals (Miller 2008). As Miller (2008) further explains, there is an enormous complexity of minerals as they occur in various foods, and he further points out that the forms of minerals in most foods are poorly known. This is especially the case with one of the most important minerals, iron. Iron, like most other minerals, is not present in foods or diet ingredients in its elemental form, but rather as a compound or ion linked to other substances.

The tremendously complex interactions of iron and other diet components are discussed in Chapters 3, 7, and 8, including its powerful effects as a pro-oxidant. Likewise, several other minerals can act as pro-oxidants in diets and within the bodies of the target insects, including copper, zinc, and manganese. An important action of chelating agents is to sequester such inorganic components, removing them from portions of the diet matrix where they can cause oxidative damage. This point pertains to both those that are deliberately added to diets, such as EDTA and citric acid, and ones that are added for other reasons but serve as chelators, such as proteins.

Also, many enzymes that occur in diets, especially in raw ingredients, have as cofactors certain minerals that help promote reactions that are often undesirable. For example, iron is a cofactor for polyphenol oxidase, which promotes nonenzymatic browning reactions, which can be destructive to proteins (or to the availability of the proteins in the insect's digestive system). A large variety of minerals are known to act as cofactors for enzymes that occur in foods (Whitaker 2008) and diet ingredients. For example, zinc (Zn^{++}) is known as a cofactor in a great number of enzymes, including alkaline phosphatase, and chloride (Cl^-) is a cofactor to amylases in many foods (Whitaker 2008) and diet ingredients.

It is also important to realize that the solubility of various minerals is profoundly influenced by pH. As discussed elsewhere in this chapter (on pH) and in the chapter on sources of complexity (Chapter 8), many minerals, most notably calcium, are much less soluble at basic pHs than they are under more acidic conditions. Mittler (1972) presents an excellent anecdotal discussion of the inadvertent removal of calcium salts by precipitation reactions that took place as pH was being adjusted with added bases. Mittler also explained the insidiousness of the reactions that caused cholesterol and other diet ingredients to be lost from the diet due to unrecognized precipitation reactions mediated by calcium in a context of increasing pH.

Although the chemistry of all the nutrients discussed here was for clarity considered in a piecemeal fashion, it must be emphasized that the real dynamics of the diet components can only be understood in the context of their organizational or structural matrix. This reiterates a point that was made in Chapter 4 and again in Chapter 8. This point is intricately tied to the concept of complexity that characterizes insect diets.

CHAPTER 6

Dealing with Changes

6.1 INTRODUCTION

Once a good, working diet has been developed and adapted to the insectary, it is risky to make changes. In nearly three decades of experience with many insectaries, I have seen countless problems that came from unplanned or poorly designed changes. Often, such problems began after a considerable period of relative colony stability and successful rearing. Certainly, such problems sometimes came from sources other than diet deviations but stemmed from microbial contamination (Chapter 13); environmental changes (Chapters 1 and 8) (e.g., unnoticed temperature, humidity, or light changes); genetic problems from inbreeding (Chapter 1); accidental introduction of parasitoids or unwanted commensal species, such as mites and lice; inadvertent introduction of pesticides; and other hidden contaminants (fumes from newly painted rooms or from other stray sources).

Unplanned and undesired results can come either from deliberately or inadvertently changing various diet ingredients, using different amounts of the same ingredients, using ingredients that have become degraded during storage, changing processing techniques, and changes in the handling of the diet after it has been prepared. All of these factors are discussed further in Chapter 8 (on complexity in insectaries).

Most of these factors are related to the fact that diet ingredients are processed materials. By the very nature of the manufacturer's processing, variations creep into the diet-making process. Variations are common in fresh produce and meats, even those purchased from the same store. A navel orange from one batch may be sweet and tasty, while the same variety from another batch (sometimes from the same tree) can be sour and flavorless. The agar, carrageenan, or pectin that is purchased from the same distributor with the same product number can have dramatically different gel strengths, water binding quality, or mineral contents. For example, four bottles of citrus pectin with the same product number from the same, very reliable chemical company can have different methoxy contents resulting in quite different digestion rates for pectinases used by insects to liquefy the pectin. The methoxy groups are methyl esters (–O–CH_3 groups) that are attached to the subunits that are linked in long linear and branched structures that make up the macromolecules that we call pectins. The differing methoxy contents will also lead to differences in the gel strength of the diets where these pectins are used, as well as numerous physical and chemical characteristics that differ according to methoxy numbers. A whole cascade of unexpected consequences may result from these differences in what should be a nutritionally inert diet component that is added to some diets mainly to provide a desirable texture. There are batch-to-batch differences in the strength, mineral content, and related outcomes in all types of gelling agents that are used in diets. In fact, because of these differences, it is strongly recommended that gelling agents be tested with every new batch that is delivered to an insectary (discussed in Chapter 11 on process control and quality control).

Various products with identical names and manufacturer or distributor assigned product numbers may be derived from different sources or from different processing conditions. All of these differences can lead to variance in the product. This is especially applicable to products that are of complex composition (complex foods or staples) such as various kinds of meal, flour, meat products, or other sparsely refined materials and to very pure materials that require multiple steps in purification. The impurities in these products stem from sources such as genetic variation, phenology of parent plant material at the time of harvesting, or physiological changes that take place as a result of stress or other environmental circumstances. For example, soy beans will differ chemically according to their genetic makeup, when they were harvested (unripe versus ripe pods), the fertilization and watering history of the parent plant, and various differences in processing treatments such as types of dehusking and grinding equipment, whether or not a roasting (or toasting) process was used (and if so, the duration and temperature of the roasting process), storage conditions, and packaging conditions. These factors are but a few of all the possible variations that can characterize what seems to be such a simple, stable diet component such as soy flour. Also, various foods, including soy flour, may contain pesticides (herbicides, miticides, insecticides) that were used on the parent crop material or that drifted from nearby fields where the crop was harvested. Even small traces of these compounds can have an adverse effect on the target insects.

6.2 CONFUSION OVER PRODUCT NAME DIFFERENCES

There is often confusion about products with names that are close or ambiguous. For example, wheat germ, wheat bran, and wheat flour are sometimes confused, and at least one insectary that nearly lost its entire stock of insects by mistakenly substituting whole wheat flour for the wheat germ called for in the diet formulation. Table 6.1, where two kinds of wheat germ and a whole wheat flour are compared, reveals the magnitude of the differences between these two milled wheat products. Wheat germ has nearly twice the protein content and more than four times the lipid content of wheat flour. Wheat flour has about 1.5× the carbohydrate found in the wheat germ, and wheat germ has a higher vitamin and lipid content than does wheat flour. The differences that do not show up in composition breakdowns such as Table 6.1 are those that include profiles of the kinds of proteins and their relative amounts in the two kinds of wheat products.

Also soy protein and soy flour are sometimes confused. Although soy flour is rich in protein (see Table 6.1), it contains many other components that make it a different factor in a diet that calls for soy protein, and if soy protein is called for in a diet formulation, there will be a substantial difference in the amount and kinds of protein present if soy flour is inadvertently substituted. Conversely, if soy flour is called for, it can be a problem to substitute soy protein. An aliquot of 100 g of soy protein has more than 90 g of protein, and soy flour (defatted) contains about 51% protein (USDA database). Even more dramatically, full-fat soy flour contains only about 38 g of protein (USDA database), less than half the amount called for by the diet formula that calls for "soy protein." Most soy protein preparations contain mainly the soy beans' storage proteins, but soy flour proteins include whey proteins as well as storage proteins, and whey proteins include biologically active factors such as protease inhibitors, lectins, and a host of other antinutrient factors. In contrast with the biologically active proteins, storage proteins are highly nutritious in terms of profiles of amino acids and digestibility (Fukushima 1991, Endres 2001).

Lastly, raw soy flours have different compositions and different nutritional qualities than those that are heat-processed. An example is seen in Table 6.1 comparing full-fat raw soy flour (Nutrient Data Base No. 16415) and full-fat roasted soy flour (Nutrient Data Base No. 16416), with the former having a fat content of 20% and the latter about 22%. The protein percentages and activities of the heat-processed versus the raw flour also differ sharply. Most of the differences between the percentage of protein, lipid, and carbohydrate result from the fact that the roasted soy flour has a slightly lower percentage of water than does the raw soy flour. However, the differences in the character of the proteins after they have been heat-processed are more profound and far-reaching than just

Table 6.1 Nutritional Components (Amount per 100 g) of Two Forms of Wheat Germ (Roasted and Raw) and Two Forms of Soy Flour (Roasted and Raw)

Component	Wheat Germ (Toasted) (NDB # 08085)	Wheat Germ (Crude, Raw) (NDB # 20078)	Soy Flour (Full-Fat, Roasted) (NDB # 16116)	Soy Flour (Full-Fat, Raw) (NDB # 16115)	Wheat Flour, Whole-Grain (NDB # 20080)
Water	3	11	3.8	5.2	10.3
Energy (kcal)	372	360	441	434	339
Protein	26	23	35	38	14
Total lipid	8	10	22	21	1.9
Carbohydrate	58	51	34	32	73
Fiber	10	13	9.7	9.6	12
Ash	4	4	6	5	1.6
Minerals					
Calcium	50	39	188	206	34
Iron	8	6	6	6	4
Magnesium	272	239	369	429	138
Phosphorus	1011	842	476	494	346
Potassium	964	892	2041	2515	405
Sodium	11	12	12	13	5
Zinc	14	12	4	0.4	3
Copper	0.72	0.9	2	2.9	.3
Manganese	16	13	2	2.3	4
Selenium	0.06	0.08	0.08	0.08	0.07
Vitamins					
Ascorbic acid (vitamin C) (mg)	0.0	0.0	0.0	0.0	0.0
Thiamin (mg)	1.3	1.9	0.4	0.58	0.47
Riboflavin (mg)	0.7	0.5	0.9	1.2	0.22
Niacin (mg)	4.7	6.8	3.3	4.3	6.4
Pantothenic acid (mg)	1.1	2.3	1.2	1.6	1.0
Vitamin B-6 (mg)	0.5	1.3	0.4	0.5	0.3
Folate (mg)	0.333	0.28	0.23	0.345	0.04
Vitamin B-12 (mg)	0.0	0.0	0.0	0.0	0.0
Vitamin A (IU)	96	0.0	110	120	0.0
Vitamin E	22	0.0	0.0	0.0	1.2
Lipids					
Saturated fatty acids	1.3	1.7	3.2	3.0	0.3
14:0	0.01	0.01	0.06	0.06	0.003
16:0	1.2	1.6	2.3	2.2	0.271
18:0	0.04	0.06	0.9	0.74	0.015
Monosaturated fatty acids	1.04	1.37	4.8	4.6	0.232
16:1	0.03	0.03	0.06	0.06	0.013
18:1	1.08	1.33	4.8	4.5	0.219
Polyunsaturated fatty acids	4.8	6.01	12.3	11.7	0.779
18:2	4.1	5.29	10.9	10.3	0.738
18:3	0.52	0.72	1.5	1.4	0.038

Source: Anonymous, USDA Nutritional Database for Standard Reference, Release 27 (accessed February, 2015), 2013.

a reduction in water content. Also, there is a decrease in the vitamin content of the heat-processed soy flour with a greater than 20% decrease in the niacin content and similar decreases in riboflavin, thiamine, and total folate (USDA Nutrient Data Base).

6.3 UNAVOIDABLE CHANGES IN DIETS AND OTHER COMPONENTS

Changes, deliberate and inadvertent, are discussed as factors that lead to unexpected and usually undesirable results. One of the unavoidable changes that are characterized in that chapter is the *cycle of ingredient replacement*. As they are used up, diet ingredients must be replaced, and it is commonly practiced by diet and rearing professionals to uphold a steadfast and highly conservative effort to purchase the ingredients from the same supplier from whence the components had been previously obtained and to further use the precaution of using the same product number or catalog description. However, changes in rearing materials are as inevitable as "death and taxes." Even nondiet materials such as organdy, flannel cloth, plastic sweater boxes, vials, glue, light bulbs, and paper towels (to mention only a few) can become unavailable in the form or composition that was previously used. Even more insidious are the ways that diet components come to change. The changes that take place due to aging and mistreatment in storage and handling are discussed in detail in Chapter 5 on the chemistry of diet components and Chapter 12 on diet-processing equipment. However, these components are also variable as they come fresh from the supplier or manufacturer.

These changes are generally not documented in the literature on diets and often go unnoticed but are likely to be responsible for many unwanted changes in insect quality and production outcomes. For example, gel strength can be greatly affected by the quality of the agar that is used to form the gels. It is not uncommon for gel strength to vary by 30% when agars from the same supplier and under the same product number are compared. Likewise, lipid profiles, potency of vitamins, and purity and profile of purified proteins such as soy, casein, and albumen can all vary from batch to batch, even when these components are obtained from the same supplier and manufacturer. As discussed in the chapters on food chemistry and physics (Chapter 5), complexity (Chapter 8), and diet processing (Chapter 12), the differences in such products occur as a result of changes in the source of raw materials and in processing protocols. For example, the characteristics of gelling agents such as agar and carrageenan can vary greatly according to the species of algae from which these products are derived as well as how they are extracted (BeMiller and Huber 2008). The other side of unwanted change stems from purchase mandates where a supplier no longer carries a product line that had been used, and it becomes necessary to find a suitable substitute. This point is extensively discussed in Chapter 8 on how disorder enters the rearing domain.

6.4 CHANGES IN PRODUCTION PROCEDURES

It is a common problem in insectary operations for equipment to wear out and for other necessary changes to be made in types of equipment and other processing procedures. Even such a simple and seemingly innocuous action as the replacement of an older 3 speed blender with a newer 10 speed model can cause changes in mixing characteristics of the incipient diet. If the diet formulation called for mixing for 2 m at speed 3, what is the appropriate speed on the new blender? If the insectary staff were trying to scale up the diet volume of 0.5 L mixed in a household blender (with a 1 L capacity) to 3 L mixed in a 4 L commercial scale blender, would the mixing and chopping profile be the same? These same types of questions can be asked for each implement and procedure of diet making, and the magnitude of potentially significant variation increases as diet volume scale-up is attempted by introduction of large-scale commercial equipment.

For example, in the food industry, there are more than 10 different heat-processing techniques and equipment associated with those procedures (Fellows 2000). He also lists more than 20 modifications of these techniques. Also there are many types of size reduction equipment, purification equipment, preservation equipment, and a wide variety of other equipment and techniques that are used in food processing. As is discussed in Chapters 8 and 12, many of these techniques have been applied or can, in the future, be applied to processing and improving insect diets. However, there are unexpected changes that can be brought into the rearing setting from making seemingly innocuous changes in various processing techniques.

6.5 WHAT TO DO IF YOU MUST MAKE CHANGES

The procedures discussed here can be incorporated with the quality control techniques suggested in Chapter 11. What follows is a recommended plan for development of a product cycling regimen designed to deal with inadvertent or cryptic changes in ingredients. The plan includes ways of dealing with ingredients that were changed either because of mistakes in purchasing or changes made by manufacturers or distributors. While some changes were made without the knowledge of the insectary staff (such as the form of iron in the Wesson salt mixture discussed in Chapter 8), this section deals with changes that are made deliberately by the insectary personnel. This section is the basis for the development of a strategic plan for maintaining consistency and quality in the insectary, even in light of an ever-changing source of dietary components and personnel who deal with these components.

It is strongly recommended that the types of changes described here are done with full documentation, record-keeping, and authorization. In terms of the integrity and quality of the colony, one of the most insidious and subversive events that can take place in a rearing program is a change in some procedure, which has not been at first validated/supported, through a controlled test, as being a positive change. The people who use the insects being reared, the customers, depend on and expect a high degree of uniformity of the insects that will be applied to some end-point purpose. Just as supplier changes in the nature of a diet ingredient is undesirable, the insectary customer has the expectation of insects whose quality and nature are consistent from batch to batch.

Any change can have one of the three consequences: it may make no difference, it may make things better, or it may make things worse. Deliberate (conscious) changes may be made with noble intentions: to save time, to make things safer, to make things more comfortable for the insectary workers, to save money, and to make things better for the insects (i.e., to increase production and/or quality). Regardless of the outcome (neutral, negative, or positive), it is important that the change(s) be recorded and communicated by the management and the personnel who carry out the rearing diet-handling duties. One of the most important duties of both those managing and those being managed is to communicate that changes are to be done ONLY after there has been complete communication about those changes and that there is an appropriate test procedure that validates the change. These practices are well documented as the basis for process control and problem-solving in Chapter 11.

6.6 MAKING CHANGES: DEVELOPING STRATEGIC PLANNING SYSTEMS

Communication regarding changes and gaining cooperation in implementing changes are probably the most difficult issues of insectary management. When any changes are to be made, all relevant personnel who are responsible for the rearing system should be involved. One of the most common sources of problems is the failure to gain buy-in of all relevant staff. As is well established in modern systems of total quality management, the team is encouraged to help develop and maintain channels of communication. The personnel who implement the change must be alerted to the exact nature of the change and the reasons behind it. The insectary manager should be a part of the change process, and possibly the customers who use the insects may need to be notified. These comments are based on many experiences where changes or deviations from SOPs were made locally, without the knowledge and buy-in from the diet team, often excluding the management and diet experts.

6.7 TESTING CHANGES: THE HALLMARK OF STABLE REARING PROGRAMS

Using such meticulous practices as those described here will ensure a high level of stability that will more than pay for itself in many insectary systems, especially in large-scale operations. The rule of thumb is simple: *all changes that are to be made must be subjected to controlled testing.* A controlled test is simply a well-designed scientific experiment that contains a control and a variable. In the case of the insect-rearing setting, the control is the exact set of protocols and components that have been used, and the variable is the exact same setup with the exception that the new, single protocol or component is used in place of the time-tested one that is to be replaced. As is the case with any scientifically conducted experiment with good design, the interpretation of the data resulting from the tests should be objectively evaluated, preferably with a suitable statistical test. In the rearing community, there are two fundamental types of tests to conduct: *the bioassay* and the *materials test.* The bioassay is the most important aspect of development and maintenance of quality rearing systems, and it is probably the most underestimated and underused tool. It will be discussed in more detail later in the chapter. The materials test, typically used in various industries and laboratories, is a measurement of the physical or chemical qualities of components or protocols.

Some examples of suggested procedures will clarify this concept. If a new source (a new brand, new batch) of wheat germ is being used in a diet, a rearing test (a bioassay) with two batches of diet, one that contains the wheat germ from the old source (control) and another batch that contains the new source (variable), needs to be performed. This assay should be conducted with both batches of diet side by side and with all other components and protocols being the same for both batches of diet. It is further advisable to repeat these tests with at least three separately mixed batches of each type of diet (the controls and the variables). This is recommended because variable results (error) can be better evaluated if multiple batches of each diet are prepared. For example, a component may have been improperly weighed or inadvertently eliminated in one diet preparation cycle—giving false-negative results with the diet formulation in question. There are dozens of examples of this type of diet production error. Statisticians tell us that such operator error is a prominent reason for using replicated tests.

Once the tests are performed, the evaluations are made based on comparisons of some key and telling biological features that reflect possible differences in the responses of the insects to the test diets. The bioassay can be no better than the parameter being evaluated as an indicator of potential problems. To evaluate possible differences in the diets being tested, a biological parameter must be measured that is sensitive to the differences being tested. As discussed in several other places in this book, especially Chapters 4, 9, and 11, the most commonly used biological parameters and some of the most robust are body weights and size measurements of various life stages, survival rates, reproductive rates (fecundity), egg hatch rates, development rates (often expressed as reciprocal values in order to make shorter development periods positively correlated with the quality of diets), sex ratios, and longevity (Singh 1977, Cohen 1992). Nutritional indices are sensitive tools for the assessment of changes (Waldbauer 1968), but parameters such as growth indices and relative consumption rates can be labor intensive.

There is no substitute for a clear empirical basis for selecting or developing bioassay parameters, that is, developing a strong base of understanding of the target species through careful observation and experience. Becoming familiar with the target species is essential as some of these parameters are ambiguous, deviating from one species to another. A safe approach is to use multiple parameters so that at least three different kinds of measurements can be used to detect subtle but important deviations in the quality of diet components. Once the bioassay protocols are chosen and worked into a routine system, the diet professional should adopt a proper statistical system for measuring possible deviations (also discussed in Chapter 11).

Besides using biological tests (bioassays), materials can be tested using physical or chemical tests. The literature in food science is an excellent source of literally hundreds of different physical and chemical tests for every type of component that may be present in insect diets. An excellent source that provides a detailed discussion and methodology for many of these tests is Nielsen (1998); however, this topic will be briefly covered here. The tests suggested and discussed here were selected because they meet two characteristics: (1) Each provides a robust assessment of whether or not a given diet component is different from the standard version of that component and (2) each is relatively easy to perform requiring only a moderate background.

Water content: One of the most robust and simple tests that can be made of completed diets and diet components is simply the percentage of water contained in the material in question. The only instruments required to measure water content are a reliable balance and a drying oven (any oven that can hold a temperature of 65°C, or 105°C). The material to be tested is preweighed, then placed in the drying oven at either 65°C (a fairly low temperature that is often used when concerns about large portions of volatile substances are present other than water, e.g., ethyl alcohol), and reweighed every few hours until an equilibrium weight (no further weight loss with subsequent reheating) is reached. The final weight is divided by the original weight to obtain a value for the percent dry weight. If multiple batches of diet are tested this way or if past records are kept meticulously, statistical comparisons can be made to determine whether or not any deviations in the percentage of dry weight (or water content) have occurred. The tests of dry weights of the complete diet can indicate an error in the diet processing. Each insectary must determine its own allowable standards, but a good rule of thumb is that a greater than 1% deviation is unacceptable. The dry weight test of certain individual ingredients can be a good indicator of whether or not those ingredients had been properly stored or whether they may have gotten stale. Water-absorbing materials such as flours and meals, gelling agents, salt mixtures, and vitamin mixtures are susceptible to gaining water from the surrounding air. As discussed in Chapters 3 and 5, these increases in water content also mean an increase in water activity, opening the door to the degradation by microbes, reactive oxygen species, and enzyme activities.

A simple and robust test is one that can be done to determine the potency of vitamin mixtures, using ascorbic acid or total antioxidant potential as a measure. This test, called the FRAP assay, is described more fully in the section on quality control in Chapter 11. Other tests that can be done with fairly limited biochemical or food chemistry background include gel strength tests, gel temperature tests, lipid peroxidation tests, and starch or sucrose tests. Several of these tests are available as simplified kits. These and other quality control measures are further discussed in Chapter 11.

6.8 USING THE INGREDIENT CYCLE CONCEPT

One of the most useful and conservative (though laborious) systems of organization that can give stability to colonies is application of the "ingredient cycle concept." This concept is a powerful safeguard against inadvertent changes in the ingredients that could result in negative consequences. This system prescribes that all ingredients be assigned a replacement date that takes into account the shelf life of the ingredient and allows for procurement of its replacement prior to the expiration of the ingredient's shelf life. It should be noted that a "lead time" of at least one generation of the target insect is needed with multiple generations lead time as the better choice. This assures adequate time for each ingredient to be tested, preferably by the bioassay technique, prior to the complete depletion of the ingredient in question. For example, the current batch of vitamin mixture would used as a control that is tested side by side with the new, replacement batch of vitamins. If there were a problem in the newly purchased replacement vitamin mixture, there would be a high likelihood that the problem could be detected with enough time left for replacement with a satisfactory vitamin mixture.

Ingredient cycling requires development of an organized schedule of ordering and testing each material in the diet or at least the components that are most likely to present problems. The ingredients that are most likely to cause problems are those that are perishable and those that are processed by rather elaborate procedures. This includes substances such as the gelling agents (e.g., agar and carrageenan), vitamins (especially vitamin C), flours or meals (e.g., wheat germ or soy flour), and antimicrobial substances. Chapter 5 covers the chemistry of various diet components and explains the basis of deterioration or degradation that takes place in various diet substances; however, the most common causes of deterioration or other ways that an ingredient can become unacceptable are through contamination, uptake of moisture, oxidation (especially of fatty acids and vitamins), or substitution of sources or processing methods of materials. Examples of each of these are discussed in Chapters 5, 8, and 13, respectively, dealing with the chemistry and physics of diet components, disorder in insectaries, and microbial contamination.

Finally, it is re-emphasized that most insectary workers and insect diet specialists are diligent people who care about their insects and want to provide a wholesome setting to rear the healthiest possible specimens. The factor most salient in the deviation from optimal conditions is the lack of understanding or education about how and why certain practices must be followed. The combination of the diligence of insectary staff and continuous improvement through training and open communication will lead inexorably to higher-quality products of rearing systems based on artificial diets.

CHAPTER 7

Insect Feeding Biology (Digestive System Anatomy and Ultrastructure, Physiology and Biochemistry of Digestion, Feeding Behavior, and Sensory Systems) and the Logic of Metabolic Systems

7.1 INTRODUCTION AND OVERVIEW OF INSECT FEEDING SYSTEMS

One of the most important steps in developing and using a successful artificial diet for an insect is to understand thoroughly the target insect's feeding biology. "Feeding biology" is the composite of these features: mouthparts (biting, piercing/sucking, lapping, or some combination of these types), factors that induce or sustain feeding responses (token stimuli, nutrients that are phagostimulants), digestive enzymes, optimal gut residence time of suitable foods, absorption characteristics, nutritional requirements, and characteristics of egestion (excretion and elimination) of waste products. For those insect diet professionals who do not develop diets but who use them, it is also useful to understand how the target insect feeds and utilizes the diets that we provide them. Too often, incorrect assumptions about the insect's feeding targets or feeding mechanisms have led researchers to fail to develop excellent or even suitable diets. Also, underestimating the complexities of insect feeding systems and how these systems are matched to their food has led to faulty decisions about diet ingredients or procedures (Chapters 6 and 8).

For example (as further discussed in Chapter 4), early researchers in diet development for tarnished plant bugs and western tarnished plant bugs assumed that these mirids were sap feeders or "plant fluid feeders," and accordingly, they provided liquid diets whose nutrient composition was inherently dilute and especially depauperate in lipids (Cohen 2000a). It was not until Debolt (1982) developed a slurry diet rich in proteins and lipids that western tarnished plant bugs could be reared en masse and in continuous generations. Further work (Cohen 2000b) on slurry diets with high nutrient concentrations allowed researchers and commercial insectaries to produce these plant bugs with inexpensive diets composed of low cost, undefined components.

Similarly, the assumption that predatory Heteroptera feed on the prey liquids (especially hemolymph) had led to misplaced efforts to nurture these important insects with liquid diets. As a case in point, my early efforts to develop artificial diets for big-eyed bugs using hemolymph as a model proved to be futile. It was only after I paid careful attention to exactly *how* the predators feed and exactly *what* they feed on that I succeeded in developing a diet for these and other predatory insects that share the feeding style with the big-eyed bugs (as chronicled in the following section). Therefore, this chapter, a survey of the feeding mechanisms of insects, can serve as a useful introduction to the salient features of how insects deal with their foods.

7.2 INSECT FEEDING HABITS

7.2.1 Liquid versus Solid Feeding: A Case Study

A personal, anecdotal example of difficulties in developing and using artificial diets will serve as a good illustration of the kinds of frustrations and misdirection that can result from failure to understand and respect the complexity and true character of the feeding biology of our target insects. My first mission in the Agricultural Research Service was to develop an artificial diet for the predaceous hemipteran, *Geocoris punctipes* (a big-eyed bug from the family Lygaeidae, now called Geocoridae). As a member of the Hemiptera, *G. punctipes* is said to be a liquid feeder. It has piercing and sucking mouthparts as illustrated in Figure 7.1 showing *G. punctipes* feeding on whitefly nymphs, with the predator's labium and stylets in contact with a whitefly nymph. This feeding apparatus prompted many entomologists to perpetuate a concept that this insect feeds on "prey juices," especially hemolymph (insect blood). It has been repeated, incorrectly, in dozens of publications that this liquid mode of feeding characteristic of Hemiptera, larval Neuroptera, many larval Coleoptera, and parasitic Hymenoptera targets only the liquids present in prey, hosts, and other food sources; however, these insects, along with many other species of arthropods, target solid foods, which they liquefy with digestive enzymes and specialized mechanical actions (Cohen 1995, 1998c, Wu et al. 2000). This is discussed further in the last part of this chapter. Details of the feeding anatomy of a neuropteran chrysopid (*Chrysoperla rufilabris*) are presented in Figures 7.2 and 7.3. Like the predaceous members of the Heteroptera, the predaceous Neuroptera are now known to use extraoral digestion (EOD) to convert the solid materials in their prey into suspensions known as slurries (Cohen 1998b). The use of EOD is also practiced by plant-feeding insects as well as other trophic guilds (Cohen 2000b), and the stylet-type mouthparts are of a typical plant-eating heteropteran are illustrated in Figures 7.4 and 7.5. The homopteran head and mouthparts are shown in Figure 7.6.

Accepting the published accounts of strict liquid feeding, my first 4 years of efforts at providing artificial diets for *G. punctipes* were confined to formulation and presentation of liquid diets that were modifications of the liquid formulations described in the literature and ones modeled after the reports from the literature and my own analyses of hemolymph of known prey species. I tried

Figure 7.1 *Geocoris punctipes* feeding on whitefly nymphs, with the predator's labium in contact with the prey and the stylets penetrating the carcass of the whitefly nymph.

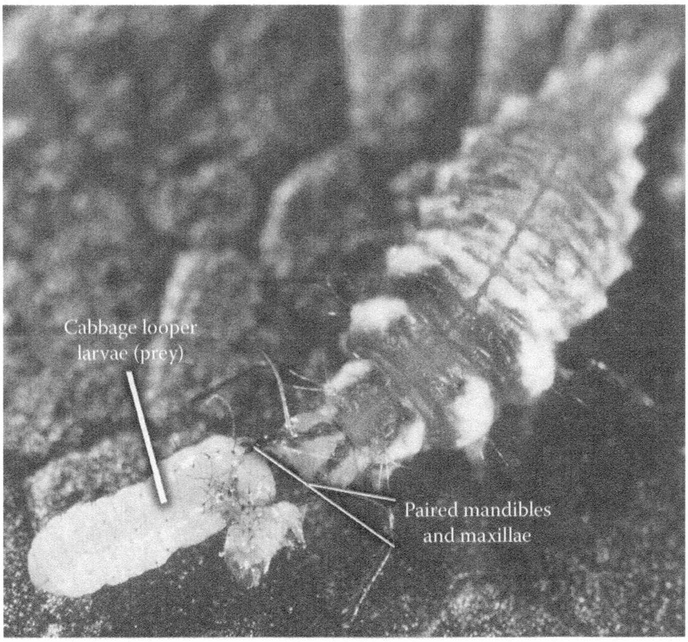

Figure 7.2 Feeding anatomy of a neuropteran chrysopid (*Chrysoperla rufilabris*).

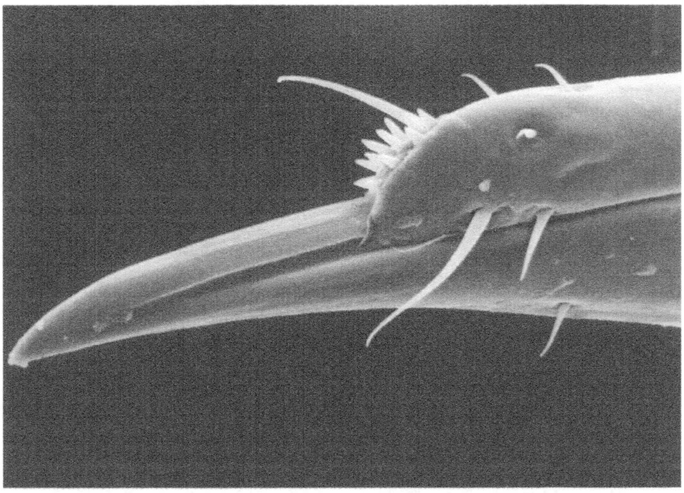

Figure 7.3 Detail of the mouthparts (mandible and maxilla) of a neuropteran chrysopid (*Chrysoperla rufilabris*).

faithfully to include all nutrient classes in an aqueous-based diet. This meant exhaustive efforts to blend lipids with the water base, as well as dissolving amino acids, various proteins, carbohydrates, vitamins, and minerals. Of course, due to solubility problems, very little lipid could be combined with the water, and I was constantly haunted with the suspicion that I was not affording the insects adequate amounts of sterol or other lipids. Furthermore, it was often futile to try to mix in mineral mixtures with the other components, because certain minerals, especially calcium compounds, were apt to precipitate out of the solution carrying with them other nutrients. To get mixtures of these ingredients, I tried blending, triturating, sonicating, homogenizing, and using ultrahigh-speed devices as well as various forms of heating. All of these efforts met with failure.

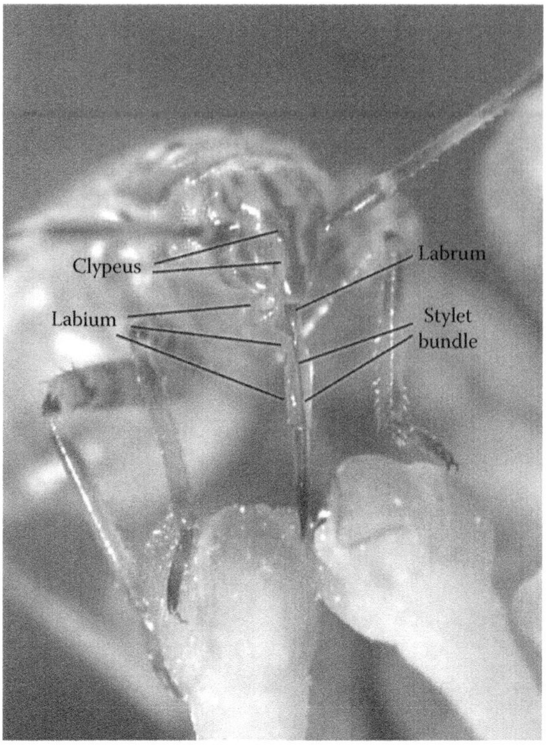

Figure 7.4 Stylet-type mouthparts of a typical plant-eating heteropteran.

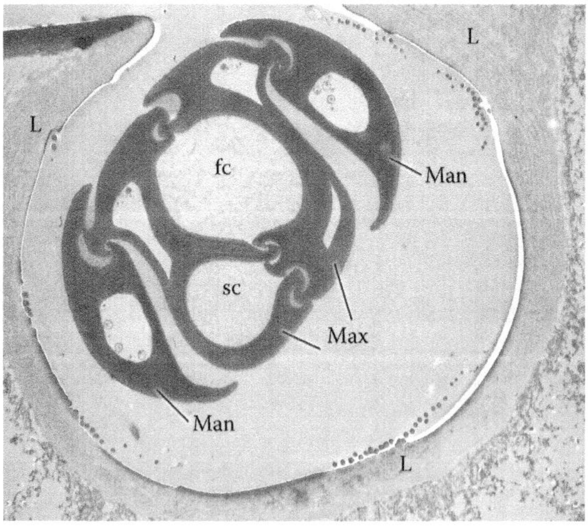

Figure 7.5 The food canal (fc), the salivary canal (s), the mandibular stylets (Man), the maxillary stylets (Max), and the surrounding labium (L) of a typical plant-eating heteropteran.

Discouraged by failures, I became convinced that there was no way to provide predators with an artificial diet that would be adequately nutritious to support robust development over multiple generations. As a last resort (realizing that the predators were thriving on live and heat-killed prey), I turned to careful observation of how the insects fed on their natural prey. Could I have missed something important regarding the nature of the big-eyed bugs' feeding process? Asking this

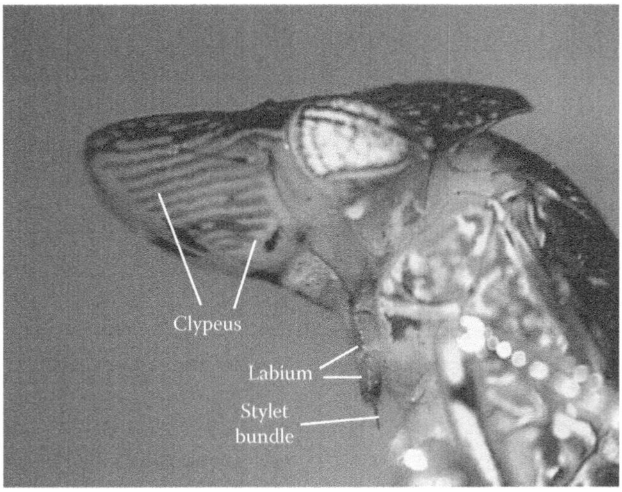

Figure 7.6 Homopteran head and mouthparts. Note the pronounced clypeus, which is linked with the massive muscles that power the cibarial pump.

question, I made careful observations of the big-eyed bugs during their feeding activities. At first, it appeared that the feeding process of these and other hemipterans was not very dynamic. The labium (Figure 7.1) seemed to rest on the surface of the prey's cuticle. However, once I started using dissections and lighting strategies along with a video filming system to allow detailed study of the feeding process, I began to realize that the mouthparts were far from static and that inside the prey, they were probing extensively and even more dramatically, they were mechanically and chemically macerating the prey's solid structures (including fat body, which is extensively distributed throughout the prey's body cavity, muscles, reproductive structures, neurons, and most of the prey's other organs and tissues). Seeing this type of feeding, I realized both that the predators were using what is called EOD to predigest their meal of solid nutrients from the prey's carcass and that any artificial diet modeled after hemolymph composition was far too dilute to fill the nutritional requirements of such a predator. These concepts are documented in several publications (Cohen 1985a,b, 1993, 1992, 1995, 1998c). The failures that I witnessed that resulted from a lack of understanding of the basics of feeding structure and function and the success that came from taking these factors into consideration prompted my enthusiasm about having a thorough understanding of the integrated nature of feeding mechanisms in any target insect. The remainder of this chapter is an effort to establish that kind of groundwork.

7.2.2 Regulation of Feeding and Sensory Mechanisms

A good general background in the factors involved in regulation of feeding under natural circumstances will help in efforts to develop artificial diets and to troubleshoot problems in an existing diet. Numerous research publications and several texts (e.g., Chapman and de Boer 1995) reflect the complexity of the feeding response, especially the wide range of feeding choices and feeding adaptations found in insects.

7.2.3 Survey of Insect Mouthparts

Although there are numerous mouthpart arrangements found among the Insecta, generally insects are adapted to feed on (1) strictly liquid diets, (2) strictly solid diets, or (3) a mixture of liquid and solid foods. Examples of strictly liquid feeders are found in the Homoptera, in general, including aphids, cicadas, leafhoppers, and scale insects. Many of the Homoptera feed on xylem sap or phloem sap, with some members feeding on cell sap. Most adult Lepidoptera (butterflies and moths) are nectar feeders. The various groups that fill the niches of vertebrate blood feeders (Siphonaptera = fleas, some Heteroptera = certain assassin bugs and bed bugs, and Diptera = flies) are true liquid feeders, although their food is a slurry of blood cells suspended in a matrix of plasma. A few Heteroptera (true bugs) feed on plant saps that are originally liquids (most notably the chinch bugs, family: Blissidae), but most of the true bugs, as well as many flies, beetles (Coleoptera), Neuroptera (e.g., lacewings and ant lions), and Hymenoptera (ants, wasps, and some bees) feed on materials that are originally solids that are converted into a liquid slurry before ingestion—solid-to-liquid feeding or EOD (Cohen 1995, 1998c). For the vast majority of the remainder of Insecta, solid-feeding via biting and chewing mouthparts (or biting and gulping) are the norm. Comparisons of representative mouthparts of insects with piercing and sucking feeding habits are shown in Figures 7.1 through 7.6, and ones with chewing habits are shown in Figure 7.7a and b showing the chewing mouthparts of a tobacco budworm larva (Lepidoptera) and a boll weevil adult and Figure 7.7c showing the boll weevil's maxillary palp with its chemical receptors. The labium of the tarnished plant bug contains the stylets, which penetrate the food source, delivering saliva that helps to chemically macerate the foods that this insect consumes as slurry.

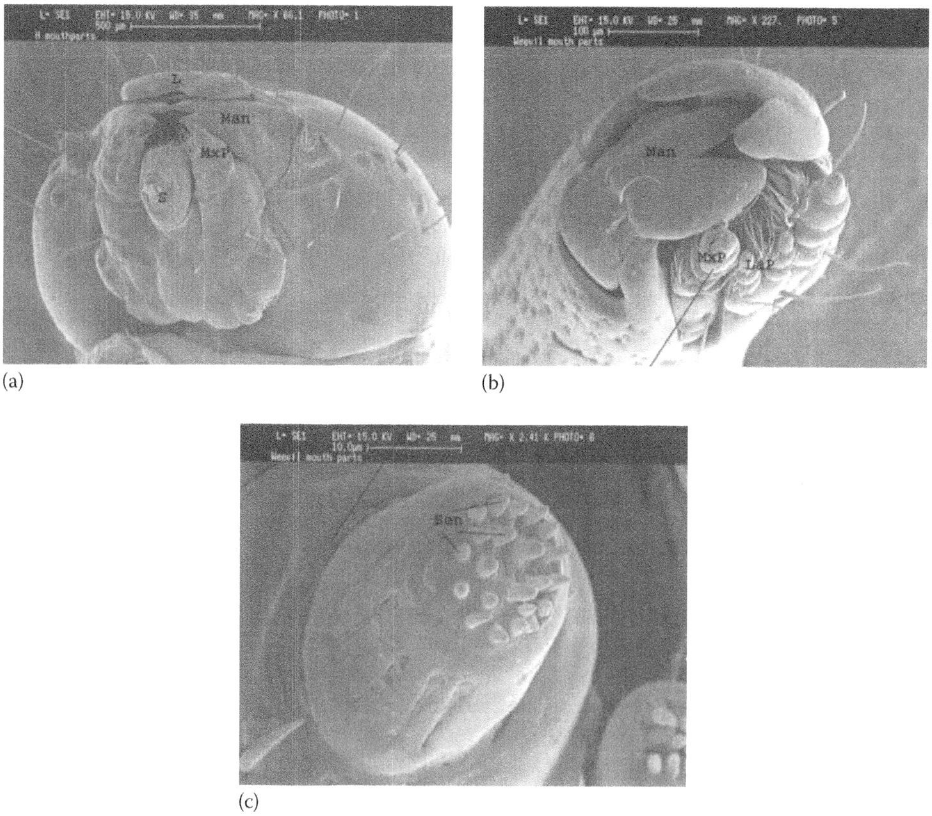

(a) (b) (c)

Figure 7.7 The type of feeding apparatus adapted to chewing habits: (a) a tobacco budworm larva (Lepidoptera); (b) a boll weevil adult. Note the prominent mandibles (Man) of these insects. (c) The boll weevil's maxillary palp (MxP) with its chemical sensillae(s).

7.3 PREINGESTION AND POSTINGESTION PROCESSING

Food processing is a major component of evolutionary success in insects as well as in other groups of animals. We humans, for example, use technology to process our foods in amazingly diverse ways. Arguably, the advent of agriculture is one step in food processing, where we have selected for cultivation the most nutritious and palatable foods and we have developed a huge cadre of techniques that guarantee availability of foods, and certainly, the genetic selection of productive varieties has been a hallmark of successful food production.

7.3.1 Insects' Food Preparation (Preingestive and Postingestive Processing)

The characteristically large and diverse insect populations are supported, in large, by the successful processing of the myriad foods used by the most diverse group of arthropods. There are the more spectacular or exotic examples of insects' habits of food preparation that include the leaf-cutting ants that cultivate microbes whose processing renders unpalatable leaves and their antinutrients into nutritionally useful forms of what was previously unpalatable or nonnutritious foods (Cherrett et al. 1987) or the ant species that tend aphids, exploiting their honeydew. Bees are legendary for their production and storage (Romoser and Stoffolano 1998) of honey and bee bread from raw materials of pollen and nutrient-dilute nectar. The process of producing honey from nectar is physiologically and biochemically very complex and very beneficial to the bees (and the other animals that exploit the bees' handiwork). No less spectacular than the bees, which produce wax, are the wax moth larvae that sustain themselves largely on the waxes that are remarkably intransigent to most animals' digestive processes. As Romoser and Stoffolano (1998) point out, termites are well known for their exploitation of symbionts that allow them to utilize nutrient-poor plant materials (wood, grasses, and other cellulose-rich and nutrient-poor substances). Clothes moths are noteworthy for their ability to utilize the keratin that is the fundamental substance in wool by a combination of production of potent digestive enzymes coupled with setting up a remarkably low oxidation-reduction potential in the gut (Gilmour 1961). The biochemical and physiological complexity characterizes the feeding adaptations of the large number of dipteran and heteropteran specialists at blood feeding. Indeed, the elaborate series of salivary strategies to ensure that blood pools in feeding sites and that it continues to flow are remarkable (Chapman and de Boer 1995). And among the most remarkably specialized feeding adaptations are those of the many homopterans that make their entire living by extracting and filtering the dilute saps of plants—especially xylem sap!

However, the main discussion in this chapter is devoted to the more common processing that insects do to condition their foods and render them into usable nutrients. It includes selection and ingestion or food components, digestion, absorption/transport of components, metabolic processing of components, elimination of nonusable components, and excretion of metabolic waste products. In the human feeding domain, with which we are most familiar, there is a preingestion processing—usually of a technological character. Then there is ingestion, digestion, absorption, transport (to the body's metabolic sites such as tissues, cells, and subcellular levels of organization = organelles), and elimination/excretion. In insect feeding processes, sometimes the order of ingestion and digestion is partially reversed. Otherwise (minus the mechanized technology), the processes are fairly parallel but with some very interesting twists.

7.3.2 Ingesting Solids: Using Chewing Mouthparts

The most common feeding strategy in Insecta is that of biting and chewing (or biting and swallowing pieces, a process called "piecemeal feeding" by Cohen 1998c). The typical mouthparts used in this type of feeding are observed in the major orders: Ephemeroptera, Phasmida, Orthoptera, Blattaria, Isoptera, Dermaptera, Plecoptera, many Neuroptera, Coleoptera, Trichoptera, Lepidoptera,

Table 7.1 Survey of the Mouthparts of Insects in the Most Prominent Orders

Order	Mouthpart Type[a]	Typical Foods	Artificial Diets?
Protura	C	Detritus	No
Collembola	C	Detritus	No
Thysanura	C	Detritus	No
Thysanoptera	S	Plants (+insects)	No
Dictyoptera	C	Detritus, etc.	Yes
Orthoptera	C	Plants	Yes
Homoptera	S	Plant saps	Yes
Heteroptera	S	Mixed	Yes
Siphonaptera	S	Blood	Yes (limited success)
Mallophaga	C	Vertebrate detritus	No
Ephemeroptera	C	Detritus	No
Plecoptera	C	Detritus in fresh water	No
Neuroptera	C + S	Insects	Yes
Coleoptera	C (+S)	Mixed	Yes
Lepidoptera	C	Plants	Yes
Diptera	S	Mixed	Yes
Hymenoptera	C	Mixed	Yes

[a] C = chewing and S = sucking.

and Hymenoptera (Table 7.1). Typically, insects with chewing mouthparts have mandibles, which are the principal biting structures. The food is manipulated by the labrum (upper lip) and the labium (lower lip) and reoriented by the maxillae. Often, chewing insects have sensory structures including labial or maxillary palps that contain external sensillae, all of which can be seen in Figures 7.6 and 7.7, showing the anterior, ventral head region of a typical lepidopteran larvae and an adult beetle.

7.3.3 Ingesting Liquids: Sucking and Lapping Mouthparts

Liquid ingestion is less common than chewing but still widely distributed. In some insects, the foods are either already liquids before ingestion has been undertaken, or they may be solids that are converted externally into liquids. The former case includes the sap-feeding insects, which ingest xylem, phloem fluids, or nectars (floral and extrafloral gland exudates, all of which are simple, aqueous solutions) and the hematophages, insects that ingest blood from vertebrate hosts. The latter feeding type includes insects that use EOD or solid-to-liquid feeding as described by Cohen (1995) and Cohen (1998c). There are important differences between the adaptations of the insects that consume foods that are already liquids (original liquid feeders) and those that convert solids to liquid (EOD feeders).

The insects that are "original liquid feeders" do little beyond concentrating nutrients to process the foods prior to their being ingested. However, this feeding process is not simple or unsophisticated. For example, the plant sap feeders use complex stylet movements and behavioral steps in the location of appropriate xylem or phloem bundles. The complexity of this feeding behavior is chronicled in Ellsbury et al. (1994). There are complex behavioral and biochemical actions involved in the movement of the stylets to the vascular bundles and the formation of the salivary sheath that covers the stylets throughout their course. Likewise, the location and selection of a capillary in a vertebrate host by a biting fly such as a mosquito or a haematophagous hemipteran involves use of anticoagulants, chemicals that stimulate localized blood flow, and a complex of behavioral actions that promote the blood-feeding process. In both plant sap-feeding and blood-feeding processes, the processing is aimed at the circumventing of the hosts' barriers or defenses against exploitation. In simple nectar feeding, there is less complex feeding action, but there is also a higher risk of failure to locate an adequately filled nectary. It must be noted and remembered that simple nectar feeding does not provide a complete nutritional package, and for complete nutrition, nectars are invariably substituted by other nutrients such as pollen, fungi, or carryovers from larval feeding. It can be noted in Figure 7.6 that the true fluid feeder (the glassy winged sharpshooter) has a pronounced clypeus, a head region anterior to the relatively small labium, which houses the remarkable bundles of muscle that make up the cibarial pump, which must create powerful negative pressures to allow this insect to ingest xylem sap, which is under a negative pressure in the plant. In contrast, the tarnished plant bug has a long labium, allowing extensive probing for nutrient-rich solids that it can liquefy with its use of extraorally delivered salivary enzymes. Its clypeus is much less pronounced, indicating that it is not taking in foods that are under negative pressure, and its food canal (Figure 7.4) has a substantial diameter of about 10 μm, compared to 1 μm diameter food canals in true fluid feeders such as whiteflies and leafhoppers (Cohen et al. 1996).

In summary for liquid feeders, liquid and slurry foods are taken into the mouth only if they pass the inspection of the sensory apparatus. The "decision" to ingest a given liquid involves a complex reflex system involving sensory neurons, interneurons, and motor pathways of the nervous system. Once the food has been selected and/or prepared by the preingestive or extraoral processes, it is taken into the mouth or buccal cavity where many insects have chemoreceptors that are further involved in the decision as to whether or not to swallow the food (Ma 1972, Schoonhoven 1972). After an allotment of liquid or slurry food has been swallowed, it goes through a process that may be similar to that of solid foods, or it may be processed in a very special manner characteristic of insects that make their living by feeding on nonslurry liquids (simple liquids). This feeding strategy is specialized for concentrating very dilute nutrient solutions. It is not suited to particulates or slurries as are the systems of blood feeders and filter feeders (e.g., aquatic, larval Diptera). The insects that feed exclusively on dilute liquids such as xylem and phloem saps have remarkably complex digestive tracts specialized to remove excess water concentration of dissolved nutrients such as amino acids, sugars, organic acids, minerals, and vitamins. Many of the dilute liquid feeders have a specialized complex in their digestive systems, known as a filter chamber (Figure 7.7a and b, for high magnification view of a filter chamber's structure). The filter chamber consists of a complex of anterior and posterior gut structures, especially Malpighian tubules and hind gut or posterior midgut, which double back and form a tightly wrapped complex that picks up water from the anterior region and shunts the water to the hindgut, where it can be excreted. Meanwhile, the concentrated liquids can be more efficiently handled in their lower volumes and higher nutrient concentrations by the midgut, where ingestive surfaces are present. The distinction between the gut structures of insect that are true dilute liquid feeders and the slurry feeders is clear from observation of the overall gut structure of the two kinds of insects, the gut of a glassy winged sharpshooter and the tarnished plant bug (Figure 7.8), where it is evident that the sharpshooter's

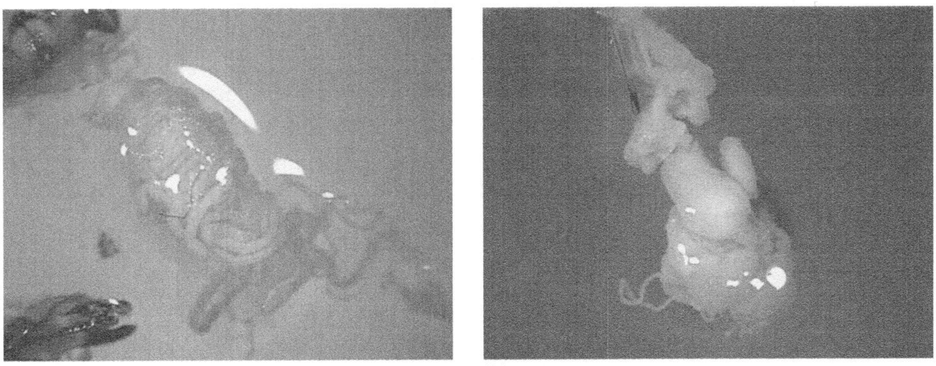

(a) (b)

Figure 7.8 Gut of a glassy winged sharpshooter (a) and of the tarnished plant bug (b).

gut is very long, highly complex, convoluted, and arranged to suit this insect' special dietary problems. Furthermore, sharpshooters, as is the case with other plant sap feeders, possess specialized bodies of cells that contain packages of microorganisms that serve as symbionts that help the host insect deal with the nutrient-poor diets that characterize this type of feeding. In contrast, the gut of the tarnished plant bug is simple and relatively short, clearly lacking a filter chamber or any liquid-filtering network. It is clear from the figures that the tarnished plant bug could not handle the throughput of dilute liquid that would be required to extract adequate amounts of nutrients from such a nutrient-poor diet such as xylem sap, phloem sap or other plant saps. Also, there is no evidence that this insect or its mirid relatives possess the microbial flora that would supplement nutrient-poor diets (Wheeler 2001).

7.4 LIQUIDS AND SOLIDS AND SLURRIES, OH MY!

It is important to distinguish between the various states of foods eaten by insects. There has been confusion throughout the history of insect feeding studies regarding the terms "solid foods" and "liquid foods." These deceptively simple-seeming terms and the various verbal modifications such as semisolid, soft solid, and semiliquid, as well as intermediate states of matter such as gels, thick versus thin slurries and suspensions all have some value in describing the state of various foods. Part of the ambiguity and difficulty stems from the fact that most foods and food components (and insect diets and diet components) are generally not simple solutions or pure, crystalline solids (such as soda pop or sugar crystals) but rather exist as complex (nonuniform), amorous matrices. To use some familiar examples from human foods, ice cubes or popsicles are crystalline solids, but meats, fruits, stems, leaves are solids, but not crystalline. Cooked oatmeal can be so loose that it readily pours from its container, or it can be so firm that it can be molded into a fairly stable shape. Orange juice can be virtually free of suspended solids or so full of pulp that it is difficult to pour. Is the orange juice in the viscous, slow pouring state a liquid, solid, gel, or slurry?

Often, we think of "true liquids" as seeking their own level and conforming to the shape of their container, but there are states of foods that are on the border of this definition, with the flow being so slow that it may take hours or a slight warming to allow them to seek their own level and to conform to the shape of their container. Gels offer special problems in interpreting liquid versus solid state, and this is very important to insect diet professionals, because many insect diets are gelled. The diffusion of solutes through a gel is often much faster than it would be in a "true solid," but not as rapid as it is through a simple, nongelled liquid.

The importance of this in the nutritional value and the stability of diets are discussed in Chapters 3 through 5. One of the most important characteristics of slurries that makes them dramatically different from true solutions is that the particulate nature of the slurries makes them potentially much richer in terms of nutrient density. For example, lipids, which will not dissolve readily in aqueous solvents, can be suspended as lipid micelles and chylomicrons—both of which can confer very high lipid density to a slurry. Sparingly soluble proteins and other nutrients can also be made to be more concentrated in suspensions than they can in solutions or even colloidal dispersions. All these issues are tied intricately with the structures and functions of the target insects' digestive system.

7.5 INSECT GUT: A STUDY IN COMPLEXITY

In my more than 30 years as a student of vertebrate and insect physiology and biochemistry, I have come to believe firmly that the digestive system of animals in general and especially in insects is the most complex and underestimated functional system. Most systems (e.g., integument, locomotion/support, neuroendocrine, reproductive, excretory, and gas exchange) have a single function or limited number of functions. The digestive system functions in defense against microbial and chemical would-be invaders, the site where digestion occurs, the complex and highly selective region of absorption of nutrients, the region of storage for many nutrients, the beginning of the transport process for nutrients, a site of water regulation and salt balance, and a principal organ of excretion. It is the first line of defense against invasive organisms and toxins, using a variety of mechanisms to detoxify antinutrients and to kill or suppress growth of microorganisms (while simultaneously serving as a refuge that harbors symbionts), and setting up barriers that filter out invasive chemicals, microbes, and mechanical threats. It is charged with the enormously complex job of digestion of foods; it must set up conditions hospitable to the selective absorption of the foods and regulating that absorption process in light of a host of defenses on the part of the organisms that are serving as foods.

To function properly, digestive systems must regulate pH and redox potential within the gut lumen. Both pH and redox potential must be at once hospitable to digestion and absorption, yet inhospitable to fostering growth of harmful microbes. The digestive system must set up the transport situation that allows the nutrients to be sent on to their appropriate targets after absorption has occurred, for example, the packaging of lipids into lipoprotein complexes, chylomicrons, or micelles that can be transported by a lipophobic hemolymph medium. The digestive system also plays a major role in excretion of wastes and elimination of indigestible food components. This system is the primary component in the regulation of salt and water balance. This same system is also modified in several taxa to produce silk or other exported products for a variety of functions from pupation arenas to glues involved in food capture. In terms of biomass, the digestive system generally exceeds that of most or all other systems, and it is often the most demanding system in terms of energy and material turnover. There is mounting evidence that the digestive system in insects and other animals has several endocrine functions (Stevens and Hume 1995). Finally, the digestive system shares with the cuticle the role of first line of defense against microbial attack.

7.6 MEAN RETENTION TIMES AND DIET COMPOSITION

Food materials are moved through the digestive system by a coordinated series of movements or contractions of the circular and longitudinal muscles that surround the gut, from the extreme anterior to posterior regions. Once the foods have passed the esophagus, they are moved posteriorly by waves of *peristalsis* and anteriorly by reverse waves of *antiperistalsis*. The coordination of both flux and reflux efforts results in a churning of the food material and the digestive enzymes (collectively making up the food bolus). This churning provides a mechanical energy to force mixing of food components and enzymes and other digestive components such as pH and redox (reduction and oxidation) control factors, and this intense mixing greatly increases the efficiency of digestion. Both peristalsis and control by valves regulate the rate of passage of foods through the digestive tract in vertebrates (Stevens and Hume 1995), and these factors are probably responsible for rates of movement in insects and other invertebrates. Stevens and Hume (1995) reviewed numerous studies that showed that in most (but not all) vertebrates, different materials passed through the digestive systems at different rates; most notably, liquids had lesser mean retention times (MRTs) than did solids, and smaller solid particles had lower MRTs than did larger particles of the same composition. There is further evidence that lipids have higher MRTs than do carbohydrates or proteins (Stevens and Hume 1995). Although the information is far less abundant for insects than for vertebrates (especially mammals), Chapman (1998) discussed the fact that different materials pass through insect (grasshopper and cockroach) digestive systems at different rates, that is, they have different MRTs, depending upon their physical state and chemical composition.

The importance of MRT cannot be overemphasized with regard to the amount of gut residence time required for digestion and absorption to occur with maximum efficiency and how much the mechanical action of gut movement contributes to digestion and absorption. This raises questions about the ability of insects to use chemically refined and defined components that (1) are in a chemical form that is simpler than what is present in the insects' natural food, (2) are in concentrations that are greater on an instantaneous basis than those of the nutrients that are "time released" or "stepwise released" by the normal digestive process, and (3) are present in solutions that remain in the gut for shorter intervals than the insoluble or sparingly soluble components. This is where the differences in the organization of highly complex natural foods and that of the simplified components of highly refined, well-defined artificial diets come into play. The question raised here is this: to what extent are the normal digestive and absorption processes affected by the churning and refluxing of foods whose lipids are hidden in recesses and whose proteins are tightly interwoven and often cross-linked with other proteins, lipids, carbohydrates? Does a certain amount of churning influence the production and release of digestive enzymes? Does the churning and mixing of food influence the hunger response so that the ingestion amounts and rates of the next meal are influenced by the dynamics of the current meal within the gut? Are gut receptors and cellular transport channels affected by the MRT or other aspects of the digestion dynamics? Does absorption of certain nutrients get bypassed or circumvented because of a low ration of gut receptors or specialized transport channels? If the last statement is answered affirmatively, does the insect miss absorption of key, limiting nutrients because of abbreviation of the MRT?

Does the "all at once" presence of simple nutrients (as opposed to macromolecular components that are released stepwise) violate the all-important rule regarding the ratio of nutrient concentrations required for efficient absorption at specific absorption sites? This last statement can be visualized by imagining a guarded gate that has an inspector checking for proper identification to determine whether each person who is trying to enter is authorized for entry. Now, if a crowd of would-be entrants appear, pressing at the gate and the gatekeeper, blocking the entrance, struggling for the gatekeeper's attention, the orderly process of permitting entry is disrupted by the bottleneck effect. If such a bottleneck occurs at the limited, specialized receptor sites in the gut, key nutrients will be eliminated without being absorbed. How much these questions may explain the inability of diets made from highly purified, simplified ingredients to support normal, robust growth of most insects remains to be answered, but it is a plausible hypothesis about why many diets fail, despite the fact that they contain all of the required minimal nutrients.

7.7 REGULATION OF DIGESTIVE FUNCTION

The many complex processes in the digestive system must be coordinated and controlled to serve within the context of the insect's other functional systems. Feeding and processing foods and their nutrients must be coordinated with growth requirements, storage, diapause, and reproduction. Along with these day-to-day operations that must be coordinated, many insects have other very complex issues that they deal with in regard to special environmental demands. Those insects that thermoregulate by producing their own body heat (e.g., bees and some moths) must have access to large amounts of fuel that is available at appropriate times, and insects that migrate face special demands of storage of flight fuels for their long distance journeys. Many insects feed intermittently and have extended periods when they are either anticipating a scarce or unreliable food source.

It is not only inefficient for insects to produce digestive enzymes during periods of nonuse, but it is also potentially harmful to the insect's internal structures, which active digestive enzymes may attack. Therefore, digestive enzymes are generally present only during the periods when food is present in the gut and during brief periods before and after the digestion is in progress. For insects that feed infrequently such as mosquitoes, the production of digestive enzymes is induced only after the meal is present in the gut (Chapman 1998). In other insects that are more or less continuous feeders (such as grasshoppers and lepidopteran larvae), the digestive enzymes are almost constantly present and active, except for periods immediately prior to and after molting.

The factors that trigger the induction or production of digestive enzymes are not entirely understood, but there are two fundamental mechanisms that are thought to regulate enzyme production: the secretagogue control mechanism and the humoral (or hormonal) mechanism. The secretagogue hypothesis holds that the presence of some food component in the gut stimulates the production of an appropriate digestive enzyme. Under this hypothesis, the presence of a blood protein such as hemoglobin in the gut of a biting fly serves as a direct stimulus to induce gut cells (mediated by specialized gut cell-surface receptors) to induce the production of trypsin, chymotrypsin, or some other appropriate proteolytic enzyme that will execute the digestion of the inducing protein (hemoglobin). In the alternative hypothesis, the sensing of the meal or the anticipation of the meal starts a neuroendocrine cascade of events that result in the release of a hormone (humor) that in turn triggers the production and release of the various digestive enzymes. The basis of control of appetite for specific foods and the digestion of those food components may well be under the control of either or both the secretagogue and the humoral systems of control (Cohen et al. 1988, Chapman 1998).

7.8 STRUCTURE AND ORGANIZATION OF INSECTS' DIGESTIVE SYSTEMS

The digestive systems of insects have several structures that are common to virtually all insect species, and built into these commonalities are countless variations on their theme. Insects generally have three major gut regions: foregut, midgut, and hindgut. The foregut and hindgut are usually lined with a chitinous or cuticular lining or *intima*. Generally, the mouth cavity is connected to the esophagus via a muscular pharynx. The pharynx is the structure that drives swallowing, and once swallowed, foods move toward the insect's posterior through the esophagus, which like the other gut structures employs circular and longitudinal muscles that are responsible for peristalsis. Although the esophagus is usually a simple tube, some insects such as sawflies have outpocketings of this structure called *diverticula*. The esophagus sometimes articulates with a crop, which is a part of the foregut, lined with a cuticular intima. Crops generally do not have a digestive or absorptive function, though there are some reservoirs where some digestive processing occurs (Chapman 1998). The predominant function of diverticula is to harbor symbiotic microorganisms that contribute to the nutritional well-being of the insects by producing substances that are essential or useful to the insects but are not present in the diet (Campbell 1990). Some digestion and absorption may also take place in the diverticula. An overview of a generalized insect digestive system is shown in Figure 7.9.

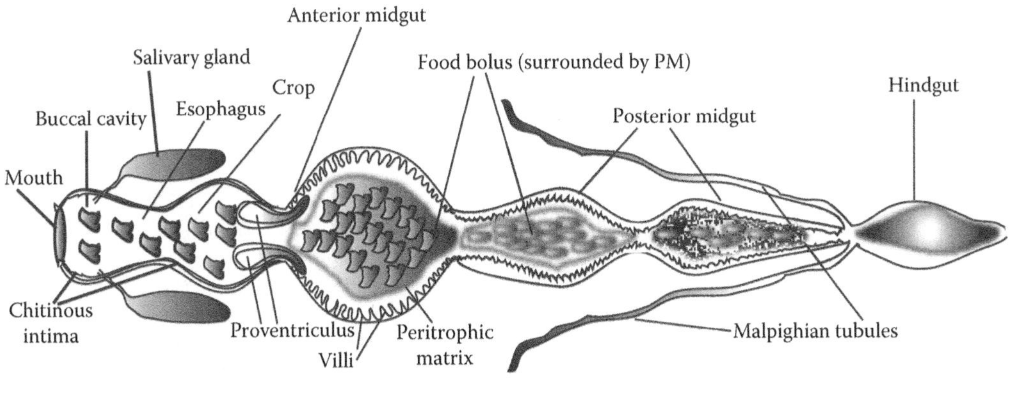

Figure 7.9 Overview of a generalized insect digestive system.

Across the class Insecta, the digestive enzymes are generally produced in the salivary glands (or modifications of these glands known as mandibular glands, maxillary glands, or hypopharyngeal glands) and the midgut (Chapman 1998). The preingestion application of digestive enzymes often involves the salivary gland secretions and is known as EOD (Cohen 1995, 1998c). It is important to realize that the salivary secretions that are mixed with food externally are recaptured and ingested along with the food materials with which they are mixed, and the digestion that results from this mixing process often continues in the gut long after ingestion has occurred (Cohen 1993, 1995, 1998c). In some respects, it appears that the contribution of the salivary glands to the overall digestive process is as important to insects that use EOD as the pancreas is to vertebrates in their digestive process. On the other hand, some insects such as the blood feeders derive no (or almost no) digestive activity from their salivary secretions (Ribeiro 1995). In these blood-feeding insects, there are defensive or other than digestive functions of the salivary secretions, and recent work (Musser et al. 2002) has shown that saliva from phytophagous insects such as *Helicoverpa zea* also serves defensive functions in counteracting plant defenses. A list of the digestive enzymes commonly found in saliva is presented in Table 7.2.

Table 7.2 Digestive Enzymes in Insect Salivary Glands and Digestive Systems

Digestive Enzyme	Substrate	Notes on Specific Enzyme
Proteases/peptidases	Proteins and peptides	May be in these classes: serine or alkaline proteases (SP), cysteine proteases (CP), metalloproteases (MP)
Trypsin	Proteins (endoprotease)	Attacks basic amino acid sites in proteins (lysine or arginine) (SP) Present in lumen
Chymotrypsin	Proteins (endopeptidase)	Attacks aromatic amino acid sites (tyrosine, phenylalanine, tryptophan) (SP) present in lumen
Elastase	Proteins (especially elastin-like proteins) (endopeptidase)	Attacks nonpolar amino acid sites (valine, leucine, isoleucine) (SP) present in lumen
Cysteine protease (thiol-protease)	Proteins (endopeptidase)	Acid proteases (present both in lumen and within gut cells)
Collagenase	Collagen (or collagen-like proteins)	Various collagen-type proteins
Aminopeptidase	Peptides (exopeptidase)	Attacks amino-terminal end of peptides (usually present at gut cell surface)
Carboxypeptidases A and B	Peptides (exopeptidase)	Attack carboxy-terminal ends of peptides (usually present at gut cell surface
Lipases	Various types of lipids	
"Lipase" (triacylglycerol lipase or triglyceride lipase)	Triglycerides	Attacks fatty acid/glycerol esters, especially those in 1- and 3-position
Phospholipases A, B, C	Phospholipids or various polar lipids	Attack fatty acid esters in polar (phosphor-) lipids
Carbohydrases	Various types of carbohydrates, including oligosaccharides and polysaccharides	
Amylases	Starches	Attack glycosidic bonds in starch, rendering the huge polysaccharides into the disaccharide maltose
Cellulases	Cellulose	Attack β-glycosidic bonds in cellulose, releasing cellobiose
Polygalacturonases	Pectin	Endo- and exopectinases attack the interior and end
α- and β-glycosidases	Maltose, sucrose, cellobiose, raffinose, trehalose	Attack the α- or β-glycosidic linkages that form di-, tri-, and oligosaccharides, releasing simple sugars
Nucleases	DNA, RNA	

Once food has been ingested and moved, via peristalsis, through and past the esophagus, it may either enter a storage organ known as a crop (in insects such as honeybees, many flies, and butterflies) where the food is stored until the midgut is prepared to receive it, or it may go directly into the anterior midgut as is the case with many beetles and most members of the Heteroptera. In certain insects that have specializations for using dilute liquids, such as many members of the Homoptera, the foods move from the esophagus directly into a filter chamber, which is a muscular structure that contains the posterior portions of the digestive system that serve as liquid concentration devices, using countercurrent exchange, active transport of key solutes, and an overall bypassing of the midgut by the watery material that is being removed from the insect so that nutrients can be more efficiently absorbed (Chapman 1998). Although the crop is not generally a major region for nutrient absorption, some insects such as cockroaches have long been known to absorb lipids or other key nutrients in this region (Chapman 1998). As food is moved either directly from the esophagus into the midgut or through the crop and then into the midgut, it must pass through a structure that is often very complex both in form and function: the proventriculus. The proventriculus (as Chapman 1998 mentions) can be a simple valve, a filter that permits a one-way flow of solids in a posterior movement but a two-way flow of liquids (e.g., in many species of ants), or it can be a grinding apparatus (Figure 7.10) containing cuticular plates or teeth (e.g., some species of beetles and many grasshopper species) (Figure 7.10).

Once the food material, now properly called a bolus, moves past the proventriculus, it is in the midgut proper (Figure 7.10). For those insects that do not use EOD (Cohen 1995, 1998c), this is where most of the chemical aspects of the digestive process take place. In most insects, the first stages of chemical digestion (a series of hydrolysis reactions, as described in Chapter 5) are aimed at reducing the complexity of and increasing access to the macromolecules and macromolecular complexes. Carbohydrases, proteases, lipases, and peptidases are the predominant enzymes in these processes. Those insects that use EOD begin these reactions outside of their bodies—within the food source itself. In the organisms that use EOD, the food has been efficiently premixed with the full complement of extraorally secreted digestive enzymes. For most of the species of insects that do not use EOD, saliva is mixed with the food in the mouth and pharynx, but in these instances,

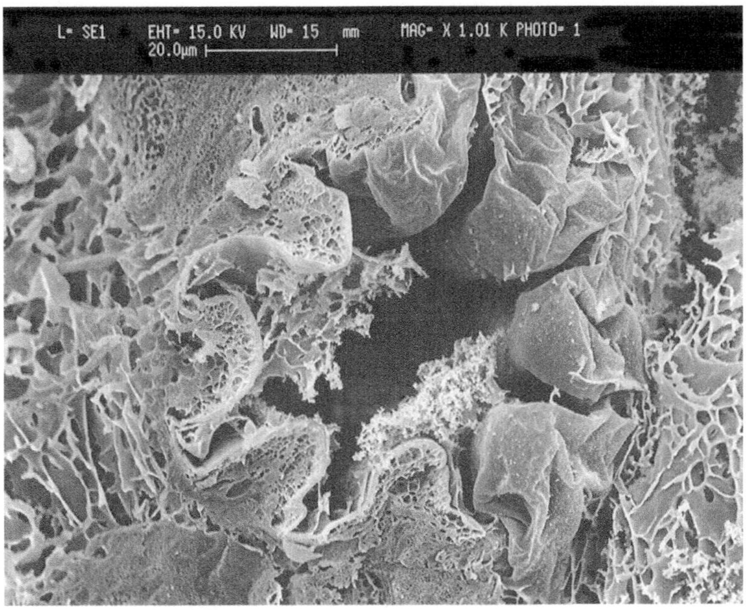

Figure 7.10 Proventriculus of the coffee berry borer (Coleoptera).

the salivary enzymes are not efficiently distributed and the mixing must take place in postingestive phases of digestion. The importance of the mixing process, achieved mainly through peristalsis, is underappreciated by many students of insect feeding.

In the midgut (which is in many insects divided into two anatomically distinct sections, the anterior midgut and the posterior midgut (Figure 7.11a through c), which show freeze-fractured sections of the midgut regions in *Lygus hesperus*, the western tarnished plant bug viewed by scanning electron microscopy), the macromolecule-digesting enzymes begin or continue their actions, reducing larger molecules and complexes of these molecules into smaller subunits. In the midgut, several other significant activities take place. The mixing and churning in this region are intensely active processes, and a great deal of mechanical reduction of the food bolus is accomplished here, as well as the chemical maceration of the targeted food stuff. The midgut cells in most species also secrete many of the digestive enzymes. Evidences of this function of midgut cells are seen in Figure 7.12 (midgut cells from *L. hesperus*), where the gut cells' rough endoplasmic reticulum (RER) is abundant along with the ribosomes (R) and mitochondria (M), which furnish the ATP for fueling the extensive synthetic activities that correspond with protein synthesis and with transport actions across the gut cell membranes. Figure 7.12 shows a portion of a cell in the posterior midgut of *L. hesperus*, where several food-containing vacuoles are evident. Also evident in these figures are the microvilli (MV), which increase the gut cells' surface area, increasing their effectiveness at absorbing nutrients. Figure 7.13 is a diagram of a receptor complex hypothesized to be present on the surface of the MV.

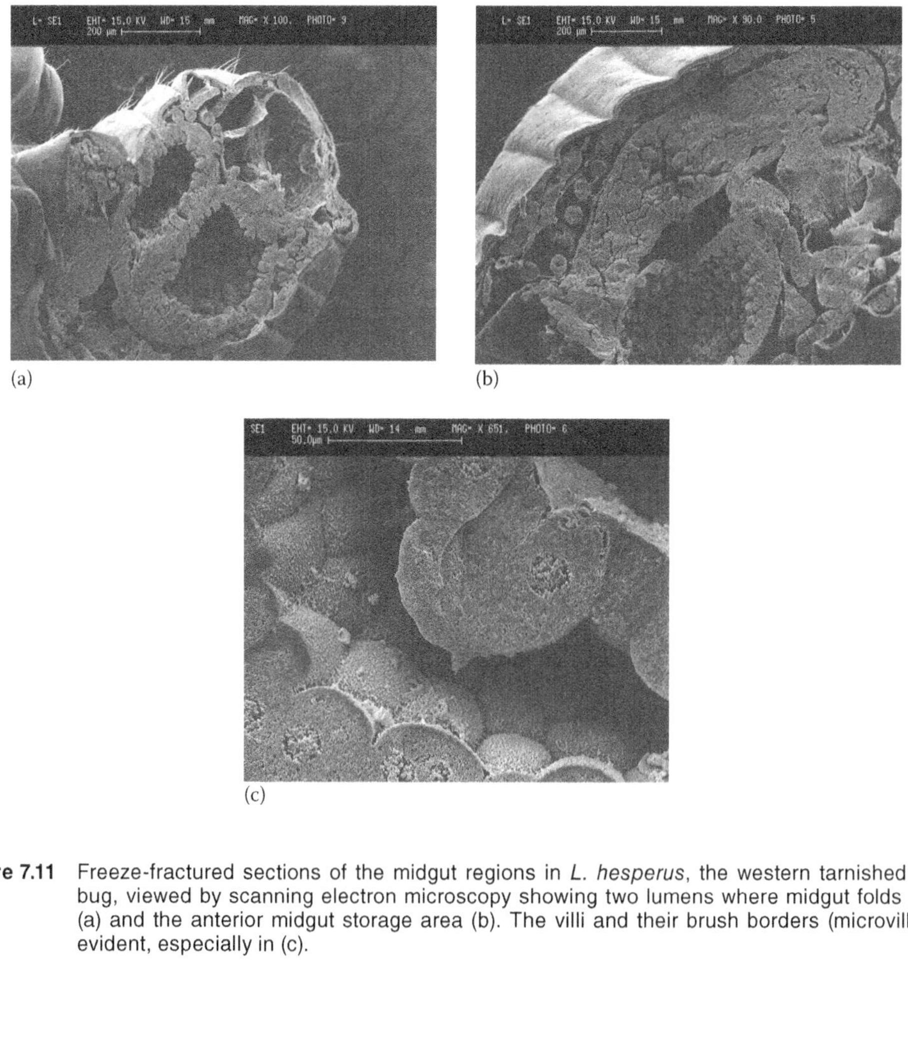

Figure 7.11 Freeze-fractured sections of the midgut regions in *L. hesperus*, the western tarnished plant bug, viewed by scanning electron microscopy showing two lumens where midgut folds occur (a) and the anterior midgut storage area (b). The villi and their brush borders (microvilli) are evident, especially in (c).

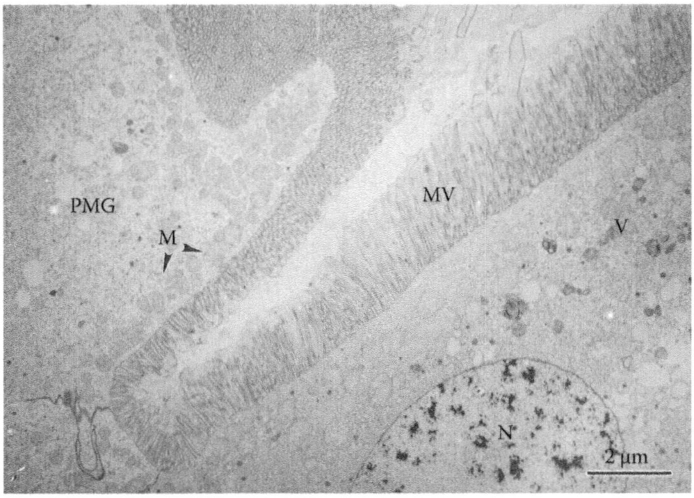

Figure 7.12 Portion of a cell in the posterior midgut (PMG) of *L. hesperus*, where several food-containing vacuoles are evident. Also evident are the microvilli (MV), mitochondria (M), vacuoles (V), and the nucleus (N).

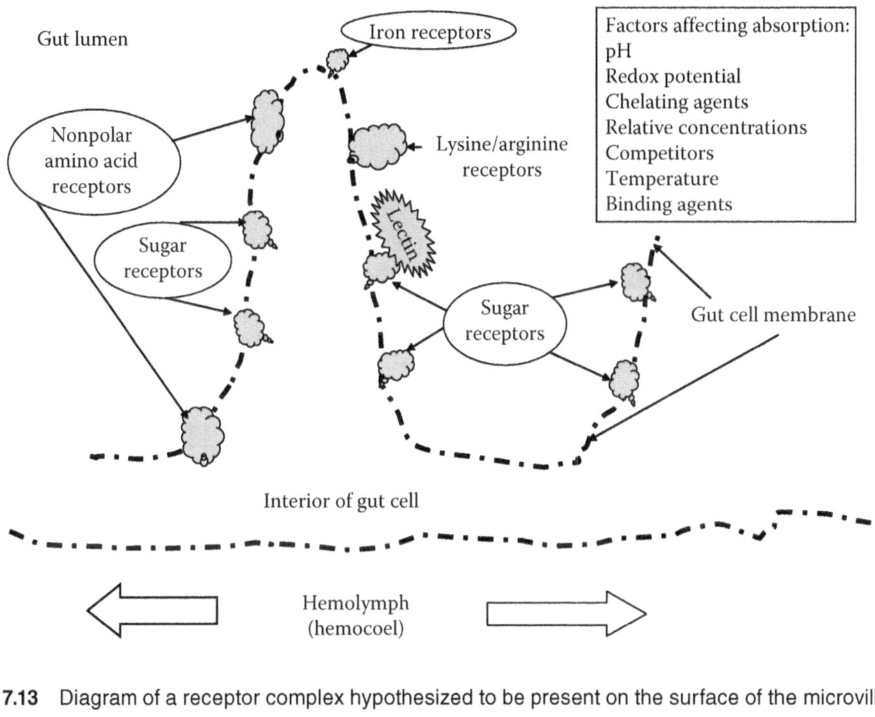

Figure 7.13 Diagram of a receptor complex hypothesized to be present on the surface of the microvilli.

Another feature of the midgut of many species of insects (including Dermaptera, Orthoptera, Blattaria, Lepidoptera, Coleoptera, Neuroptera, Plecoptera, Hymenoptera, and many Diptera) is the presence of the peritrophic matrix (PM) (Figures 7.14 and 7.15). The PM is also known as the peritrophic membrane or the peritrophic envelope. It is composed of chitin, several kinds of proteins, proteoglycans, and glycosaminoglycans. The PM has several important functions that are centered on protection and organization of digestion. Early interpretations of the PM as a protection of delicate gut cells from mechanical damage from sharp or abrasive foods, though probably correct in

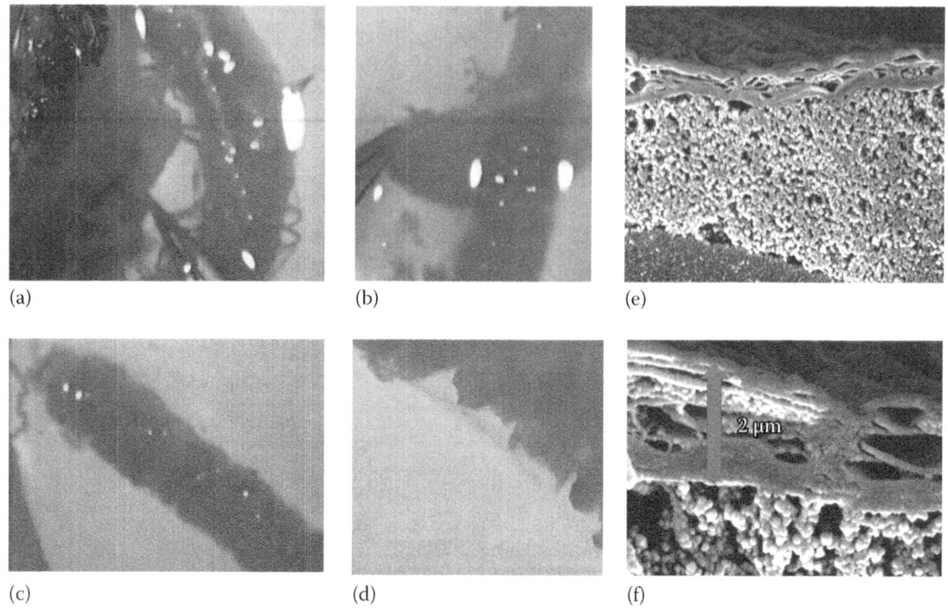

Figure 7.14 **(See color insert.)** Views of the peritrophic matrix or PM (peritrophic membrane or peritrophic jelly) of *Heliothis virescens* after feeding on plant materials. (a–d) show dissections of gut: (a) Malpighian tubule attached to gut, tracheae and cuticle being removed on the left; (b) the gut being cut and pulled away from the PM; (c) gut contents being held in place by the PM; and (d) shows the edge of the PM (at 100×) as a sharp line: note the contents of the bites of plant material. (e) and (f) show scanning electron microscope images of the PM at (f) 4,500× and (f) 14,000×. Note the multiple layers of PM, especially visible in (f).

part, are too limited to explain the full importance of this remarkable structure. More recent evidence indicates multilateral functions of the PM, including protection from microbial invaders and a large array of potentially damaging chemicals (Terra 1990, Lehane 1997, Chapman 1998). The organizational functions of the PM is especially useful in characterizing the intricacies of the digestive system and how the PM's special chemical and physical nature increases the efficiency of digestion by the formation of specialized compartments inside the PM (endoperitrophic space), outside the PM (ectoperitrophic space), and even within the PM itself (Terra 1990). For example, the pore size of the PM is a barrier to the escape of macromolecules such as larger proteins but is permeable to smaller molecules (including certain digestive enzymes such as trypsin, chymotrypsin, and elastase—all of which are small enough to pass through the PM's pores). Thus, the proteases diffuse into the endoperitrophic space where they attack food proteins, rendering them into smaller subunits (peptides) that diffuse out of the PM and into the ectoperitrophic space where they can then be further attacked by exopeptidases, such as aminopeptidases and carboxypeptidases, which reside on the surface of midgut cells, within the fluids of the ectoperitrophic space, and perhaps on the outer surface of the PM itself. This compartmentalization sets up an orderly sequence of digestion as follows:

7.9 MACROMOLECULAR COMPLEXES—MACROMOLECULES—LARGE SUBUNITS—SIMPLE MOLECULAR SUBUNITS

Once the simple molecules such as individual amino acids, free fatty acids, and simple sugars are broken free of their more complex parent materials, they can be absorbed or otherwise transported into the MV in the midgut—mainly the posterior portions of this organ. Along with the

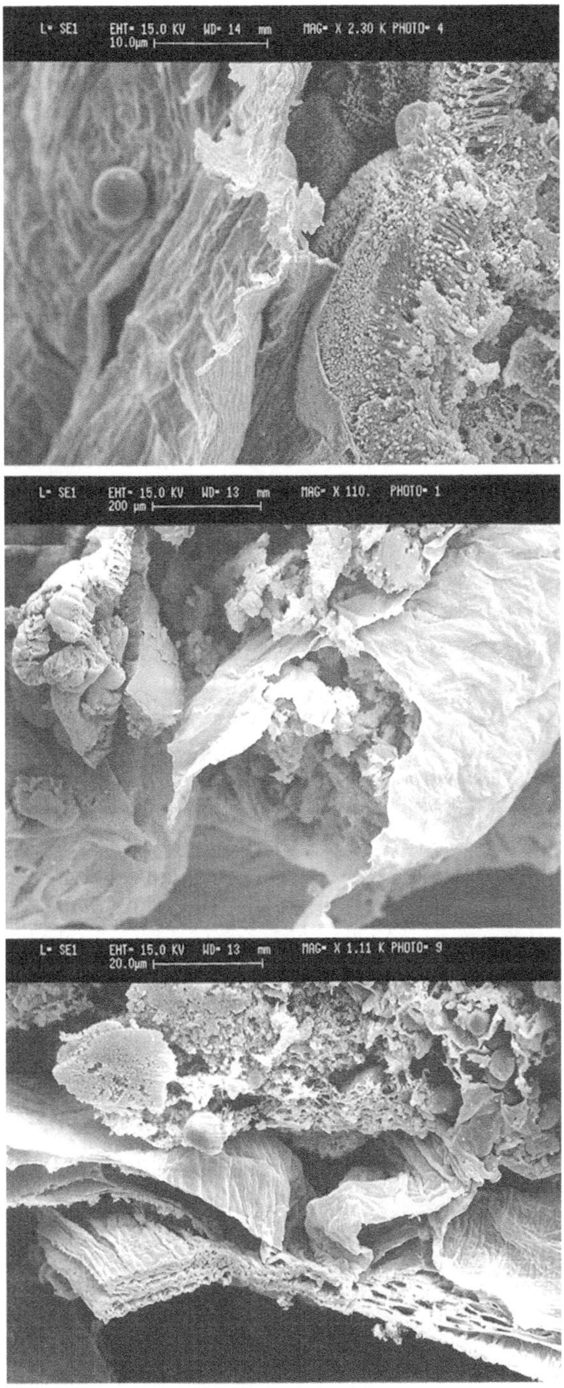

Figure 7.15 Three views of a peritrophic matrix (PM) in the gut of a *Heliothis virescens*.

products of digestion, minerals, vitamins, and various other substances present in foods (including some toxins) may be absorbed once they make contact with the appropriate sites on the surface of the MV (07–10a, 07–10b, 07–10c, and 07–11). The *decision* about whether or not a substance will be absorbed is based on several features of that substance including size, charge (positive, negative, or neutral—nonpolar), and shape. Many materials that are absorbed by the gut surface regions fit specialized receptors that are regulated by the cells upon whose surface they reside and are influenced by the milieu of the extracellular space immediately surrounding the receptor. For example, it is well documented that pH strongly influences the absorption of certain minerals such as zinc and iron, lower pH, favoring rapid absorption. The presence of ascorbic acid in the vicinity of iron or zinc greatly enhances the bioavailability (or rate of absorption) of these minerals (Gregory 2008).

Many other interactions at the surfaces of the MV are instrumental in governing rates and percentages of absorption of all diet components (which equals the bioavailability of various diet components). The presence of competitors for absorption sites can inhibit the uptake of key components. Often such competition takes place between components whose structures are relatively similar in size, shape, and charge. So calcium and iron would interfere with one another (Miller 2008) as would zinc and manganese, potassium and sodium, arginine and lysine, and probably a whole host of other substances. It is also crucial to recognize that certain naturally occurring materials such as phytic acid and oxalic acid (both present in several plant species) can sequester components such as iron and zinc, reducing their bioavailability just as other compounds (such as citric acid and ascorbic acid) can enhance the absorption kinetics of key components. Undoubtedly, the highly complicated issues of iron availability in Gypsy moth larvae (O'Dell et al. 1997, Keena et al. 1998, Willis and Allen 1999) are related to the diet and gut matrix of the iron as much as they are to whether the iron is present in the crystalline or amorphic form of ferrous phosphate. It is also worth noting that the intricate interactions of receptors on gut cell surfaces (MV surfaces) and the milieu surrounding the receptors will have profound bearing on how putative toxins that are delivered via foods will affect a target insect.

Gut lumen pH and redox potential exert major influence over digestion and absorption in the midgut. They are also influential factors in the susceptibility of the insect to microbial invaders that enter per os. Many insects, especially phytophages, have very high (basic) pH, sometimes exceeding 10.0 or 11.0 (Chapman 1998). The pH of the gut is not only influential in terms of absorption kinetics but also as a modifier of digestive enzyme activity. So for example, those insects that use trypsin as a major proteolytic enzyme must accommodate this enzyme by keeping the local conditions where the trypsin is acting at a pH above 7.0 or even higher than 9 or 10, the optima for this alkaline protease. Also, the reduction/oxidation (redox) potential greatly influences the stability of food substances in the gut (Gilmour 1961) as well as the digestive potential of many digestive enzymes (Johnson and Felton 2000). Gilmour (1961) explained that the very low redox potential (–300 mV) in the midgut of wool-eating species is largely responsible for the ability of these insects to digest otherwise indigestible keratin, the protein that composes wool and hair. Chapman (1998) describes the high metabolic cost of maintaining special pH and redox conditions, explaining that at least 10% of the metabolic energy used by some lepidopteran species is dedicated solely to this function!

Yet another aspect of midgut physiology that is often overlooked or underestimated in terms of its overriding importance in normal, efficient digestive processes is peristalsis. Although no estimates are available of the amount of energy devoted to maintaining peristalsis, this process must cost most insect species a great proportion of their energy budget. Students of gut physiology are often surprised at the high degree of peristaltic activity in freshly sacrificed insects. The rapid and vigorous gut movements are important beyond moving food through the digestive system. It is an oversimplified interpretation of peristalsis to consider this dynamic process as being limited to propulsion of food to the posterior portions of the insect's gut. The active churning that can be observed, especially in the anterior midgut, must be taken to function as a means of mixing digestive enzymes into hard-to-reach portions of the foods that are being targeted for digestion. Also the intense mixing and the hydrostatic forces must help energize the movements of nutrient subunits, enhancing their absorption. For the many species of insects that possess a PM, the pressures and movements generated by peristaltic actions may help to force digestive enzymes into the PM lumen and digested products out of the PM.

Finally, once the food materials are converted from macromolecular complexes into simple organic subunits and free minerals, the insect can absorb the nutrients, either by passive means (osmosis for water, passive diffusion of lipophilic substances, or facilitated transport across specialized receptors) or by active (energy-requiring) means, sometimes known as active transport. Many of the nutrients are taken up into gut cells via specialized absorption sites that are known as receptors.

Once the insect has fully processed the food bolus and extracted all the nutrients that it can derive from the food, the nondigestible materials are passed into the hindgut. The hindgut is also important as the major site of water resorption and salt regulation in most insect species. At the juncture of the posterior midgut and the anterior hindgut, the Malpighian tubules empty their contents, which they have absorbed from the hemocoel or directly from other organs and tissues (e.g., water from the midgut in species that must remove copious amounts of water such as xylem sap, phloem sap, and blood feeders). The Malpighian tubules also carry into the hindgut nitrogenous wastes, excess minerals, and various toxins (usually ones that were exogenously derived such as metabolites from foods). The urine that is deposited into the hindgut from the Malpighian tubules is added to the undigested food material to form a pre-egested material that is finally processed for water removal, in the case of insects that tend to go into negative water balance (many terrestrial species) or for water addition, in the case of insects that must remove excess water in order to achieve water balance. Examples of the latter category are the sap feeders. Finally the end product, the final egestate, is expelled from the posterior hindgut through the anus.

In summary, the total process of food handling is a highly complex and well-coordinated interplay between various insect systems and various organs. In the overall treatment of food, insects of various kinds perform processing steps that render nutrients as usable to the insect, often having to overcome antinutrients and concentration differentials that render many foods as what would be expected to be suitable sources of nutrition. The preoral (extraoral) processing in some insects does an incredible amount of processing and selection of nutrients against a background of toxins and concentration issues, much of the process being accomplished by application of potent and diverse digestive enzymes (Cohen 1995, 1998c). The postingestive processing of foods by other insects is no less remarkable, with the PM, diverticula (containing sometimes most exotic and fastidious microbes), and liquid-filtering systems for concentrating what would otherwise be completely useless saps into suitable levels of nutrition. Truly, insect feeding systems must be considered among the most amazing entities in the biological world: the proof of this statement is to be found in the long-standing evolutionary and ecological success of insects!

7.10 METABOLIC LOGIC: WHAT HAPPENS TO FOOD COMPONENTS AFTER INSECTS CONSUME THEM?

After ingested foods are digested to the point that macromolecules have been reduced to their simple subunits, they are subject to absorption. Absorption takes place in the specialized gut regions known as MV, and depending upon the chemical nature of the substances to be absorbed. Lipids are generally able to cross the cell membrane because of their nonpolar nature and the fact that the cell membrane is predominantly lipid. This follows a general axiom, "like dissolves like." As for the more polar substances such as sugars, amino acids, and most vitamins and minerals, crossing the cell membrane barrier is a matter of transport via a specialized receptor discussed in the following text and elsewhere in this chapter.

7.10.1 Transport of Materials after Absorption

Once food materials have been digested and absorbed, that is, after they have crossed the insect gut, they are transported by the hemolymph (the insect blood system) to target tissues. In contrast with vertebrates, which have a closed circulatory system, insects have open circulatory systems. The insect's hemolymph system, being open, directly bathes the tissues, rather than being confined in blood vessels. The hemolymph is moved by a dorsal blood vessel (a heart) throughout the body cavity (hemocoel) with some help from movements of other body parts such as legs, wings, and gut movements. The cells of structures such as gonads, fat body, Malpighian tubules, and the gut in Figures 7.12 and 7.16 can be seen to have a conspicuous nucleus, and cytoplasm. It is noteworthy that the close proximity of many of the organs (Figure 7.11a through c) suggests that nutrient uptake from the gut may be direct from gut to target tissue, rather than solely via hemolymph mediation. Many components such as sugars, amino acids, most minerals, and most vitamins can be carried readily by the hemolymph by virtue of the fact that they are water soluble and are therefore very compatible with the hemolymph, which is primarily an aqueous matrix. In contrast, the lipid-soluble components such as sterols, fatty acids, and carotenoids must be treated with another strategy carried out by the insect—this process involves the use of carrier molecules (Shapiro et al. 1988), especially *lipoproteins*, lipid *micelles*, and *chylomicrons*.

These lipid-bearing components and the lipids carried by the lipophorins are carried from the gut to the fat body, muscles, neurons, reproductive tissues, glands, the cellular portions of the cuticle, and the Malpighian tubules. The cells in the tissues of the organ systems mentioned have, on their surfaces, specific receptors that recognize and bind the various nutrients, which are then taken into the cells. Once taken into the appropriate cells, the nutrients can be shuttled to the sites where they will participate in the processes that the cells are specialized to carry out.

Once the various nutrients reach their target cells where they are to enter the metabolic pool, they can be absorbed or transported through the membrane of the appropriate cell. The transport

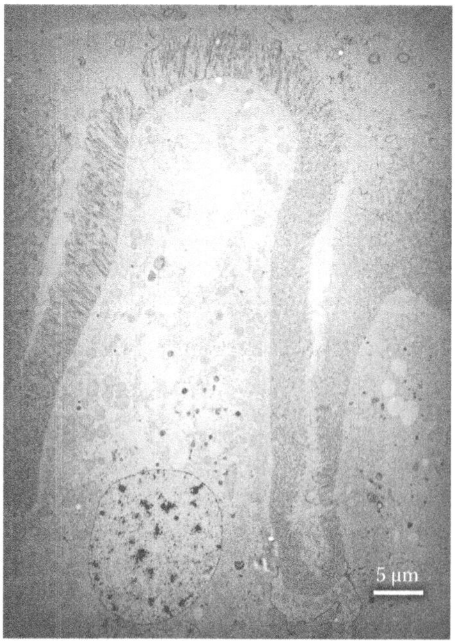

Figure 7.16 The function of the cell is contained in the blueprints specified by the strands of DNA that constitute the genes that make up chromosomes in the nucleus.

across the membrane may be simple osmosis (in the case of water) or diffusion, which is gradient-driven. It may be a receptor-mediated diffusion (such as glucose absorption). Or it may be active transport, where the cells must utilize a special set of receptors and mechanisms of ATP-driven active transport. Ion pumps (which maintain membrane homeostasis, osmotic gradients, and which drive excitable tissues such as neurons and muscle actions). Regulation of pH and sodium/potassium pumps are common examples of this kind of active transport system.

The various kinds of nutrients will be accounted for here with a brief explanation of the typical processes that these nutrients undergo. The nutrients that will be tracked include amino acids, lipids, carbohydrates, vitamins, and minerals. Also, a brief accounting will be made for various toxins that are also ingested.

7.10.2 Getting inside of Cells of Target Tissues

Cell membranes are mosaics of bilipid layers interspersed with proteins and glycoproteins. Because lipids are soluble in these membranes, they can readily penetrate cell membranes, and water molecules can penetrate readily because of their small size. Water molecules move, by *osmosis*, in and out of cells, whose membranes serve as semipermeable gateways. However, most other nutrients and raw materials can get into cells only by special gateways. Some substances such as glucose (or other substances such as amino acids) enter cells by a *facilitated transport* mechanism, where a specialized glucose (or amino acid)-recognizing protein serves as a channel of glucose passage (or amino acid). Facilitated transport depends on a gradient or higher concentration of the substance in question outside of the cell than there is inside of the cell. Yet another way that substances may enter the cell is through *active transport*, where cellular (ATP) energy is expended to import substances. The most thoroughly understood active transport system is the sodium/potassium pump. Another important cell surface phenomenon is cell surface receptor–mediated activity. This is the process where some substance such as an insulin molecule docks at an insulin recognition site, triggering a cascade of events inside the cell and pertaining to sugar regulation. Recent research has been revealing instance after instance of cellular activities that are regulated by surface receptors and their interactions with specialized triggering molecules. This is a domain where the complex communication systems inherent in glycoproteins are being implicated as the mechanistic basis of the cell surface receptor complex. The unique combinations and arrangements of sugars are what make possible the high degree of specificity that characterizes these sites.

7.10.3 What Happens inside Cells?

The complex logic of cellular operations can be conveniently simplified with a few generalizations. The function of the cell is contained in the blueprints specified by the strands of DNA that constitute the genes that make up chromosomes in the nucleus (Figures 7.12 and 7.16). When appropriate signals are sent to the genetic material that lies organized within the cell's nucleus (in the context of the cell type and the messages from the cell surface), the appropriate string of DNA (gene complex) becomes activated and along with the enzymes that guide the process, the DNA unravels and becomes *translated* into a sequence of messenger RNA (mRNA), which is then transported to the cytoplasm. Here the mRNA finds an appropriate element within the complex of the RER, where the ribosomes *read* the message carried by the mRNA as to the prescribed sequence of protein that was encoded into the original strand of DNA that was the gene that carried the template for the protein's specific sequence of amino acids. In summation, (1) the sequence of amino acids confers the fundamental and specific nature of the protein made of that sequence and (2) the proteins, especially enzymes, are the genes' way of expressing themselves as what we often call *traits*.

Once a cell has produced the series of specific proteins that its genes have called for, that cell has special properties (structures and functions) that confer upon its special functions. Further, the nature of cells is the basis for the nature of the tissues that the cells make up, and the nature of organs (fat body, muscles, nerves, testes, ovaries, etc.) is based on the types of tissues making up the organ. At the next higher level of organization, the organs work in a unified way to serve a specific purpose (or set of purposes, e.g., reproduction, movement, digestion, endocrine activities), and the organ systems work in a coordinated way (known as homeostasis or steady-state regulation) to make the whole insect function adaptively as a living organism.

The sum total of all the biochemical processes that organisms perform is what we call their metabolism. The synthetic (anabolic) reactions and the degradative (catabolic) reactions are the two major facets of metabolism. Also, the cell's environment stimulates the transfer of information from DNA to messenger RNA (mRNA), the mRNA leaves the nucleus and moves into the RER, where ribosomes *read* the mRNA message and translate the order of the message from RNA base sequence to protein structure. Large concentrations of ribosomes, associated with RER, are evident in the salivary glands of *L. hesperus* (Figure 7.17). The ribosomes have the catalytic ability to bring appropriate amino acids (carried on special transfer RNA = tRNA molecules). Once the proteins are synthesized (i.e., the message is translated from nucleic acid information to protein information) in their preliminary form, they are further processed in the Golgi apparatus (a process called posttranslational processing). This processing often involves the addition of carbohydrate groups (glycosylation), which helps mark the protein to guide the cell to send the protein to its site of activity. The fate of the protein depends upon its structure and the function of the cell in which it is being produced. It may be retained in the cell as an enzyme, a membrane component, a structural component, replacement for other such components, or exported from the cell to serve one of these or some other function. So for example, if the cell is a fat body cell whose function is to produce vitellin, the imported amino acids from the diet, after they had moved through the hemolymph and were transported into the RER, the amino acids may then be synthesized into the vitellin via the joining of amino acids via peptide bonds. Once the vitellin has been synthesized, it may be posttranslation-processed by addition of sugars and lipids. The now completed protein can be stored in the fat body cell's vacuoles, awaiting export into the hemolymph, which will carry it to the ovarioles, where the vitellin will be absorbed and deposited into the forming eggs.

Nutrient amino acids are transported to a region in the cytoplasm called endoplasmic reticulum where ribosomes are involved in protein synthesis. At the endoplasmic reticulum, through the genetic instructions that had been transferred from DNA to messenger RNA, the amino acids are fit into a synthesis of specific proteins that the cell involved in this process is specialized to produce. So, for example, if the cell is an endocrine gland cell or a neurosecretory cell that is specialized at synthesizing neuropeptide hormones such as prothoracicotropic hormone (PTTH), eclosion hormone (EH), or ecdysis-triggering hormone (ETH), the amino acids will be synthesized into the appropriate sequence that results in the formation of the specific peptide. That peptide can then be secreted from the neurosecretory cell as that cell performs its function.

As an example, when the brain of a locust senses that the insect requires fat mobilization, it signals the corpora cardiaca to secrete into the hemolymph adipokinetic hormone. This hormone with a simple amino acid sequence (pyroglutamic acid—leucine—asparagine—phenylalanine—threonine—proline—asparagine—tryptophan—glycine—threonine) is transported to the receptors in fat body cells that, in turn, triggers the fat body cell to activate its previously inactive lipase molecules (Chapman 1998). Once the lipase is activated, the appropriate amounts of free fatty acids are released from the fat body cells. This is a clear and relatively simple example of how the amino acids that had originated from the food source in some previous meal became incorporated into the metabolic pathways in the insect's body. In this case, the amino acids that had been taken up by corpora cardiaca cells were traced in their role as components of the peptide hormone, adipokinetic hormone.

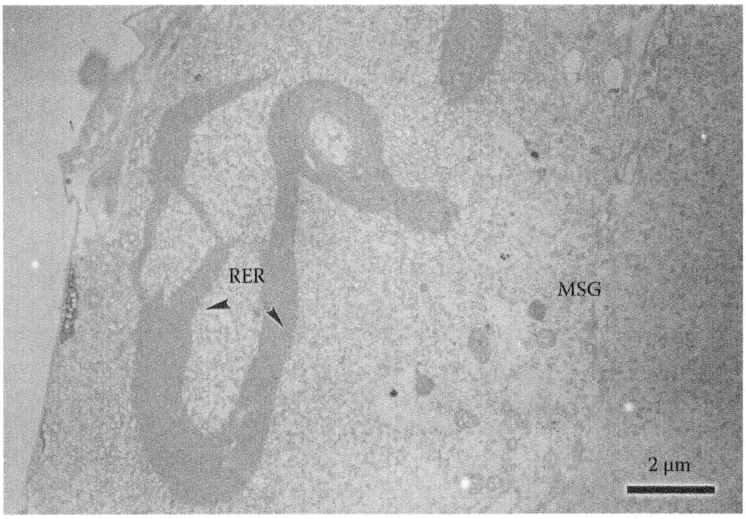

(a)

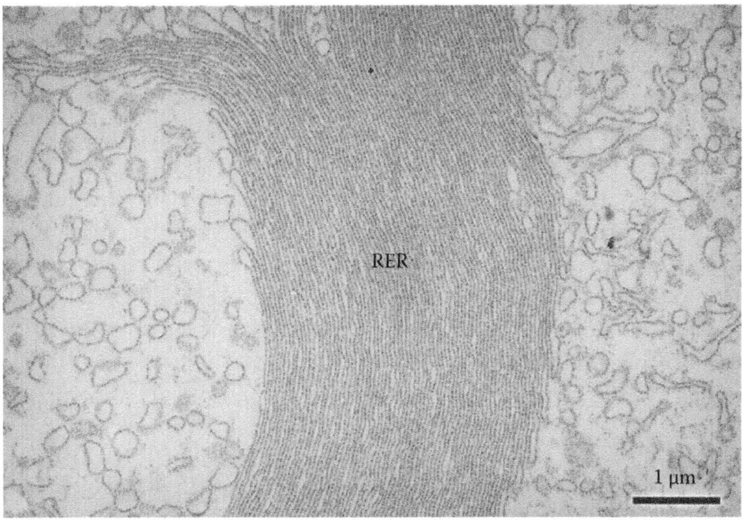

(b)

Figure 7.17 Large concentrations of ribosomes associated with RER are evident in the salivary glands of *L. hesperus.*

Likewise, fat body cells can just as well take up the amino acids in question for the synthesis of appropriate peptides or proteins such as the insect's vitellin, insulin-like hormone, or any of the other myriad proteins manufactured by these cells.

Because they are water soluble, the products of protein and carbohydrate digestion, free amino acids and simple sugars, can be carried directly by the aqueous phase of the hemolymph. The lipids and lipid-soluble factors (such as carotenes and tocopherols), on the other hand, require transport by carrier proteins known as lipophorins. The lipophorins act as a shuttle system, carrying lipids to target body cells from the gut cells where the nutrient lipids had been originally absorbed from the gut lumen. Lipids are not the only components that require special transport mechanisms. Certain minerals such as iron (Nichol et al. 2002) require special transporters or carriers (transferrins). After the nutrients that were absorbed by the gut are transported to the target cells, the nutrients must then gain entry into the appropriate cells. The target cells use a variety of means to admit the nutrients, passive diffusion, active transport, receptor-mediated passage, and *endocytosis* (the formation of invaginations into the cell membrane where masses of liquid or solids can be taken into the cells, forming a *vacuole*).

Once the nutrients have been taken up by the target cells throughout the insect's body, they can be put to work by the metabolic machinery inherent in every cell, with each cell type possessing its own characteristic metabolic pathways. These pathways are governed by the genetic makeup of each cell (which is literally a clone of the original parent cell—usually a fertilized egg that had progressed through the various developmental processes (e.g., embryogenesis, growth, differentiation, postembryonic development). Thus, the metabolic pathways are the cells' way of expressing their and their parent insect's biological fate, and the nutrients that the insect had ingested as part and parcel of its food are the raw materials that feed the process of metabolism. Now it can be fully understood how the sources of energy, materials, and cofactors interrelate within this metabolically logical framework. It can be further understood how the interplay between essential nutrients, beneficial but nonessential nutrients, antinutrients, and waste products forms the greater scheme of function to serve the overall biology of our target insect.

7.11 FEEDING SPECIALIZATION AND GROSS ANATOMY OF INSECT DIGESTIVE SYSTEMS

It is well recognized that much of the success in numbers and diversity of Insecta is related to remarkable features of their feeding system, where there is an interplay between mouthparts structures, salivary glands, and gut structure. The diversity and trophic success are evident in the coordination of structure, function, and behavior. While it is not within the scope of this book on insect diets to treat comprehensively gut anatomy and physiology, there are some broad generalizations that can help diet specialists better grasp how their target insect's feeding requirements may fit the diets they intend their insects to use. Some portions of this topic have been discussed in earlier parts of this chapter, as well as in Chapters 4 and 13. However, here, the focus is on an overall perspective about how insect guts can tell a story—provided we are willing to *listen*.

The evolutionary primitive structure of insects' guts is a simple tube. This is seen in embryos of all insects and in postembryonic stages of primitive insects (Wigglesworth 1972). The simple tube arrangement is also present in insects that feed on complex, nutrient-rich diets. The more derived characteristics of gut structure in terms of gross anatomy have evolved toward complexity (Figures 7.18a–f and 7.19), characterized by twists or convolutions, chambers with convoluted tubes, and outpocketing or evagination of gut walls in structures known as diverticula or diverticular caeca (Sections 7.8 and 7.9, and Chapter 13). Again at the level of gross anatomical organization, the convolutions are associated with (1) an increase in surface area allowing increased food residence time and opportunity for digestion and absorption (selective or due to mass action) and (2) an opportunity for countercurrent exchange to further increase the efficiency of concentration of nutrients. The range of gut structures from an ultrasimple gut in the form of a blind sac (in immature stages of many parasitoids) to the longest, most convoluted systems in xylem sap feeders is illustrated in Figure 7.19. This figure shows a model of the continuum between gut structures adapted to feed on the most nutritionally concentrated foods through the most dilute liquids. It is also intended to reinforce the gestalt concept of complexity of foods being inversely correlated with the complexity of digestive systems. This gestalt tells us that when we see a short, simple gut, we should not expect the insect to feed on simple, dilute diets. The idea that an insect with a blind sac or simple tube could "make a living" on a dilute material such as xylem sap, phloem sap, or

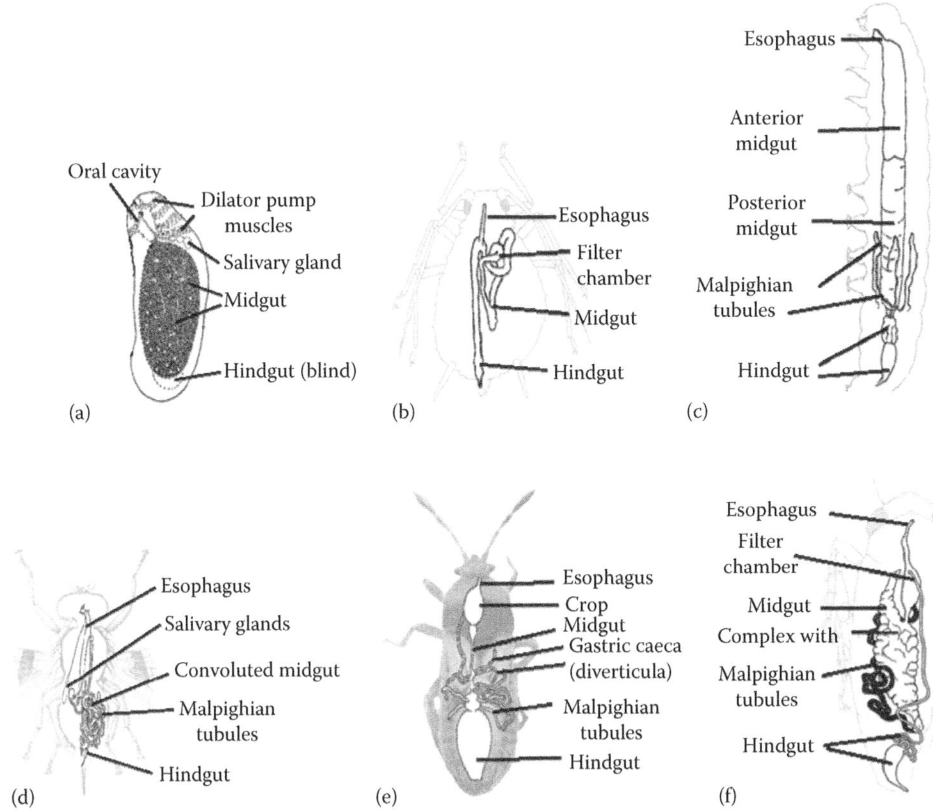

Figure 7.18 Simplified representations of feeding systems. (a) A trichogrammatid wasp larva with its blind sac structure. (From Wu, Z.X. et al., *Entomol. Exp. Appl.*, 96, 1, 2000.) (b) A lepidopteran gut representing a simple tube-type gut. (From Snodgrass, R. E., *Principles of Insect Morphology*, McGraw-Hill Book Company, New York, 1935.) (c) An aphid gut with moderate convolutions and countercurrent exchange structure. (d) A tephritid fruit fly adult's gut with moderate convolutions and intertwined Malpighian tubules adapted for fluid removal or nutrient concentration. (From Snodgrass, R. E., *Principles of Insect Morphology*, McGraw-Hill Book Company, New York, 1935.) (e) A lygaeid hemipteran adult with complex caecal diverticula, which harbor microbiota, especially bacteria. (From Snodgrass, R. E., *Principles of Insect Morphology*, McGraw-Hill Book Company, New York, 1935.) (f) A cicada gut with extensive convolutions and countercurrent systems, including a filter chamber that houses folded Malpighian tubules that are adapted for rapid removal of copious fluids. (From Hickernell, L.M., *Biol. Bull.*, 45(4), 213, 1923.) (Figures redrawn by Ms. Jackie Cohen.)

even insect hemolymph should stretch credulity and generate critical questions. Yet examples of misinterpretations of feeding biology are far too common. This point is made in the turning point in the work (Cohen 1981, 1985b) on diets for the predaceous geocorid, *Geocoris punctipes*, which was thought to consume hemolymph as its sole food source from its varied prey. After several years of failure to produce robust growth and reproduction on liquid diets that were painstakingly modeled after hemolymph from a prey insects such as *Spodoptera exigua* and *Heliothis* and *Helicoverpa* spp., Cohen had discovered through careful observation, including extensive video microscopy of the feeding process in *G. punctipes* and several other hemipteran predators, that these insects fed on virtually all tissues (fat body, muscles, reproductive organs, digestive systems) of their prey, rather than restricting themselves to sucking hemolymph from their prey and abandoning the remaining tissues. Several time lapse photos of this process taken from older videos are shown in Figure 15.2e.

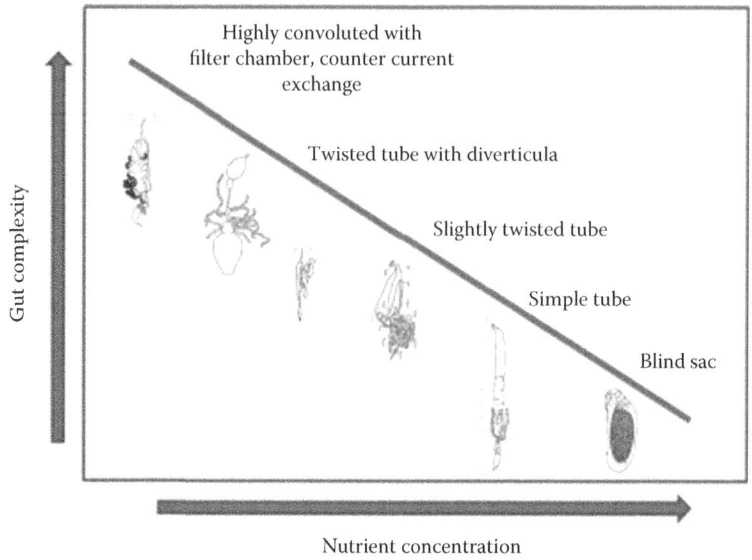

Figure 7.19 A model that summarizes the inverse relationship between nutrient concentration and gut complexity.

7.11.1 Insects with Digestive Systems That Are Blind Sacs

The larval *Trichogramma* shown in Figure 7.18a feeds on host egg yolk, which is one of the most nutrient-rich foods in nature, with high concentrations of protein, lipids, and carbohydrate and all the vitamins and minerals required to form an insect's biomass. It must be remembered that the egg contains (with the exception of oxygen) all the nutrients required for the developing embryo to complete its development (this is especially the case with cleidoic or closed egg systems of most insect eggs). Wu et al. had shown that the *Trichogramma* larvae use EOD to render egg yolk materials into a concentrated slurry, and the ingested food is readily converted into the larval wasp's biomass with little or no removal of water or need to concentrate ingested materials: this is one of the most simple possible feeding systems in terms of low requirements for processing foods to concentrate and convert ingested materials to consumer (*Trichogramma*) biomass. The occurrence of blind sac-type digestive systems is somewhat common, with the Neuroptera larvae having such systems and many parasitic Hymenoptera and Diptera sharing this characteristic. In all these cases, the natural diets of the insects in question are very concentrated, nutritionally rich materials that require little or no concentration to remove water or other diluting materials such as nondigestible fiber. It is also noteworthy that these insects have very high values for the nutrient indices approximate digestibility (AD), efficiency of conversion of ingested food to insect biomass (ECI), efficiency of conversion of digested food to biomass (ECD) as noted by Slansky and Scriber (1981), Cohen (1985b), and Cohen and Urias (1988) when compared to insects with more dilute nutrient-containing diets.

7.11.2 Insects with Digestive Systems That Are Simple Tubes

Many insects that have simple tube-type digestive systems feed on nutrient-rich foods such as other insects, seeds, and nuts. A substantial number of insects with short, simple digestive systems use EOD as a strategy that allows them to select concentrated, highly nutritious parts of their food source (Cohen 1995, 1998c). Such insects include predators from Neuroptera, Coleoptera, Hemiptera,

Diptera, and a few other orders. And for the Lepidoptera that feed on nutrient-rich materials such as seeds or developing fruit, the simple tube meets their needs without exceptional throughput. However, many Lepidoptera larvae (Figure 7.18b) feed on plant materials (leaves, stems, and roots), which are less nutrient rich than the egg yolk and yet more nutrient dense than xylem sap or even phloem sap. Some Lepidoptera larvae specialized in plant materials with dilute nutrients feed almost constantly, using a strategy of maximizing the window of opportunity for nutrient uptake by passing the greatest possible plant biomass through their digestive system. These insects along with many beetles have a very low efficiency of conversion of food to insect biomass (Waldbauer 1968 and Chapter 9), so in a sense, leaf-eating, stem-eating, and root-eating herbivores have a similar problem to that of fluid-feeders that consume dilute liquids. But instead of having lengthened and convoluted digestive systems, they simply force a large biomass through their guts.

7.11.3 Insects with Moderately Convoluted Structure and/or Gastric Cecal Diverticula

For other herbivores, including scarabaeid beetle larvae (such as Japanese beetles, *Popillia japonica*), an otherwise simple gut has three sets of gastric caeca extending from anterior, middle, and posterior midgut or ventriculus (Snodgrass 1935) and harboring microbiota that help the beetles process foods that are poorly nutritious and that contain plant secondary compounds that the microbes help to detoxify. The nature of the relationship between insects and their gut microbes is discussed in more detail in Chapter 13. The moderately convoluted guts can be found widely among diverse taxa such as Diptera, Hemiptera, and Coleoptera as seen, for example, in a representative dipteran, *Rhagoletis pomonella* (Figure 7.18d), and a hemipteran, a chinch bug *Blissus leucopterus* Figure 7.18e. Both of these insects (the tephritid fruit fly adult and the lygaeid chinch bug) feed on plant materials, and in both cases, their feeding biology is specialized for using fairly dilute fluids. This is reinforced in the huge community of mass rearing of tephritid fruit flies where billions of these insects are produced on liquid artificial diets. The tephritid fruit fly gut is especially complex midgut (ventriculus) region that can be seen in Figure 7.18d to make multiple twists and turns and to be interwoven with the Malpighian tubules, which suggests strong countercurrent exchange, which is adaptive for rapid removal of excess water. In this sense, the fruit fly gut is similar to the homopteran guts of aphids (Figure 7.18c) and the cicadas (Figure 7.18f).

The digestive systems of aphids are suited to utilize phloem sap, which is more nutrient rich than xylem sap (used by cicadas and many leafhoppers (Figure 7.8a)). The phloem sap of plants is the source of products of photosynthesis and the other nutrients (e.g., amino acids and minerals) that are translocated to the cells of higher plants for growth and storage. With this function, phloem sap can contain very large concentrations of sugars, especially during peak photosynthetic times in bright sunlight. In fact, the sugar concentrations in phloem sap are so great that along with water, excess sugars must be discarded by aphids, whiteflies, and other phloem-feeding insects, and the resulting exudate known as honeydew is characterized by its sticky character. This point is discussed earlier, but in this context of comparative gut structure, it can be seen in the aphid gut Figure 7.18c that the countercurrent exchange helps to remove the excess water along with the sugars, excess amino acids, and minerals that were present in phloem sap. In contrast to phloem sap, xylem sap is even more dilute and contains very few organic compounds and minerals.

7.11.4 Insects with Highly Convoluted and Complex Digestive Systems

The gut of the cicada pictured in Figure 7.18f is not only highly convoluted, but it also contains a filter chamber where Malpighian tubules are tightly wrapped and structured to allow countercurrent exchange for removal of water. This same type of structure is to be found in the leafhopper pictured in Figure 7.8a with extreme convolutions and part of the Malpighian tubule network in the anterior filter chamber. The head of this leafhopper *Homalodisca vitripennis* (formerly *H. coagulata*, Hemiptera: Cicadellidae) is seen in Figure 7.6 where the very pronounced clypeus is evident as a housing for the very strong muscles that power the sucking pump required to ingest tremendous quantities of xylem sap, which is under a negative pressure, which must be overcome. Exudate from *H. vitripennis* was shown to be produced at a rate of more than 100 times the biomass of the insect each day—a testimony to the remarkable concentration that is required for this insect to meet its demands for biomass accumulation (Cohen unpublished data).

7.11.5 Scope of Extraoral Digestion and Extraoral Processes

Although the topic of EOD has long been recognized in a wide range of insects (and other arthropods such as mites, spiders, and centipedes), it is often overlooked as an important feeding mechanism or misunderstood. Wigglesworth (1972) provides a useful though brief synopsis of some of the dramatic observations of EOD, in insects that have salivary glands but also by many (such as beetles) that lack salivary glands. His review of what he calls "extraintestinal digestion" spans herbivores, entomophagous insects, and even saprophages. The source of the digestive secretions—especially enzymes—is a rather complex issue, and it is sometimes difficult to evaluate. In some cases, such as feeding by reduviids on nontransparent prey, the salivary glands pump several kinds of hydrolytic enzymes, especially proteases (Cohen 1993) into their prey; after allowing the enzymes to liquefy the contents of the prey, they ingest the partially digested slurry (Figure 7.20, top). It is nearly impossible to determine the extent of

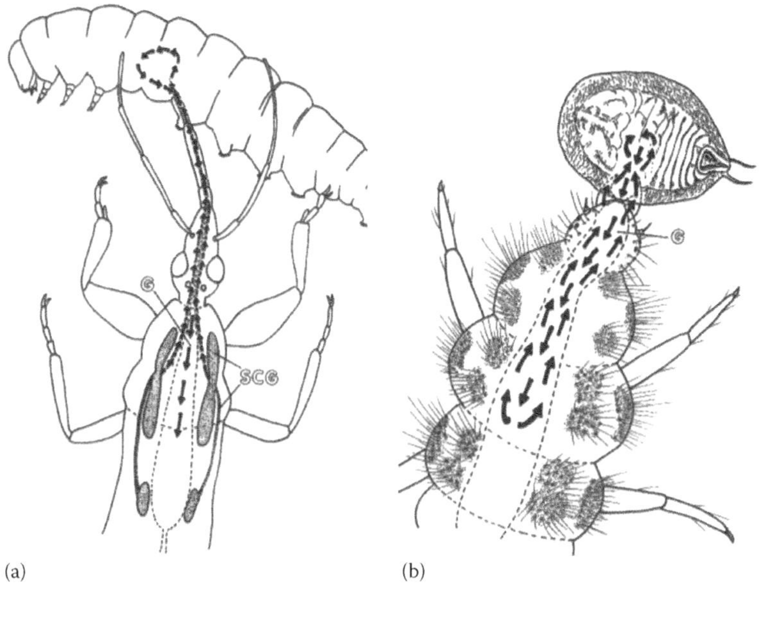

Figure 7.20 Two kinds of extraoral digestion (EOD): (a) Nonrefluxing EOD (by a reduviid hemipteran, *Zelus renardii*) where the digestive enzymes are secreted from the salivary glands, pumped into the food source (a lepidopteran larva as prey), and then the saliva and the digested internal structures of the prey are ingested as a concentrated slurry and further digested in the simple tubular gut. (b) A larval lady beetle (*Serangium pracesetosum*) uses refluxing EOD by pumping digestive secretions from its gut into the prey (a whitefly nymph), and the partially digested contents of the whitefly are ingested as a slurry, which is rapidly digested and absorbed by the simple tubular digestive system of the predator. (From Cohen, A.C., Biochemical and morphological dynamics and predatory feeding habits in terrestrial Heteroptera, in *Predatory Heteroptera: Their Ecology and Use in Biological Control*, M. Coll and J.R. Ruberson, eds., Thomas Say Publications in Entomology, Entomological Society of America, Lanham, MD, 1998a, pp. 21–32.)

the digestive process in such conditions, whereas with transparent prey, the digestive action can be directly observed. In the case of beetles, which lack salivary glands, the gut apparently provides the EOD fluids, and they are refluxed as illustrated in Figure 7.20 (bottom). While the process of EOD is clearly important for efficient consumption of complex foods with macromolecular contents, there are other functions of the extraorally delivered enzymes and possibly other secretions: detoxification of plant defenses (Musser et al. 2002) and disinfection of food substances (Eichenseer et al. 1999).

7.11.6 A Case That Illustrates Complex Interactions of Feeding System Components: *Arundo* Wasp Feeding Biology

Figure 7.21a–f shows several aspects of the feeding system of a plant-parasitic insect the *Arundo* wasp, *Tetramesa romana* (Hymenoptera: Eurytomidae). The Arundo wasp is a gall-former, which was imported into North America from Europe and North Africa to control an invasive weed, *Arundo donax*, which chokes waterways and disrupts natural ecosystems associated with streams and rivers (Moran et al. 2013). The galls and the galleries within the galls caused by the wasps are seen in the top frames of Figure 7.21 (a and b). The question of determinants of gall-formation remains controversial (e.g., Hartley 1998, Schonrogge et al. 2000), where questions of microbial involvement versus insect-produced factors remain unresolved. However, the nature of galls is becoming better understood, and their chemical composition is now known to be far more nutritionally complete and more nutrient-concentrated than the nongall tissues of stems or leaves where the galls are formed. In fact, the reduction in the amounts of secondary compounds (such as the alkaloids that prevail in *Arundo donax* tissues: Khuzhaev et al. 2004) and the increase in nutrients certainly render gall tissues as a far higher quality food for gall-forming insects than the nongall plant materials (Hartley 1998, Schonrogge et al. 2000). This makes efforts to develop an artificial diet for gall-formers more problematic than it is generally for simple feeding on leaves, roots, fruit, or seeds.

The overall feeding process of *T. romana* is a complicated phenomenon with many "moving parts" to be understood. The mandibles are small, weak structures that are clearly not adapted to feed on tough tissues in ways that leaf-eating Lepidoptera larvae or boll weevils are adapted to use (Figure 7.7a and b). The head of *T. romana* has a peculiar bulbous region known as a prementum and postmentum (Moran et al. 2013), and the gut is a simple tube, which seems to end posteriorly as a blind gut. Larvae of *T. romana* can be observed through their transparent cuticle to maintain almost constant peristalsis and antiperistalsis where gut contents are continually being churned and moved alternately from the anterior to the posterior gut and back to the anterior. The larvae, when feeding on either artificial diet or tissue-cultured callus (Figure 7.21) frequently produce droplets of exudate (from either salivary glands or gut contents), allow the droplets to grow to a maximum size on the prementum/postmentum bulb, and then they dab it onto the surface of the feeding site where it is thought (Cohen unpublished observations) to soften the food to allow the weak mandibles to rasp the pablum consistency material and ingest it.

So where does this leave us in the gestalt understanding of the anatomical basis of *T. romana* (or other gall wasp) feeding biology? The short, simple, blind-ended gut is suited only to foods that do not require concentration (dilute foods). The weak mouthparts indicate that only soft foods can be consumed, and the use of EOD indicates that the food (gall material or artificial substitute for gall material) needs to be digested, probably with the help of tissue-macerating enzymes such as pectinase (Cohen and Wheeler 1998, Moran et al. 2013). This narrative about understanding the feeding biology of *T. romana* does not, in itself, provide an automatic success in diet development. As discussed in Chapter 10 (on diet development protocols), this is only a base of understanding that may help lead diet development specialists to start with a more proximate point in successful diet formulation.

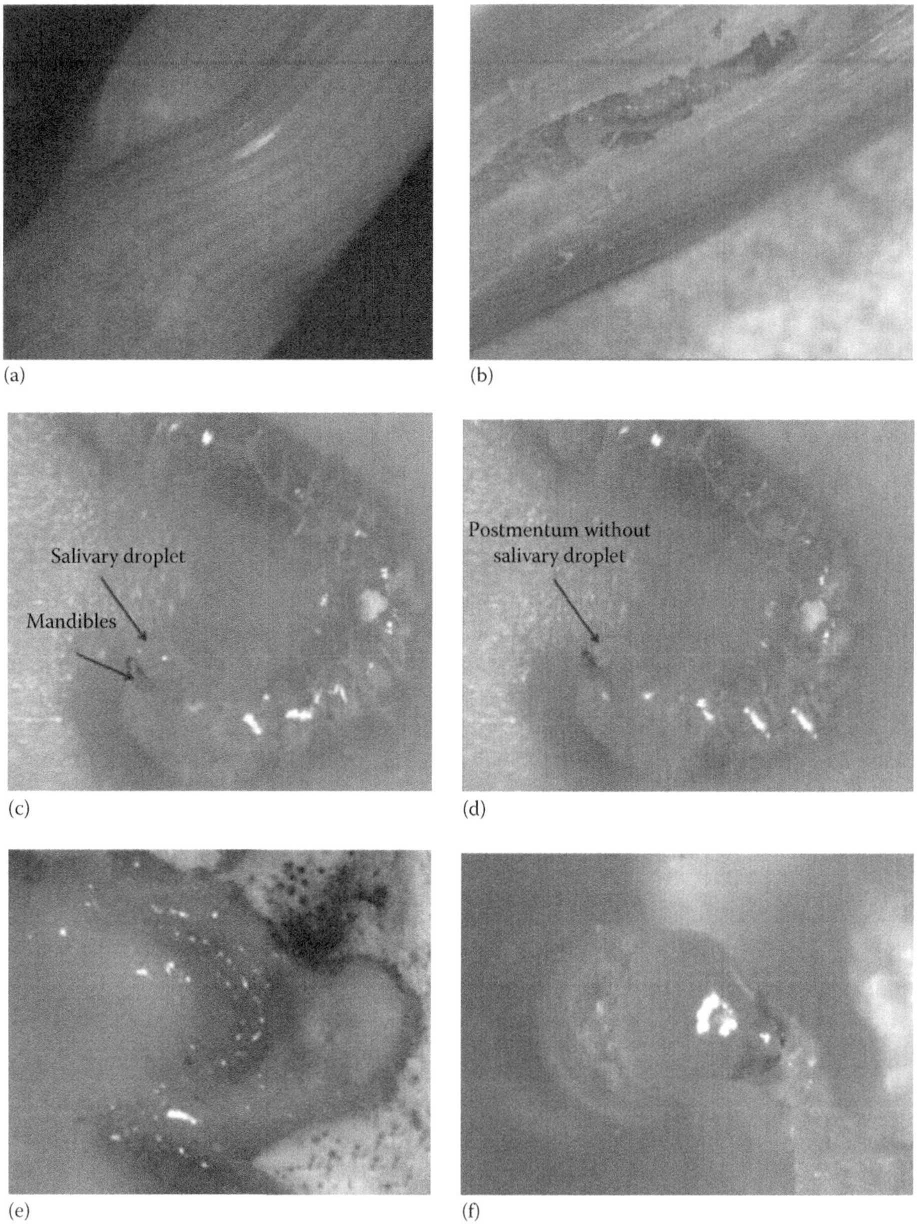

Figure 7.21 **(See color insert.)** Top two images are an external view of a gall on the invasive weed *Arundo donax* (top left) (a) and a view of a larval *Tetramesa romana* (Hymenoptera: Eurytomidae) the Arundo gall wasp in its gallery within a gall (b). The second row shows a *T. romana* larva with a droplet of saliva that it dabs onto gall material to digest it by EOD (c), then the larva uses its weak mandibles to rasp the softened gall tissue (d). This process was observed when larvae fed on galls, artificial diet, and callus from tissue culture. The third row is a *T. romana* feeding on artificial diet (left) (e) and on *Arundo* callus tissue (right) (f) (grown by Dr. Sergei F. Krasnyanski of the Plant Transformation Laboratory in the Department of Horticulture at NCSU). These processes are discussed by Moran et al. (2013).

7.11.7 Digestive System of Termites: Gestalt and Empirical Science

One of the most remarkable achievements in nature is the success of termites in their ability to use for their food one of the most stable, intractable, and nonnutritious materials—wood. In the spirit of the discussion thus far in this chapter, the question that can be asked is this: if we were to examine the gut of a termite (and for a helpful clue the mouthparts, too), could we tell that it is adapted to consume and utilize a material such as wood? For guidance to the answer to this question, please see Figure 7.22 for a comprehensive characterization of termite mouthparts, gut structure, and microbial components. The interactions of these subsystems are evident in this figure. This is in contrast with the compact mycetomes of whiteflies—shown in Figure 7.23, which feed on phloem sap, a material lacking in amino acids that are produced by mycetome-bound microorganisms. Clearly, this question is simplistic, but there are some useful conclusions that can be reached with the gestalt approach: first, the mouthparts of worker termites consist of very strong, highly sclerotized mandibles powered by strong muscles that help to account for the large head: along with a head and mouthparts that are clearly predisposed to biting hard materials, and the digestive system, which has both gastric caeca diverticula (on the ventriculus or midgut) and an unusually long hind gut (in comparison to most other insects) with a greatly distended pouch (sometimes called a "paunch" to emphasize its bulbous nature). It is the pouch that harbors the remarkable microbiome, which is composed of hundreds of species of bacteria, archaea, and protists (Carpenter et al. 2013). These authors write, "Bacterial diversity in lower termite hindguts is very high, with as many as 700 phylotypes representing 15 bacterial phyla commonly, Spirochetes, Firmicutes, Bacteroidetes, Elusimicrobia present in a single hindgut (Ohkuma and Brune, 2011)." And finally, the question of

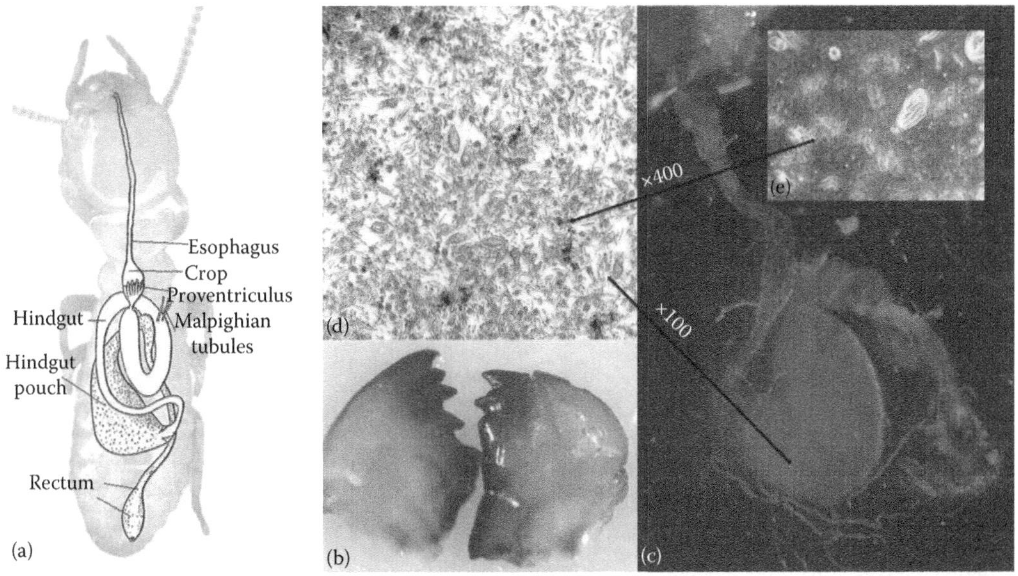

Figure 7.22 **(See color insert.)** Termite digestive system (a, c), mandibles (b), and endosymbionts from the enlarged pouch (d, e). The drawing (a) (from Wigglesworth 1974) and the image of the entire digestive system (c) shows that the termite has a relatively simple, tube-like digestive system, except for the notable enlarged pouch (sometimes called a *paunch*) where dozens of species of symbionts (bacteria, Archaea, and protozoans) abound. The mandibles (b) in this composite image are seen as very strong, hard structures whose highly sclerotized dentition is responsible for the termites' ability to break up wood into pieces small enough for the microbiome to access and digest. The microbiome images are magnified 100× (d) and 400× (e). It should be noted that the smaller bacteria and Archaea are not visible under these magnifications.

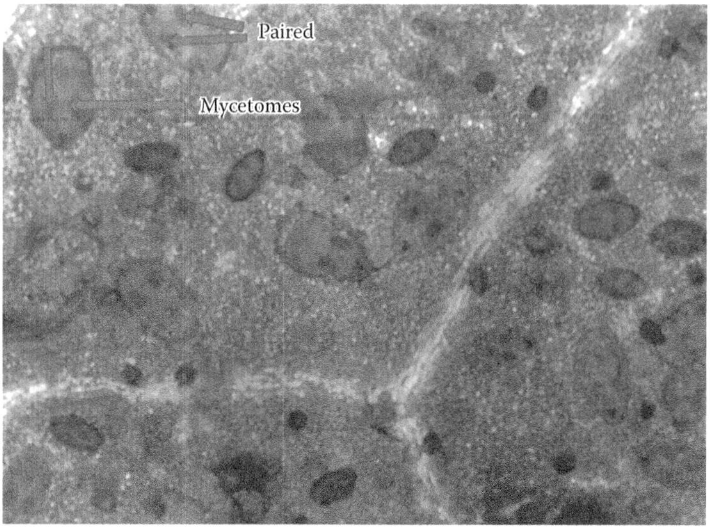

Figure 7.23 **(See color insert.)** Immature whiteflies (nymphs) on cotton leaf that has been partially cleared. The paired mycetomes in each individual are visible as dark spots that are associated with the digestive system of the whiteflies. The mycetocytes and their symbionts are transferred prior to oviposition by a transovarial process from the female to all her offspring.

how we can understand more fully the mind-bending complexity of the roles and interactions of all these microbes, Carpenter et al. (2013) wrote about their adoption of direct imaging systems:

> To show how the ecological roles of protists and nutrient transfer with symbiotic bacteria can be elucidated by direct imaging, we combined stable isotope labeling ^{13}C-cellulose of live termites with analysis of fixed hindgut microbes using correlated scanning electron microscopy, focused ion beam-scanning electron microscopy FIB-SEM, transmission electron microscopy, and high resolution imaging mass spectrometry NanoSIMS. We developed methods to prepare whole labeled cells on solid substrates, whole labeled cells milled with a FIB-SEM instrument to reveal cell interiors, and ultramicrotome sections of labeled cells for NanoSIMS imaging of ^{13}C enrichment in protists and associated bacteria. Our results show these methods have the potential to provide direct evidence for nutrient flow and suggest the oxymonad protist *Oxymonas dimorpha* phagocytoses and enzymatically degrades ingested wood fragments, and may transfer carbon derived from this to its surface bacterial symbionts.

Clearly, the nature of the feeding systems of the numerous of species of insects is highly complex, almost to a degree that could paralyze efforts of diet specialists to use mouthparts, digestive system structures, and microbial and metabolic analyses to understand them. However, it is the point of this chapter and the ones on diet chemistry (Chapter 3), microbial relationships (Chapter 13), and diet development (Chapter 10) that using a systematic approach to diet development, including basic understanding of structure and function in the feeding systems of the insects we target, will lead to enhanced success with the science and technology of insect diet development and utilization.

CHAPTER 8

Insectary and Rearing System as a Context for Diets

Order and Complexity in Nature and in Diet-Based Rearing System

8.1 ORDER AND UNPREDICTABILITY: AN OVERVIEW

Population crashes, fluctuations in developmental periods, shifts in body weights, and reduction in fecundity in the target insect are examples of sudden and undesirable outcomes that can occur in a rearing facility. Such events are reminiscent of the unexpected results manifested by the dinosaurs that ran amok in Michael Crichton's *Jurassic Park* and the intellectual construct, the so-called butterfly effect, which suggests that if a butterfly flaps its wings today in Peking, it may cause changes in the weather systems next month in New York (Gleick 1987). The most salient feature of the butterfly effect is the amplification of some seemingly minor factor that triggers complex or unpredictable results. Because diets are complex, unpredictable outcomes in insect-rearing facilities are always possible. This chapter examines some seemingly innocuous factors that can become amplified and lead to undesired outcomes in insect-rearing facilities.

8.2 ORDERLINESS OF SYSTEMS IN NATURE

Efforts to develop strategies to improve insect diets and rearing systems underscore the extent to which we place insects in environments that are alien to their natural habitats. Target insects in natural and agricultural ecosystems thrive in a complex and orderly world. The regularity and predictability in nature are remarkably precise. This order is both spatial and temporal. The regular seasons in all life zones provide insects with cues for entering and leaving diapause, mating, egg production, when and where to oviposit, number of generations per year, finding food, and myriad other signals that inform insects about their environments. Changes in day length, temperature, and other regular factors that characterize the environment feed information to the insects' biological clocks and centers of physiological regulation in the form of visual, olfactory, auditory, and thigmotropic (touch) cues. Often the female's choice of oviposition site determines the location of the food source for the insect. One cue or a combination of several cues known as releasing stimuli trigger the selection. Although the predictability of the type of food may vary from one environment to another, the food source is generally reliable. Figure 8.1a and b illustrate the diversity of food choices available to insects such as leafhoppers and whiteflies in terms of light, temperature, and humidity gradients, and different kinds of specialized plant tissues. Clearly, such choices are not available in our rearing settings.

The regularity or order becomes evident upon careful analysis of the various foods that insects eat in nature. The pink bollworm *Pectinophora gossypiella* (Saunders) (Lepidoptera: Gelechiidae),

(a) (b)

Figure 8.1 (a) Flowering ornamental cherry tree and (b) potted cotton plants are examples of the kinds of environmental and nutrient gradients that are available to insects in nature and in agricultural ecosystems.

for example, does not feed indiscriminately on leaves, stems, roots, or flowers of the cotton plant but rather on specific tissues within and around the seeds in the cotton boll. Most insects feed on such narrowly circumscribed choices as xylem sap, phloem sap, blood drawn only from capillaries, and specific tissues in prey insects. The nutritional composition, structure, and other chemical and physical characteristics of specialized feeding targets are predictable in terms of specific feeding cues or stimuli.

In contrast, rearing systems inherently limit food selection choices. For example, when the female beet armyworm moth, *Spodoptera exigua*, lays eggs on a cotton leaf, the neonate larvae find themselves on a suitable host, so that all they need to do is start eating the leaf tissue that surrounds them. Likewise, when a green lacewing adult, *Chrysoperla rufilabris*, lays her eggs near a group of aphids or other suitable prey, the neonate larvae only need to crawl a short distance to find an abundant food source. Such food sources are of consistent nutritional composition, and the location of the feeding target is clearly demarcated with chemically appropriate surface stimuli such as waxes, aphid-derived honeydew trails, volatile chemicals, and surface structures such as protrusions or indentations that help the insects orient themselves to feeding targets. The wax particle exudates from whitefly nymphs illustrated in Figure 8.2a appear as concentric circles. Whitefly predators and parasitoids use such wax trails to locate their target nymphs. All these foods contain additional internal cues the sleuthing insect reads until it reaches the feeding target. That target is generally adequate for each feeding insect in terms of nutritional profile and required amounts of material.

This is not to say that there are never mistakes and faulty choices on the part of the insects or that nature does not force hardships upon the insects. Each species has programmed into its population dynamics the inevitability of imperfect food supplies and other difficulties (such as host plant tissues drying out, disease, and inclement weather) because it produces offspring beyond whatever number could remain within the carrying capacity of its ecosystem. However, the highly abstracted, simplified conditions that we use for artificial diet-based rearing systems are much more limited and confined, and more greatly influenced by small changes than those conditions encountered in nature. Figure 8.2a and b show views of the structure of leaves and stems of plants being attacked by whiteflies and glassy-winged sharpshooters (GWSSs), a phloem-feeder and xylem feeder, respectively. Figure 8.3 shows a view of the internal structure of an egg of an *Lygus hesperus* nymph that serves as a target for predators or parasitoids that feed on nymphal plant bugs. The internal organization of the great variety of tissues that are available to the parasitoid comprises a series of sequential feeding sites. In contrast to this

(a) (b)

Figure 8.2 (a) Leaves with whitefly waxes (concentric circles) and (b) hibiscus leaves with GWSS adults and the watery exudates (Ex) on the leaf surfaces.

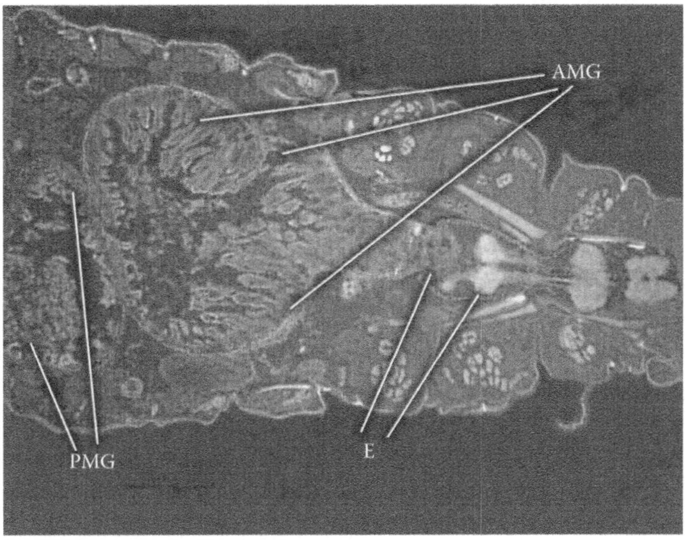

Figure 8.3 The variety of internal structures in *L. hesperus* nymphs. These structures represent a wide range of targets with various nutritional and textural qualities for predators and parasites of these plant bugs and other insects.

diversity, complexity, and orderliness of the internal structure of the host plant and prey are the artificial diets used in typical rearing settings for *S. exigua* and *C. rufilabris*, as shown in Figures 4.1 through 4.4. An examination of the differences between feeding opportunities of *S. exigua* and *C. rufilabris* that consume natural diets compared to captive specimens reared on artificial diets illustrates this point.

S. exigua larvae are commonly kept in small containers such as 8 cc plastic cells. *C. rufilabris* are even more restricted, usually reared in 1 cc containers. Typically, these insects are provided with artificial diet that is far less compartmentalized and diverse than the natural diet. The natural diet is abundant in components of different chemical makeup, offering a range of cues that help the insect locate compartments containing optimal food. For example, Figure 4.5 shows that the cotton leaf contains a remarkably orderly set of structures, lower and upper epidermal cells, a palisade layer, spongy mesophyll cells, vascular bundles that are differentiated into xylem and phloem, companion cells, and the apoplastic compartment. These compartments are further subdivided into subcellular components that contain a considerable number of organelle types, each of which has its own biochemical and nutritional character. For an insect, a visit to a leaf such as this is the equivalent of a visit to a supermarket where each aisle (compartment) offers its own special resources. For example, cell membranes are rich in phospholipids, chloroplasts are rich in proteins and lipid-soluble antioxidants, and various storage granules or vesicles are reservoirs for starch, protein, or triacylglycerols. The phloem-feeding specialists, such as aphids and whiteflies, can locate the phloem sap. The xylem specialists, such as certain leafhoppers, can locate the xylem sap. Leaf miners can find ample mesophyll tissues between the layers of epidermis. Even insects that are generalists can pick and choose from the various compartments and self-select the nutritional profile that best fits their needs. In contrast, artificial diets consist of size-reduced nutrients that have been stripped of nearly all compartmental integrity.

8.3 FACTORS THAT INFLUENCE DIET COMPLEXITY

The deconstruction of compartments is a characteristic feature of artificial diets. For the convenience, stability, and economic benefits of using artificial diets, we have sacrificed the organization and predictability of structure and placement of nutrients (and antinutrients) inherent in living or highly fresh foods. Because it is costly and under some circumstances impossible to maintain supplies of living plants at their appropriate phenologies (growth stages) or living prey and hosts for predators and parasites, we have replaced these natural nutrients with mixtures of stored products. We have added to the disorder by deliberately implementing size-reduction, mixing, and temperature processing practices to ensure that the mixtures that we call artificial diets are fully available to our target insects.

Often minerals such as iron, copper, zinc, and manganese, all known to be destructive to polyunsaturated fatty acids in dietary lipids, are put into contact with those lipids in the diet-making process—a practice that greatly increases the likelihood of lipid destruction. Refined sugars and proteins are mixed and heated, forming indigestible complexes called Maillard products (see Chapter 5). Destructive enzymes are freed from the safety of their natural compartments, inducing oxidation or hydrolysis of nutrients, and often inadvertently extracting antinutrient substances that can wreak havoc in our diets.

8.4 PARADOX OF NUTRIENTS AND ANTINUTRIENTS

Along with nutrients, foods, especially those derived from plants, contain substances known as *antinutrients* that deter potential consumers. Nutrients are part of the living systems of parent organisms, but the biomass that is nutritious to plant consumers is there for the well-being of the plants, for their own metabolic functions, storage, or reproduction, not to serve as foods for other organisms. The presence of these nutrients and the ongoing competition for scarce resources in nature make plants targets for consumers. In response to the selective pressures imposed by these potential consumers, plants have evolved defenses that protect them from attack. While some of these defenses are physical (thorns, spines, hairs, hard coverings for seeds), the most elaborate and universal defenses are chemical. Some of the chemical defenses are geared to defend against feeding (feeding deterrents), but many others are adapted to reduce the potential to use the nutrients at hand. Some chemical defenses are effective because they impart a noxious taste that discourages feeding (feeding deterrents). Many other chemical defenses, however, act by preventing digestion, absorption, or utilization of nutrients (antinutrients).

How can substances with antagonistic purposes exist side by side in the same organism? The answer is *compartmentalization*. The defensive chemicals and the nutrients that they defend are often organized into spatially distinct regions, or they may be organized by temporal (time-based) separation. An example of temporal separation is unripe reproductive structures in plants. These structures are well defended until it is time for seed dispersal by an animal. As part of the ripening process, the feeding deterrents or antinutrients are degraded, and the fruit becomes palatable to the same animals that served as seed dispersing agents.

Physical compartmentalization is another common means of separating potentially self-destructive antinutrients from remaining metabolic and structural components. Figure 9.1 shows an example of the compartmental organization wherein antinutrients are sequestered in the gossypol gland of a cotton leaf. The gland is set apart from the other cotton leaf cells as a site of synthesis and storage of gossypol, a terpenoid, antinutrient compound toxic to most species that discourages feeding by many insect species. As discussed in more detail in Chapter 12, a significant source of potential adverse outcomes in using artificial diets stems from destruction of the compartments that separate nutrients from antinutrient substances.

Plants are not the only organisms that use an antinutrient strategy for self-protection. Chicken egg yolk contains an extensive complement of nutrients such as B vitamins, including biotin. However, egg white contains a protein called avidin, which is known to bind biotin in tight complexes that make absorption of this vitamin impossible. Avidin is thought to protect the developing chicken embryo from microbial attack by removing biotin from the potential nutrient pool of microorganisms. Because the yolk and white are in separate compartments in an undisrupted egg, the avidin never gains access to the yolk biotin used by the developing chick embryo. However, when the egg is broken and these two major compartments are mixed, the destructive potential of the avidin is released, not only binding the biotin in the yolks but also removing any extra biotin present from other sources (e.g., from vitamin mixtures that might be added to diets). The biotin-binding characteristics of avidin are completely destroyed by denaturation such as thorough cooking, but the heating must be done *before* the avidin/biotin-binding reaction has taken place.

The nutritional value of eggs (Table 3.4) and the nutrient availability of egg proteins (PER values) enhance insect diets that contain chicken egg yolks or whole eggs (Cohen 2000c, 2003b). Included in their potential value to insect diets are the functional properties of eggs that contribute to the emulsification, water-binding capacity, gel formation potential, and fat- and flavor-binding qualities. These properties are reviewed by Damodaran (2008). While much of the value of the egg, as described earlier, resides in the yolk, the white is almost entirely protein. Therefore, egg white would seemingly increase the protein value of an insect diet. However, before eggs can be used to their full potential as insect diet components, the avidin/biotin issue and the digestive inhibitor problems must be addressed.

This paradox of cooccurrence of nutritional and antinutritional components is common in a wide variety of other foods and may be a source of many problems when diet components are not thoroughly understood and not properly handled. This has become a well-established concept in vertebrate nutrition. For example, biotin deficiencies are rare in mammals (Lehninger et al. 1993) because of the widespread occurrence of biotin and because biotin is produced by bacterial flora in the intestine. These authors note that biotin deficiencies are generally confined to cases where large quantities of raw eggs have been consumed. Not long ago, bodybuilders and athletes seeking to build large muscle mass were encouraged to consume large quantities of raw egg milk shakes to obtain the large amounts of protein needed for such growth. However, these athletes inadvertently created a biotin deficiency that was at odds with the desired outcome. Many other kinds of antinutritional factors can become problematic in diets, including phytic acid, which acts as a powerful and intractable chelating agent (discussed further in Chapters 3 and 5).

8.5 UNEXPECTED CHANGES AFTER MANAGEMENT DECISIONS

It would be helpful for current and future workers in insect diet science and technology to have a complete, published profile of instances where unexplained outcomes in rearing facilities are documented, chronicled, or otherwise explained in honest and complete detail. Unfortunately, no such body of information exists. People do not write papers on mistakes that were made in their insectaries, nor are editors of scientific journals eager to publish such accounts, illuminating as they might be. Therefore, I offer my experiences and anecdotal information acquired over the years. The validity and universality of these anecdotes are constantly confirmed by other rearing specialists who share similar experiences. I have kept the situations general and avoided specifics that might appear to be recriminations. In every case, the personnel involved had the best of intentions:

- *Substitution of carrageenan (a gelling agent) from a different source.* The original supplier stopped making the carrageenan used in a diet. The product of a new supplier resulted in a dramatic population decline in the insectary that was caused by a change in seaweed species from which the carrageenan was derived. Specifically, the sodium content was found to be elevated in the new material, and this increase reduced the palatability of the new diet, which was otherwise identical to the old diet.
- *Substitution of low-fat ground beef for high-fat ground beef.* An effort was made to improve a predator diet by substituting the original 21% fat ground beef with a low-fat (<17%) beef product. The colony declined and was nearly lost before ground beef with the original fat concentration was reinstated. More recently, ground beef with more than 30% fat has replaced the original 21% with excellent results.
- *Substitution of water source (from a sponge to sealed-in Parafilm).* In an effort to reduce the labor needed to supply free water to a colony of predators, water was sealed in packets made of a laboratory wrap (Parafilm). The predators all died, evidently due to desiccation, because they were unable to detect or ingest the water in the sealed packets.
- *Changes in frequency of diet changes.* In an effort to reduce labor cost, feeding packets were distributed once every 4 days rather than every other day. This resulted in a precipitous decline in colony numbers. In an effort to reduce labor cost, feeding packets were replaced once every 4 days rather than every other day as described in the original report on the diet for big-eyed bugs (Cohen 1985a). This resulted in a precipitous decline in colony numbers.
- *Changes in the chelating agent in salt mixture.* Abnormal performance syndrome (APS) in rearing production of a moth larva was linked to changes in the chelating agent and the state of the iron (Willis and Allen 1999). The bioavailability of the iron chelate was completely lost when the crystalline form of iron replaced the amorphous form (Willis and Allen 1999).
- *Changes in the diet quantity being autoclaved (time of heating or degree of heat processing).* In a plant bug–rearing facility, efforts to reduce labor by scaling-up of the diet production led to the doubling of the standard batch (1 L) of diet for plant bugs. Bioassays indicated that the scale-up that involved 2 L batches autoclaved for the same 20 min period at 121°C at 15 PSI produced no noticeable adverse effects. Based on the assumption that 2× amplification was a satisfactory way of producing diet, the rearing staff went to a 3× amplification (3 L batches). This led to a rapid decline in the portion of the colony of plant bugs that had been fed this diet.
- *Changes in the length of time a diet was heat processed and the temperature of diet to which formalin was added.* Efforts to increase the heating time to more fully heat process plant bug diet resulted in unacceptable and irreversible clumping of diet, especially pronounced at the sides of the container. Efforts to reduce the time delay before adding formalin resulted in the rapid evaporation of the formalin, reducing its concentration in the diet and increasing the microbial contamination, especially by fungal contaminants. The well-intentioned change to 55°C deviated from the standard (Cohen 2000c) calling for a wait until the diet cooled to 50°C, with detrimental effects on contaminant growth.
- *Changes in the source of soy flour.* Changing the source of soy flour from a company that sold high-fat soy flour to one that provided lower-fat, unroasted soy flour resulted in a dramatic plunge in quality with respect to the rate of development, biomass accumulation, and weight of individuals.

All of these changes resulted in unexpected outcomes and a decline in or complete loss of a colony. The outcomes and the rationale behind the problems are explained in the following sections.

8.6 CONSCIOUS DECISIONS AND HIDDEN FACTORS

In every step of the rearing process, management and rearing staff make conscious decisions. The rearing staff members control choice of products to purchase, time of purchase, storage method, and every step in the diet production process (measuring ingredients, mixing, heating, delivery steps, storage). At any stage in diet production, a decision may be made that will impact the quality of the diet.

Starting with the purchase of materials, decisions are made as to how often to make the purchase, what quantity to buy, how to have the material shipped, which supplier to use, and which product of the choices listed to select. To demonstrate the complexities in these decisions, let us take a couple of simple innocuous-sounding diet ingredients such as sucrose and asparagine. If the diet formulation calls for sucrose, the first decision may be where to buy it. Sucrose is table sugar, which can be purchased from a local market. The purchaser must decide whether sucrose from a local store is of adequate purity to serve the purposes at hand or if a chemical supplier such as ICN Pharmaceuticals, Inc. or Sigma Aldrich Fine Chemicals should be the source. I use the 1998–1999 catalogs from these two companies to illustrate my points. Should the purchaser choose to use a chemical supply company, the cost will increase from less than \$5/kg to at least \$20/kg. Is the higher price or quality necessary?

The two company catalogs list more than 10 products under the sucrose heading. Some of those listed can be excluded from consideration because they are radioactive or sucrose compounds that are irrelevant for our purposes. However, whether these differences are obvious to a purchaser depends on the background of the purchaser. Sucrose synthase, for example, may sound as applicable as ACS reagent sucrose or one of the other products listed. The purity of the sucrose also must be decided. For example, the ICN catalog describes Product A as crystalline and suitable for density gradient studies with RNA, Product B as a cell culture reagent, and Product C as slightly less pure than Product A and B, will not reduce Fehling's solution, and suitable for most routing reactions. To further complicate the situation, Product C is priced by the pound rather than by metric weight as are Products A and B, and Product C is considerably less expensive than Products A and B (even after English to metric conversions are made). The choices in the Sigma Aldrich catalog are equally complex.

The next question is the amount to order. Assume that the purchaser decides on Product A, the ICN sucrose listed as suitable for density gradient studies with RNA. A 500 g amount would cost more than \$33/kg plus shipping, compared to \$26/kg for the 1 kg listing, or about \$15/kg for the 5 kg listing. Ten separate shipments of 0.5 kg would cost more in the long run than a single shipment of 5 kg because of shipping costs. Frugality and convenience of recordkeeping would favor the purchase of larger shipments. Having a large amount on hand would reduce the possibility of running out of sucrose and having to wait for a new shipment. A subtle benefit of having large batches of diet ingredients is that certain materials are variable according to idiosyncrasies of the lot of the parent materials. With sucrose, batch-to-batch idiosyncrasies would be of far lesser concern than for less processed and more inherently variable components such as agars, carrageenans, various flours, fresh meats, produce, and dairy products. Almost everyone has experienced the variations in taste in samples from the same variety of fruits and vegetables, and anyone who has used fresh produce to feed their insects has probably experienced die-offs attributable to pesticide contamination (Debolt 1982).

The drawbacks to large batches include requirements for extra storage space in appropriate places (freezers, refrigerators, cool, dark rooms, laboratory shelves) and the likelihood of product degradation over prolonged periods of storage. The caveats about moisture and oxidative degradation associated with prolonged storage are covered in detail in the discussion of food chemistry (see Chapter 5) and microbial degradation (see Chapter 13). Along with the increased risk of degradation by chemical and microbial means, stored products are also subject to contamination by absorption of local contaminants such as ammonia, carbon dioxide, or other airborne gases. Vermin are found in laboratories and storage facilities. I have found insect (fly) excrement and rodent droppings in what appeared to be closed packages of diet materials. On several occasions, I have found extensive carpets of fungi growing on the outsides of sacks of soy flour and wheat germ in a cool, dry storage room. Was the reduction in cost of large batch purchases truly economical if they had to be discarded (or worse yet, if they were used in the diets of insects without notice or regard to their potential subversion of the diet and the colony that it supports)? These points are further discussed in Chapter 12.

Diet-based rearing problems often result from handling and storage abuses such as when a product is kept well beyond its shelf life. In one instance, 50 lb sacks of wheat germ had been stored in a walk-in refrigerator for more than 5 years. The purchase of the wheat germ was a large order of several pallets of the meal. In this case, a large purchase was made because funds were available through a reimbursement transaction, and it was considered prudent to make a single purchase of product that would supposedly last for several years. The assumption behind the purchase was that wheat germ is very stable and relatively nonperishable and that even if it had to be kept for several years, it would not deteriorate. Although the thinking was reasonable and the intentions were honest, the outcome was a large batch of spoiled material with mold infestation that could contaminate the entire rearing facility.

8.7 CHANGES IN THE ORDER OR NATURE OF PROCESSING STEPS

Although it has been known for decades that the order of mixing ingredients when processing diets greatly impacts the degree to which components remain in solution and how they interact, changes are made frequently (Dadd 1968, Mittler 1972). Such changes probably assume that as long as all required ingredients are included, the diets should be satisfactory. However, altering the sequence of processing steps can result in pH-based precipitation of minerals and other sparingly soluble components such as cholesterol (Mittler 1972). Also, untimely addition of difficult-to-mix ingredients can cause uneven distribution of these components. In my opinion, uneven distribution of components is often the reason for diet-based rearing failures.

Even what appears to be a minor change in processing may have a profoundly adverse effect on diet. For example, I have witnessed an increase in the contamination of a diet for tarnished plant bugs that was finally attributed to a change in the timing of the addition of antimicrobial agents. To save time, formalin and antibiotics (both heat-labile compounds) were added to the diet mixture when it had cooled to 55°C–60°C after autoclaving, rather than waiting until the diet had cooled to the recommended 50°C. As documented in the Merck Index (Windholz 1983), the boiling point of formalin is 96°C, and the flash point is 60°C. In retrospect, it became obvious that much of the antimicrobial activity of the additives was lost due to the exposure of these heat-sensitive components to high temperatures.

The more automated processes that are usually associated with large-scale production tend to deviate less from the standard, well-proven order of processing than do the smaller-scale practices. Steam kettles or autoclaves are common heating implements in the moderate-scale settings. For most diets that include specialized gelling agents (usually agar or carrageenan), the solution and heat activation of the gelling agent are the first steps in the mixing process. Once the gelling agent is dissolved and heat activated, the other ingredients are mixed into the molten gel solution, often with the more heat-labile ingredients added last, after the solution has cooled. Beyond lowering labor costs, these processes used in larger-scale, automated diet production systems often produce a product superior to those produced more manually, and the mechanization lends itself to better sanitation (Chapters 12 and 13).

8.8 IMPORTANCE OF IRON IN INSECT DIETS

Possibly no interactions better illustrate the theme of diet complexity than those interactions involving iron. While iron is one of the most important, multifunctional nutrients, it can also be a component in highly destructive oxidative degradation reactions. Adding to the complex character of this element is that the form of iron and the diet's chemical matrix determine the nutritional value, bioavailability, and toxicity of the iron. The many important functions of iron in insects are explained in Chapter 3, and several aspects of the complex chemistry of iron in diets are discussed in Chapter 5. This section here is included to underscore the degree of complexity of diets that must be understood to better manage diets. What amplifies this complexity is that iron reactions are subject to various kinds of synergistic interactions. In this section, we emphasize that (1) iron has many functions in insects and many different reactions in insect diets, (2) dietary iron comes from many different sources, (3) iron has both beneficial and destructive roles, and (4) the bioavailability of iron is very dependent on the chemical environment (diet matrix) that surrounds the iron atoms.

8.8.1 General Nature of Iron

Iron is a mineral that is required by all living things. It is the fourth most abundant element on Earth, but it is frequently lacking in the nutrition of many organisms (Mertz 1981), including plants, humans, and insects. In insects, iron is an essential component of the cytochrome system, crucial in energy metabolism. The oxidation of biological fuels is impossible without iron. Iron is also a cofactor for several enzymes, including catalase and peroxidase. Iron is also involved in detoxification processes, DNA and RNA synthesis, nitrogenous waste product formation, and synthesis of essential molting and growth hormones (Locke and Nichol 1992). Early reports that certain insects did not require iron must be taken as erroneous (summarized in Gilmour 1961).

8.8.2 Forms of Iron

The types of compounds and matrices in which iron occurs in natural diets are varied and poorly understood. Iron can be present in one of several organic forms in several inorganic compounds: ferric form such as ferric phosphate, ferric nitrate, ferric sulfate, and ferric chloride (all of which contain iron that is in the Fe^{3+} or oxidized state) or ferrous form such as ferrous chloride, ferrous sulfate, and ferrous phosphate (all of which contain iron in its Fe^{2+} or reduced state). It may also be present in one of several chelated forms associated with organic ions such as citric, oxalic, and fumaric acids, which are among the many organic compounds that bind (chelate) iron.

8.8.3 Sources of Iron and the Issue of Bioavailability

Several foods such as fish, poultry, and meats are known to enhance bioavailability or absorption of iron in vertebrates as do certain compounds such as citric acid, ascorbic acid, and EDTA (Miller 2008). Conversely, other substances are known to act as inhibitors of iron absorption, thus decreasing the bioavailability of iron, including polyphenolic compounds such as tannins from tea and legumes, oxalic acid, some plant proteins (especially from legumes such as soybeans, chickpeas, and lima beans, as well as whole grain cereals), calcium, organic phosphates such as phytic acid (Miller 2008), and phosphorylated proteins such as the egg yolk protein phosvitin. Iron is most readily absorbed in its heme form or one of the other organic forms. Absorption is strongly pH dependent, with acid pH favoring absorption kinetics (Miller 2008).

8.8.3.1 Case Study: How Iron's Complexities Caused a Major Problem

The complexity of iron is exemplified by a well-documented case known as the APS or what was initially termed the *straggling syndrome* in the New Jersey Standard Strain of gypsy moths, *Lymantria dispar* (L.). The insects in question were reared at the Animal and Plant Health Inspection Service (APHIS) Methods Development Laboratory (in Cape Cod, MA) and the Forest Service Insect Rearing Facility (in Hamden, CT) (O'Dell 1992, O'Dell et al. 1997). The gypsy moths were used for sterile release programs, virus production, and extensive research programs on various aspects of biology of this species, including genetics, nutrition, behavior, ecology, and physiology. According to O'Dell (1992), the straggling phenomenon first reported by APHIS laboratory staff around 1980 caused reduced virus production and problems in research and other programs involving these moth colonies. APS was neither confined to the insects reared in the Otis facility nor confined to insects that had undergone extensive laboratory in-breeding (O'Dell 1992). The problem was so serious and affected so many programs that in 1989, a workshop was held in which 50 scientists from various disciplines were assembled to determine the cause. After ruling out environmental, genetic, and microbial causes and a variety of nutrient and other diet factors, the problem was traced to an iron deficiency that stemmed from changes in the iron formulation of the Wesson salt mixture used in the diet (O'Dell et al. 1997, Keena et al. 1998).

Through a complex series of very careful investigations, the researchers determined that one of the manufacturers of Wesson salts had substituted a crystalline form of ferric phosphate for the conventional amorphous form (Keena et al. 1998, Willis and Allen 1999). For reasons that are not clear, the amorphous form of ferric phosphate apparently has a higher degree of bioavailability than the crystalline form.

8.8.3.2 Iron Economy in Gypsy Moth Diets

A batch of gypsy moth diet with a fresh weight of 10,810 g contains 1,550 g of dry ingredients, 1,200 g of which is wheat germ, 250 g casein, and 80 g Wesson salt mix. The wheat germ contributes about 75 mg of iron, casein about 5 mg, and Wesson salts about 444 mg, or a total of 524 mg of iron (nearly 5 mg/100 g diet), about 84% of which comes from the salt mixture. This is a substantial amount of the total iron, and one may wonder how this compares with the iron availability in the natural foods of gypsy moths—foliage from several species of trees. Although no specific figures are available for the natural foods, it is known that leafy foods such as spinach, cabbage, and lettuce have 0.5–2.5 mg of iron/100 g fresh weight (USDA database); it seems unlikely that oak leaves and other gypsy moth hosts have a higher iron content than a relatively high iron food such as spinach. This raises several questions about the nature of iron balance in gypsy moths vs. the balance of this mineral in other phytophagous species.

Is there a reason why gypsy moths, more so than other phytophagous species, would be more sensitive to deviations in the amount or the form of iron in their diets? Is the iron that is present in the wheat germ unavailable to the gypsy moth larvae, or is it simply not an adequate amount to fill the dietary requirements of this species? Interestingly, a survey of 100 papers on artificial diets randomly selected from Singh's (1977) compendium revealed that more than 50% of all these diets did not contain a salt or mineral mixture. The percentage of diets with no salt mixture was even greater when defined diets were removed from the count. This means that for most species, the form and amount of iron inherently present in components such as soy flour and wheat germ were adequate. A pinto bean diet (Shorey and Hale 1965) and the many derivatives that use pinto beans, lima beans, or other legume seeds perform excellently with no added salts. In the Shorey and Hale (1965) diet, the two sources of iron, pinto beans and yeast, contribute a total of 1.0 mg of iron/100 g of diet (compared to the gypsy moth diet described by Bell et al. (1981), which contains ~5.0 mg iron/100 g of diet). At this point, there is no clear reason why gypsy moths might require five times as much iron as the dozens of species reared on the diets with no added salt mixtures. Gypsy moth rearing teams practice excellent quality assessment and extensive tracking; this raises the question of whether or not the problem of iron deficiency has cropped up in other species of insects but was not diagnosed because of incomplete recordkeeping.

8.8.4 Synergistic Complexities of Iron in Diets: The Potentially Destructive Character of Iron

Discussions of iron dynamics make it clear that there are complex circumstances and interactions regarding dietary iron and physiological complexities including absorption, availability, transport, storage, metabolism, and excretion (Dadd 1985, Mittler 1972, Locke and Nichol 1992, Nichol et al. 2002). Despite its essential, life-supporting role, iron is potentially destructive as a prooxidant and toxin. This paradoxical nature of iron is expressed in the following statement: "Intrinsic interest in iron lies in the fact that it is both an essential nutrient and a potent toxin. In the presence of oxygen, iron can catalyze reactions producing free radicals that are damaging to membrane components, nucleic acids, and other essential biological materials" (Nichol et al. 2002). Iron must be in an appropriate form that enables it to be absorbed or maximizes its bioavailability. Iron can also be destructive in diets by causing free radical generation, lipid peroxidation, deterioration of other nutrients sensitive to oxidative degradation, and other chemical reactions adverse to the nutritional value of the diet.

Locke and Nichol (1992) discuss the paradoxical role of iron as a component of the enzyme catalase, which breaks down hydrogen peroxide (therefore playing an antioxidant role), and as a prooxidant in the form of free iron in its reduced (Fe^{2+}) form, causing hydroxyl radicals (•OH) to form. Free radicals are destructive to many essential components of living systems, especially cell membranes and nucleic acids. The hydroxyl free radicals (˙OH) form by a process known as the Haber–Weiss reactions (Locke and Nichol 1992):

$$Fe^{2+} + O_2 \rightarrow Fe^{3+} + O_2^-$$

$$2O_2^- + 2H^+ \rightarrow H_2O_2 + O_2$$

$$Fe^{2+} + H_2O_2 \rightarrow \cdot OH + OH^-$$

If catalase is not present in the diet and iron is present, the Haber–Weiss reactions will generate free radicals that participate in chain reactions that result in the degradation of various susceptible diet components, including the nucleic acids mentioned earlier, vitamins such as α-tocopherol and β-carotene, as well as lipids, especially lipids that contain unsaturated sites (palmitoleic, oleic, linoleic, linolenic, and arachidonic acids), causing a rancidization of these nutrients. It is important to recognize how destructive chain reactions can be if they are allowed to proceed unchecked. Once the free radicals have been set into action in the diet, the bulk of the lipids can be degraded. This has a double-negative effect because it not only removes important nutrients from the nutrient pool, but also creates toxic by-products that inhibit feeding and are harmful once they are ingested. Free radical generation can indirectly cause malnourishment, reduced feeding rates, and increased toxicity. In addition to the free radical oxidation reactions within the diet (and within the insects that ingest free radical–containing diets), there are many other iron-related reactions that may occur both in diets and in the insects after ingestion of the diets in question.

8.8.5 Bioavailability of Iron and Its Various Forms

The amounts of iron reported in Table 8.1 or the USDA Nutrient Database are derived from analyses of iron that was recovered from total heat or acid-based destruction of the organic components in a given food. This process does not tell us anything about the form of iron that is present in each food. As stated elsewhere in this chapter, iron in the heme form generally has higher bioavailability than iron in inorganic forms—the predominant type found in plant materials. There are different degrees of bioavailability within the range of inorganic forms, phosphates, oxides, carbonates, chlorides, or organic chelates. The pH of the diet and the gut lumen can have a profound effect on absorption. An acid pH favors absorption of iron, and a less acidic or basic pH decreases the effectiveness of iron uptake (Miller 2008). The molecular biology and biochemistry of the iron receptors and absorption sites in the guts of insects and other animals remain unexplained; however, the pH-related effect on absorption is well documented, especially with regard to increases in absorption instigated by ascorbic acid and citric acid.

It is also becoming clear that several naturally occurring compounds decrease the absorption of iron, reducing bioavailability to levels below thresholds of nutritional minima. For example, phytic acid has been shown to greatly reduce iron absorption in mammals, and destruction of the phytic acid by appropriate heating or fermentation reverses the reduction in iron bioavailability (Bhatia and Khetarpaul 2002). *Heliothis virescens*, adept at using plant tissues that contain high concentrations of phytic acid (e.g., fruits and seeds), are less susceptible to the adverse effects of this iron-chelating compound than insects that feed on leaf tissues (*Trichoplusia ni* and *Depressaria pastinacella*) (Green et al. 2001). The explanation of these differences is that *H. virescens* is better adapted to extract iron from phytic acid than the latter two species.

Mammalian systems can serve as instructive models for understanding the dynamics of insect systems because the complexity of interactions with iron in mammals has been quite thoroughly explored. A complex interplay between iron and other trace minerals such as zinc, copper, and calcium and proteins, citric acid, ascorbic acid, and phytic acid has been demonstrated (Jovani et al. 2001). Phytic acid was found to reduce absorption of iron and zinc, but the addition of ascorbic acid to the diet (infant formula) increased iron absorption, reversing some of the effect of the phytic acid (Jovani et al. 2001). It was found unexpectedly that the same benefit of ascorbic acid did not apply to zinc and that endogenous protein decreased iron absorption but did not affect calcium absorption (Jovani et al. 2001). This research demonstrates the unpredictable nature of interactions between iron and other trace minerals and dietary components such as proteins, phytic acid, citric acid, and ascorbic acid.

Table 8.1 Iron Content of Various Foods

Food	Iron Content (mg/100 g)	Water Content (%)	Iron Content/100 g Dry Weight
Rice flour, white	0.35	12	0.40
Rice flour, brown	1.98	12	2.25
Wheat germ, crude	6.26	11	7.04
Wheat germ, toasted	8.05	3.4	8.35
Broccoli flower clusters, raw	0.88	91	9.7
Egg yolk, dried	5.42	3	5.59
Egg yolk, fresh	3.53	49	6.92
Beef liver	6.82	69	22.0
Wheat flour, whole grain	3.88	10	4.3
Soy flour, full fat, roasted	5.82	4	6.06

Source: USDA Nutrient Data Base, 2002.

The picture becomes even more confusing when we consider common food processing treatments of insect diets such as cooking, soaking of legume seeds, extraction with acids or bases, or other processing steps. Soaking legume seeds in a mild (0.5%) sodium bicarbonate solution decreased the amounts and activities of antinutrients, most notably phytic acid, tannins, trypsin inhibitor, and lectin (hemagglutinin) activity, thus increasing the digestibility and bioavailability of key nutrients, including iron and other trace minerals (El-Adawy et al. 2000). It is becoming clear that fermentation greatly influences the nutritional quality of other diet materials in soy products including the destruction of phytic acid by an enzyme that is found in a number of microbial species, including *Aspergillus* spp. (Fukushima 1991).

In all cases described here, seemingly simple, innocuous changes led to unexpectedly large, far-reaching, and generally undesirable impacts on the target insects.

Thus, the complex interactions of diet components with one another, with microbes, and with the diet processing steps often generate outcomes that require special attention if diet and rearing specialists are to have genuine control over their production systems.

8.9 BACK TO THE CONTEXT OF THE INSECTARY AND REARING SYSTEM

Thus far, we have discussed the context of insects in their natural environments or at least in agroecosystems. We contrasted the gradients in nature (light, humidity, temperature, and even nutrient gradients). Almost always in natural systems, insects have choices, and they can use the adaptations that have accrued over hundreds of millions of years of evolution to select favorable living conditions. We have also pointed out in Chapter 1 that domestication versus *natural* population genetics constrains insects in ways that nature overcomes with diversity that has an enormous scope in natural ecosystems. And when we force a small population of insects through the restrictions of rearing, and we simplify the environmental conditions as treated in Chapter 1, where physiological ecology was discussed, we end up with a population of insects that are in a precarious state of survival and population steady state or homeostasis. The take-home message of all this discussion (and really for the entire book) is that rearing professionals, including diet specialists, must remain ever vigilant to guard against the forces that can turn a smooth-running rearing operation into chaos.

CHAPTER 9

Nutritional Ecology and Its Links with Artificial Diets

9.1 INTRODUCTION TO NUTRITIONAL ECOLOGY AND ARTIFICIAL DIETS

A significant portion of the progress in the science and technology of artificial diets has come from fundamental studies of the composition of natural foods of insects and in understanding how insect feeding mechanisms deal with the nutrients and how they cope with the antinutrients that are often found to be present in the natural foods. A large part of the current knowledge of these insect feeding dynamics and mechanisms is derived from the field known as *nutritional ecology*, a hybrid of nutrition and ecology. Although nutritional ecology of insects had its origins in such early works as those of Fraenkel throughout the 1940s and 1950s, Scriber and Slansky (1981) used this term to compare feeding efficiencies of insects from various feeding niches. Taken in an even broader sense, nutritional ecology, as it pertains to insects, is the study of natural foods of insects and the interplay (or reciprocal interactions) between insect feeding systems, including nutritionally related aspects of metabolic pathways and the foods insects consume in nature (i.e., outside of the laboratory).

Besides the classical studies of feeding efficiencies articulated almost simultaneously by Waldbauer (1968) and Gordon (1968), there has emerged an extensive literature on feeding efficiencies with respect to different diets and various diet components. Another subset of nutritional ecology that has helped to advance the scientific understanding of insect diets is the study of plant (and other food source-derived) secondary substances. Attention to this approach was pioneered by G. Fraenkel in a series of now classical papers (1959 and Hsiao and Fraenkel 1968) in which key components from plants were characterized in their roles as regulators of insect feeding responses. And, of course, the overall nutritional composition of insects' natural foods has been a long-standing mainstay of information that has been adapted by researchers in the field of artificial diets.

Attention to the gross nutritional composition of natural foods clearly helps us understand the needs of an insect targeted for rearing on artificial diet. So, for example, early workers on diets for pink bollworm needed to know the gross composition of cotton bolls with regard to water, protein, carbohydrate, and lipid content so that these profiles (at different stages of development of the boll) could be mirrored in the incipient artificial diets (Vanderzant and Adkisson, etc.). Similarly, early progress in pioneering diet work with silk moths (Ito 1961a, b, Ito et al. 1975) and other lepidopteron larvae was facilitated by these early workers having gotten a grasp of broad or gross nutrient profiles of the natural foods of their targeted subjects. Although progress was sometimes made despite the lack of information about the nutritional needs of the target insects and of the nutritional composition of diet components, undoubtedly, progress would have been faster and more comprehensive if these basics were better understood. Also, it would have been useful if the insights of the pioneers and some of their later counterparts had been published in detail, explaining the rationale for including or excluding certain components or the rationale for certain processing steps had been more fully explained. An excellent example of such explanation is found in a chapter written by Mittler (1972) on the interaction of diet components and the rationale for procedures that Mittler himself and several of his notable colleagues had discovered. Unfortunately, even when authors are able and willing to provide rationale for choices of components or procedures, journal editors often discourage such explanations. This point is discussed in detail by Cohen (2001).

9.2 NUTRIENTS AND ANTINUTRIENTS IN THE FOODS OF INSECTS

Although there has long been recognition that ecological interactions are governed by nutritional relationships, the question of how nutrients and the plant secondary compounds (many of which are now called antinutrients) influence the choices made by insects was articulated by Fraenkel (1959). Soon after Fraenkel provided a *raison d'etre* for the many nonnutritional secondary chemicals from plants, an important new movement began, one in which a great number of plant chemicals were identified and tied to plant/insect interactions (Table 9.1).

The surge of research activities probing the intricacies of plant/insect interactions included revelations that the chemicals of plants influenced multiple trophic levels as elegantly demonstrated by Brower (1969) and Brower and Brower (1964), who showed that toxins known as cardenolides and cardiac glycosides from milkweed plants eaten by monarch butterfly larvae served to deter predators of the adults, which had the toxins sequestered in their bodies. Birds that had a bad experience with the foul tasting and toxic orange and black butterflies avoided further consumption of similarly colored insects. In fact, the orange and black coloration is so widely associated with noxious tasting and toxic species that there is a predisposition among potential predators to avoid potential prey that advertise their unpalatable character with warning colors, known as aposematic coloration.

Interestingly, some insects that are distasteful and that have warning coloration manufacture their own toxins while others derive their toxins from their foods. The blister beetles (family Meloidae) are examples of insects that produce their toxic component, cantharadin, using their own metabolic pathways. However, a huge variety of insects, including many members of Lepidoptera, Coleoptera, and Hemiptera, use defensively toxins from their foods. Still other insects consume toxic or generally unpalatable foods and use various mechanisms of avoidance of the toxins or the metabolic process of detoxification. The assessment of the fate of toxins and palatability-reducing substances is too large to be treated in detail here, but several reviews and reports survey this topic (Feeny 1970, Rothschild 1972, Duffey 1980, Harborne 1988, Rosenthal and Berenbaum 1992, Bernays and Chapman 1994).

What emerges from the many studies of plant secondary chemistry and its relationship to insect feeding and especially to use of artificial diets is that many insects use plant secondary metabolites

Table 9.1 List of Several Secondary Compounds Known to Have an Influence on Plant/Insect Interactions

Insect Affected	Compounds (Plant Source)
Aphid, *Brevicoryne brassica*	Glucosinolate: sinigrin (cabbage)[a]
Beetle, *Diabrotica undecimpunctata*	Triterpenoids: cucurbitacins (watermelon)[a]
Butterfly, *Papilio ajax*	Essential oils (fennel)[a]
Moth, *Bombyx mori*	Flavenoids and essential oils (mulberry)[a]
Ant, *Atta cephalotes*	Monoterpene: limonene (citrus fruit)[b]
Beetle, *Leptinotarsa decemlineata*	Alkaloid: demissine (*Solanum demissum*) (in the nightshade family)[b]
Beetle, *Scolytus multistriatus*	Quinone: juglone (*Carya ovata*)[b]
Moth, *Spodoptera ornithogallii*	Sesquiterpene lactone: glaucolide-A (*Vernonia glauca*)[b]
Moth, *Heliothis* (*Helicoverpa*) *zea*	Terpenoid: gossypol (cotton)[b]
Beetle (boll weevil), *Anthonomus grandis*	Terpenoid: gossypol (cotton)[a]

Source: From Harborne, J.B., *Introduction to Ecological Biochemistry*, 3rd edn., Academic Press, Inc., London, U.K., 1988.

[a] Indicates that the compound is a feeding attractant.

[b] Indicates that the compound is a feeding deterrent.

as guides to selection of what is for them a suitable host plant. In effect, the insects are using the plant chemicals as taxonomic guides, and when such insects are targets for our rearing programs that are based on artificial diets, it may be important to recognize the plant chemicals that stimulate normal feeding and metabolic responses.

This concept is important in the development and provision of artificial diets for many specialists such as the insects that feed on weeds that we are anxious to control. For example, Blossey et al. (2000) were careful to include host material from purple loosestrife when they developed an artificial diet for *Hylobius transovitattus*. Blossey et al. (2000) discussed the rationale for including host material in the artificial diet, largely as a means of providing the chemical feeding stimuli that help *H. transovitattus* recognize the artificial diet as a suitable food source. This is an excellent example of the use of token stimuli that are used to *fool* the insect into accepting what would otherwise be a suitable nutritional mixture but which lacks the specific chemicals that characterize the actual host plant. The Blossey et al. (2000) study was not aimed at isolating the specific chemicals that characterize the purple loosestrife, but other nutritional ecology/physiology studies have been aimed at determining the specific chemical agents that insects use as cues or "taxonomic identification tags." For example, the nature of the activity of plants in the mustard family (*Brassica* spp.) has been extensively characterized such as in Ma's (1972) and Schoonhoven's (1972) classical studies of sinigrin [a glucosinolate containing a thiocyanate (R=N–C=S) and glucose]. Extensive discussions of the ecology and evolution of the nature of various plant secondary compounds are to be found in Bernays and Chapman (1994) (Table 9.2).

Bernays and Chapman (1994) point out that many of these secondary chemicals act in concert with other dietary substances to enhance feeding responses. For example, sinigrin and sucrose together induce more feeding than sinigrin alone. While it is clear that the responses of some insects to these chemicals is the eliciting of feeding, for most insect species, these compounds are feeding deterrents, rather than feeding stimuli (Bernays and Chapman 1994). Therefore, it can be generalized that while these and other plant secondary compounds are used in a manner that is called token stimuli (sign stimuli, according the usage of Bernays and Chapman), the general function of these chemicals is as a protection against herbivory on the part of the plants that manufacture them (Fraenkel 1959). This point is emphasized by an extensive literature that deals with experiments and observations that explain the mechanisms of the plant chemicals' activity and the means by which certain insects can bypass the toxic effects. In this vein, Feeny (1970) characterized the plant defenses as being either "qualitative" for "quantitative," the former expression meaning that the plant chemicals were found in small amounts but that they had acute or very strong toxicity. The plants that produce cyanide are good examples of this and are represented by the species, *Lotus corniculatus* (family Fabaceae), and the cyanogenic compounds from this plant are linamarin and lotaustralin (Bernays and Chapman 1994).

Table 9.2 Plant Secondary Compounds That Positively Influence Feeding Behavior in Insects

Chemical	Insect Species Affected	Plant Group of Compound's Origin
Caffeic acid	*B. mori*	General
Chlorogenic acid	*L. decemlineata* and *B. mori*	General
Rutin	*Schistocera americana*, *Heliothis virescens*, *Plagiodera versicolora*	General
Tannic acid (tannins)	*Anacridium melanorhodon*, *Lymantria dispar*	General
Gossypol	*A. grandis*	Cotton
Morin	*B. mori*	Mulberry
Salicin	*P. versicolora*, *Laothoe populi*	Willow
Sinigrin	*B. brassicae*, *Pieris brassicae*	Cabbage family

Source: After Bernays, E.A. and Chapman, R.F., *Host-Plant Selection by Phytophagous Insects*, Chapman and Hall, New York, 1994.

Another familiar example of the qualitative defenses includes the milkweed family with its acute poisons, the cardenolides or cardiac glycosides, described earlier and in the works of Brower (1969) and Brower and Brower (1964). The quantitative defensive compounds are present in large amounts (sometimes as much as 40%) of the dry weight of the plant material (Feeny 1970) and act by digestive interference or disruption of the feeding process, rather than as a potent metabolic poison. Terpenoid and phenolic compounds are examples of the so-called quantitative factors, and examples are found in oaks, walnuts, and many other species of plants that have a conspicuous and long-term presence as opposed to short-term, ephemeral plants, which often exhibit the presence of the metabolic poisons (Feeny 1970). From a perspective of the totality of insect/plant interactions, considering the behavioral and biological chemistry of nutrients, secondary substances, and inert materials, what emerges is a model of complex interplay between all of these plant components and the insects that respond to them. From this, the emergent principle is the fact that insects use certain features, especially chemical cues, as stimuli to undergo their normal feeding responses. Having knowledge of this stimulus–response-driven phenomenon, insect diet researchers should have a strong basis for developing artificial diets that both serve the insect's nutritional needs and draw the feeding responses that make diets suitable for production of robust insects.

9.3 PLANT SECONDARY COMPOUNDS, FEEDING, AND ARTIFICIAL DIETS

Once it is understood that certain chemicals other than the standard, well-accepted nutrients play a role in feeding by target insects, this knowledge can be applied to the development or enhancement of artificial diets for those target species. It is clearly a worthwhile strategy to consider incorporation of the exotic chemicals that are used by monophagous or oligophagous insects to distinguish the plant species for which they are specialized. Monarch butterflies (*Danaus plexippus*) select host plants in the genus of milkweeds, *Asclepias* spp., using the cardenolides as stored chemicals that confer upon the larvae and adults of this insect species a protection from predation (Brower and Brower 1964). Therefore, incorporation of cardenolides in the diet of monarch larvae would seem to be a rational approach to the development of artificial diets for this species. However, care must be exercised in adding cardenolides (and other host plant extracts) to the diet, because as Bernays and Chapman (1994) point out: there is a critical level of cardenolides that elicits feeding. An excess of the critical amount will inhibit feeding and will, in turn, subvert the desired effects aimed at in adding these compounds. Also, the ratios of the different kinds of cardenolides must be reflective of what is found in the original host plant, as profiles of secondary compounds are an important signature of the host plant. Furthermore, there is the matter of optimization of combinations of phagostimulatory nutrients such as sugars and amino acids with appropriate amounts of secondary chemicals. And finally, the matrix of the natural versus artificial diets must be taken into consideration—the point of the following paragraph.

There are nuances of the potential use of secondary substances as additives that further complicate the potential practice of using plant signature chemicals to induce or improve feeding responses. Gossypol, for example, is a terpenoid compound that has been shown to act as a feeding deterrent to first instar larvae of *H. virescens* and *H. zea* (Parrott et al. 1983). At the same time, gossypol is also a feeding stimulant for boll weevils (Bernays and Chapman 1994) and in low concentrations may also evoke feeding by *H. virescens* and *H. zea*. Therefore, in developing or improving diets for these various cotton-feeding species, it is rationale to include cotton plant extracts or flours, meals (as used by Sterling and Adkisson 1966), or other solids such as cotton square powder (as used by Oliver et al. 1970). However, while diets for boll weevils were improved by utilization of the gossypol-containing cotton plant solids, the cotton solid–containing diets for boll worms proved to reduce feeding rates compared to diets free of gossypol (Oliver et al. 1970). These results must be interpreted with caution for several reasons. First, the amounts of gossypol were not controlled in these experiments,

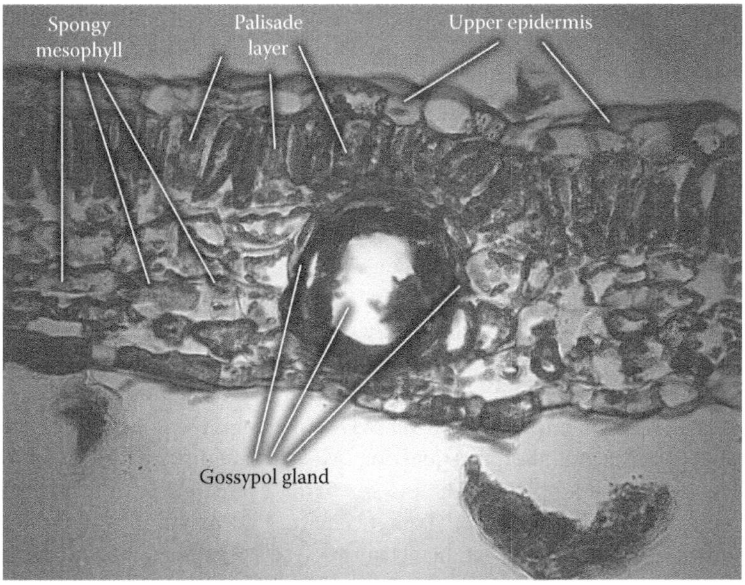

Figure 9.1 Plant secondary chemicals are often localized in glandular regions or other compartments such as the gossypol glands, which also contain dark-staining anthocyanins.

and the proportions of this secondary chemical may have been inappropriate (too little or too much), especially in relationship to the other diet components. Second, the distribution of gossypol in the diet was not accounted for and was, in all likelihood, released from the glands where it is normally confined in the plant (Figure 9.1). Furthermore, the granules or compartments of anthocyanins (Figure 9.1) were also likely to be disrupted, releasing these compounds from the sites where they are normally sequestered. Thus, the addition of unpurified extracts or powders may unleash a much more complex chemistry than was anticipated. These points are in concert with the discussion of order and chaos in diets in Chapter 9. The concept of adding plant extracts and solids to artificial diets is discussed further in the section on development of diets (Chapter 10).

9.4 EFFICIENCY INDICES

A large, important sector of nutritional ecology that has considerable significance in the development and assessment of artificial diets is the domain of nutritional indices or efficiency indices as defined by Waldbauer (1968). Broadly, these indices define or describe the efficiencies of digestion or utilization of diets or diet components; that is, how easily the insect can convert foods to their own biomass. The most commonly used indices are approximate digestibility (AD), *efficiency of conversion of ingested food to insect biomass* (ECI), *efficiency of conversion of digested food to biomass* (ECD), *relative growth index* (RGI), and *relative consumption index* (RCI). The formulae for these indices, taken from Waldbauer (1968), are listed and explained here:

$$\text{Approximate digestibility (AD)} = \frac{\text{Weight of food ingested} - \text{Weight of feces}}{\text{Weight of food ingested}} \times 100$$

The AD takes into account the amount of food consumed for a given period of time and the amount of fecal matter (egested material) that was eliminated and that was associated with the feeding cycle of the insect being tested. As is the case with all of the indices discussed here, the evaluation is more accurate if the materials are dried prior to measurement, and the measurements are further refined if allowances are made for natural losses that might take place as a result of microbial decomposition and evaporative losses of volatile components (Waldbauer 1968). Finally, if nitrogenous waste products are measured (or estimated) and their mass subtracted from the fecal biomass, the estimates of AD are more reflective of true digestibility of the foods in question. Clearly, this measurement can reflect a very important aspect of diets—their digestive availability to the insects in question. Any changes in the diet that increase the AD would be improvements in the diet in question. For example, Brewer and King (1979) demonstrated that Heliothine larvae had higher ADs when corn cob grits were added to the diets. This very telling finding indicated that somehow corn cob grits, which are thought to be nonnutritive, increased the ability of Heliothine larvae to digest and extract a higher portion of the diet's components than could be achieved without the corn cob grits. The same logic could be repeated for tests of any diet component, and wherever ADs were improved, that treatment would likely represent an improvement of the diet in question. Measurement of AD can be achieved without destruction or major interference with the insects being evaluated. It must be remembered that such indices are species specific, as well as diet specific.

As Waldbauer (1968) has pointed out, in measuring AD, it is preferable if the insect has an empty gut, meaning that the test should be conducted immediately after hatch or after a molt, given that insects empty their guts of essentially all of their contents prior to ecdysis. After a preset period of feeding (based on and appropriate to the life history of the insect), the diet, which had been weighed prior to allowing the insect to feed, is dried and weighed to measure the postfeeding dry weight. The prefeeding dry weight must be determined from weighing, drying, and reweighing fresh diet, using samples not used in the feeding experiments. When final calculations are made, the dry weight of the remaining diet is subtracted from the estimate of the weight of the original diet aliquot. The reason for using dry weights is that they are the least biased with error that derives from differences in loss of volatile components—especially water. It is intuitively evident that freshly made diet will have a higher percentage of water than a diet that has been held in the rearing container for several days at rearing room temperatures and less than saturation humidity.

The second index that is widely used to evaluate insects' responses to diets or diet ingredients is the efficiency of conversion of ingested food to insect biomass (ECI) whose formula is as follows:

$$\mathrm{ECI} = \frac{\text{Weight gained}}{\text{Weight of food ingested}} \times 100$$

The ECI is also known as the *growth efficiency* index. The measurement of the two components in this index should, like the AD, be measured with dried materials (meaning that the insect must be destroyed in order to obtain an accurate dry weight). The live insect in each test must be preweighed, and its dry weight must be calculated from the dry weight measurements of individuals of similar age and physiological condition. The test insects must be sacrificed after the feeding trials so that their dry weights can be measured and from which the original dry weight is subtracted. The difference in the dry weight of the food eaten is divided into the weight difference of the insect and multiplied by 100 to convert values into percentages. The ECI is a robust index for showing the overall ability of the insect to use the food in question for building biomass. An appealing factor in ECI measurement is that accurate estimates can be made without recourse to collecting the fecal materials—a matter of complexity and difficulty in working with insects whose frass is difficult to collect, especially those that are liquid feeders and which produce liquid frass.

The third index, efficiency of conversion of digested material to insect biomass (ECD), is calculated with this formula:

$$\text{ECD} = \frac{\text{Weight gained}}{\text{Weight of food ingested} - \text{Weight of feces}} \times 100$$

ECD is also known as metabolic efficiency, because it takes into account already digested food (the weight of the food ingested minus the weight of feces). The measurement of ECD provides a resolution of the question of the food's overall nutritional value once the nondigestible materials are eliminated. It is a good tool for assessing the nutrient balance within the diet's digestible components. For example, if a diet contains an excess of carbohydrate to the exclusion of lipid or protein (all three of which are usually highly digestible), the insect will not gain as much weight from a given digested portion of that diet as it would from an equal amount of digested diet that had better nutrient proportions. Once a low ECD has been detected in a feeding experiment, the researcher has a clue that the balance of nutrients in the diet is substandard. From this, the diet researcher can make more directed efforts to pinpoint the imbalance.

Although there are some limitations inherent in the difficulties of conducting quantitative feeding experiments, studies of feeding efficiencies can reveal very useful information about the feeding biology and feeding fitness of the insects being studied or reared for various purposes. The indices can also provide high-resolution information on whether food components or whole foods are negatively or positively impacting the insects. For example, recently, Chang et al. (2000) and Pechan et al. (2000) demonstrated that a cysteine proteinase from maize plants caused a decrease in the efficiency of digestion and absorption in fall armyworms fed with natural and artificial diets containing this substance. This led to a further work (Pechan et al. 2002) which clarified the mechanism of faulty digestion and absorption—damage to the peritrophic matrix caused by an enzyme in the resistant corn.

In an earlier study, Cohen and Patana (1984) showed that even after more than 25 years of being reared on artificial diet, members of a colony of *H. zea* had higher ECI, ECD, and AD when fed a natural host (green beans) than they did when fed artificial diet upon which their ancestors had been reared for more than 200 generations. These very detailed and sensitive tests showed that laboratory rearing did not reduce or restrict the fitness of these insects in terms of ability to utilize their natural foods. In another study incorporating artificial diet, Cohen and Urias (1988) found that food utilization indices showed that the plant secondary compound rutin (a flavenol glycoside) did not adversely affect the utilization indices of predatory lygaeids, (*Geocoris punctipes*), even when as much as 1.0% rutin was added to the diet. The high level of resolution available from the food utilization studies such as this one provided a very sensitive assessment of the potential impact of the plant compound outside of the context of a natural host plant with its complex and uncontrolled variables (Cohen and Urias 1988). The differences between utilization efficiencies can be illustrated graphically in making comparisons of different diets or diets with different additives that are being compared. Such a diagram is illustrated in Figure 9.2.

The consumption index (CI or more properly, the *relative* consumption index) is useful in performing an overall assessment of the palatability of a given food or an extract of a food or putative toxin. The index

$$\text{CI} = \frac{\text{Weight of food eaten (fresh or dry)}}{\text{Time (days or hours)} \cdot \text{Mean weight of insect(fresh or dry)}}$$

allows for the weight-specific and time-corrected assessment of consumption. The mean weight of the insect during a given period of feeding is measured at regular intervals, and an average is taken,

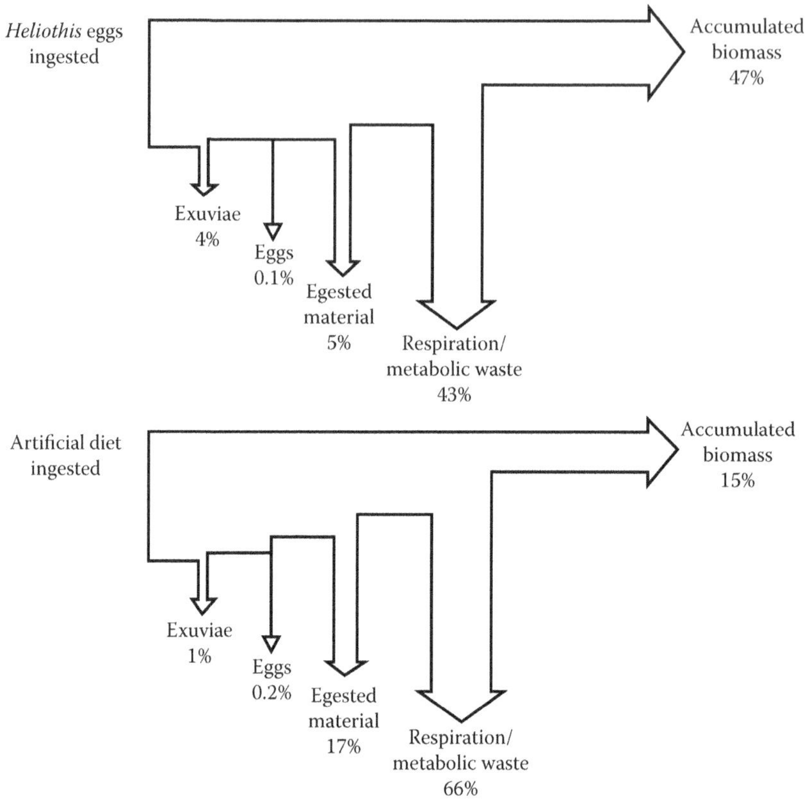

Figure 9.2 Diagram illustrating differences between utilization efficiencies of different diets or diets with different additives.

or if the growth curve is irregular, the mean value must be calculated in a more complicated way: by taking the area under the curve of the insect's weight plotted over the full time interval when the feeding measurements took place.

An index that is similar to the CI is the relative growth rate (RGR) that is measured as follows:

$$\text{RGR} = \frac{\text{Weight gain of insect (fresh or dry)}}{\text{Time (days or hours)} \cdot \text{Mean weight of insect (fresh or dry)}}.$$

As was the case with the CI assessment, the mean weight is estimated either as an average for the feeding time interval or by taking the area under the curve. The value of using this index is that it allows the researcher to evaluate how well a diet is utilized by a target insect in terms of actual biomass accumulation. While the CI gives a good picture of how palatable the diet is, the relative growth index allows interpretation of the actual nutritional quality of the diet. Measuring both parameters can allow the researcher to distinguish between the questions: how palatable is the diet? and how nutritious is the diet?

The accuracy of the nutritional indices is dependent upon several precautions that an experimenter should heed. First, it is advisable to use dry weights instead of fresh weights. Fresh weights are variable according to the conditions of humidity, length of time between sample weighing, and microbial decay of such materials as fecal matter (frass) and diets. In this vein, it is also useful to use a correction factor for what Waldbauer (1968) calls "natural losses." Correction for natural losses can inject a discouraging complexity to the nutritional index experiments, but especially if the experiment is going to last for several days to several weeks, accuracy demands that the researcher use some kind of correction. The correction begins with using a control where the insect is not present so that the various changes in food weight, other than that imposed by the insect, can be determined. The formula suggested by Waldbauer (1968) is as follows:

$$\text{Corrected weight of food eaten} = [1 - a/2]\ [W - (L + bL)]$$

where

a is the initial weight of the food aliquot
b is the ratio of loss to the final weight of the aliquot
W is the weight of the food introduced
L is the weight of the uneaten food

A final precaution suggested here and by Waldbauer (1968) is that whenever measurements of frass are included in estimates of nutritional efficiency, investigators should measure the nitrogenous waste products (such as uric acid). Subtracting the biomass of nitrogenous wastes will restrict measurements of frass to reflect only the amount of undigested food, rather than food plus metabolic wastes.

These types of inquiry are very robust ways of measuring the quality of diets in relationship to the specific target insects. Although they require extra steps beyond simple measurements of weights or linear dimensions of a given stage of insect, they yield a very rich amount and quality of information about how diets are working or failing to work. They allow diet professionals to assess a differential between the suitability of a food or food component in terms of palatability versus digestibility versus absorptive potential versus toxicity. Collectively and with the experiments done properly, the indices can provide a wealth of information and feedback about diets, both in terms of their development and any changes that are being considered. As such, the nutritional indices should be considered as tools for quality control and diet development programs. One last point, a recent study by Zha and Cohen (2014) shows the effects (generally harmful ones) from antifungal agents as indicated by changes in the nutritional indices. This is further discussed in Chapter 13 where antifungal agents are discussed in detail. However, for the present discussion, it is important to understand that the nutritional indices can be used to dissect the effects of potential toxins, even ones routinely used in diets such as antimicrobial agents.

9.5 SIFTING THROUGH THE FUNCTIONAL ROLE OF COMPONENTS

Finally, it is valuable to sort out the issues of whether given diet components act as nutrients, phagostimulants, phagodeterrents, or as functional diet components that do not provide one of these three functions. So to take the *Hylobius* diet as an example, it would be useful to know which part of the purple loosestrife powder that was added to the diet actually was a token stimulus. Should this be known, that chemical could be added to an otherwise completely artificial diet, and the presence of the token stimuli could then be used to promote feeding by an insect that is a fastidious specialist on purple loosestrife. If caution is taken to determine, via extraction/purification/feeding experiments, that it is either a single compound or a complex mixture of compounds that signal *Hylobius* to feed on the loosestrife, then use of these compounds could be used not only to stimulate feeding, but also to "keep the target insect honest" about remaining a specialist on the loosestrife. It is essential to a program of weed biological control to retain, fastidiously, the specialization for which the weed-eating insects were chosen in the first place.

Next, staying with the example of *Hylobius* and its purple loosestrife-containing diet, it would be very useful, especially in terms of the economy of diet production, to determine whether or not the loosestrife-derived powder contained any nutrients that were not present in the nonloosestrife portions of the diet. In other words, did the loosestrife powder contain some amino acid, carbohydrate, lipid, vitamin, or mineral that is essential to the insect but not present in the other ingredients? Should this be the case, then special efforts could be made to find an alternative and less expensive source of that nutrient or nutrient complex. Such determination would require biochemical analysis of the purple loosestrife powder, possibly coupled with bioassays.

9.6 ARTIFICIAL DIETS AS DELIVERY SYSTEMS FOR TESTING ANTINUTRIENTS AND TOXINS

One important use of artificial diets is as a delivery system for a variety of substances that researchers wish to test against target insects (Cohen 2001). For example, artificial diet was used to deliver a plant secondary compound, xanthotoxin, to test ascorbic acid and uric acid responses to this putative toxin against generalist and specialist herbivores (Timmermann et al. 1999). Likewise, Forcada et al. (1999) used artificial diet to deliver Bt toxin to test the effects of this substance on *H. virescens*. In these cases and in numerous others, the test substances are added to the diet either during the mixing process or afterward. Sometimes, the test substance is sprinkled or poured onto the completed diet without efforts to thoroughly mix it for homogeneous distribution.

The rationale behind the use of artificial diets as delivery systems is that diets are simpler and more convenient means of administering per os (orally) in a medium that has been proven to be suitable to the target insect or to a trophically similar species. The underlying assumptions behind this rationale are that (1) the substance is not affected chemically by the diet context, (2) the diet is as attractive and palatable with the substance as it is without it, (3) the substance cannot be avoided by specialized feeding mechanisms (such as extraoral digestion), (4) the ingredients in the basic diet formulation do not mask or enhance the effects of the substance being tested, and (5) the diet is fully suitable to support and maintain healthy subjects. However, these assumptions are often tacit and unsupported by careful testing and documentation.

In assumption (1), it is important to recognize that diet ingredients are not necessarily inert. For example, the predator diets developed by the present author (Cohen 1985a, Cohen and Smith 1998) contain raw beef liver, which contains various active enzymes, including a suite of detoxification enzymes that could very possibly serve to modify putative toxins. Such biochemical modifications would certainly compromise the accuracy of tests that are aimed at determining whether or not a given toxin is harmful to predators that consume these diets. In order to reduce concerns about assumption (1), a minimal demonstration would include a biochemical or chemical test demonstrating the persistence of the putative toxin throughout a reasonable exposure/incubation period. So, for example, if researchers were testing a Bt toxin against predators considered to be nontargets of the toxin, they would be well served to include the toxin in the diet and perform sequential tests such as electrophoresis to assure the preservation of the toxin. In concert with this, a bioassay employing an insect known to be susceptible to the toxin could be fed with the diet or extracts from the diet. Furthermore, it should be recognized that diets, in general, contain numerous other very active components that could impact the chemistry of putative toxins, including iron and ascorbic acid, which can act in concert to alter the chemistry of lipids via peroxidation reactions (Chapter 5). Finally, diets such as those mentioned earier often contain microbial contaminants, which could also profoundly alter the putative toxin.

For tacit assumption (2), the relative attractiveness of the diet with and without the test substance can be measured by employment of a good feeding assay. Several assays that were covered earlier in this chapter are appropriate, but it is essential that demonstration of a negative effect of the test substances be interpreted properly and not interpreted as toxicity if it is really a matter of feeding deterrence imposed by the nature of the substance or the presentation procedure. For example, if the substance is mixing in with an already gelled diet, did the mixing disrupt the gel structure and the physical characteristics of the diet that helped make it palatable to the target insect? On the other hand, if the substance is mixed early on in the diet production procedure, did the heating or other aspects of processing (formation of protein/carbohydrate complexes, chelation, or pH effects) alter the toxicity or overall chemistry of the test substance as described in the earlier paragraph?

Regarding assumption (3), it was thoroughly discussed in several places in this book, especially Chapter 7 on feeding mechanisms, that many insects have very specialized feeding mechanisms that allow them modify foods or to be highly selective of the foods that they ingest. Insects that use extraoral digestion, for example, ingest only what their feeding process allows them to mobilize and leave behind intractable components. It is also noteworthy that application of the toxins to the surface of a diet may not deliver a significant dose of the substance, considering that the insect is feeding on components deep within the diet. Insects have an incredible capacity to be selective in what they ingest. Therefore, any heterogeneity in their diet provides an opportunity to avoid toxic materials.

Assumption (4), masking effects of diet ingredients, is another wrinkle in the diet as a delivery system that is hard to assess. However, careful consideration should make it clear that the potentially adverse effects of a substance such as a protease inhibitor can be masked by providing a diet that contains ample and high-quality protein. Likewise, chelating agents in the diet (additives and natural agents) could mask a potentially adverse effect of any number of potentially harmful substances such as heavy metals. The masking effects can also complicate interpretation of multiple effects of diet additives. In conclusion, it is evident from these considerations that the chemical and structural dynamics of insect diets complicate their use as delivery systems for substances to be tested. This caveat is not intended to discourage or preclude the use of artificial diets as delivery systems for testing various additives; but instead, it is meant to guide researchers through planning high-quality, meaningful bioassays.

9.7 SPECIAL EXAMPLES OF NUTRITIONAL ECOLOGY APPLICATIONS

The scope of nutritional ecology is the examination of our organism of interest in the context of its trophic interactions from an environmental context. This means that all factors that influence the success or failure of target species in a feeding system context must be considered. Although this subject reaches into several other topics in this book, there are some excellent examples of ecological relationships that have nutritional implications that reach into our efforts to rear target insects. One more important point: although nutritional ecology is a large field encompassing all aspects of consumer/food interactions, the discussion in this chapter has been limited to aspects of nutritional ecology that have strong bearing on or relationship to insect diets. Two examples provided here: (1) a specific case of efforts to develop an artificial diet for hemlock woolly adelgids (HWAs), which met with many obstacles that stemmed from lack of an adequate understanding of the nutritional ecology (host seeking, host acceptance, and mechanisms of nutrient extraction by the insects in question) and (2) the broad ecology of behavioral and nutritional interactions between drosophilids and the environmental yeasts—a study in reciprocal interactions.

9.7.1 Efforts to Develop an Artificial Diet-Based Rearing System for Hemlock Woolly Adelgids

The first example of this topic is culminated in a paper by Oten et al. (2014). Some years ago, a group from North Carolina State University's forest entomology laboratory led by Dr. Fred P. Hain were charged with discovering ways to manage the exotic, invasive pest HWA, *Adelges tsugae* Annand (Hemiptera: Adelgidae) through a host plant resistance approach. It was clear that having an artificial diet-based rearing system for the adelgids would be helpful for a program of discovery of potential mechanisms of resistance to be used against this pest. However, early on in the project, it became apparent that with the exception of a few studies such as that of Young et al. (1995), little was known of the feeding habits of HWAs. The work of Young et al. was based on microscopy (mainly scanning and transmission electron microscopy), and it revealed the HWA targeted xylem ray parenchyma as feeding sites—based on direct observations of stylet placement in the needle petiole of hemlock trees.

However, for the purpose of developing an artificial diet-based rearing system, the type of tissue (xylem ray parenchyma) was only a partial clue as to how to proceed with diet development. The tasks that needed to be resolved were: (1) What specific kinds of nutrients was the insect finding and using from the parenchyma cells? (2) What physical form (consistency: see Chapter 15) characterized the food: liquid, slurry, semisolid? (3) What cues was the insect using to locate a starting point for its stylet probes? (4) Could the conditions (food characteristics and presentation of foods) simulate adequately the natural feeding process (the nutritional ecology features) to allow artificial production of the insect?

The paper by Oten et al. (2014) does not cover all the heuristic and analytical research that went into this tremendously complex problem, but many diets were tried (without success), ranging from liquids to semisolids, especially slurry diets that contained suspended yeasts, algae, conventional materials such as wheat germ, soy, and rice bran (to mention a few), and the research also involved diet presentation systems such as parafilm-covered diets, wax-coated liquid diets (Cohen 1983), and even systems where diets were designed to flow through feeding arenas (designed by Mr. John Strider). None of these systems were successful at even getting the HWA to feed sufficiently to progress through a single developmental stage. The intransigence of this problem and the targeted insect informed the research team that too little was understood of the feeding biology (nutritional ecology) to elicit normal feeding.

Many of the findings that help to fill in the gaps are presented in Oten et al. (2014), including the characterization of several trophic enzymes (protease and amylase for cues to targeted nutrients: protein and starch) and antidefensive enzymes against plant defenses (polyphenol oxidase and peroxidase). Also, the study by Oten et al. (2014) shows the extremely regular, pin-pointed path of the HWA's stylets on the hemlock surface, including the angle of attack and stylet insertion. The stylet sheath imaging showed that the HWA uses surface features such as the following observation: "The sheath material adheres to the plant, beginning distal to the stem in reference to the abscission layer, then travels across the abscission layer to the insertion point on the pulvinus..." (Oten et al. 2014). Therefore, the surface configuration seems important as part of the insect's interpretation of where it must probe: the crease or gap of the abscission layer and the swollen node-like structure—the pulvinus—at the base of the needle. This raises the question about the possible efficacy of structuring artificial diet presentation systems to contain the crease and the swelling at appropriate distances from one another to help stimulate feeding by HWA. Efforts to use chemical simulation of host plant surface and internal chemistry by adding extracts from exterior surfaces and interior contents failed to induce feeding responses, so a physical model of hemlock needle surface may offer more success.

9.7.2 Microbially Based Tritrophic Interactions

The topic of microbially based interactions between certain insects is enriched by the longest history of rearing system and artificial diet science and technology. Although this topic is covered in other parts of this book (Chapters 4, 7, 13, and 16), it deserves special mention in a section on nutritional ecology. There is probably no insect/food system in nature that does not have some mutualistic or at least commensal microbial interaction between insects and their microbiome. The symbioses include endosymbionts (internal), ectosymbionts (external), and even "agricultural systems" that specialists like leaf-cutting ants and ambrosia beetles use. When ecologists provide examples of the shaping of ecosystems by keystone species of insects (such as termites in the savanna or pampas of South America or ants in tropical rainforests), they are never really talking only of the insects and how they shape ecological communities by recycling nutrients and providing a base of food that fuels the system as a whole, but they are really holding up examples of the insect/microbe relations that underlie these relationships. Furthermore, we know that whatever aphids and whiteflies do as pests in cropping (or natural) systems, they can only accomplish with their bacterial symbionts in their mycetomes. This point goes on with termites and their plethora of gut microbes (Chapters 7 and 13), and with wood-eating beetles and their fungal partners that they carry in their special mycangial pouches or recesses. Even lice, bedbugs, and other parasites depend upon internal microbial flora to help them obtain nutrients from foods that are deficient in certain vitamins or amino acids.

Of all these many stories of partnerships and their ecological consequences, the one with the longest history of study and interpretation is the *Drosophila*/yeast interaction. The first publication that reported that *Drosophila* consumed yeast specifically—not just rotting fruit—was Guyenot (1913b). Although Guyenot had explained the yeast-eating habits of *Drosophila*, Loeb (1915) performed an experiment where he tried to determine the minimal requirements of *Drosophila* by providing it with ammonium compounds, sugar, and minerals. In that report, Loeb interpreted his results of growth on media that he had sterilized to mean that the flies were growing aseptically. However, in response to Loeb's 1915 and a later report by Loeb, Baumberger (1917) wrote the following:

> In 1915 Loeb raised Drosophila on a synthetic medium consisting of inorganic salts, sugars and ammonium tartrate. He believed the synthetic power of the fly to be as great as that of bacteria. Later, however, he pointed out that the medium used 'is a well-known culture medium for certain microorganisms; e.g., yeast cells which are capable of synthesizing their proteins and other complicated organic compounds from ammonium salts.'
>
> It is very evident from my experiments that the food of Drosophila larvae is yeast.

And Baumberger further wrote the following:

> From the foregoing experiments we may conclude that yeast is not present in the eggs or pupae of Drosophila. However, a loose symbiosis exists between the two organisms. As mentioned above, surface fungus growths disappear in the presence of larvae which often seemed to be more numerous at this point. From these observations I inferred that the larvae fed upon the microorganisms present.

And Baumberger, in conclusion, made these statements, which can be thought to be the foundation of the nutritional ecology in the context of insect/microbial relationships:

> It is probable that the food relations pointed out in my experiments are common to a number of organisms. The house fly, for example, oviposits only in the presence of the odor of fermentation, always has a certain form of bacteria on its body in great abundance and the larva is unable to survive in garbage which gives an acid reaction. Those insects which live in strange media such as strong salt water and petroleum may be associated with microorganisms which have unusual power of oxidation.

We now realize that the scope of the insect/microbe relationships is not confined to oxidation, but the scope of this statement about insects and their interdependence on microbes is a masterful insight. Once this foundation was laid, the studies of the nutritional ecology of *Drosophila* have become the basis for many elegant studies of the nutritional requirements of these cosmopolitan flies, including a landmark paper by Dobzhansky et al. (1956) (Differential attraction of species of

Drosophila to different species of yeasts.), where the mutual interactions of *Drosophila* as a selective force for yeast colonization and yeast species set the stage for selection of *Drosophila* species. Although Theodosius Dobzhansky is best known for his work with genetics and evolution, he and his colleagues clearly understood the importance of understanding the genetics and evolution in the concept of the "ecological theater" in which the "evolutionary play" is performed (a reference to G. Evelyn Hutchinson's series of essays: Hutchinson 1965).

In the context of these interactions, Starmer and his colleagues (Starmer et al. 1976, 1977, 1986, 1990, Ganter et al. 1986, Starmer and Fogleman 1986, and Ganter 2006, to cite a few) produced an extensive body of literature on the evolution and ecology of the trophic interactions between Drosophila and the yeasts that illustrate profoundly the dynamics of the Hutchinsonian "ecological theater" and "evolutionary play." A centerpiece of much of this work was done in the relationship between several *Drosophila* species, several yeast genera and species, and several cacti in the Sonoran Desert and later in other cactus habitats. However, the body of the *Drosophila*/yeast interactions involves many other situations where rotting plant materials play a role, including agricultural systems (such as guava orchards). Besides involving evolutionary ecologists, many (scores) of these studies included the great yeast specialist and the late Professor Herman Jan Phaff whose incredible understanding of yeast diversity and biology gave these studies an added scope and competence for recognizing the complexities of the interactions between the insect and yeast component of this tritrophic system. An important note is that the Phaff collection of yeast is maintained at the University of California by the curator of yeasts Dr. Kyria Boundy-Mills, and one of the valuable services of this collection is available as a resource for research into yeasts.

Finally, it is noteworthy that the tradition of the works cited has been continued with continued research on *Drosophila*/yeast interactions. An example of this work is a recent paper by Stamps et al. (2012) where it is shown how specific *Drosophila* are in their choices of yeasts and how the insect helps shape the community profiles. This is discussed further in Chapters 4 and 13 where the role of yeasts in insect diets is developed.

CHAPTER 10

How Artificial Diets Are Developed and Improved

10.1 DIFFICULTIES IN DIET DEVELOPMENT METHODOLOGIES

Diet development is a process of invention. When we set out to develop a diet for an insect that has not been reared previously on artificial diet, we are immersing ourselves in a new realm, and it would be helpful to have a set of rules that would assure us of success. It would be convenient to have a set of rules such as Koch's postulates in microbiology where the causality of a disease is determined for a putative microbial species. Such a set of "rules" can be considered an algorithm where a logical set of steps are taken in a certain order (study feeding dynamics → analyze natural food → develop model of diet components → test components (bioassay) → return to model in incremental steps of optimization). Unfortunately, no such rules exist nor do tested and validated algorithms exist, though several researchers have tried to develop such a model (e.g., Cohen 1992). Therefore, the process of diet development remains in a fairly chaotic state where we depend on chance more than established, vetted principles.

The intention of this chapter is to provide a review of the salient points of the past 100 years of efforts to develop artificial diets for insects and to highlight the features that were useful in the successes and the features that undermined the diet development (and improvement) process. Just as in any invention, the successful outcomes are more likely to come from researchers who have knowledge of the subject. Like all inventions, diets do not "spring forth full-blown like Pallas Athena from Zeus's head." Rather, diets always come from some previously existing base of knowledge. Therefore, a familiarity with the diet development processes that worked and those that failed is an important starting point.

This leads to a summary of the kinds of approaches used by researchers in diet development and improvement for insect rearing. First, researchers have started with diets that were *derived* for other insects or for other organisms such as microbes. Second, diet researchers have used *heuristic* techniques (discussed in more detail in the following text) where a process of trial and error is applied. It is important to recognize that in heuristic systems, the trial and error or intuition-based procedures are not wild guesses, but rather decisions that stem from experience-based judgments. Third, diet developers have used the process of studying the target insects to get cues and clues from the *natural feeding biology* that were applicable to artificial diets. Fourth, researchers have used *analysis of the natural food and/or analysis of feeding/digestive mechanisms* to build models of the incipient artificial diet. Fifth, some researchers have used *mathematical models* (e.g., linear models, geometric models, statistical systems of mixture analysis), mainly to improve diets or to better understand the ecological impacts of nutritional components. Finally, some researchers have tried to use combinations of two or more of these approaches in what can be called the *synthetic or eclectic approach.*

A major difficulty in prescribing (or describing) diet development methodology stems from the fact that few scientists who developed successful diets explained the scientific or technological basis for their contributions. Inclusion of *rationale* in the use of various procedures and materials in papers on diets has not been encouraged and in fact has been generally discouraged by editors vigilant to reduce printing costs (Cohen 2001). Review of hundreds of papers on artificial diets for insects reveals amazingly little about how the diets were formulated or the rationale for improvements in materials or methods. The lack of stated rationale and methodology is unfortunate, because it forces all other researchers to start from scratch to develop methodologies, repeating the same mistakes that their predecessors had made. This chapter is an effort to reconstruct the implicit and explicit rationale behind development of insect diets. Much of the effort to explain rationale is also treated in the explanation of why certain diets or diet components are successful and why others are not (Chapter 4).

10.1.1 Historical Perspectives

As exceptions to the trend of not fully stating rationale behind diet components, a few older papers in diet development do include some excellent explanations of rationale that are based on careful observations by pioneering researchers into diet methodology. These include some of the earliest publications on diet development where the philosophy prevails that understanding the effects of artificial diet components helps us better understand the biology of the target insect. So, for example, Baumberger (1917a,b) discussed intricacies of the feeding biology and biological responses of *Drosophila melanogaster* to diet components such as agar, yeast, and bananas. Baumberger's works followed the studies of Loeb (1915) who tried to rear *Drosophila* on a simple medium of ammonium tartrate, sugar, and inorganic minerals that he sterilized. He reported that the flies grew suitably on this minimal medium, not realizing until 2 years later that the flies had inoculated the media with their own yeast and/or bacteria. But these works (of Loeb and Baumberger) and those extensive studies by Guyenot in the nineteen teens set the stage for insect diet development as an outgrowth of the practices of the burgeoning field of microbiology—both in terms of the conceptual framework and the techniques. During this period of insect rearing infancy, a few researchers directed their attention to other insects such as flour beetles (*Tribolium*) and blowflies (*Lucilia sericata*).

Michelbacher and his associates helped further set the stage of rearing protocol based on careful observation—similar to natural history observations. Also, Michelbacher et al. (1932) did an admirable job of providing the rationale for including materials such as casein (for the first time in insect diet history), "fish juice," egg yolk, egg whites, meat, and numerous other materials in their original diet for flesh flies (*L. sericata* Meig.).

The philosophy behind understanding such rationale is best stated by Mittler (1972), regarding understanding mechanistically the nature of and interplay between diet components, he wrote "Until we have this awareness, some aspects of diet formulation will remain a mysterious art, only achieved by those who inadvertently or intuitively do things in a certain way—often without realizing the significance of their actions and hence without describing them adequately in the literature." This statement followed a sage and insightful discussion of the interactions of components in aphid diets (Mittler 1972), including compelling insights on how diets or mixtures of diet ingredients such as amino acids change during storage and how minerals and other diet components can be made to be more readily absorbed. The remarks of Mittler, although they were written in a context of aphid diets, which are very specialized liquid media, apply to diets in general and should be heeded by insect nutritionists and others setting out to develop artificial diets for insects in general.

10.2 STARTING OUT: THE FIRST STEPS IN DIET DEVELOPMENT?

The first tenet of diet development strategy is to answer the question, "what is the purpose of the diet?" Is the diet being used to produce a reliable supply of research subjects? Is it intended for mass rearing and continuous generations of the target insect? Is the diet intended only to allow a temporary presence of the target insect or as a makeshift food to keep the target insect alive until its natural food becomes available? (This very practical problem is faced by groups such as the teams charged with rearing insects such as specialist predators of the hemlock woolly adelgid *Adelges tsugae* (Hemiptera: Adelgidae), which can only complete their life cycle when fed with their natural prey. In the adelgid predator rearing programs, the practice is for personnel to go into the field and collect adelgid-infested hemlock twigs and provide them to predators in the laboratory. However, this cumbersome practice is frequently hindered by the mismatch of the prey availability and the predators' developmental stage where food is needed immediately. This situation has led to development of "rescue diets" or temporary stopgaps (Cohen and Cheah 2011).

Answering this question about the diet's purpose will help with the next question, "will I be developing a holidic, meridic, or oligidic diet?" The next step is to review the literature on other existing insect artificial diets to determine whether or not diets for the target species or closely related species, or species with similar feeding characteristics, already exist. If such diets exist, it is sometimes fruitful to use the existing diets as a starting point. For example, once the wheat germ diet described by Adkisson et al. (1960a) was published, numerous "spin-off" diets were successfully developed, based on wheat germ. According to Adkisson et al., their wheat germ diet was derived (except for the wheat germ component) from a casein diet described by Beck and Stauffer (1950), who acknowledge several works of Fraenkel and Blewett as having paved the way for development of their diet. Other examples of "spin-off" diets based on one core diet abound in the literature (Singh 1977, Singh and Moore 1985, Cohen 2001).

As the diet researcher approaches this early point in developing a diet, this question is raised: can the target insect be cultured on its natural food under the same conditions that are to be used for an artificial diet? *Implicit in this question is the basis of the scientific method, and this is really a question about* "controls" *and* "variables". The control in this case is the process of rearing the insect on its natural diet or a factitious host. The variable is the substitution of the artificial diet for the natural one. If the researcher plunges into efforts to rear the insect without knowing that the insect has hospitable conditions of environment, the cause of any failures to support the insect's growth on artificial diet cannot be attributed unambiguously to the diet itself. Implicit here too is the concept of feedback that one is getting from the bioassays that he/she is conducting to evaluate changes in the diet. The protocols for bioassays and evaluation of changes through the feedback process will be discussed later in this chapter. These protocols are linked with the concepts explained in the chapter on quality control. It is also crucial for the diet researcher to have confidence that the rearing conditions, other than the diet, are all suitable for rearing the target species.

This point is exemplified by recent efforts to develop an artificial diet for the glassy-winged sharpshooter *Homalodisca coagulata* (Say) (Homoptera Cicadéllidae) (Cohen unpublished data). Rearing this insect, even on some of its natural host plants and on factitious hosts, has proved difficult. When this insect was brought into the environment of Petri dishes, incubators, and parafilm-based packaging for artificial diets, it was exasperatingly difficult to balance the humidity and temperature needs with inclinations for mold to form and for a host of other environmental conditions to present barriers to survival. It became almost impossible to separate mortality factors and to attribute them properly to diet versus environmental inadequacies.

Early on in the diet development process, the researcher must come to terms with understanding, as completely as possible, the feeding mechanism used by the target insect. Too often failures in diet development programs have resulted from trying to feed insects totally wrong physical forms of food. For example, several researchers working over a span of more than two decades failed to develop satisfactory diets for *Lygus* spp., because they continued to offer these plant bugs aqueous solutions, thinking that these insects fed on "plant juices" (Cohen 2000c). It was not until Debolt (1982) offered *L. hesperus* very concentrated, solid-rich slurry that this insect could be sustained on artificial diet (discussed further in Chapter 4). Following the model of using slurry diets to feed *Lygus*, *Geocoris*, *Chrysoperla*, and other insects that feed similarly, Cohen (1985a, 1998c, 2000c) has succeeded in developing artificial diets for numerous species of insects. Conversely, trying to feed solid diets or slurries to strictly liquid feeders such as phloem sap–feeding aphids or xylem sap–feeding leafhoppers will be equally disastrous.

Once it is decided what purpose the diet is intended for (e.g., small research colonies for testing nutritional requirements versus mass rearing for a sterile release technology), once the feeding mechanism is understood, and once a rearing program is in place with natural or factitious foods, the next step is presenting to the target insect an artificial diet known to be suitable for insects with similar feeding mechanisms. In embarking on an effort to develop a diet for a xylem sap–feeding insect, one would be better served to begin with an already established diet for another xylem sap feeder (Hou and Brooks 1975) rather than making efforts with a diet devised for phloem sap feeders. This raises a point expressed by Beck (1972) that the beginning process of diet development is complicated by the fact that if there is no knowledge of a suitable feeding system (presentation methodology) and the most rudimentary physical and chemical requirements, one cannot get enough of a start in feeding the target insect to even begin to get feedback to correct minor faults. Fortunately, after a century of experiments on presenting artificial diets to insects and an even greater period of observation of feeding habits, there is, for a wide variety of species, a substantial base of information that could be applicable to diet development for species of target insects that have not been previously used in programs based on artificial diets. However, it is essential that early on the diet researcher develops a thorough knowledge of what is the exact feeding mechanism and target of the insect for which the diet is to be developed. The more exact this understanding, the better are the chances of developing a successful diet.

10.3 USING DIETS DEVELOPED FOR INSECTS WITH SIMILAR FEEDING HABITS, INCLUDING DERIVATIVE DIETS WITH AND WITHOUT HOST EXTRACTS

Applications of diets that are successful with closely related insects have a better chance for success than those diets designed for species with disparate feeding habits. The diet of Adkisson et al. (1960a) is an excellent example of this "spin-off" phenomenon, where many other insects were found to be supported excellently on the parent diet. For example, Berger (1963) showed that the pink bollworm diet of Adkisson et al. (1960a) was an excellent medium for rearing *Heliothis* species. A diet for a generalist leaf-feeding insect such as the cabbage looper is more likely to be suitable to another leaf feeder such as an armyworm than would be a diet for a carnivore or a phloem sap eater or even a specialist on other plant tissues such as the one that consumes fruits or seeds. This is in accord with what was explained in Chapter 4 on why diets are successful. The point was made there and bears repeating that one of the most important factors in predicting or explaining the success of diets is that they have a general composition that parallels that of the natural food. For insects that are adapted to use a natural food that is solid and that has a water content of 90% and a 3% protein content (such as broccoli and spinach), an artificial diet that contains 70% water and 20% protein will almost certainly be unsuitable. An insect on such a diet will be constantly battling water deficit and nitrogen excess problems. The other factor that goes hand-in-hand with the general composition profile is the physical texture replication that makes the artificial diet comparable to the natural diet.

An example of the employment of this technique is found in the report (Blossey et al. 2000) on the development of an artificial diet for the weevil *Hylobius transversovittatus* (Coleoptera: Curculionidae), a biological control agent for the weed known as purple loosestrife (*Lythrum salicaria*). Blossey et al. used as a basis for their diet for *H. transversovittatus* an existing diet (Hunt et al. 1992) for two species of weevils that were also in the genus, *Hylobius*, *H. radicis* and *H. pales*. It was found that the existing diet required modification by adding natural host material and alteration of the salt mixture and moisture content (Blossey et al. 2000).

Similarly, many other species of insects, such as silk worms (Ito 1961a,b, Ito et al. 1975), coffee berry borers (Brun et al. 1993), and boll weevils (Lindig and Malone 1973), have been reared successfully on diets to which some natural host material was added to otherwise conventional nutrients such as soy flour, wheat germ, or casein. However, it should be noted that there are many instances where insects that are considered specialists on a given host species (cotton) or host family (e.g., the mustard family) can thrive on diets that contain no host plant components or extracts. This was the case with pink bollworms and the diets by Adkisson et al. (1960a,b). Interestingly, a diet for the same species of weevil mentioned in the previous paragraph (*H. transversovittatus*) was developed by Tomic-Carruthers (2007, 2009) with no host materials from the purple loosestrife weed to which the weevil is known to be restricted in its host range.

However, there are some very well-documented cases of rejection of diets that do not contain some fraction of the specific host plant. One such case that is very compelling in regard to host plant specificity is the saga of silkworm diet development. In the silkworm *Bombyx mori*, there has been an extensive body of work that has shown that the larvae strictly require at least some form of mulberry leaf material. Singh (1977) has reviewed the papers on silkworms, which are numerous (with 50 papers alone coming from T. Ito's group). In all of these reports and dozens of others since the Singh (1977) review, there have been several failed efforts to rear silkworms on diets void of mulberry components. In all of Ito's diet studies, regardless of the other diet components, the larvae would only accept diets if they contained powdered mulberry leaves (e.g., Ito 1961a,b, Ito et al. 1974). The conclusion seems to be a clear indication that silkworms require some kind of mulberry signature, even if it is only the purified extract component, morin (an isoflavonoid).

There are several reasons for including portions of the natural host in artificial diets. First, the natural materials contain a complex of chemicals that may serve as token stimuli. It may be impossible to get target insects to feed on what would otherwise be perfectly nutritious and suitable diets without the presence of these token stimuli (discussed in Chapters 3, 7, and 9). Besides token stimuli, there may be cryptic nutrients or suitable proportions of nutrients that make natural host plants desirable in a diet. The third point is the desirability of reinforcing the insect's specialized feeding habits that steer it to certain plant species (Blossey et al. 2000). For example, the target of the rearing program described by Blossey et al. (2000) is mass production of weevils that retain their predisposition toward the consumption of purple loosestrife so that when the laboratory-reared insects are released into the field, they will do the jobs for which they were reared: destruction of a weed. The danger of breeding insects in the laboratory that have lost their natural inclinations toward predation or other desirable attributes has been extensively discussed (Bartlett 1984, Blossey et al. 2000, Cohen 2000c).

It remains a fine line and a great challenge to insect diet and rearing professionals to perform the balancing act of keeping the insect well nourished and reproductively capable in the laboratory while not diminishing its robust qualities as a field insect. Finding the right mixture of natural foods in relationship to the "artificial ingredients" is the guiding force in this quest. This is further complicated by the difficulty in finding the appropriate balance of token stimuli and nutrients, and by the economics of mass rearing, where cheaper diet components and reduced labor in processing are always a driving force (Figure 10.1).

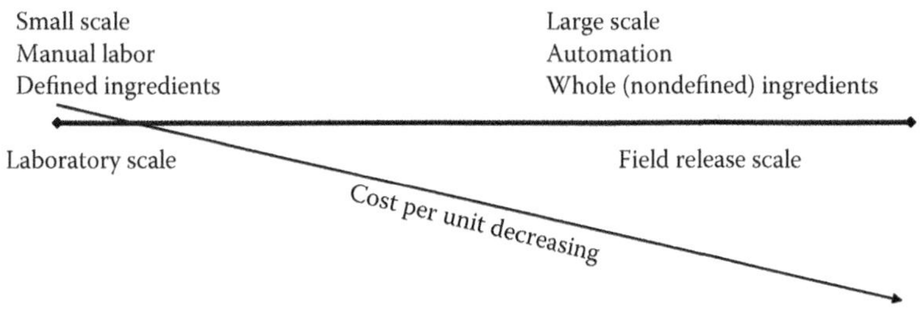

Figure 10.1 Model of the continuum of diet scale, ranging from small-scale, developmental phases of diet formulation to production.

Finally, the strategy for adding extracts is a very important factor in providing a foundation for a successful diet. If we can assume that the major nutritional and textural features of an incipient diet have been established in what we can call a "base diet," then what we would expect from the extract is that it would serve as a token stimulus or possibly a combination of token stimulus and nutritional supplement, supplying small amounts of some unknown factors. The well-established diet components (discussed in much more detail in Chapter 4) include the staples or complex nutrients such as soy flour, wheat germ, cottonseed meal, yeast, and supplements such as proteins (e.g., casein or soy protein isolate), sugars, starch, lipid complexes, mineral mixtures, and vitamins. Also, the base diet would include other functional components such as gelling agents, antimicrobial agents, and other preservatives.

The procedures for adding plant materials have been nearly as varied as the species of insects targeted for extract-based diet development. The work on silkworm predominates the efforts (because of their tremendous economic importance and the desired outcome to produce diets that are less expensive and more convenient than the use of fresh plant materials in a seasonal host [mulberry]). Despite all this research, the bulk of silkworm rearing, for several reasons, remains a system of providing fresh mulberry leaves. Therefore, the example that is provided here comes from the recent research of Allen Cohen and Dr. Carole Cheah (unpublished information) where the imported weevil *Rhinoncomimus latipes* (Coleoptera: Curculionindae) was the target for diet development for the USDA, Forest Service and the New Jersey Department of Agriculture efforts to mass-rear the insects for control of the invasive weed, mile-a-minute (*Persicaria perfoliata* [syn. *Polygonum perfoliatum*]) (Figure 10.2).

Figure 10.3 shows a pictorial flow of the steps used in the process of making extracts and adding them to diets being developed for *R. latipes*. As seen in the figure, there is a considerable space in the greenhouse at the PABIL (the Philip Alampi Beneficial Insects Laboratory in Trenton, NJ)

Figure 10.2 Processing plant extracts for inclusion in diet. (a) The greenhouse at the PABIL laboratory in New Jersey showing the growing setup for mile-a-minute weed to be used as host plant for the mile-a-minute (*P. perfoliata* [syn. *P. perfoliatum*] weevil *R. latipes* [Coleoptera: Curculionindae]). (b) Removal of leaf, stem, and nodes of the plant being sorted (shown in c). (d) Freeze-drying process and (e) grinders (cutting mill and hammer mill) for making a powder of the dried plant materials. (f) Soxhlet extraction apparatus used to make ethanolic or petroleum ether extracts of the plant and (g) adult weevils and a larvae (inset) consuming the diet that contains 1% extract. The process is further explained in the text.

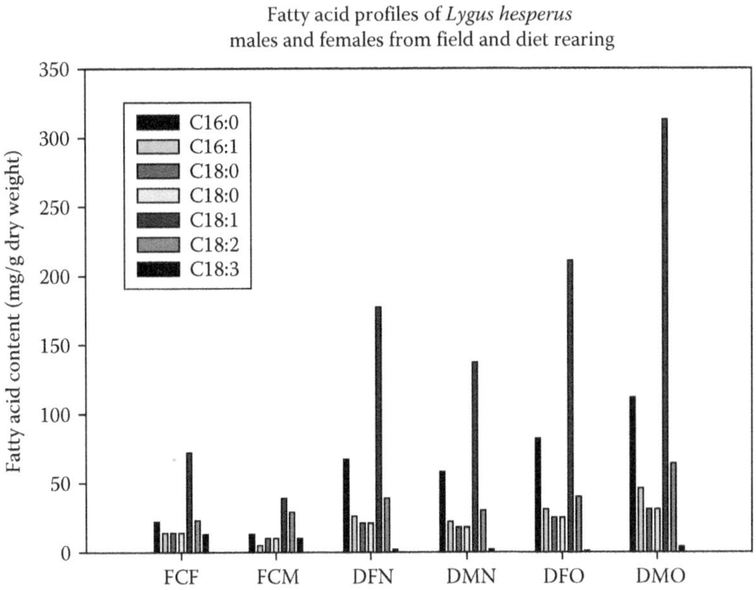

Figure 10.3 Distribution of fatty acids in the whole carcass of the western tarnished plant bug (*Lygus hesperus*) from various diets and from the field. FCF, field-collected females; FCM, field-collected males; DFN, diet females newly eclosed; DMN, diet males newly eclosed; DFO, diet females, 5-day post-eclosion; DMO, diet males, 5-day post-eclosion. (Adapted from Cohen, A.C., *J. Entomol. Sci.*, 25, 75, 1990b.)

where the weevils are mass-reared. The rationale behind the effort to develop artificial diets for weed insects such as for *R. latipes* is to reduce the burden (in terms of space, personnel, and materials) of growing huge numbers of plants to use as living hosts. The underlying assumption for the artificial diet in lieu of host plants model is that using artificial diets would require possibly 1/10 to 1/100 of the number of plants needed to produce an equal number of insects. It is important to understand that the most expensive part of the live-host rearing system is the host plant, so it becomes a question of dilution. The advantage of using an artificial diet, then, is that for the same amount of greenhouse space now used to rear 10,000 weevils, production could be increased to between 100,000 and 1,000,000 weevils.

We proceed in this work with this underlying premise: we can find an amount of extract that would, when added to the artificial diet components, elicit normal feeding, growth, development, and reproduction. The next questions are (1) what base diet will be used? (2) What extraction technique will meet these needs? (3) How much of the extract will need to be added? (4) How will we add the extract? In the previous example of the Blossey diet for purple loosestrife predators, nearly 25% of the artificial diet's biomass was root materials from the host plant, added as a ground-up meal or flour. In the experiments with *R. latipes*, a base diet was made of wheat germ, soy flour, sucrose, mineral, and vitamin mixtures added as supplements and carrageenan gelling agents. To this base, extracts were added in the following forms: ground, fresh *P. perfoliata* was added, powdered and freeze-dried *P. perfoliata* was added at concentrations from 1% to 10% of the total diet mixture, and extracts were made with cold water, hot water, methanol, ethanol, hexane, and petroleum ether. The liquid extracts were dried under a stream of nitrogen and then were mixed with the dry diet ingredients or mixed with hot, molten diet. The dried extract was mixed with the dry ingredients by hand-stirring dry diet in containers that were coated with extract residues, allowing a visually determined complete uptake of the extract onto the dry diet. The soy flour was a full-fat soy, which helped take up the extracts because of soy product's somewhat sticky character (attributable to the emulsifying nature of lecithin and soy oil in the flour). The diet being eaten by the weevils in Figure 10.3 was the best diet in terms of survival and longevity (more than 3 months on the diet alone), and it was the ethanol extract diet with 0.1% dry residue from the *P. perfoliata* in 99.9% diet.

The assumption for the base diet formulation is that all or nearly all of the nutritional needs of the insect are being met by this portion. Then the extract only needs to stimulate robust feeding as a "token" stimulus. This means that the extract is present only to serve as a "signature" to "fool" the insect into mistaking the diet for its actual host. The strategy that we are undertaking is the dilution of host materials (let us use host plants as examples) so that only a small fraction of the diet is derived from the host plant, and most of the diet is from materials that are more convenient and economical to use. Clearly, more work is needed at the basic level for determining optimal feeding stimuli from host plants (or hosts of entomophagous insects), and applied studies are also needed for optimizing extraction of key materials that are void of possible toxins that can be amplified by extraction techniques. And with this information, it will then be necessary to optimize strategies for adding the extracts (powders, liquids, or residues) to the other diet ingredients.

10.4 USE OF FOOD ANALYSIS AS A BASIS FOR DIET DEVELOPMENT

Another method, "the analytical approach," is the analysis of natural foods of the target insects and using the resulting nutrient profile as a model for incipient diets. In this procedure, the first step is learning the exact feeding choices of the target insect. Once this is achieved, the diet researcher determines the chemical profile of the specific foods by published accounts or by direct analysis (Cohen 1985a, 1992). However, erroneous assumptions about the specific feeding targets can be misleading as can be faulty chemical analysis. These points are discussed in Chapters 4 and 7, where anecdotal information on the methodology is explained, with special reference to the development of diets for several species of predators—especially *Chrysoperla* spp. and *Geocoris* spp. Briefly, these accounts explain misdirection that stemmed from assuming that these species of predators were strict liquid feeders, which consumed hemolymph and other prey body fluids. The benefits and pitfalls of this method are also discussed in Chapters 4 and 7 in regard to the development of plant bug diets.

There are two major steps in the analytical process once the correct feeding target is determined: the gross analysis of the insect's food and the more specific analysis of the major food components. In the gross analysis, the water, protein, complex and simple carbohydrates, lipid content, and overall mineral (ash) content are measured. For most species of insects, this level of organization is mainly at the macromolecular level. For certain highly specialized feeders such as xylem sap feeders or phloem sap feeders, the amino acid or small peptide contents should be analyzed as well as physical/chemical parameters such as the osmotic potential, water activity, and pH. Also, it should be understood here that analysis is not confined to nutritional components of the diet, but also structures that the insect is observed to consume (based on microscopy of uneaten and eaten natural foods, frass analysis, or direct observations of the feeding process—such as the visualization of extraoral digestion in the development of predator diets by Cohen (1985a, 1998a)). Also, a type of analysis that can be very productive in helping the diet researcher to understand the feeding biology of the target insect is biochemical measurements (or genetic determinations with cDNA) of the digestive enzymes.

Gross analyses may be circumvented for some insects if their feeding is known to be restricted to foods used by humans and whose nutritional composition is well documented in the literature. For example, for insects that thrive on broccoli florets or on spinach leaves, the information on nutrient content of these foods can be obtained from the USDA nutrient database (http://www.nal.usda.gov/fnic/cgi-bin/nut_search.pl). References to this database are made at several points in this book, and several tables summarizing data on specific foods are also presented. This database is especially useful when the researcher knows the nutrient content of the insect's natural diet and he/she wishes to match the natural food profile with available foods that can be used as substitutes. It should be further noted that the USDA nutrient database is constantly growing where they add new foods and the results of expanded analysis of components such as phytosterols, flavonoids, and various kinds of antioxidants. This was the procedure used by Cohen (1985a, 1998c) and Cohen and Smith (1998) as a major guideline in developing successful diets for predatory insects and for plant bugs.

Analysis is next conducted at the submacromolecular level of organization. This includes individual amino acids, fatty acids, sterols, simple sugars, and minerals. At this level of organization, researchers also test potential token stimuli—ones that are difficult to pinpoint or to extract in clusters or aggregates. For example, lipid-soluble components that are putative nutrients or feeding stimuli may be extracted with acetone or other organic solvents. This extract may then be added to the other nutrients with care to remove all traces of the solvent. Similarly, water-soluble extracts or the remnants after organic solvent and aqueous extraction may be added to the artificial nutrients. If a certain major fraction (organic, aqueous, residue) is found to stimulate feeding and/or growth and development, the researcher may decide to further purify and isolate subfractions such as a lipid extract that can be further separated by thin layer chromatography (TLC), solid extraction chromatography, high-performance liquid chromatography (HPLC), or other such extraction/isolation procedures (Cohen 1992).

Although vitamins are essential nutrients, few analyses have been conducted to determine their contents in foods of insects, and therefore, the amounts and kinds of vitamins added to insect diet formulations are arrived at by little more than approximations. The reasons for this shortcoming are that (1) vitamins are present in trace amounts that push the limits of accurate quantitative analysis, (2) they belong to diverse chemical groups (Chapters 3 and 5), which require separate kinds of analysis for each vitamin (rather than a composite analysis of the 20 amino acids from protein hydrolyzates), and (3) there are often several forms of the same vitamin, further compounding the difficulty of analysis. Some of the modern analytical techniques for vitamins are improving in detection and quantification thresholds, so future progress in this important area is promising. However, anyone who has tried to obtain a clean, reliable sample of a material such as the phloem sap in cotton plants realizes that a broad-spectrum and reliable analysis of this type of material is still a difficult and daunting task, even for the most experienced analytical chemist.

This raises another point about the efficacy of analysis of insect foods to serve as a basis for diet development. Several researchers, including this author, expressed the hypothesis that if a complete picture of nutrients could be known of the natural foods of target insects, these nutrient profiles could be satisfactorily simulated by the addition of a few complementary compounds to an otherwise inexpensive foodstuff. So for example, if a target predator's natural prey (such as *Heliothis* eggs) contains more methionine and threonine than soy protein, one could bring the soy protein up to the level of quality of the prey (*Heliothis* egg) protein by adding appropriate amounts of the two mentioned amino acids. The argument behind this hypothesis is that what makes the artificial diet containing soy protein inadequate compared to the insect eggs is the apparent deficit of methionine and threonine. Unfortunately, it is apparently not merely the difference in a few amino acids or other such simple nutrient differences that render unsuccessful diets as failures. It is unlikely in this author's current opinion that any diet that contains a complement of typical proteins will be so inadequate in one or several amino acids that an insect with its resourceful metabolic pathways cannot work its way around such minor deficiencies. It is much more likely that the problem in most diet failures stems from violations of the general tenets of diet success as stated in Chapter 4 (feeding stimulant deficiency—including digestive system retention/gut mobility problems, bioavailability failure, inappropriate matrix, causing degradation or inaccessibility of digestive enzymes, presence of antinutrients, and lastly, a missing simple nutrient). This is not to say that good, comprehensive analytical work on the natural and artificial diet cannot be an excellent tool in both diet development and in quality control. But the analyses must be put into a context of good overall practices that include bioassays and feedback about how the insects deal with its foods.

10.5 USE OF WHOLE CARCASS ANALYSIS IN DIET DEVELOPMENT (AN ALTERNATIVE ANALYTICAL APPROACH)

It is appealing at first to hypothesize that the composition of an insect's body reflects its nutritional needs (Rock and King 1966). The advocates of whole carcass analysis subscribe to using the profile of the various nutrients in a target insect's body as the template for artificial diets. The underlying assumption behind whole carcass analysis is that what an insect needs to consume, nutritionally, is reflected in that insect's body composition. For example, if an insect that weighs 100 mg contains 20 μg of methionine, then it requires at least this amount in its diet to build its 100 mg of healthy biomass. This idea is reinforced by the fact that by definition, essential nutrients such as methionine must be obtained from the insect's feeding sources, and also by definition, the insect must acquire at least the needed dose of each nutrient in its diet. However, one of the failings of the whole carcass analysis hypothesis is that animals have complex metabolism and behavior, and the natural "wiring" or inclinations of animals is to consume foods that have excesses of some components and deficits of others. The logical extension of the whole carcass approach would lead us to conclude that cannibalism would be the optimal nutritional strategy. As is the case with detailed analysis of target insects' foods, whole carcass analysis may provide some useful adjunct information, but as a sole means of diet development, this procedure is an oversimplification of the enormous complexities of feeding strategies, behavior, and metabolic pathways that characterize most insect species. Figure 10.3 shows the distribution of fatty acids (from Cohen 1989) in the whole carcass of the western tarnished plant bug (*L. hesperus*) from various diets and from the field. It is evident from this figure that both the overall lipid profile and the specific fatty acids in the whole carcass reflect diet profiles.

10.6 RADIOISOTOPES AND DIET DELETION TECHNIQUES

A very powerful approach to the development of understanding of insect nutrition has been the use of isotopes (usually unstable or radioactive isotopes) to determine the ability of a subject to produce a given compound. A pioneer and leading figure in this area (as well as whole carcass analysis and the "eclectic approach" discussed in the following) was the great insect physiologist and biochemist G. C. Rock. For example, Rock and Hodgson (1971) used a combination of radioisotopes and dietary deletion to determine the essentiality of several amino acids in the *Helicoverpa zea* (formerly *Heliothis zea*) in what now must be considered a classical study. In this approach, the researcher provides a diet that contains all of the known protein amino acids except for one (dietary deletion). Also included in the diet is a radioisotope of either carbon (^{14}C) or hydrogen (^{3}H) either in acetic acid or a sugar, and after the insect has had a chance to consume and metabolize the diet ingredients, the carcass is hydrolyzed (digested to its simple molecular components), analyzed by conventional techniques (ion exchange chromatography, TLC, GC, HPLC, or paper chromatography), and the separated components are examined separately for the presence of the radioisotope. If, for example, lysine was omitted from the diet but was found to be present and labeled with the isotope, it is concluded that the insect (or one of its possible symbionts) was able to produce the amino acid de novo from the precursors provided in the diet.

However, the radioisotope has not been the only method of determining dietary requirements. Prior to the widespread use of isotopic determination of metabolic pathways, the dietary deletion method was used productively. If a defined diet that is nutritionally sufficient to support robust development, it could be used as a basis for employing the deletion method. If a substance that was deleted from the diet appeared in the carcass after a reasonable period of feeding and metabolic activity, it could be concluded that the insect or a symbiont was responsible for the synthesis of the substance.

However, most defined diets fail to support robust growth and prolonged maintenance of the target insects, so recourse is needed to a more sensitive method such as the radioisotope method. The deletion/isotope method can be used with the deletion of multiple amino acids, the concept behind this being that if the pathways for synthesis of these amino acids are present in the insect, it can then use another source of nitrogen and a carbon source to synthesize the species of amino acids suspected of being essential. This can be a broad-spectrum (shotgun) approach to gaining information about which amino acids are essential. For example, if lysine, arginine, threonine, and leucine were deleted, but if lysine and arginine were found to have ratio-tracers and threonine and leucine were found to have none, it could be confidently concluded that the insects lacked the pathways for the latter two compounds but possessed the pathways for the former two amino acids. It must be noted further that the ambiguity of the results of these studies always exists raising the specter of microbial involvement in the synthesis of the putative essential amino acids. The removal of the symbionts by antibiotics introduces the possible variable that intolerance to the diet in question may be a result of direct toxicity of the antibiotics coupled with the restricted palatability and overall nutritional value of the defined diet. Obviously, the difficulty of dealing with all of these variables has made progress in nutrition in general and insect nutrition specifically very difficult and laborious.

10.7 USE OF DIGESTIVE ENZYMES AS AIDS IN DIET DEVELOPMENT

The logic behind this approach is that the presence of a given digestive enzyme is an indication that the insect is adapted to use the substrate that this enzyme hydrolyzes (breaks down). The ability to hydrolyze a substrate such as a protein, carbohydrate, or lipid means that the insect is prepared to use the food material in its diet, further meaning that the material is likely to be a useful part of the insect's nutritional capacity. A teleological but convenient way of regarding the presence of specific enzymes is that they express the insect's "expectations" about its food. The fact that *L. hesperus* possesses salivary α-amylase is an indication that (1) these insects regularly encounter starch (the substrate for α-amylase) in their diet, (2) they are capable of breaking down the starch into smaller units that can readily be ingested, and (3) they can use starch as a nutrient if it is included in their diet (Agusti and Cohen 2000, Zeng and Cohen 2000c). Likewise, the presence of the proteolytic enzyme trypsin in the salivary glands indicates the ability and predisposition of this insect to use proteins in its diet. In contrast, the lack of a β-galactosidase would indicate the lack of ability of this species to utilize lactose sugar or other dietary β-galactosides. This factor has been demonstrated (Gingrich 1972) to be responsible for the poor performance of screwworm larvae on diets containing milk.

The application of such basic techniques as analysis of digestive enzymes has proved to be useful in the development or improvement of artificial diets for several species of insects (Gingrich 1972, Cohen 1992, 2001). However, this and other basic science techniques that help us understand the feeding adaptations and characteristics of the specific insects that we are trying to rear have not been as fully exploited as the potential of these techniques promises.

10.8 NUTRIENT SELF-SELECTION, GEOMETRIC ANALYSIS, AND STATISTICAL OPTIMIZATION APPROACHES

In a sense, the basis for nutrient self-selection and the thinking behind this concept came from earlier works such as those of Professor Vincent Gaston Dethier. Over his half century of studies, Dethier made tremendous contributions to the physiological basis for behavioral and ecological choices by insects. His main model organism was the blowfly, though he did make substantial progress in helping us understand phytophagous insects, especially with respect to the influences of plant secondary compounds on feeding choices. But Dethier's work is probably best known for the simple, yet profound design of the proboscis eversion response to different tastes. An example from Bowdan and Dethier (1986) explains this fully:

> The interactions of excitatory and inhibitory neural input controlling feeding behaviours of flies *Phormia regina* Meigen, at three different levels of deprivation were assessed by studying the proboscis extension reflex, volume of sucrose imbibed and the responsiveness of the contact chemosensory neurons. The responsiveness of the sensory cells tested did not change with respect to length of deprivation. The two known inhibitory systems consist of stretch receptors associated with a) the foregut, and b) the abdomen. These two systems were demonstrated to act at different times. The effect of the foregut stretch receptors was shown to be delayed, cumulative and slowly decaying. The abdominal stretch receptors were demonstrated to act immediately and to be important in ending a drink.

This discussion explains the complexity of feeding or drinking responses and characterizes the exquisite intricacy of regulatory system that underlies feeding choices, including the history of insect, especially its physiological state. In fact, a large part of Dethier's works and those of his associates were designed to tease out these feeding determinants (Dethier 1941, 1976, Dethier and Chadwick 1947, Dethier et al. 1956, Dethier and Bodenstein 1958, Dethier and Yost 1979). The issue of selection and its fundamental determinants is expressed further by Dethier et al. (1956):

> Diet selection and preference are commonly evaluated in terms of quantity of food consumed; however, measurements of intake alone give little information concerning the degree to which different factors participate in the regulation of ingestion. It is clear in the case of insects that a sequential contribution by various stimuli governs the finding of food, the initiation of biting or sampling, the continuance of feeding, and the termination of feeding. It is believed by some (e.g., Dethier, 1953; Fraenkel, 1953) that stimuli which initiate sampling and which drive continued feeding are neither necessarily nor invariably correlated with nutritional values. Other workers, notably Kennedy (1953), believe that there is an important causal relationship between stimulating and nutritional characteristics. The present study is intended as a step toward the ultimate clarification of this problem.

These intricacies of feeding responses are the mechanisms that are involved in the choices made by insects in the nutrient self-selection experiments, and the nuances certainly deserve further attention when broad generalizations are made about carbohydrate versus protein selection and optimization by insects fed with chunks of diet containing agar/casein and agar/sucrose choices. This is the case for a wide variety of insects, including honeybees and butterflies, and it explains a great deal of the basis of feeding responses (Minnich 1921, Wykes 1952).

Like the approaches describe thus far, the nutrient self-selection technique has a much greater potential than has yet been exploited in the development of a base of information for diet development. The underlying assumption behind nutrient self-selection is the concept that insects (and animals in general) come equipped with an innate (inborn) "wisdom" about what is healthy or nutritionally sound. The idea goes on to suppose that if an insect is given a proper set of choices, it will select a food or combination of foods that completely fills its nutritional needs. The foundation behind these concepts in Insecta is derived largely from the works of Vincent G. Dethier, Stanley Beck, and Gottfried Fraenkel who were pioneers in efforts to understand the relationships between feeding behavior, feeding ecology, and nutrition.

Early work in the investigation of nutrient self-selection by the Waldbauer group (Waldbauer et al. 1984) showed that *H. zea* was capable of selecting an optimal mixture of protein and starch when offered diets with these components that were spatially separated. In its most fundamental way, self-selection experimentation allows the researcher to develop a profile of major nutrients (such as gross protein, carbohydrate, and lipid content) in an ideal diet. In this vein, it is useful to recognize that diets are heterogeneous enough that the insect gets to make choices about the components of the diet that it is accepting. For example the lipids in the NI diet (Cohen 2000c) are in compartments of wheat germ cells (Figures 4.1 through 4.4), and the *Lygus* bugs that feed on this diet can pinpoint the lipid-containing compartments with their stylets. Likewise, tobacco budworm larvae showed the ability to do pinpoint selection of desirable nutrient components (Parrott et al. 1983).

The statistical, geometric, or mathematical models are derived from a whole cadre of systems of analysis and interpretation of mixtures and complex systems described by Lee et al. (2004, 2008), Simpson and Raubenheimer (1995, 2001), Soran and Raubenheimer (2008), Raubenheimer (2011), and Ojeda-Avila et al. (2003). The geometric approach is explained by Raubenheimer (2011):

> The problem thus arises of how to capitalize on the benefits of graphical analysis when multidimensional data entail only compositions, and not amounts or rates. In one approach, the mixture triangle (also known as triangle plots, ternary graphs, simplex plots, or de Finetti diagrams), three-dimensional compositions (e.g., protein, lipid, and carbohydrate) are depicted as points within an equilateral triangle.... Mixture triangles have, however, seen surprisingly little service in nutritional biology; they have occasionally been used in animal production nutrition (Emmans 1991, Ruohonen et al. 2003, Lapointe et al. 2008), and only rarely in nutritional ecology and related disciplines.

One of Raubenheimer's citations is Lapointe et al. (2008), which is a paper on optimization of a diet for a root weevil *Diaprepes abbreviatus*. In this paper and also in Lapointe et al. (2010a,b) present information on optimization of an existing diet, showing that some of the components such as wheat germ and cellulose were redundant. The model that they used was a mixture triangle where the components are laid out in a triangle representing the test mixtures with each side of the triangle representing the concentrations of each component. Several statistical packages such as JMP (SAS Institute JMP, Version 11) provide software for such analyses, giving the users choices of each of the plots mentioned by Raubenheimer (2011) plus several other kinds of mixture optimization procedures. In another treatment of optimization with statistical procedures, Tan et al. (2013) show efficacy of these techniques for an improved diet for the predator *Orius sauteri* (and anthocorid hemipteran).

In the application of the geometric model, a common methodology of providing food choices with simple choices of carbohydrate and protein provided as separate aliquots with the nutrients in question dissolved or suspended in agar. The weight of the foods consumed is measured by difference after an allotted time, and the results are graphed to help visualize the optimization. These methods are promising as tools toward diet development, because they help to isolate variables that are attractive to the insects, but even with multivariate analysis, statistical mixture analysis, linear or nonlinear methods, they may not capture or help explain the complexity of the physical and chemical interactions that underlie the food choice patterns that the researchers are trying to grasp.

An example of an optimization experiment is provided as a hypothetical model in Figure 10.4. The flow of this experiment would be (1) testing some mixture variants of the Yamamoto diet (for example) where three key ingredients are tested in varying proportions and (2) weights of the insects subjected to the various diets are measured (or other biological parameters such as rate of development or body fat content). The power of the mixing programs is that they allow a small number of mixtures to be tested, and they use mathematical (linear) programs to project other predicted outcomes. This is seen, for example, in the Gaussian curves in Figure 10.4. It is axiomatic that in the mixture problems, the sum of the components tested in different relative proportions always adds up to 1.0; so 0.7 wheat germ, 0.2 casein, and 0.1 torula yeast would equal 1.0, or 0.6, 0.2, and 0.2 of the respective components would also equal 1.0. For simplicity, the other components of the diet (salt, vitamins, agar, etc.) would remain fixed, so the total of 1.0 is only based on mixtures of the three components in question.

The outcome of these experiments is expressed as a geometric shape: in this case, the shape is a triangle (Figure 10.4, top portion), and the outcomes can be visualized as a projection of weight

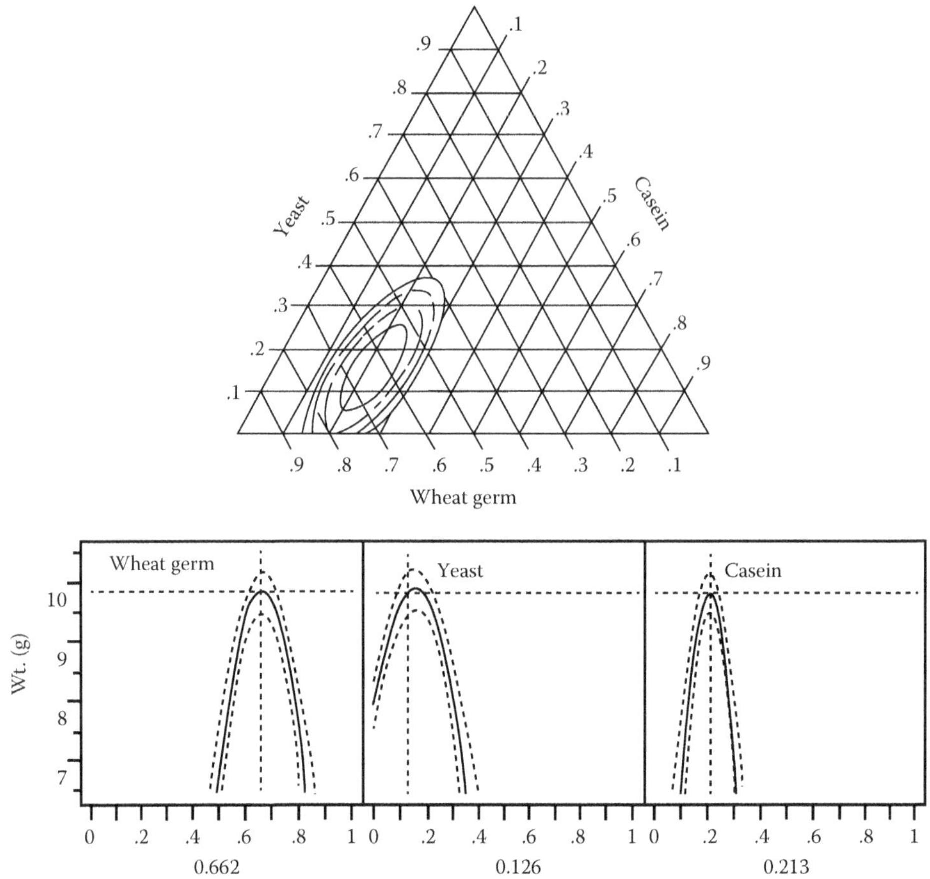

Figure 10.4 Hypothetical optimization triangle developed and modified from JMP simplex mixture program. The bottom graphs represent weights of a larva after treatment with Yamamoto diets with varied proportions of wheat germ, torula yeast, and casein. (From SAS Institute, JMP®, Version 11, SAS Institute Inc., Cary, NC, pp. 1989–2007.)

versus proportions in the geometric space. The oblong shapes represent the combination of actual proportions and their outcomes as well as the projected outcomes where the model has "filled in" the unknown results by projecting from the model's calculations.

In a simple variation of this type of test, a component such as casein can be tested as to its importance in the diet by substituting another material such as soy protein in an equal amount to the casein. In this case, 45 g of soy protein isolate was put in the diet instead of the 45 g of casein, and the results (shown in Figure 10.5) clearly indicate that the casein was superior to the soy in terms of its ability to produce a faster biomass accumulation in *Manduca sexta* from the same batch of eggs and held under identical rearing conditions (both groups within the same container). If these two kinds of experiments were conducted in simultaneously or in tandem, they could be very useful in (1) optimizing proportions of diet components and (2) helping determine optimal substitutions of components either to improve diets or for economic advantage (or possibly some other effort to improve or understand the insect/diet relationship). These rigorously experimental approaches in conjunction with analytical-driven approaches can be useful and systematic ways of improving diets. These approaches may be used in conjunction with the heuristic approach, or they may be replaced by it to meet certain research expectations.

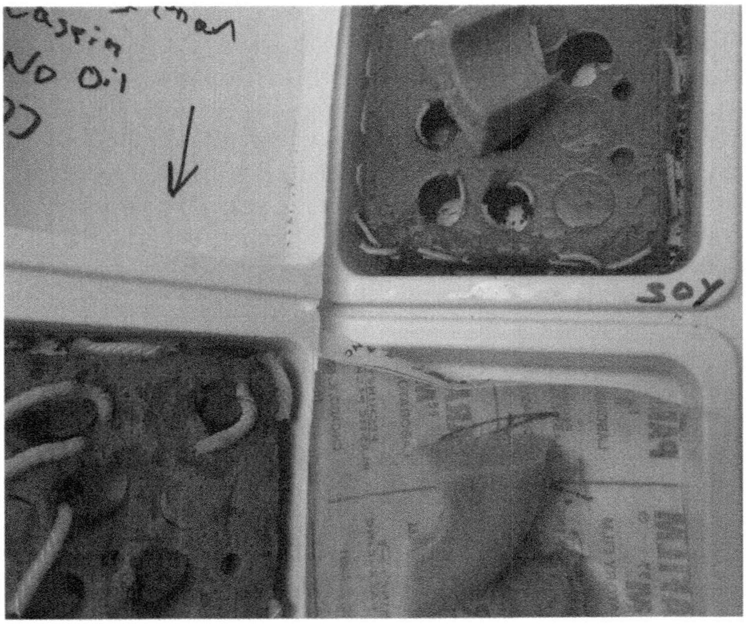

Figure 10.5 *Manduca sexta* (6 days, post-hatching) on modified Yamamoto diet with either casein (lower left) or soy protein (upper right). Although larvae were kept under otherwise identical conditions including the fact that aside from the type of protein supplements all diet components were identical, those from the casein-containing diet had accumulated more than five times more biomass than did the soy protein–treated insects.

10.9 HEURISTIC APPROACH

There is a considerable base of literature on the concept of heuristic philosophy of understanding. Heuristics is a term that is used in mathematics (Pólya 1945), computer science (Nielsen and Molich 1990), psychology (Kahneman et al. 1982, Kahneman 2011), and biological taxonomy (Nogrady 1998). Heuristics can be thought of as a trial and error approach or in the sense of Kahneman's text *Thinking Fast and Slow* (2011), it is the "fast thinking" while analytical thinking is the "slow thinking" approach. So a heuristic approach is intuitive and experience enhanced. In a sense, a heuristic approach to diet development is contrary to the scientific treatment that is advocated in this book and contrary to much of the suggested methodology that is rigorously systematic and mechanistic. Yet there is much to be said about the success that heuristics has had where it is applicable, and it seems that if rearing and diet specialists develop a rigorous and systematic understanding of the multidiscipline nature of rearing situations and dynamics of diet function, they can advance the field greatly, especially if they approach the diet development, diet improvement, and the grasp of rearing context with an awareness or self-consciousness of what they are doing.

10.9.1 Example of Heuristic and Analytical Diet Development

After working for several years on boll weevil and pink bollworm diet, in 1959, Dr. Erma Vanderzant (1920–1999) used wheat germ in a diet for boll weevils (Vanderzant et al. 1959), and she found that although the wheat germ was not very impressive as a nutritional component, it did elicit oviposition response from the weevils. This showed that for oviposition, boll weevils did not need a cotton component in their oviposition medium. This was the first use of wheat germ reported for any insect diet. The next year, Dr. Perry L. Adkisson and Vanderzant (Adkisson et al. 1960a) used wheat germ in a diet for pink bollworms, and they found that the diet was a remarkable breakthrough (which is now understood to have revolutionized insect rearing). This topic is covered in Chapter 4 and is touched on in several other places in this book, but the question here is what motivated Dr. Vanderzant to use wheat germ in the first place (in the boll weevil diet)? In both the boll weevil diet and the pink bollworm diet, there were derivative components such as casein, sucrose, corn oil, Wesson salts, vitamins, and agar. The two papers mentioned here do not contain a rationale for using wheat germ (nor does Vanderzant's later treatment of wheat germ in several papers), so it can only be guessed what motivated her toward wheat germ. She was a biochemist by training and had a very strong background in nutritional biochemistry, but it seems unlikely that her biochemistry coursework would have brought wheat germ to the forefront of her academic experience. Therefore, the interpretation offered here is that something from her day-to-day experience led her to come up with wheat germ. Of the thousands of products in the human food domain, she could have tried any number of nutritious foods, but somehow, her scientific background and her life experiences led her to wheat germ. This seems to be a heuristic leap.

10.10 ECLECTIC APPROACH: CONSTRUCTING A DIET DEVELOPMENT ALGORITHM

This leads to the final and most robust approach to diet development: the use of multiple strategies or the "eclectic approach." The information in the literature, though spread out among many sources and often cryptically nested within publications on topics other than diet development (Cohen 2001), does contain very useful guidelines, especially regarding ingredients that have worked to feed and nurture various species of insects. This gives diet developers several good starting points with respect to ingredients, methods of presentation, and other aspects of rearing. If the diet professional, either through direct analysis or from reliable sources in the literature, goes on to develop a good grasp of the basic food compositional profile (percent water, protein, total lipids, carbohydrates, minerals, and nutritionally "inert" ingredients such as fiber), he/she can use these percentages as guidelines for proportioning the time-tested "standard" ingredients, especially with the aid of a guide to the composition of commonly used ingredients such as that listed in the USDA nutrient database (2002). As emphasized earlier, these combined approaches are more likely to work if the diet professional has a clear grasp of the feeding mechanisms of the target insect. Understanding the feeding biology (including mouth parts, specific feeding targets, digestive enzymes, and digestive system morphology) will help guide the diet developer to avoid mistakes such as trying to force a form of diet that is alien to the target species. Examples of failures to obey the evidence from thorough understanding of an insect's feeding biology are all to common, including several from the author's works (e.g., Cohen 1981, 1985a).

Once some degree of successful feeding has been established, it can then be determined if the new diet is inadequate if the insects are not completing their development on the new diet, or if other biological parameters are indicating a problem such as lower fitness as indicated by substandard fecundity, fertility, or body weight (among many other biological parameters). The reliance on the bioassay places a significant burden on diet experts to develop what can be considered confidently as "high-resolution assays." Learning several weeks after a nutritional inadequacy was created that such a failing exists is too late to make the changes that must be made to correct the problem. Using a parameter such as oviposition rates or adult feeding or mating behavior is too remote to allow timely assessment of a nutritional deficiency that actually occurred in the larval or early nymphal stages. This amounts to a development of a finely tuned feedback system—the essence of good quality control and quality assessment systems. In fact, the entire process of diet development is a matter of using finely tuned feedback to determine nutritional and dietary efficacy.

The eclectic approach may be considered the basis of the model or algorithm of diet development provided as a starting point for development of a fully validated model. The algorithm presented in Figure 10.6 summarizes the approaches used in diet development and improvement or optimization, but one feature of this model emerges with paramount importance: it is an axiom of diet development that if the target insect cannot be reared under laboratory conditions and if the rearing conditions are not firmly established, it should not be expected that an artificial diet can be used successfully. There are several reasons for this axiom: (1) If environmental conditions that are suitable for the insect have not been established, no amount of diet quality will make up for inappropriate temperature, humidity, light conditions, etc. (2) If the insect is newly collected from the field, it has not passed through the "bottleneck" or winnowing effects described in Chapter 1. The comments of Yamamoto (1969) regarding selection (previously discussed in Chapters 1 and 4) are an excellent example of how domestication shapes reared insects. Yamamoto explained that his *M. sexta* colony was highly selected to remove insects that were more occupied with defensive behavior than with feeding and reproduction. A similar discussion is found in the study of diets presented by Shorey and Hale (1965). This is a domestication phenomenon that is unfortunately overlooked or underappreciated by many rearing specialists and their customers. (3) In using insects recently collected from the field, the diet researcher is unable to discriminate which failures in the rearing process are responses to diet problems and which are related to failure to meet environmental requirements.

The algorithm in Figure 10.6 is an attempt to provide a blueprint for students of diet development and improvement to get started. The purpose of the diet must be established as this will shape the approaches and outcomes. This point is discussed at the beginning of this chapter. A second point that is of tremendous importance is that for any hope of success, the diet development team must be able to rear the insect in the laboratory using natural or factitious host materials. If, for example, the insect is a corn (maize) consumer, and if we cannot rear the insect on this host plant, why would we expect that we will do better with an artificial diet? If we do not know the environmental conditions that the insect demands, if we do not have appropriate containers, or if any other shortcomings are at hand, how can we expect that an artificial diet will fill in the gaps?

This leads to another point: do we understand the feeding biology of the target insect. If we are not sure (or have no idea) what the insect eats, how can we provide a diet that even begins to approximate the natural food? The example from the experience with rearing big-eyed bugs (*Geocoris punctipes*) on artificial diet (Cohen 1981, 1985a) is a perfect example of failing to develop a diet until an understanding of the actual feeding system was achieved. It is also worth

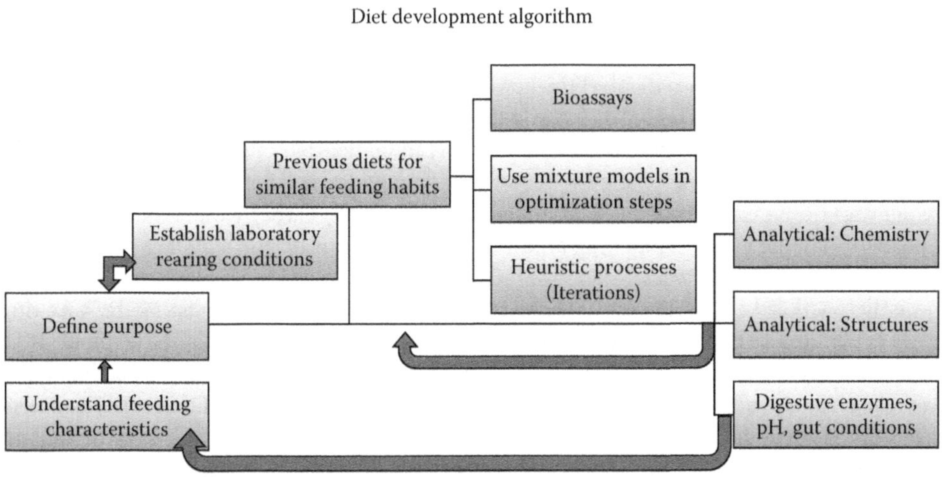

Figure 10.6 Flow chart suggesting an algorithm of diet development: an eclectic approach.

emphasizing that the process of understanding the insect and making improvements on the artificial feeding system is a continuous improvement/iterative exercise. As the understanding of the feeding system is improved, the growing understanding informs the diet developer about nuances that will help with diet formulation and presentation. The idea that the mouthparts are adapted to feed through a membrane (plant surface waxes and epidermis or insect cuticle) can help guide the researcher to use parafilm, wax coating, or some other membrane that permits or even stimulates feeding action.

It should also be noted that direct observation of the insect while feeding can be of vast importance and help in diet development. The words, "know your insect," apply here. The more you know about the feeding process and how that ties to reproduction, diapause, and other aspects of the insect's biology, the more likely the researcher can meet with success. This point is reinforced again with the example from *G. punctipes* research, where the literature was fairly abundant and inaccurate about how the predators "penetrated their prey with their piercing and sucking mouthparts and sucked the hemolymph." It was not until the actual feeding mechanism was understood that the diet was successfully developed, the fact being that liquid diets that simulated hemolymph were not nutrient concentrated enough to support complete growth and development. This story is repeated frequently in literature on diet development (whenever authors are allowed the luxury of explaining how they arrived at their favorable results).

Once the diet development team gets a good sense of the insect's feeding habits, it can attempt to use an existing diet for insects with similar feeding habits. If two insects, closely related taxonomically or ecologically, feed on seeds of certain legumes (for example), it is more likely that the new target insect will be able to use the same diet (or similar diet) as the seed-feeder's currently useful artificial diet. So the combination of knowing how an insect feeds and using a diet for an insect with similar feeding habits should take us to a place where the insect is at least trying to feed on the established diet. And of course, the process demands that bioassays be conducted of the diets in question, and if these diets are not a complete success, then optimization/improvement measure can be made.

If the target insect feeds on any of the formulations of existing diets, but if they fail to thrive in the ways indicated on the bottom of Table 4.1 ("robust feeding," "normal growth," and "normal reproduction"), then the algorithm should lead the researchers to analysis of specific feeding targets (if possible) and/or to heuristic techniques where a rationale-driven set of choices sets the next formulation in motion. One kind of rationale that has been used with some success is finding components or processes used with other successful diets and introducing these factors into the incipient diet. So, for example, if the diet was started with corn meal and brewer's yeast, but it did not fully succeed with the new target insects, the researchers may try to substitute wheat germ or soy flour entirely or in part for the corn meal. If the newly tried components improve the results of the bioassays, then they may be interpreted as leading the researchers to a "trail" that is closer to an optimal diet. The rationale for trying the new components is more heuristic than analytical. However, if a good analysis has been done where the researcher can, with confidence, conclude that the proximate analysis of the natural food is 85% water, 3% protein, 7% carbohydrate, 3% ash, then finding components that match (or formulating mixtures that match) this profile may be a good guide based more on analysis than heuristics.

Finally, once a diet has been established that takes the researcher most of the way through Table 4.1, the geometric optimization (mixture optimization statistics) may be applied. It seems that if diet developers become competent with the mixture statistics, they can take advantage of the power of such analyses to reduce the number of trials of diets with combinations of key components. But thus far these methods have not been used to make major contributions to diet formulation, possibly because of the complexity of diets and possibly because of a lack of expertise in the kinds of complex mathematical modeling that make up these approaches.

10.11 DEVELOPMENT OF MINIMAL DAILY REQUIREMENTS

Authorities in human nutrition (and in mammalian nutrition, in general) have devoted themselves over the past half century or more to the establishment of minimal daily requirements (MDRs) for all nutrients and recommended amounts of the various food groups. Although the absolutely definitive MDR and recommended amounts of food groups (sometimes stated as a food pyramid) have been elusive and controversial, there are certainly some good guidelines available on human food needs. Despite the efforts of several pioneers in insect nutrition, no such MDR or food group pyramid has been established for any insect. Furthermore, as rearing and diet development expands to include hundreds of species of insects that have not been previously cultivated, there is an expanding body of undefined characterization of MDR and food pyramid models. In this author's opinion, there are several factors that conspire to make the worthy goal of establishing such standards and they are (1) the inherent difficulty of the task (especially in light of the difficulties pointed out in Chapter 4 on using defined diets and on understanding the complexities of matrix interactions discussed in Chapters 4 and 8), (2) the failure to draw into the field of insect nutrition and insect dietetics a strong body of researchers who are committed to answering these types of questions, and related to (3) the paucity of funding that is available for performing labor-intensive studies that must be conducted with a proper mixture of insect behavior, nutritional chemistry, and insect physiology/biochemistry. In Chapter 15, the education for such researchers is discussed as part of the visions of the future of insect diet developments and rearing systems that are based on artificial diets.

CHAPTER 11

Fundamentals of Process Control, Quality Control, and Problem Solving

11.1 INTRODUCTION TO PROCESS CONTROL, QUALITY CONTROL, AND INSECTARY PROBLEM SOLVING

This chapter integrates ideas of statistical process control (SPC) with the principles of quality control (QC) and systematic problem solving (Figure 11.1). It is crucial to understand that these three concepts or constructs are interconnected and are seamless outgrowths of one another. While it is a useful metaphor for this three-faceted system to be thought of as organically interconnected, it will be necessary for clarity and practicality to dissect these components. This chapter is intended to convey to readers who are either not versed in engineering (systems control) statistics or applications of control theory to rearing systems a practical understanding of these concepts. It is further intended that this chapter provides rearing professionals (especially diet-making personnel) a blueprint for developing an SPC system to their rearing system. It is a crucial point that several large-scale rearing systems, especially ones overseen by FAO, IAEA, and/or USDA, have profited from the development of QC applications. In these systems, there has been impressive attention to questions of the fitness or usefulness of the product-insects produced by the systems in question.

The excellent attention (whose history and basic tenets are reviewed in the following text) resulted in some exquisite tests of quality ranging in complexity from quantitative or qualitative measurements of sex ratios and pupal weights to genetic makeup and complex behavioral screening for flight capability, mating propensity, and other high-level measurements of the rearing products' characteristics. However, there remains a serious gap in controlling these outputs. The recognition that a factor such as pupal weight can have powerful predictive value as to whether an insect intended for sterile release technology will be able to do the job for which it was intended (such as being competitive with wild or feral males in the field so that it can displace its competitors and prevent a fertile mating event) does not explain *how* the pupa came to have a lower than acceptable weight. This disconnect between effects and their causes in the rearing process results in the *control* part of QC to be a matter of removing the individual or batch of pupae from the population to be released. Or there may be some other strategy such as increasing the number to be released to make up for the predicted incompetence of the individuals of lower fitness. This strategy is better than blindly releasing inferior products (unfit sterile adults) with no regard to their potential to do their job, but it is much less efficient than the alternative possibility of controlling the process of production in such a way that the pupae (and the adults that will emerge from the population of pupae) are large, in robust health, and can be predicted to do their job in a fully satisfactory way.

This connection between the causes of fitness failures and the failures themselves is the basis of process control (PC). If we knew, for example, that a stale diet ingredient such as rancid soy flour or wheat germ causes a decrease in pupal weight and the concomitant fitness problems, then measures that assure fresh, wholesome soy flour or wheat germ could prevent the failures described here.

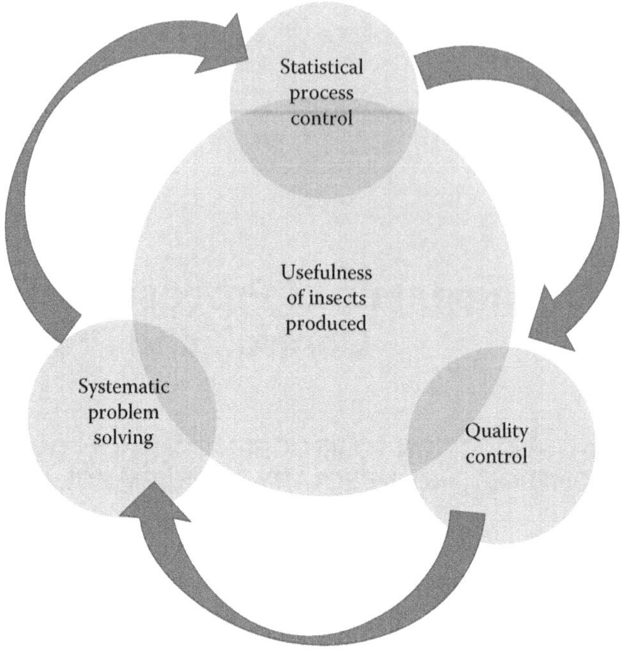

Figure 11.1 Interconnections of statistical process control, quality control, systematic problem-solving techniques, and the outcome: useful insects that meet the needs of the end-user.

Similarly, variations in raw materials of diets such as lower than normal protein content in wheat germ, soy, or yeast or insufficient lipid content in a raw material such as wheat germ can result in degraded fitness syndrome (low pupal weight, poor flight capability, insufficient stored glycogen in fat body and flight muscle, etc.). The outcomes that are considered low quality in the rearing products could be averted by testing the raw materials prior to their being used in diets and then replacing the defective raw material or otherwise remedying the inadequacy could save a large amount of money in wasted materials, and labor. Therefore, the remainder of this chapter and much of the other parts of this book are intended to help identify and treat the problems that occur in rearing systems, especially those pertaining to diets.

11.2 RELATIONSHIPS BETWEEN PROCESS CONTROL AND QUALITY CONTROL: HISTORICAL PERSPECTIVES AND AN OVERVIEW

When diet/rearing personnel make a batch of artificial diet for an insect, they are going through a process. If they do not use the tenets and protocols for PC, or more pertinently, SPC, they invite problems that inherently creep into all manufacturing efforts. Almost all manufacturers of nearly every product use data-based, statically driven PC techniques to assure reliability of their processes and quality of their final product. However, insect rearing, with a few notable exceptions in screwworm and fruit fly mass-rearing facilities, has lagged behind other industries in utilization of quality and PC. This chapter is a blueprint to help rearing specialists develop and adopt PC into all aspects of their diet-making procedures. Also, very importantly, most procedures that have been developed are not published or otherwise publicized. Hopefully, this chapter will fill in many gaps in the conceptual basis for PC and the practical issues of putting data-based (statistical) PC into a useful context. In fact, the intention behind this chapter is to provide an operational plan for developing a practical SPC plan for small, medium, and large rearing operations.

A comprehensive QC document is the report from the Joint FAO/IAEA Division group meeting in Vienna (May 5–9, 1997), listed in the References under Anonymous (1999b). The standards in this document are those used for the Mediterranean fruit fly *Ceratitis capitata*, the Caribbean fruit fly *Anastrepha suspensa*, the Mexican fruit fly *A. ludens*, and various *Bactrocera* species. The suggested standards are routine measurement and statistical treatment of pupal size, percent emergence, flight ability, longevity under stress, sex ratio, mating compatibility, dispersal and survival, mating tests, pheromone compatibility tests, a sterility test, and counts of egg numbers in sterile females. This document is based on these background publications: Boller and Chambers (1977), Boller et al. (1981), Brazzel et al. (1986), Calkins et al. (1994), Churchill-Stanland et al. (1986), and SPSS (1996).

Although the quality of the science is tacitly accepted as part of doing research (and the technology development that stems from such research), the entomological community—with some notable exceptions mentioned in the following text—has been somewhat lethargic or myopic about adopting standards and practices of quality in insect rearing. This statement has several important exceptions with some of the pioneers in mass rearing such as Boller and Chambers (1977), Chambers (1977), Chambers and Ashley (1984), Leppla and Fisher (1989) and later Calkins et al. (1994). As might be expected based on economic forces that demand quality and the efficiency that attends QC, mass-rearing facilities have been the most conspicuous leaders in advances in quality practices. In keeping with this premise, the published literature on quality in rearing has been promoted by organizations such as IAEA, FAO, and large government organizations such as USDA, APHIS.

Although there were allusions to or implications of the role of process in QC, the earlier publications on quality were more concerned with the product and how it should be measured and treated than with the process. The paper by Leppla and Fisher (1989) was an important departure from this emphasis, and as can be seen from Figure 11.2, the concept of process is central in the total quality model that these authors presented. In the model and in the discussion by Leppla and Fisher, the point is made that raw materials and processes (including what they call "production control," which includes environmental factors and practices that include routine maintenance) are to be controlled (or at least given attention) (Figure 11.3).

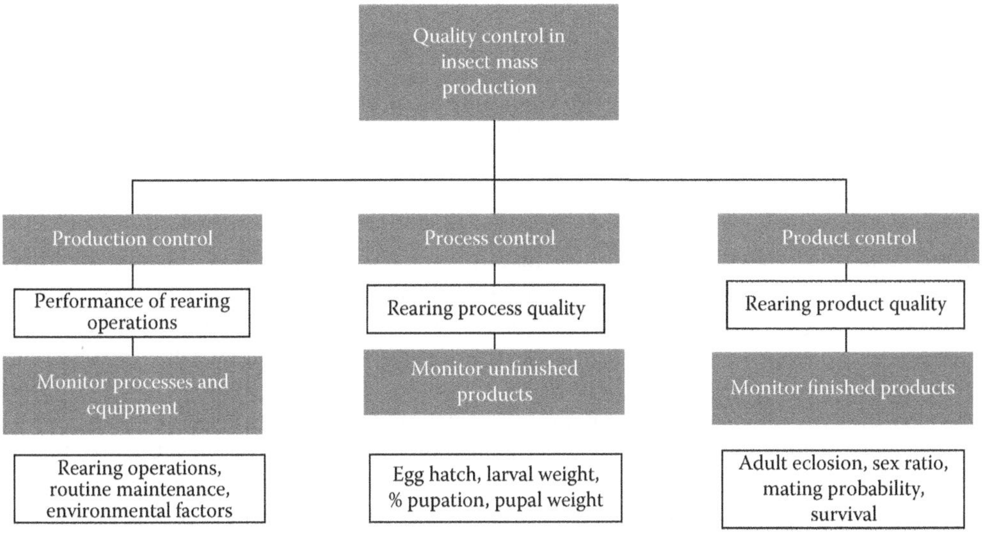

Figure 11.2 Model of relationships between process and quality. (Adapted from Leppla, N.C. and Fisher, W.R., *J. Appl. Ent.*, 108, 452, 1989.)

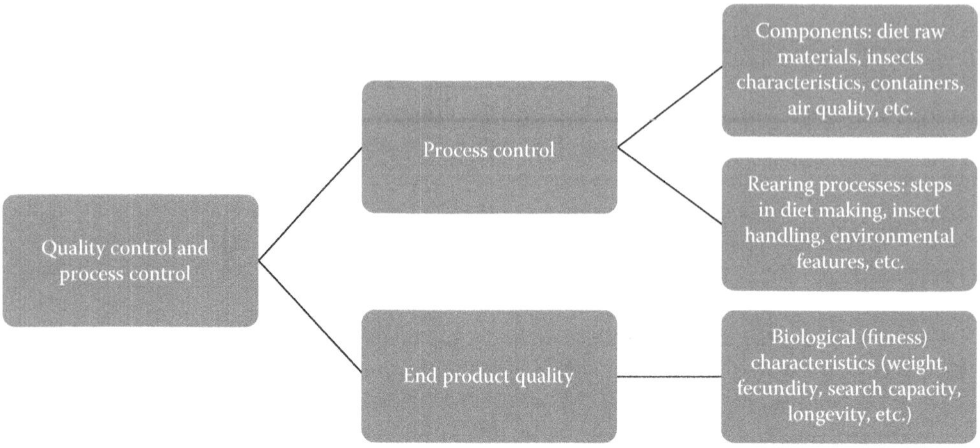

Figure 11.3 Model of the direct interrelationship between process and product. This model shows a more direct relationship between process and product than the model in Figure 11.2.

In all of these publications on quality, the emphasis was on QC of the end product, and the arguments pertained to establishing outputs (insects that were the products of rearing operations), which would do the job of serving as determinants of sterility in the wild or feral populations that were targets of control through the sterile insect technique (SIT). Other aspects of product quality (QC) could include voracity or search capacity for parasitoids or predators in rearing systems whose purpose is production of insect for biological control. In other instances, the quality may be interpreted in terms of nutritional value of insects being used as foods for other organisms or the ability for insects intended for research purposes (such as testing genetically modified crops) to compare favorably with field (wild or feral) counterparts. It is a basic tenet of QC engineers that the customer or end-user of a product should be consulted to help establish the decision process for standards of quality. These engineering standards became the motivating force behind adoption of standards of automobile production, for which the Japanese auto industry became famous in the 1970s and 1980s, forcing many of the auto manufacturers in the United States to go out of business. The power of QC cannot be overestimated in any kind of production, and the relationship between QC and PC is fundamental to improvement of the quality and economics of production. However, until Lorraine and Bruzzone (1992) published their paper on PC as a promising tool for quality assurance and control, the historical emphasis was mainly on the outcomes, rather than the process.

Although it is not within the scope of this book to discuss in detail the rather extensive body of literature on QC per se of insects, it is important to consider the kinds of outcomes that may be influenced by diet or the diet context in rearing systems. The bulk of literature on insect QC is found in mass rearing of insects for SIT, and the greatest amount of attention to quality in SIT programs has been in tephritid fruit fly rearing. For example, two publications that are a basis for much of the QC testing are FAO/IAEA/USDA (2003) and FAO/IAEA/USDA (2014), both dealing with a variety of fruit flies such as several species of *Ceratitis*, including Mediterranean fruit flies *C. capitata* (Wiedemann), as well as, *Anastrepha ludens* (Loew), *Anastrepha fraterculus* (Wiedemann), *Anastrepha obliqua* (Macquart), and *Bactrocera cucurbitae* (Coquillett), to mention a few of the tephritid used in SIT systems. In each of the QC systems designed for these insects, several direct and indirect parameters were adopted for QC evaluation, including pupal weights, sex ratios, emergence period, flight capability, longevity of adults, egg production (fecundity), and egg hatch (fertility). Such testing can be expensive and time consuming, so there have been questions raised about the efficacy of these tests, including questions of their potential relevance and possible redundancy.

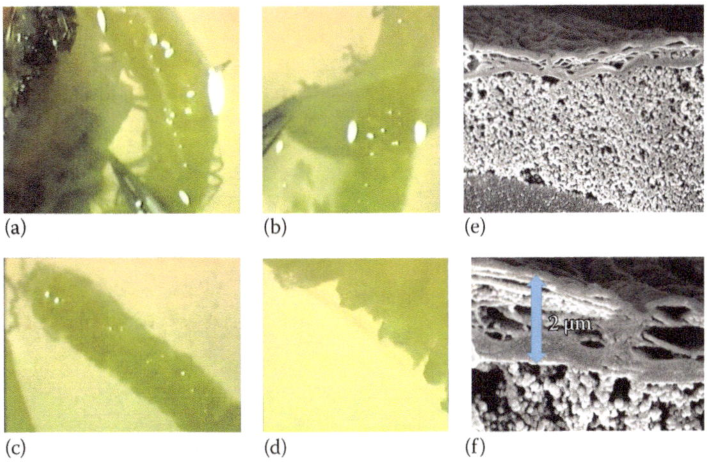

Figure 7.14 Views of the peritrophic matrix or PM (peritrophic membrane or peritrophic jelly) of *Heliothis virescens* after feeding on plant materials. (a–d) show dissections of gut: (a) Malpighian tubule attached to gut, tracheae, and cuticle being removed on the left; (b) the gut being cut and pulled away from the PM; (c) gut contents being held in place by the PM; and (d) shows the edge of the PM (at 100×) as a sharp line: note the contents of the bites of plant material. (e) and (f) show scanning electron microscope images of the PM at (f) 4,500× and (f) 14,000×. Note the multiple layers of PM, especially visible in (f).

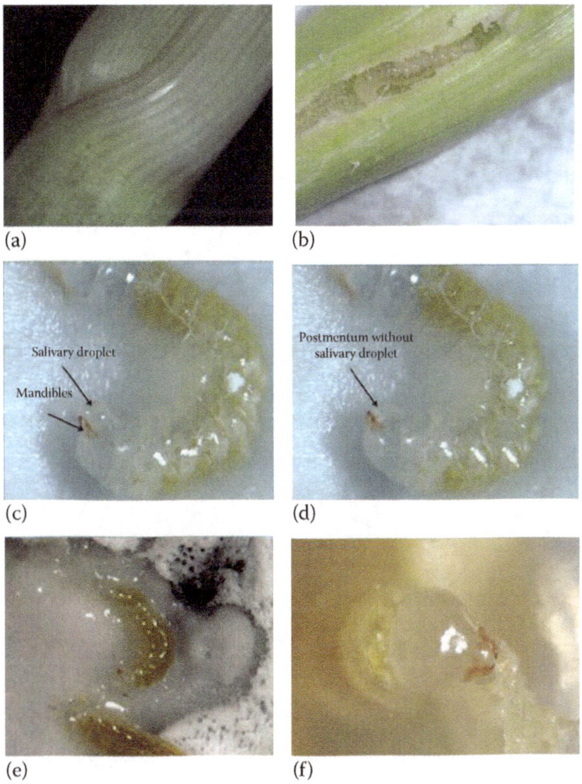

Figure 7.21 Top two images are an external view of a gall on the invasive weed *Arundo donax* (top left) (a) and a view of a larval *Tetramesa romana* (Hymenoptera: Eurytomidae) the Arundo gall wasp in its gallery within a gall (b). The second row shows a *T. romana* larva with a droplet of saliva that it dabs onto gall material to digest it by EOD (c), then the larva uses its weak mandibles to rasp the softened gall tissue (d). This process was observed when larvae fed on galls, artificial diet, and callus from tissue culture. The third row is a *T. romana* feeding on artificial diet (left) (e) and on *Arundo* callus tissue (right) (f) (grown by Dr. Sergei F. Krasnyanski of the Plant Transformation Laboratory in the Department of Horticulture at NCSU). These processes are discussed by Moran et al. (2013).

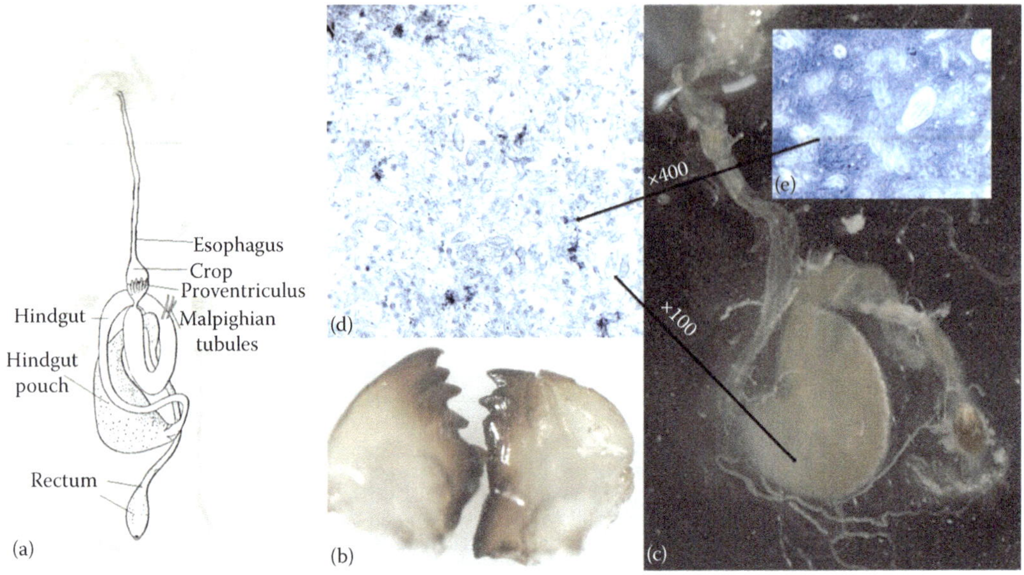

Figure 7.22 Termite digestive system (a, c), mandibles (b), and endosymbionts from the enlarged pouch (d, e). The drawing (a) (from Wigglesworth 1974) and the image of the entire digestive system (c) shows that the termite has a relatively simple, tube-like digestive system, except for the notable enlarged pouch (sometimes called a *paunch*) where dozens of species of symbionts (bacteria, Archaea, and protozoans) abound. The mandibles (b) in this composite image are seen as very strong, hard structures whose highly sclerotized dentition is responsible for the termites' ability to break up wood into pieces small enough for the microbiome to access and digest. The microbiome images are magnified 100× (d) and 400× (e). It should be noted that the smaller bacteria and Archaea are not visible under these magnifications.

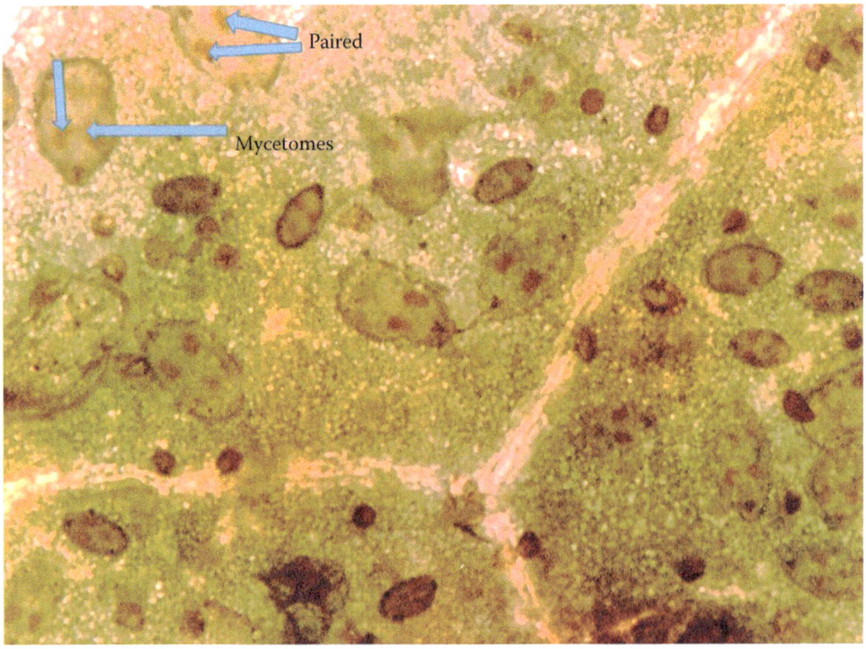

Figure 7.23 Immature whiteflies (nymphs) on cotton leaf that has been partially cleared. The paired mycetomes in each individual are visible as dark spots that are associated with the digestive system of the whiteflies. The mycetocytes and their symbionts are transferred prior to oviposition by a transovarial process from the female to all her offspring.

Figure 11.18 The overall experimental setup in 32-cell and 8-cell rearing containers (a), healthy larvae (b), improper humidity (c), larva infected by pathogens (d), properly managed frass (containers and diet permitted larvae to either remove frass from container or to pack it away from diet) (e and f), improper humidity and larval density (g and h), and mold outbreak (i).

Figure 15.1 (a through e) Liquid and slurry feeders. (a) Whiteflies (*Bemisia tabaci*, Aleurodidae: Hemiptera) feeding on hibiscus phloem sap. (b) Hemlock woolly adelgids (*Adelges tsugae*, Adelgidae: Hemiptera) feeding on cell contents of xylem ray parenchyma. (c) *Euthyrhynchus floridanus* L. (Hemiptera: Pentatomidae) feeding collectively, via extra-oral digestion, on a mealworm pupa (*Tenebrio molitor* L. Coleoptera: Tenebrionidae). (d) Tarnished plant bug *Lygus lineolaris* (Miridae: Hemiptera) feeding on extra-oral digestion products from broccoli florets. (e) *Sasajiscymnus tsugae* (Coccinellidae: Coleoptera) feeding on slurry-type artificial diet. (f) *Geocoris punctipes* (Hemiptera: Geocoridae) using extra-oral digestion to liquefy and consume prey contents.

Figure 15.2 (a through e) Showing insects that feed on solid foods. (a) Silkworm larvae, *Bombyx mori* L. (Bombycidae: Lepidoptera) feeding on mulberry leaves. (b) Confused flour beetles *Tribolium confusum* (Tenebrionidae: Coleoptera) feeding on wheat flour and wheat bran (not tunnels). (c and d) Termite colony feeding on wood, showing galleries and frass produced by termites (c) and mandibles of termite, capable of breaking down wood. (e) Stop action images showing movements of mandibles of a typical chewing insect (*Manduca sexta*, Lepidoptera: Sphingidae).

In a recent paper, Fanson et al. (2014) present the results of a study that provided analysis of data collected from 1996 to 2012, from the mass-rearing facility at Camden (New South Wales, Australia) where they had been producing Queensland fruit flies, *Bactrocera tryoni* (Froggatt). They had questioned whether datasets that they had been collecting contained redundancies that could be eliminated while maintaining the quality standards. Interestingly, they found (using statistical techniques such as principal component analysis [PCA]) that there were redundancies with respect to several parameters. In discussing their QC analysis from data collected between 1996 and 2012, they wrote:

> During this time, the facility has regularly recorded fly quality parameters, creating a unique dataset that provides an invaluable opportunity to evaluate the interrelationships among standard quality control (QC) parameters and test for redundant QC variables. Here, we conducted an exploratory data analysis to reveal relationships among the QC parameters. We found that pupal weight, adult lifespan, and longevity under nutritional stress (i.e., survival duration without food or water) had distinct monthly trends, suggesting that these QC parameters are sensitive to seasonal conditions. Furthermore, emergence percentage, flight ability, and adult lifespan were adversely affected by the dyeing/handling/irradiation process associated with sterile insect releases. Using a multivariate approach and controlling for monthly and yearly patterns, we showed that pupal weight and egg hatch are consistently negatively related and that percentage male and emergence rates are consistently negatively related. These results suggest that these correlation pairs measure similar quality information and hence one QC variable from each pair could be dropped. Flight ability was not strongly correlated with any of the QC variables, suggesting that this QC variable remains a useful QC metric. Finally, the longevity under nutritional stress QC appears to be fairly insensitive to QCs and we suggest that it should be replaced by the standard mortality under stress test.

It is noteworthy that they used multivariate statistics and controlled for monthly and yearly variability due to changes in personnel and other insectary conditions, and also that they used PCA to determine which parameters were redundant, such as pupal weight and egg hatch or percentage male and emergence rates were also redundant (both as negatively correlated parameters). Furthermore, using the system of statistical analysis of long-term data collection, they were able to show that the processes of dyeing, handling, and irradiation adversely affected "downstream" parameters of fitness and quality such as emergence percentage, flight ability, and adult lifespan. Finally, but also noteworthy, the authors point out that through the design of their statistical analysis, they were able to demonstrate that quality parameters such as pupal weight and longevity of adults were affected by season, with lowest adult weights and highest longevity occurring within the midwinter (which is June, July, and August in the southern hemisphere).

The longitudinal scope of the data analysis by Fanson et al. (2014) is very extensive and useful as a model for relationships of rearing conditions and quality. One of the most useful features of this work is the data treatment, which should serve as a model for insectary workers who are trying to tie quality of products (insects being produced) with the quality of inputs or PC. A few other studies such as the *Trichogramma* study by Cerutti and Bigler (1995) also reports use of advanced statistical methods such as PCA, eigenvalues, and multivariate statistics to associate and evaluate QC outcomes. But the rearing systems and rearing personnel who have done the most advancement in tying *PC* and *QC* are clearly the pioneers in the mass rearing of insects for SIT. This point is developed in the following paragraphs.

In providing a template for application of PC to mass rearing of tephritid fruit flies, Lorraine and Bruzzone (1992) stated the benefits of using PC techniques used in industry in their abstract:

> Results of a pilot project to apply advanced rearing techniques to the mass production of sterile medflies (*Ceratitis capitata* Wied.) indicate that industrial process-control procedures can provide program administrators and production managers with a multifaceted management tool. Quality control of process can be used to (a) control variation in the production process in order to identify and correct

problems before they become crises, (b) generate the information needed to estimate the tradeoffs made when procedures are modified, (c) identify parameters for program evaluation that appropriately guide and motivate the behavior of program implementers, (d) identify new research questions, and (e) provide a formalized structure to direct and facilitate the adoption of new technology.

Lorraine and Bruzzone (1992) go on to explain:

The application of industrial quality-control techniques to rearing sterile medflies was first suggested by Chambers 1977 and Boller and Chambers 1977, leading to the development of standardized procedures for measuring and accessing fly quality (Boller et al. 1981 and Brazzel et al. 1986). *Measurements of fly quality, however, are only half of the total quality-control package. Product quality measurements provide only retrospective and very indirect information about the consistency, precision, and reliability of the production process.* Potentially, every aspect of the production process can influence fly quality in some way. If the sterile insect technique (SIT) is to achieve its potential as a fully developed industrial production technology, *managers must be able to distinguish between systematic and random variation in the production process and also be able to identify, measure, track, and correlate the impact on fly quality of all key production parameters.* To help focus the attention of researchers, agency administrators, and program managers on this neglected half of total quality control, Chambers and Ashley again borrowed from industrial quality-control concepts and procedures to identify three distinct components of the total quality-control package required for sterile insect production: process control, production control, and product control (Chambers and Ashley 1984). (emphasis is the author's)

The crucial point of Lorraine and Bruzzone's paper is that the quality of the insects, in terms of their usefulness for a given purpose, must be related back to the elements of the rearing process. It follows from this that *deviations* known as *errors* in the process (special causes of error, random error, or gross operator error) must be determined and linked to the deviations from set points or quality goals. This is illustrated in Figure 11.5 and is further discussed in detail later in this chapter and elsewhere in this book, but as a simple example: if an event where biomass of the output insects is lower than expectations (based on past measurements), it would be most useful (and correctable) if it could be shown that the lower quality (weight) is attributed to lower protein concentrations in the wheat germ used to make the batch of diet for the subpopulation with the lower quality. The all-important connection, therefore, is the linking of the deviation in diet quality (protein content from wheat germ) with the insect quality (weight of pupae).

In discussing QC in codling moth, Dyck (2010) makes an excellent point:

A very important aspect of QC is routine monitoring of quality parameters (Huettel 1976, Hathaway et al. 1973, Rogers and Winks 1993). "However, monitoring itself does not correct problems. Each rearing facility should develop SOPs for rearing operations, QC operations and finally responses to adverse QC findings" (Parker 2005).

Rogers and Winks (1993) illustrated case studies of successfully identified problems such as diet deficiency, material contamination, and a lack of response to pheromones and a larval attractant (α-farnesene). These case studies illustrated how an Insect Rearing Management system identified, solved and helped to prevent quality problems.

What Dyck is asserting here is the point about finding connections between *processes* that go into rearing systems and the quality (or deviations from quality standards) of the *products* of rearing, the insects. The point made by Parker (2005) within the quotation from Dyck (2010) is a very important, though implicit, criticism of practices that are followed in many insectaries, where staff members, sometimes assigned the role of "QC" managers, measure and/or keep records of processes and products, but often in such circumstances, the causal connections between processes or practices and outcomes are not made. This makes the efforts to measure quality into a kind of placebo effect where managers and staff feel that they are doing something helpful toward improvement of their rearing situation. The solution to this problem is to make deliberate efforts to tie processes to the characteristics and quality of the products. A suggested set of protocols is presented in Chapter 17 as well as in several places in this chapter.

Figure 11.4 offers a template for developing the system discussed in the previous paragraph. In this hypothetical scenario (based on actual insectary problems but compressed here for clarity), an insectary staff member (QC technician) is responsible for keeping records of biological features of tobacco budworms being produced in a rearing facility. Records are kept of pupal weight, development time, and percentage adult emergence from pupae, rate of egg production (fecundity), and

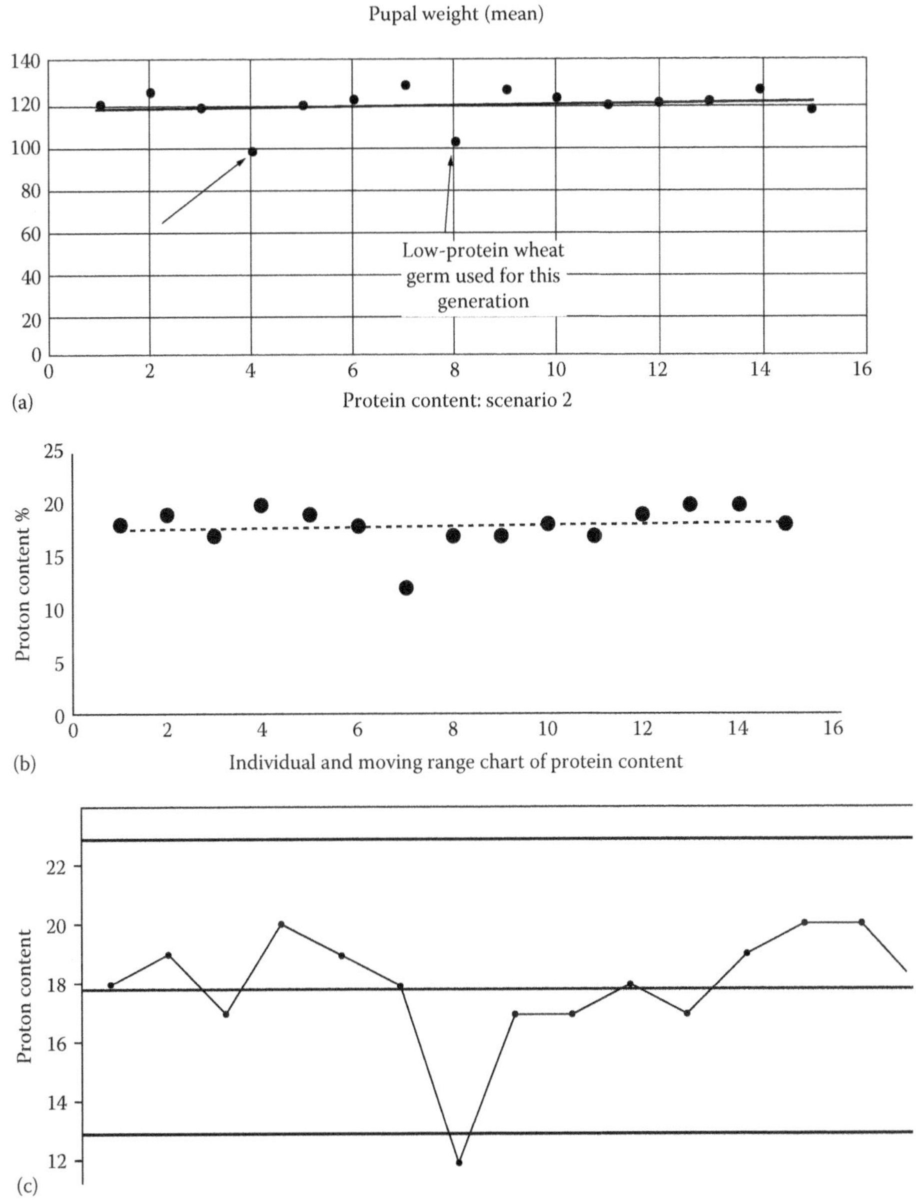

Figure 11.4 (a) Relationship between insect characteristics, (b) protein content of diet staple, and (c) a capability (process control) chart. This figure provides a model of the relationship between quality and process control.

percent egg hatch (fertility). Also, throughout the rearing operation, another person (the PC technician) makes measurements of rearing processes and keeps detailed records of such matters as dates of receipt of raw materials, special characteristics of the raw materials, chemical and physical measurements of raw materials; the QC technician also makes observations of a routine set of measurements of processes (such as notations about grinding, mixing, heating, and pouring of the diet) for each batch of diet being made. As indicated by Figure 11.4, there is a record of the protein concentration (% protein) of key raw materials such as wheat germ, soy flour, and yeast, as they are received.

Studying the top graph in Figure 11.4, it is evident that the weights of tobacco budworm pupae sampled from Rearing Batch 4 and Rearing Batch 7 are lower than the remainder of the rearing batches (rearing group cohorts). A check of the records indicates that the shipment of lidding material used for rearing trays was defective, and the larvae reared in Rearing Batch 4 were in containers that had fewer pinholes in the lidding (due to an error in the hole punch process at the lidding manufacturer's facility). Each lid was supposed to have 8 holes, but in this batch, there were only 4 holes. (Again, the reader is reminded that this scenario is adopted from an actual case where the lidding maker supplying the rearing facility at the USDA at Mississippi State University was supposed to run the lidding through a punch machine in two directions to get the full number of holes, but in this batch, the lidding was run in only one direction, resulting in half the holes being punched.) Once the new lidding arrived with the proper number of holes, the defective lidding was discarded, but in the meantime, the problem seemed to manifest itself as a reduction in pupal weight attributable to insufficient gas exchange. This hypothesis was later confirmed in side-by-side tests that showed that indeed, the defective lidding caused reductions in the survival and fitness of larvae and resultant pupae.

The second problem with pupal weight was attributable to the lesser protein content in the wheat germ. The variability of wheat germ in terms of protein content, lipid content, and other nutrient and antinutrient factors is well documented (such as by Vanderzant 1967). In this case, the batch of wheat germ that had about 12%–13% protein, rather than 18%–20% protein, was sufficiently defective to cause a decline in mean pupal weight in the subsamples from Rearing Batch 7. Once this relationship was established (cause and effect), it would be a reasonable response to continue to measure protein content (or other composition features of incoming wheat germ), and to reject batches of wheat germ that did not meet the established standards, rather than using wheat germ known to be defective in future diets.

11.3 VARIABILITY AND ERROR: CAUSES AND TREATMENTS

The kinds of problems that occur in production systems that do not use SPC involve *variability* in any or all of the system's processes. The variability falls into two types: *special causes* and *common causes* (Montgomery 1991). While *common causes* are the normal variability factors that randomly occur (and described by Montgomery 1991 as "background noise") when a system is in control (not failing), *special causes* are those deviations in the system's performance that throw the process into an unacceptable state of production of outputs. These unacceptable outputs are the results that are considered system failure. The driving forces behind special causes are (1) defective raw materials, (2) machinery not properly adjusted or malfunctioning, and (3) operator error.

Some examples of these causes will help clarify these concepts. If the agar in a new batch is defective in such a way that its normal activation temperature of 85°C has shifted to 92°C, then the normal heating to 90°C will be insufficient to cause the agar to become properly hydrated (see Chapter 5), and the diet will not gel properly. The diet produced with this new, defective agar will be mushy rather than a crisp gel, and it is likely that the insects will not feed properly, and they may possibly drown in the diet or witness other problems with collapse of the tunnels that the insects usually make in a normally formed diet gel. A second example of special causes is illustrated by a thermal control in the heating system that causes the incipient diet to heat to 83°C, 2°C below the normal activation temperature for agar in the hypothetical system. Again, this will result in production of a mushy, poorly gelled diet, with the same results described for the defective agar in the previous example. The third scenario is where the technician who ordinarily sets the thermostat for the diet kettle to 85°C fails to do so, and the kettle remains set to 80°C from a previous use of the steam kettle for another procedure. Again, the result is a mushy diet with the previously described dire consequences for the insects that depend on the diet's texture being within an established, acceptable range. These errors illustrate the three special causes: defective raw materials, machinery malfunction, and operator error. In all three cases, the colony is threatened, and even if there are survivors, their quality will be greatly compromised.

Figure 11.6 shows an overview of an insectary's production process where inputs such as raw materials for diet, containers, oviposition components, and other materials as well as processes are evaluated for their contribution to the variability in different parts of the system. Some key concepts behind this diagram are (1) that measures can be taken to reduce the variability of controllable inputs (discussed in the following text as *special causes* and *common causes*) and to develop ways to change uncontrollable inputs to controllable ones. The feedback system that the rearing specialist uses to make these changes is an orderly and carefully planned regimen of *measurement*, *evaluation*, and *control*. An example will clarify this model.

In rearing systems for many lepidopteran larvae, wheat germ is an important staple, and also a material that can be a source of a great deal of unwelcome variability. The characteristics of wheat germ are covered in detail in Chapter 4, but the kinds of variability include (1) the type of wheat germ (source or producer and whether the product is heat stabilized or raw), (2) the particle size, (3) the storage history (before and after shipment to the insectary), and this includes the method of storage (Figure 11.5 shows examples of suitable storage containers, and Figure 11.6 shows the packets that remove oxygen and moisture) (Figure 11.7). Factor (1) causes of variability include (a) differences in the wheat genetics and growing conditions, (b) differences in milling and sorting of the germ, and (c) preshipment storage. (Was it stored under cool conditions? How long was it stored? Did storage conditions include exclusion of moisture and air? Was the product vacuum packed? Figures 11.5 and 11.6 illustrate these concepts.) (2) The particle size can influence rates of degradation and mixing characteristics with other diet components. (3) Postshipment treatment includes repackaging in air-tight containers, refrigeration, or freezing, using small aliquots in each package to avoid reopening of packages and exposure to warmth, light, moisture, and air, using an oxygen scavenger and desiccant in the packaging (Figure 11.7).

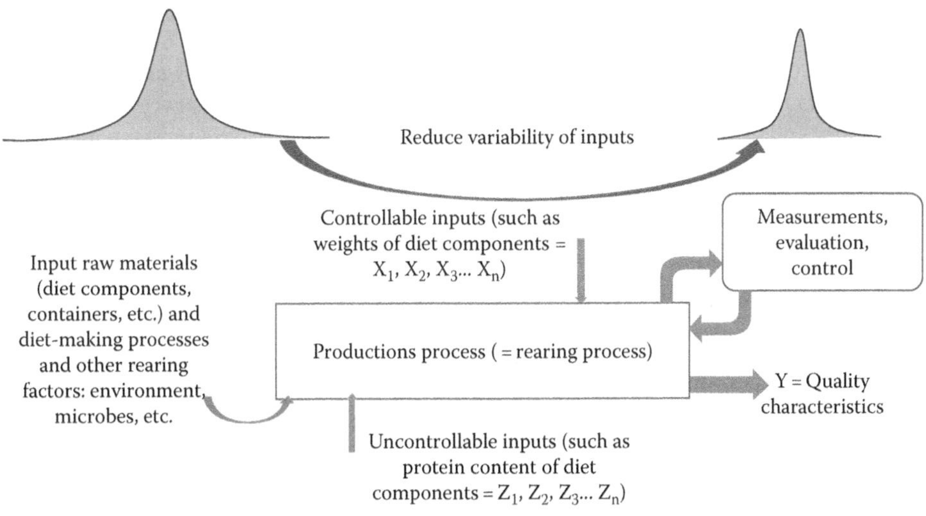

Figure 11.5 Model of system for removal of special causes and reduction of variation. (Adapted from Montgomery, D.C., *Introduction to Statistical Quality Control*, 2nd edn., John Wiley & Sons, New York, 1991.)

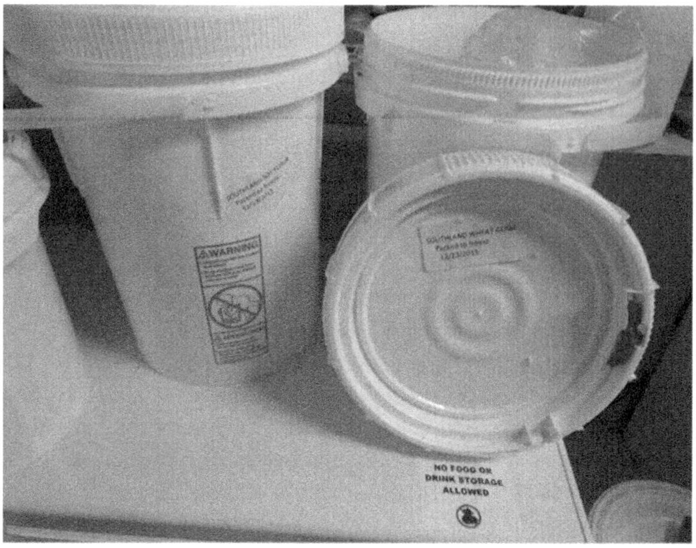

Figure 11.6 Containers for storage of wheat germ and soy flour in smaller aliquots than the original shipment. Containers can be sealed to exclude moisture and stocked with packets of oxygen absorbent, which is usually supplied with moisture absorbent.

Too often, rearing personnel are overly generous in their confidence in the homogeneity and quality of the product they receive. They assume that a product that they are obtaining from what they consider a reliable source would have passed QC standards that eliminate undesirable variability (such as moisture, lipid, protein content, or other features including minerals, carbohydrate, or vitamins). In various rearing facilities under the auspices of USDA, APHIS, there has been a recognition of variation in the diet raw material, agar. They had learned over many years of rearing that there are differences in the gel strength of agar, ranging from strengths of less than 500 g/cm^2 to more than 1800 g/cm^2 force required to fracture a 2% gel (personal communication with Dr. Fred Stewart, formerly head of the USDA, APHIS Pink Bollworm Facility in Phoenix, AZ). The topic of gel strength and other rheological properties of gelled diets is discussed further in Chapter 15, but this illustrates that raw materials, often from the same supplier, can vary greatly from batch to batch. And although wheat germ has generally not been subjected to the same scrutiny as agar, it has an equally great potential to cause variability problems in successive batches of diet. A little of this potential for rearing failure attributable to wheat germ is noted by Vanderzant (1967) and Bell and Joachim (1976) (Figure 11.6).

11.4 EXAMPLE OF GAINING CONTROL OVER VARIABLES IN WHEAT GERM

First, a controllable input would include the treatment of the incoming product. This would include collection of a subsample and analysis of that sample for water activity, moisture content, protein content, and crude lipid content. The next step would be to make aliquots of the product in small enough masses to allow production of batches of diet without thawing and refreezing or opening packages repeatedly. The third step in modifying controllable inputs would be to improve the precision and accuracy of measurements of the wheat germ by using the most precise balance for weighing batch-sized aliquots and using a properly calibrated balance. This is discussed in more detail in the section on error and variability, but briefly, this means using a balance that could weigh a 100 g aliquot to 100.000 or 100.00 g. Using a balance that reads to the nearest gram can add a 1% error on top of other errors mentioned herein.

Figure 11.7 Examples of packets containing combined oxygen scavengers and moisture absorbents. Such packets are inexpensive and greatly extend the shelf-life of diet components that are vulnerable to oxidation and moisture damages. For example, the cost of protecting a 19 L bucket of wheat germ is less then $ 1.00 U.S.

Care in selecting a container appropriate to reducing tare-based errors will help reduce variability. If the balance has a 1% error, a 100 g container can generate a 1 g error, but on the same balance, a 10 g container can generate a 0.1 g error. Of course, the container must match the material being weighed, so a small 1 g plastic weigh boat will not hold 100 g of wheat germ, but a larger 4 g weigh boat will hold the 100 g sample. However, using a large beaker (1000 cm^3) that weighs about 400 g will hold the 100 g easily, but it can introduce a 4 g error (Figure 11.8).

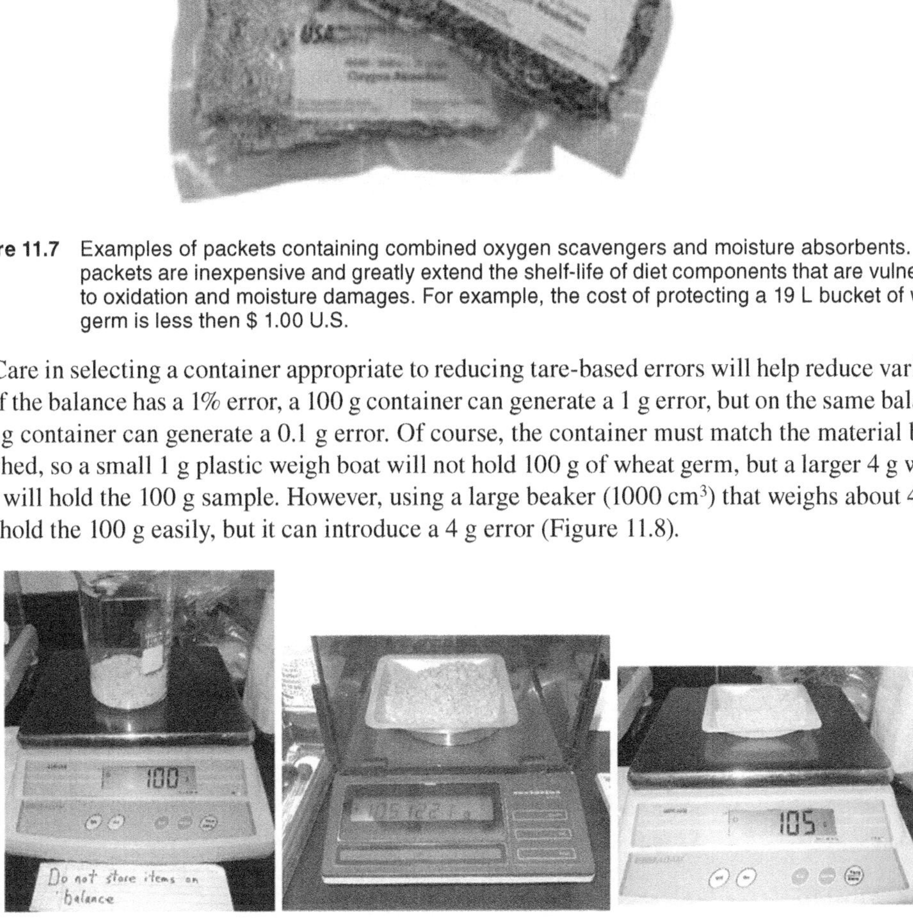

Figure 11.8 Selecting appropriate weighing equipment and supplies to reduce variability and error. The top loading balance reads to grams, and the analytical balance (in the middle) reads to thousandths of a gram. The wheat germ pictured here was weighed to 100 g in a beaker after a tare to zero on the top-loading balance. The sample was placed in a weigh boat (whose weight was 5.3649 g) and weighed on the top-loader at 105 g and then on the analytical balance where the weight was measured at 105.1221 g minus the pan, so the actual weight of the wheat germ was 99.7572 g, which is 0.24% less than the intended weight of 100.000 g. It should be noted that the error from weighing the wheat germ on the less sensitive balance could have been as much as 0.5 g above or 0.5 g below the desired weight, which is an error of 1% of 100 g. However, if the same low sensitivity balance were used to weigh a diet material where 10 g was called for, samples that actually weigh 9.5 or 10.5 g would both read "10 g," representing a 10% error.

All of this discussion has been aimed at suggesting how much can be done with simple measures of variability control (system control) such as repackaging diet materials, storing them under more deliberate and well-planned conditions, and reconsidering measurement strategies that improve precision. Now the discussion is focused on how personnel can move inputs (shown in Figure 11.5) from the "uncontrolled inputs" to "controlled inputs."

Several researchers described measures to correct deficiencies in raw materials, generally by the addition of some component to compensate for such deficiencies. An example of this is the measure taken by Bell and Joachim (1976) to add linseed oil to their diet for *Manduca sexta* in case the wheat germ was deficient in its lipid content, a flaw that Vanderzant (1967) pointed out about wheat germ, which she explained can be variable in its lipid nutrient components. Adding extra ascorbic acid or other vitamins is considered a safety factor when there is concern that heat-labile vitamins are being destroyed in diet processing or that they are otherwise unavailable in the amounts required by the target insects. Such thinking is expressed occasionally in the literature, for example, by Edwards et al. (1996), where that research team developed the procedure for applying twin-screw extruder technology to production of the pink bollworm diet. They discussed the possibility of vitamin destruction by the high temperatures, pressures, and mechanical forces of the extruder and how these conditions could destroy labile vitamins such as thiamine and how such destruction would require possible compensation by addition of extra vitamins. Fortunately, in the case of the pink bollworm diet and the extruder, the research team demonstrated that the vitamins were not destroyed. Despite this demonstration, researchers and diet-making personnel often consider using an overcompensation strategy.

However, some mechanistic thinking about such measures should raise questions about the "downstream" effects of such compensatory actions. For example, we should ask: what happens when we add 0.4% linseed oil to a batch of wheat germ that contains greater amounts of lipids than the average? Is there not a possibility of too much lipid? So this argues for some kind of measure to evaluate how much lipid is present in a batch of wheat germ, and if it is below a threshold, then and only then would it be efficacious to add extra lipid. This subject, including a treatment of the rheological effects of adding oil to diets, is treated in Chapter 15.

11.5 FOUR DIET/PROCESS CONTROL CRITERIA

Elsewhere in this book, diets are discussed from the perspectives of meeting the insects' *palatability* and *nutritional* needs as well as maintaining *bioavailability* and *stability* (Figure 1.2). Clearly, the diet-making process must result in the diet's ability to meet these four criteria. It would be helpful if the processing steps could be measured immediately and directly, but this is not possible. Therefore, steps in the diet making, storage, and presentation must be based on best estimates of key criteria, ideally based on scientific testing. A few examples and case studies should help clarify this issue.

11.5.1 Applying Process Control in Insect Diets

When we make a batch of artificial diet for our insect, we are going through a process. If we do not use the tenets and protocols for PC, we invite problems that inherently creep into all manufacturing efforts. Almost all manufacturers of nearly every product use data-based, statically driven PC techniques to assure the reliability of their processes and quality of their final product. However, insect rearing, with a few notable exceptions in screwworm and fruit fly mass-rearing facilities, has lagged behind other industries in utilization of quality and PC. This chapter is a blueprint to help rearing specialists develop and adopt PC into all aspects of their diet-making procedures. Also, very importantly, most procedures that have been developed are not published or otherwise publicized. Hopefully, this chapter will fill in many gaps in the conceptual basis for PC and the practical issues of putting a PC system into operation.

The kinds of problems that occur in production systems that do not use SPC involve *variability* in any or all of the system's processes. The variability falls into two types: *special causes* and *common causes* (Montgomery 1991). While *common causes* are the normal variability factors that randomly occur (and described by Montgomery 1991 as "background noise") when a system is in control (not failing), *special causes* are those deviations in the system's performance that throw the process into an unacceptable state of production of outputs. These unacceptable outputs are the results that are considered system failure. The driving forces behind special causes are (1) defective raw materials, (2) machinery not properly adjusted or malfunctioning, and (3) operator error.

Some examples of these causes will help clarify these concepts. If the agar in a new batch is defective in such a way that its normal activation temperature of 85°C has shifted to 92°C, then the normal heating to 90°C will be insufficient to cause the agar to become properly hydrated (see Chapter 5), and the diet will not gel properly. The diet produced with this new, defective agar will be mushy rather than a crisp gel, and it is likely that the insects will not feed properly, and they may possibly drown in the diet or witness other problems with collapse of the tunnels that the insects usually make in a normally formed diet gel. A second example of special causes is illustrated by a thermal control in the heating system that causes the incipient diet to heat to 83°C, 2°C below the normal activation temperature for agar in the hypothetical system. Again, this will result in the production of a mushy, poorly gelled diet with the same results described for the defective agar in the previous example. The third scenario is where the technician who ordinarily sets the thermostat for the diet kettle to 85°C fails to do so, and the kettle remains set to 80°C from a previous use of the steam kettle for another procedure. Again, the result is a mushy diet with the previously described dire consequences for the insects, which depend on the diet's texture being within an established, acceptable range. These errors illustrate the three special causes: defective raw materials, machinery malfunction, and operator error. In all three cases, the colony is threatened, and even if there are survivors, their quality will be greatly compromised (Figure 11.9).

11.6 LINKING PROCESS CONTROL IN DIETS TO QUALITY OF INSECTS PRODUCED

The fundamental purpose of PC is to improve the quality of the insects being reared and then to maintain that quality for the duration of the rearing operation. An important accessory purpose of PC is to maintain the quality within economical and other practical constraints. Therefore, it is of paramount importance to relate process steps to quality outcomes. Decisions and goals concerning quality are reached through an iterative dialog between the insect production team and the customers. Measures of quality must be made in the context of purpose or uses of the insects in the production system. If insects are being produced as feeders for other organisms, the mobility of the insects may not be as important as it would be if the insects were to be used for a sterile release program where males must be competitive with field-derived males. Similarly, if insects are being reared as study materials for an insect physiology course, their overall fitness may not be as crucial as it would be for predatory insects that are to be released in a biological control program. None of this discussion is intended as a suggestion that any rearing specialist should relax standards of quality, but it is intended to point out the reality and importance of linking intended purpose and quality standards. It is further hoped that this discussion will drive a standards (quality) development effort that reflects the fact that even small changes in a process of rearing, in general, and diet making can have a very important impact on production quality (Figure 11.10).

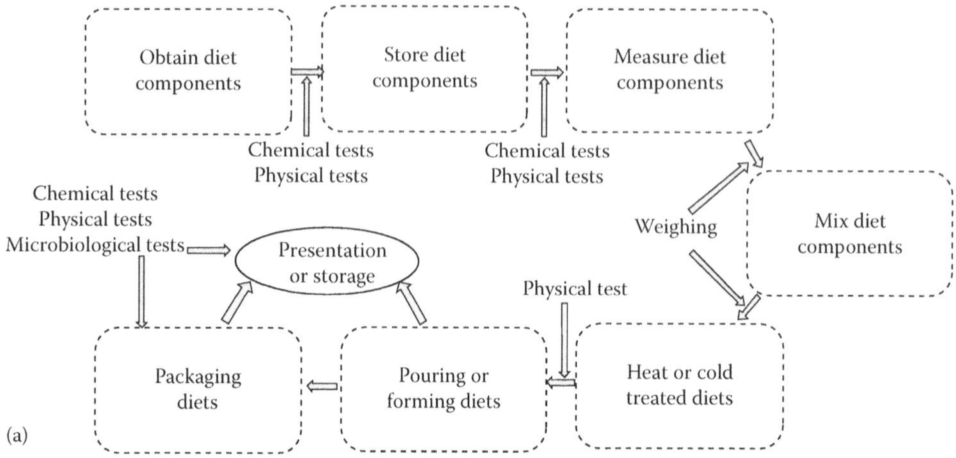

Diet process tests

- Chemical tests
  - pH
  - Proteins
  - Carbohydrates
  - Lipids (and lipid components)
  - Antioxidants, free radical scavengers
  - Antimicrobial agents
  - Other specific tests
- Physical tests
  - Water activity
  - Moisture content
  - Particle size
  - Weight/volume
  - Rheological
    - Gel strength
    - Gumminess
    - Cohesiveness
    - Stringiness
- Microbiological
- Colony counts
- Bacteria vs. fungi
- Kinds of microbes

(b)

Figure 11.9 Components of process control in production of insect diets, (a) showing the steps in diet making and (b) listing the kinds of tests that are useful and parsimonious in diet processing. This model is a basis for finding inflection points where process controls can be adopted.

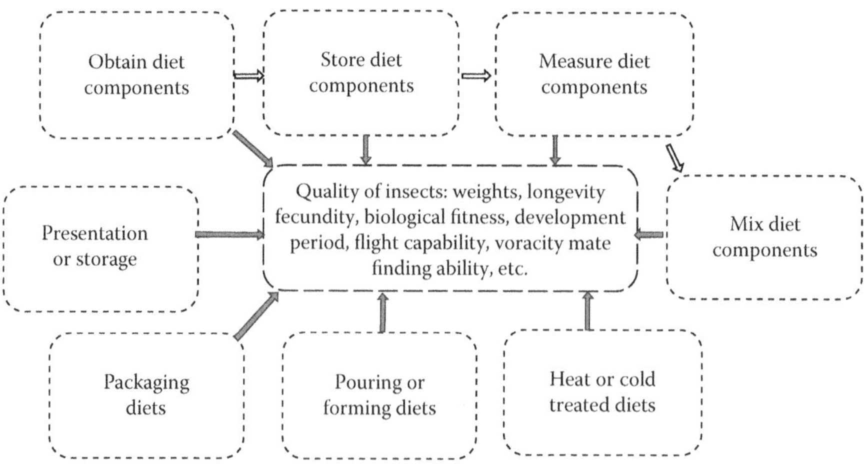

Figure 11.10 Linking the process in a diet-production system to the ultimate quality of the insects.

The kinds of issues that Figure 11.4 is intended to underscore are the relationships between factors such as protein content of a raw material and the quality factors described, the impact of storage of an oxidation-susceptible material such as wheat germ or soy oil and the longevity of the insects, or how improper mixing of antifungal agents can affect the weights or digestive efficiency of the target insects. These are only a few of the hundreds of kinds of interactions that exist but are incompletely documented. Unfortunately, the relative dearth of demonstrated cause and effect or correlation-based examples places these connections almost in the "butterfly effect" (chaotic) status where connections of distant factors become a kind of chaos theory model. However in this author's nearly four decades of rearing experience, it is a dead certainty that there are countless instances where rearing systems became derailed due to seemingly small and hitherto unconsidered violations of sound PC procedures. Often the deviations from optimal procedures (departures into the realm of *special causes*) are too subtle to capture notice, or even more often, insectary workers lack the training to observe the deviations. Also, as is the case in many administrative systems, workers are not adequately encouraged to point out deviations, especially if the culture of the system presents a risk of blame for problems that have crept into the rearing operation. These concepts are discussed elsewhere in this book, but for now, it is valuable to recognize that the philosophy and approaches to production called total quality management (TQM) as discussed by Deming (1986) and Feigenbaum (1983) are the guiding principles that help the Japanese automobile industry succeed in a remarkable way, and these principles have been applied to countless industries and systems.

Hopefully, the remainder of this book will underscore this point.

11.6.1 Linking Process Control in Diets to Quality of Insects Produced: The Four Diet/Process Control Criteria

Elsewhere in this book, starting in Chapter 1, diets are discussed from the perspectives of meeting the insects' *palatability* and *nutritional* needs as well as maintaining *bioavailability* and *stability* (Figure 1.2). Clearly, the diet-making process must result in the diet's ability to meet these four criteria. It would be helpful if the processing steps could be measured immediately and directly, but this is not possible. Therefore, steps in the diet making, storage, and presentation must be based on best estimates of key criteria, ideally based on scientific testing. A few examples and case studies should help clarify this issue.

11.6.1.1 Case Study 1: Replacement of Carrageenan (Palatability)

Prior to 1969, the Biological Control of Cotton Insects Laboratory in Tucson, AZ, was rearing several species of cotton pests on a Lepidoptera diet (Patana 1969, 1977). The diet contained carrageenan as a gelling agent and was being used in place of agar because of its lower cost and excellent palatability to each target species. After several years of successful mass rearing, there was a sudden truncation in the cotton Lepidoptera production where suddenly, the larvae failed to reach full size and demonstrated greatly reduced pupation rates, adult emergence, and egg production (fecundity). Upon careful interpretation of shipment records, the rearing staff in Tucson discovered that the supplier of the carrageenan had changed its source, but the supplier reported that the only difference between the new gelling agent and the older material was in the sodium content, which the supplier felt would be of no consequence to the customers who used carrageenan for products such as ice cream, puddings, and other dairy products that required a gelling or thickening agent. Clearly, however, the unconventional "customers" (the Lepidoptera larvae) had a different response from human consumers, and they responded to the new product with greatly reduced consumption of the diet. The problem was solved, fortunately, by locating a large supply of the older type of carrageenan, which was used until the laboratory was closed down in 1991. Other kinds of palatability problems can stem from changes in gel strength or other texture parameters, substituting a raw soy flour or wheat germ for a toasted (heat-treated form of this component), addition of too much antifungal agent, and many other factors that change the sensory qualities of a diet.

11.6.1.2 Case Study 2: Decreasing Lipid Content of a Predator Diet (Nutritional)

The diet for rearing big-eyed bugs (*Geocoris punctipes*, Hemiptera: Geocoridae, formerly Lygaeidae) was composed of ground beef, liver, and sucrose solution (Cohen 1985a). Although the diet supported continuous production of the predators, the rearing personnel in the USDA, ARS and USDA, APHIS laboratories were continually trying to improve the diet to bring biological parameters to the standards dictated by rearing outcomes with natural prey—mainly eggs and larvae of heliothine moths and aphids (Cohen and Staten 1994, Cohen 2000a). The original ground beef used in the big-eyed bug diet contained about 16%–17% fat. Insectary workers noticed that the frass of big-eyed bugs reared on the artificial diet and which was deposited on the plastic rearing containers was chronically greasy, whereas the frass from the predators, which were reared on natural prey, was decidedly free of the fatty deposits. This difference prompted the predator rearing team to find a ground beef with a lower fat content, based on the assumption that the very fat ground beef was overloading the predators with lipids, possibly contributing to the lower body weights and fecundity of artificial diet-fed predators. A diet with ground beef that was about 10% fat provided with the expectation that the fitness parameters would improve in predators fed with the new diet. Just the opposite occurred. While the frass was less greasy in the groups of big-eyed bugs fed with the lower fat diet, their survival, body weights, and fecundity also declined. The problem did not seem to be related to feeding response as the low fat-diet group seemed to accept the new diet feeding for the same amount of time. However, the fitness clearly was related to a nutritional disadvantage with the lower fat diet. The USDA website on composition of foods (http://ndb.nal.usda.gov/ndb/foods/show/7407) lists cholesterol levels that are about 15% lower in low fat beef than in the higher fat product. Also vitamin E levels and levels of unsaturated fatty acids are lower in lean beef than in high fat beef. It was never determined definitively which factor was responsible for the decline in predator colony fitness with the low fat experiment, but once the high fat diet was resumed, the colony thrived for hundreds of successive generations (Cohen 2000a).

11.6.1.3 Case Study 3: Destruction of Antinutrients (Protease Inhibitors) to Make Proteins in Plant Bug Diets More Digestible (Enhanced Bioavailability)

During the process of developing a mass-rearing system for tarnished plant bugs (*Lygus lineolaris*) and western tarnished plant bugs (*L. hesperus*) at the USDA, ARS Biological Control and Mass Rearing Unit in Mississippi (at the R. T. Gast Facility), the research team encountered many difficulties inherent in elevating a small-scale rearing system (production of hundreds or thousands of insects per week) to a truly large mass-rearing system (production of tens of thousands to millions of insects per week) (Nordlund et al. 2001). A major problem involved the increase in batch size of plant bug diet from one liter batches (Cohen 2000c) to 20–30 L, a minimal amount considered appropriate for mass rearing of insects at the Gast Facility (Nordlund et al. 2002). The first step in the scale-up procedure was enlarging batch size to 3–5 L quantities. During this process, it was discovered that the autoclave treatment of diet used for the 1 L batches resulted in a diet that was toxic to the *L. hesperus* and *L. lineolaris* (Cohen et al. 2005). Careful analysis of the diet indicated that in the larger batches, the diet was not being fully cooked due to heat transfer problems discussed in Chapter 12, and further, it was determined that the main culprit was undercooked soy flour. Soy products, while potentially rich nutrient sources, also contain a series of heat-labile antinutrients such as protease inhibitors, lectins, and phytates (which are slightly sensitive to heat processing, such as flash sterilization, autoclaving, and extruding). Therefore, the medium-scale batches of diet heated in four-liter beakers in an autoclave at 121°C and 15 psi for 20 min were not adequately cooked to destroy the antinutrient compounds, and the colony declined with symptoms of lower biomass per individual, reduction in egg production, and lowered longevity. The condition was reversed by application of adequate diet heating procedures (Cohen et al. 2005). Although there are probably multiple causes of the colony decline associated with inadequate heating, the most immediate one was the decrease in digestibility of the diet's proteins caused by the soy's high titer of protease inhibitor (serine protease inhibitor), to which *Lygus* bugs have been shown to be very susceptible (Zeng and Cohen 2000).

11.6.1.4 Case Study 4: Application of Wax (or Other Materials) to Diet Surface and Strategic Control of Diet Moisture and Softness—Increased Stability

In a large number of diets in successful rearing programs, stability is achieved in ways that may at first seem subtle or even trivial. However, the importance of stability cannot be overemphasized and can be what makes or breaks the success of a rearing program. Patana and McAda (1973) used dried diet flakes on the surface of tobacco budworm and corn earworm diet to (1) limit aggressive interactions between individuals, (2) reduce rates of diet desiccation, and (3) provide an enhanced pupation substrate. With these species and other insects that normally pupate in the soil, diet flakes serve to enhance pupation conditions. Even today, many rearing specialists allow their insects to pupate in their own frass and spent diet—a practice that may be convenient for the insectary personnel but not as sanitary as soil or spent diet. Many rearing programs use a "ballistic desiccation" process that allows the diets to dry out as the insect is completing development, resulting in a more stable diet than would be the case if the moisture content were kept at the level of initial diet synthesis—but with several weeks when antimicrobial agents could deteriorate, and microbial loads could increase.

One of the most well-designed diet stabilization strategies is found in codling moth rearing. The complex progress in diet development and diet improvement in this rearing system is chronicled by Dyck (2010). In Dyck's account of modifications in codling moth diet, he classifies the diets into two major types: those that remain soft (hydrocolloid-type diets) and those that harden and dry out (sawdust/woodchip type diets). He explains the logistics behind the two types of diets in terms of the ease of harvesting prepupae or pupae as the basis for making diet modifications. The optimization is a kind of tipping point determination where the maximum economic feasibility/survival/minimal deterioration/maximum insect recovery is achieved. The balance of these four factors that result in a system of production that meets quality demands, efficiency, and economic production makes this program a model system.

11.6.2 Diet-Making Process

The first step in diet production is procurement of the diet's raw materials (ingredients). Usually, the ingredients are stored for some period of time before being used. Some ingredients are further processed before being added to the incipient diet, including size reduction, sifting, or freeze drying. Further processing usually includes measuring (by weight or volume), mixing, and heating, and some of the postsynthesis processes include pouring, cutting, "ripening," and storage. Many diets are further processed in later stages of production by some type of packaging.

The quality-threatening variability in diets can be introduced with any of these steps. If the variability (known by statisticians as "error") is part of the normal variability, we use the term "common causes."

If the variability is exceptional, we refer to it as a "special cause." Special causes threaten the process and the outcome—possibly the entire rearing process and quality of our rearing product.

It is extremely important to recognize that variability is inherent in all aspects of the real world. Using wheat germ as an example of both common cause and special cause variability, we recognize that the nutritional/chemical content of wheat germ can vary from batch to batch and *within* batches. Also, physical and microbiological aspects of wheat germ can be sources of variation. We routinely find variability of several percentage points of wheat germ's water, protein, sugar, lipid, and vitamin content. There are several sources of variation in wheat germ: variety of wheat, growing conditions, and processing conditions (such as whether or not it was toasted, how the germ was separated from the remainder of the wheat kernel, and several other factors). Packaging and storage can influence the character of wheat germ (e.g., wheat germ used as nutritional supplement is both toasted and vacuum-packed, resulting in a more stable product). Often insect rearing standard operating procedures (SOPs) call for "stabilized" wheat germ, meaning that the product is heat treated (roasted or toasted) to denature enzymes such as peroxidase and lipo-oxygenase, which can cause changes in flavor and nutrient quality of wheat germ. Even a process such as gamma radiation that would seem at face value to be minimally influential can have important effects on the quality of the ingredient. For example, Ramadan et al (2008) showed that stabilization with gamma radiation caused an increase in the antioxidant capacity of wheat germ oils compared to heat stabilization. This means that insect diets made with gamma radiation–stabilized wheat germ would have higher free radical–scavenging capacity than the heat-stabilized raw material. Many other processes are used for various wheat products, including wheat germ—such as extrusion, which has a profound effect on the quality of wheat germ compared to conventionally processed germ.

11.6.2.1 Stability: Diet Component and Synthesized Diet Storage

Cryptic dilution: In Table 11.1, a crude analysis of soy flour composition is presented. The column with ~5% water is the analysis of the fresh flour (USDA 2013).

The remaining columns are based on calculations of the protein, lipid, carbohydrate, and ash composition after the flour has absorbed moisture equal to 7%, 9%, and 11% moisture content. Clearly, when the soy flour has been stored under conditions where it has been allowed to absorb moisture from the ambient air, the increasing moisture content displaces the other (nonwater) components and dilutes them proportionately to the amount of water absorbed. So when an insectary worker routinely adds 100 g of soy flour, the amount of nutrients in the dryer soy product is lower in proportion to the amount of water absorbed. This introduces an error in the protein content (for example) where the protein goes from almost 38% of the 100 g aliquot to less than 36%, meaning that the insect feeding on the diet is being provided with a reduction of almost 6% of the initial, intended soy protein (37.8 – 35.6/37.6 * 100 = 5.8%). There are similar reductions in the concentrations of all the other diet components. In diets where a single component such as wheat germ or soy flour is the sole or major source of protein (or other nutrients), this translates to a substantial, unplanned (and usually unnoticed) reduction of nutrients that results simply due to storage conditions.

Table 11.1 Changes in the Proximate Nutrient Profiles as Water Content Increases from 5% to 7%, from 7% to 9%, and from 9% to 11% Moisture

Water	5.2	7	9	11
Protein	37.8	37.2	36.4	35.6
Lipid	20.7	20.5	20	19.6
Carbohydrate	31.9	31.6	30.9	30.3
Ash	4.5	4.4	4.3	4.2

Note: This is an example of the complexity of mixture problems, where changing the %composition of one component affects the %composition of all components.

It is important to recognize further that with cryptic dilution, there are other potential losses in the quality of diet components that take place as results of storage, especially improper storage. Stability during storage can be greatly enhanced by use adoption of practices from the food industry such as placement of oxygen and moisture absorbents to dry stored diet components (Figure 11.6).

11.7 ATTRIBUTING QUALITY LOSS IN INSECTS REARED ON ARTIFICIAL DIETS

As discussed in Chapters 1, 4, and 10, there are several quality-deteriorating factors or stress factors that may take effect in rearing settings. Besides potential degradation of diet quality, insects may be degraded by environmental stresses, genetic truncation, or deterioration caused by microbes. As discussed in Chapter 13, microbial contamination of the diet causes both losses of nutrients and the introduction of microbial toxins. Also, even facultative, nonpathogenic microbes within insect bodies have been shown to cause reduction in pheromone production, reduction in amino acids and fatty acids in various tissues, and deleterious changes in the overall metabolism (Sikorowski and Lawrence 1994). This subject is discussed comprehensively by Tanada and Kaya (1993).

As diet components and completed diets age, they deteriorate. The causative factors are changes in water content and water activity (basically overwetting or drying out), oxidation, loss or gain of volatile components, microbial growth, and enzyme-mediated chemical changes (all discussed in Chapters 5 and 12). These changes are driven thermodynamically by entropy or decline in the orderliness of the chemical components, that is, "the ravages of time." To detect and prevent these ravages, diet specialists can adopt a well-designed QC system.

11.8 MEASUREMENT OF WHOLE DIET AND COMPONENT QUALITY

Early detection of problems is a major goal of QC teams. It would be ideal to detect the problem prior to adding a faulty component to the diet (i.e., using PC to effect QC). If the processing of the diet were the problem, rather than a faulty component, it would be useful to have early warning of the problem before the diet was used. Such early warning can be obtained from appropriate physical or chemical tests (Chapter 17). Because most diets have at least 10 components, and many have even more components, it would be helpful (in terms of saving extraneous labor) to target components that are most likely to be faulty or to become faulty. The decision about which components are most likely to go bad must, of course, be made on a case-by-case, diet-by-diet, basis, but there are a few useful generalizations.

In diets that use fresh materials (such as the Blossey et al. (2000) diet or the Cohen and Smith (1998) diet), one of the first considerations should be that perishable plants (purple loosestrife) or animal products (liver, ground beef, chicken eggs) may go bad, especially if these components have been stored for a period of time before they were used. Because of the complexity of these components, in terms of chemical diversity and structural characteristics, it is difficult to suggest simple quality tests, but sensory examination of thawed materials can be indicative, including observations of off-colors and off-odors. The destructive potential of storage is discussed in depth in Chapters 5, 8, and 13.

Next, complex foods and diet components that have been processed to increase shelf life can become degraded sooner than many insect diet specialists may suspect. Materials such as meals (especially wheat germ) and flours from plant materials are subject to several quality-eroding forces such as oxidation, peroxidation, enzymatic destruction, and microbial deterioration. Tests of these materials can include water content and water activity tests, tests for evidence of lipid peroxidation (Miller 1998, Nielsen 1998), microbe counts (Difco Laboratories 1984, Pierce and Leboffe 1999, Jay et al. 2005), and sensory qualities (visual inspection for crusting, gumming, caking, or moisture; off-odors).

The more highly purified components, which are also chemically simpler both in terms of composition and detection of quality deviation, include vitamins, mineral mixtures, sugar, gelling agents, specific amino acids, sterols (especially cholesterol), buffers, and the various preservatives, including antimicrobial agents. For the most part, there are specific tests for each of these in chemical and biochemical literature. Many such tests are listed in standard references such as Nielsen (1998) and Miller (1998). Most of the tests employ complex techniques such as chromatography and spectrophotometry.

The most likely components to become deteriorated or to have degraded quality are the vitamins (especially ascorbic acid) and the gelling agents. The ferric-reducing antioxidant power assay (FRAP) has been used as has the DPPH assay (e.g., by Cohen and Crittenden 2004, with detailed explanation in Chapter 17) to assess the total antioxidant power and the ascorbic acid content, by simply including a control that incorporates the ascorbic acid–destroying enzyme ascorbate oxidase. A more widely used and simpler test to perform (compared to the FRAP test) is the DPPH test that is fully explained in Chapter 17. However, it is useful to point out here that both the FRAP test and especially the DPPH test can be applied specifically to individual antioxidants such as ascorbic acid or as general test of a diet or diet component's antioxidant/free-radical-scavenging potential. So, for example, if the quality/PC personnel in an insectary suspect that ascorbic acid is being destroyed by diet-processing procedures (such as heating or mixing with iron compounds), they can use the DPPH or FRAP test to measure the ascorbic acid. There are several specific tests for ascorbic acid, such as HPLC methods, but the colorimetric tests can cost less and require less expertise. It should be noted that if one uses colorimetric tests (FRAP, DPPH, or a titration test), then it is essential to use a specific enzymatic destruction of the ascorbic acid to make these general color-change tests specific.

This raises the question about which materials in diets are most likely to cause variability problems. Clearly, the process/QC staff cannot measure every component—certainly not on a routine basis. So the question arises: which components are most likely to be defective, unstable, or variable? It next raises the question: which tests of these components would be most likely to reveal variability problems that would cause deterioration of quality of the insects? Once a plan of handling raw materials is devised (storage conditions), then mapping out a testing scheme is in order.

11.8.1 Raw Materials: Base Diet Components That May Cause Problems

Commercially available plant products such as soy flour, rice flour, wheat germ, wheat bran, oat flour, etc.
Processed host plant materials, processed insects (such as drone powder for entomophage diets), vertebrate blood, meats, vertebrate eggs
Specially made extracts from plant materials or other extracts (such as hemolymph or egg extract for *Trichogramma* rearing)
Yeast products such as live and killed yeasts, yeast extracts, and autolysates
Some purified components such as soy protein or casein
Antimicrobial agents such as antibacterial agents such as chlortetracycline, streptomycin sulfate, antifungal agents such as methyl paraben, potassium sorbate, etc.
Hydrocolloids such as gelling agents (including agar, carrageenan, alginates, gellan, etc.)
Vitamins/vitamin mixtures
Salt/mineral mixtures such as Wesson salts, Beck salts, etc.

An example of diets for the Mexican fruit fly (*A. ludens* Loew) (Diptera: Tephritidae) will serve as a guide to the decision-making process with a diet that is used to rear billions of this insect species per year for the USDA, APHIS sterile release program (diets in Table 11.2 include ones used in the three SIT [sterile insect technique] laboratories mentioned in this table).

Examining the list of ingredients in the Mexican fruit fly diet, the question can be raised, "Which components are the best candidates for PC testing?" In other terms, the issue is to identify the components most likely to vary in a way that significantly affects the outcome of the rearing process. Yet another valuable way to consider these components is to ask what can happen to the material (1) as it is produced, (2) as it is shipped, and (3) as it is handled or stored once it reaches the rearing facility. For every component except water, storage, especially under improper conditions,

Table 11.2 Diets Used for Rearing Mexican Fruit Flies in the USDA, APHIS Laboratories in the Methods Development Laboratory, Mass-Rearing Laboratory (both at Mission, Texas), and the San Miguel Petapa (SMP) Mexfly Production Facility in Guatemala

USDA, APHIS Fruit Fly Diet Components	Methods Development (Mission, Texas) (g)	Guatemala Mass-Rearing Lab (g)	Mission, Texas Mass-Rearing Lab (g)
Corncob grits	157.6	180.3	157.6
Sugar	82.7	87.2	82.7
Torula yeast	63	67.4	63
Wheat germ	35.5	8.9	59.1
Soy flour	35.5		23.6
Corn flour	11.8	47.6	
Sodium benzoate	1	3.6	1
Methyl paraben	2.3	1.8	2.3
HCl	9.8	8.9	9.8
Water	600.8	594.4	600.2

Source: Data were kindly provided by Dr. Hugh Conway, USDA, APHIS in Mission, TX.

can lead to the absorption of water. Corncob grits can absorb enough moisture to create a 10% error in weight. A second problem that can happen to corncob grits is that the mesh (particle size) can be different from batch to batch. A third problem can come from microbial or other biological contamination. Especially in humid environments, corncob grits can become an inviting target of mites, mold, and even mice (Table 11.2).

11.9 DEVELOPMENT OF PROBLEM-SOLVING STRATEGIES, QUALITY ASSESSMENT, QUALITY CONTROL STANDARDS, AND SOPS

Even after excellent diets and efficient rearing systems (processes) are developed and optimized, things can still go wrong. Furthermore, whatever their purpose, laboratory-reared insects, if they are to be useful, must conform to an established set of standards. To make sure that these purposes are met, rearing system teams develop systems of QC and quality assessment (QA). In well-managed insectaries, problem-solving protocols are outgrowths of the QC program, centered on the insectary's purpose: dependable production of high-quality insects. Well-conceived datasets are collected and monitored for possible deviations from QA characteristics. This point is made clearly in the report from the Joint FAO/IAEA Division group meeting in Vienna (5–9 May, 1997) on product quality for SITs used with mass-reared tephritid fruit flies (Anonymous 1999b, also FAO/IAEA/USDA 2003, FAO/IAEA/USDA 2014). With such careful record-keeping, deviations from quality standards can be diagnosed. Figure 11.15 shows an overview of all the potentially relevant components that can become problems that require problem solving.

In Figure 11.12, components of the diet-making process are presented as possible sources of failures that can be analyzed after diet synthesis is completed but before infestation. The tests include measurements of nutrients such as proteins and sugars, component distribution such as homogeneity of antifungal agents, palatability tests where rapid response bioassays are performed, texture tests, such as gel strength or texture profile analysis (Chapter 15), pH of diet surface, physical defects such as excessive bubbles from foaming, and microbial load where a rapid microbial assay is taken prior to diet infestation so that contaminated diets are not used in large-scale production efforts.

The importance of selecting appropriate parameters to measure pivotal points in the rearing operation cannot be overemphasized. Even with good data-collecting and data management systems such as those described in this chapter, the QC staff can only keep track of a limited parcel of information, so selection of the most telling information is crucial. Because of the emphasis of this book, this chapter deals specifically with QC centered on diets. However, as discussed in Chapter 1, the diet is made and used in the insectary, and therefore, environmental and microbial parameters are always relevant. In fact, problems from environmental and microbial set-points can adversely affect diets. For example, temperature deviations can accelerate the oxidation or enzymatic deterioration of sensitive diet components, and excessive microbial populations can contaminate diets, reducing their shelf-lives. An overview of diet production is provided in Figure 11.2. Chapter 17 contains a suggested protocol for developing a QC system that detects unacceptable deviations in the insectary's microbial populations, and Chapter 17 provides a basis for developing a QC system for the insectary's environment.

The diagrams in Figures 11.11 and 11.12 are abbreviated versions of a fishbone diagram. A more complete fishbone diagram is presented in Figure 11.13, where an overview of insectary problems is provided (from Cohen et al. 2011).

The fishbone diagram (Figure 11.13) is also known as a cause-and-effect diagram where in this case, the overall effect of quality loss can be contributed to or caused by diet factors, container issues, genetics, soil factors, personnel errors, environmental factors, and microbial problems. This type of diagram is inherently flexible, so that if problems can be dismissed as inconsequential, they can be removed, or if new problems become identified, they can be added. This type of diagram can lead to specialized attention to one area such as diet failures, which can become a model that focuses only on diet issues such as the model in Figure 11.14. This figure shows a more focused model of failure: in this case, problems that can be attributable to diet.

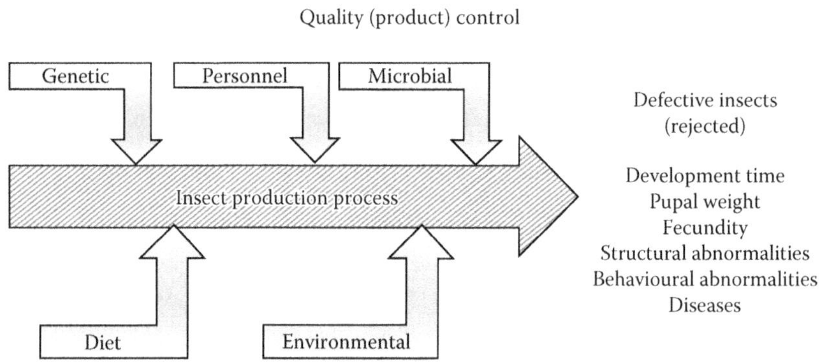

Figure 11.11 Model of relationship between inputs (simplified for clarity) and outputs expressed in terms of defective insects or insects rejected for the reasons in the list. Please note this list of reasons for rejection is also simplified.

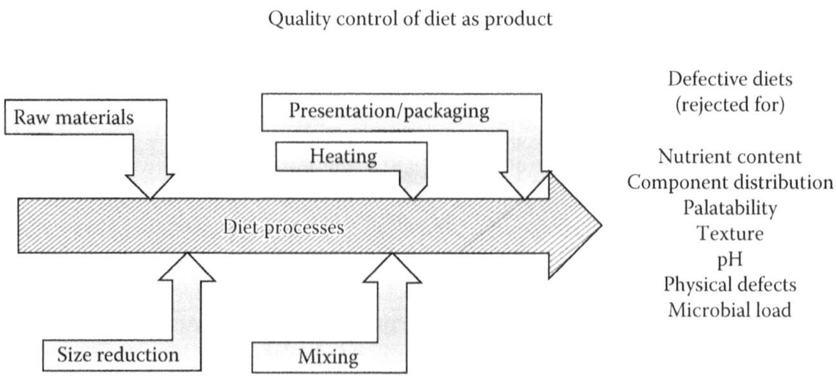

Figure 11.12 Model of the diet as a product, where quality control tests are performed on the synthesized diet. Input arrows indicate possible points in the diet-making process where failures may take place.

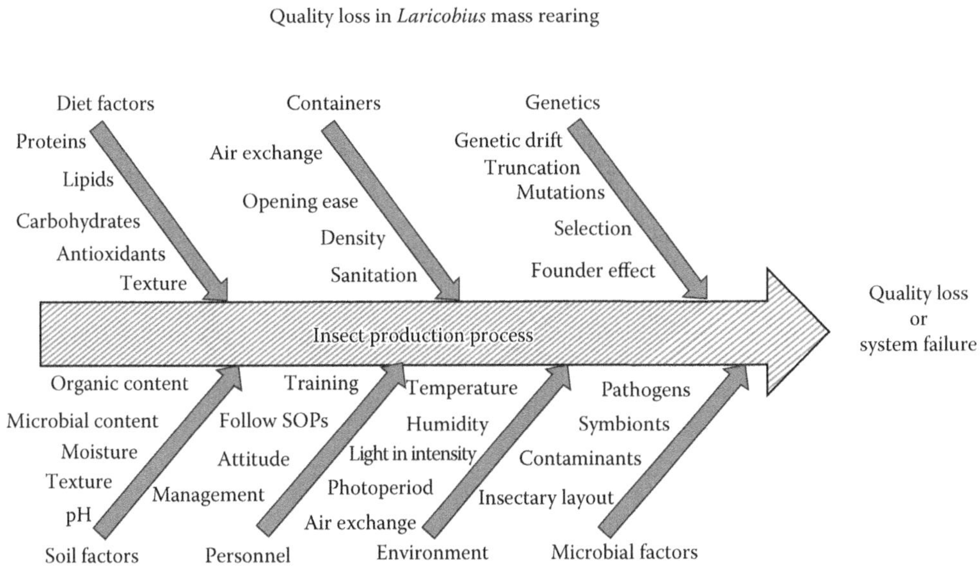

Figure 11.13 Comprehensive model of insectary components that can lead to loss of quality or total system failure. This diagram is modified from Cohen et al. (2011) on quality control in mass-rearing systems for predators of hemlock woolly adelgids.

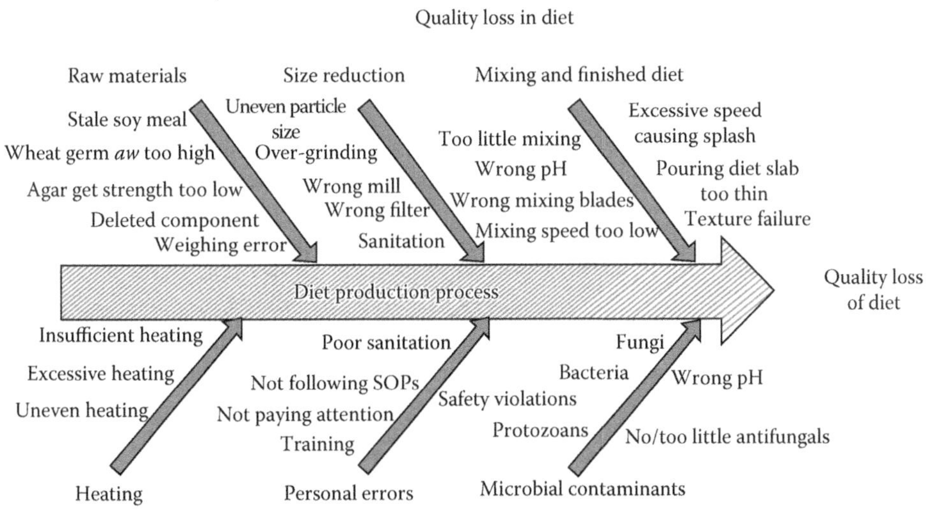

Figure 11.14 A fishbone diagram (cause and effect diagram) that focuses on diet problems as sources of rearing system failure.

The figure that follows is another way of representing potential problems, and in this model, the sources of potential failure are presented in lists of suspected or plausible causes. This type of model (like the fishbone diagram) helps to focus attention on the systems in the rearing operation. The systems approach in trouble shooting or problem solving is useful, because it allows us to rule out irrelevant factors that may be causing the problems.

The cause-and-effect diagrams and systems analysis help insectary personnel determine the causes of problems, but they do not help in the decision process of the *ranking* of these problems. This is where

Sources of insectary problems

Insectary environment
- Temperature
- Humidity
- Photoperiod
- Light quality
- Light intensity
- Air contaminants
- Noise/vibrations

Insects
- Adapted to rearing conditions
- Genetic truncation
- Microorganism relations
- Phenology/development
- Morality factors
- Density
- Exposure to waste products

Insectary staff
- Sanitation practices
- Not following SOPs
- Attendance
- Communication gaps
- Inflexibility
- Uncooperative attitude
- Lack of education/training

Diets
- Raw material purity/history
- Improper storage
- New source
- Changes in process
- Changes in equipment
- Texture "failure"
- Stability/bioavailability/palatability failure

Containers, cages, diet presentation system
- Air exchanges
- Oviposition sites
- Inadvertent traps
- Escape
- Diet accessible to all stages
- Diet dries out or fails to dry in timely fashion

Figure 11.15 Systemic model of sources of insectary problems.

the Pareto model comes into play (Figures 11.16 and 11.17). Going back to the standards of control stated throughout this chapter, it is clear that when several sources of variability or deviation from desired outcomes are present, it would be most productive to determine the extent to which each factor causes failure. We can ask the question, "Which factors (in Figure 11.15) play a role in diet-caused failure?" And we can answer this question with a fishbone or cause-and-effect diagram. However, when we ask, "Which factors will be most important in solving or reducing our problem?," we are thinking in a Pareto mode. In Pareto analysis (sometimes called the 90:10 or 80:20 concept), we try to find a major factor that accounts for the bulk of either success or failure in our system. Two examples from the mass-rearing system for *Lygus hesperus* and *L. lineolaris* will help clarify this concept.

The tarnished plant bug mass-rearing program was initiated from the diet developed by Cohen (2000c, 2003a) (discussed further in Chapter 4) and was adopted by the Biological Control and Mass Rearing Research Unit (BCMRRU) (USDA, ARS) at Mississippi State University in the late 1990s. The diet was originally developed (derived with major modifications from the Debolt 1982 diet) by Cohen at the Western Cotton Research Lab in Phoenix, and in efforts to duplicate the excellent rearing results obtained with this diet in Phoenix, the BCMRRU *Lygus* team was baffled by the decline and repeated loss of colonies. The team went so far as to hire the technician who had made the diet (Ms. Lisa Smith) to consult with the BCMRRU *Lygus* team to help ferret out the problem(s). Virtually everything in the diet made at BCMRRU was exactly like the composition and processes used in Phoenix, except for the change that was made in grinding the baby lima beans. In Phoenix, beans were ground with a small coffee grinder, but at the BCMRRU, the beans were ground in an industrial hammer mill, and the filter that was arbitrarily selected was the 0.5 mm mesh, which reduced the particle size from to about 1/6 the size of the particles from the coffee grinder.

11.10 LOGISTICAL AND STATISTICAL BACKGROUND: USING PROCESS CONTROL STATISTICS AND CHARTS IN THE QC ENVIRONMENT

Statistical treatment of key parameters is the centerpiece of every QC system, and this subject is treated in depth by Montgomery (2009), Lorraine and Bruzzone (1992) Grant and Leavenworth (1988), Ryan (1989), and the industry standard: Western Electric (1956). The discussion here is based mainly on the reference text from the statistics program, Statview® (SAS 1999). An excellent document on QC in the program on mass rearing of tephritid fruit flies (Anonymous 1999b) makes reference to an alternative QC/statistics package that also automates data management for to QC procedures (SPSS 1996). One of the most user-friendly yet powerful software packages designed for QC/PC and design of experiments (DOE) useful for diet mixing and optimization problems is the JMP®, Version 11 Software (SAS Institute Inc., Cary, NC, 1989–2007). This software allows exploration where an insectary staff member can collect a table of data, then simply plug the data into a spreadsheet, which the JMP program plots in a control chart with means, upper control limits, and LCLs automatically presented. This makes the decision process about whether the diet-making process is out of control or in danger of trending toward failure or being out of control. The processes involved in the analysis of egg failures (mortality) are illustrated in the composite of egg photographs, egg counting, and image analysis and the resulting Pareto chart is in Figure 11.16.

Several of the authors cited in the previous paragraph discuss quality (and process) control in tephritid fruit fly mass-rearing facilities. Dyck (2010) discussed the QC/PC issues in the context of codling moth rearing: the common feature of all these programs being quality in mass-rearing for sterile insect control. These authors explain that the first question that must be asked in developing a QC program is "How will it be determined if the process is in control or out of control?" The underlying assumption here is that the overall rearing procedure is a *process*. There are many sub-stages of the overall rearing or processes within the overarching process of rearing. The *product* of the overall process is the target insect population that is being produced. As the diet/rearing team develops a meaningful QC system, they are constantly aware that the purpose of these efforts is to produce high-quality, reliable products, using a reliable system whose deviations are understood and in control of the system's management.

This leads to the distinction between *PC* and *process capability*. If a process is out of control, there is a *systematic (or nonrandom) deviation* from the mainstream progress of the process. These factors are examples of systematic deviations that are out of control, processes typically found in insectaries: irreversibly degraded diet components, continuous drift in instruments (such as a balance that constantly gives readings that are higher than true sample weights or a pH meter that

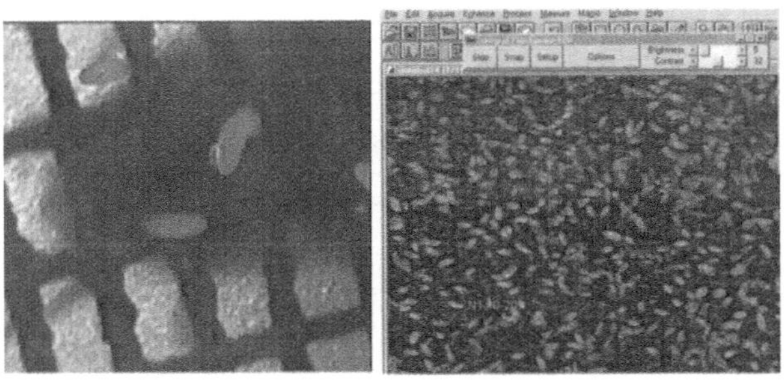

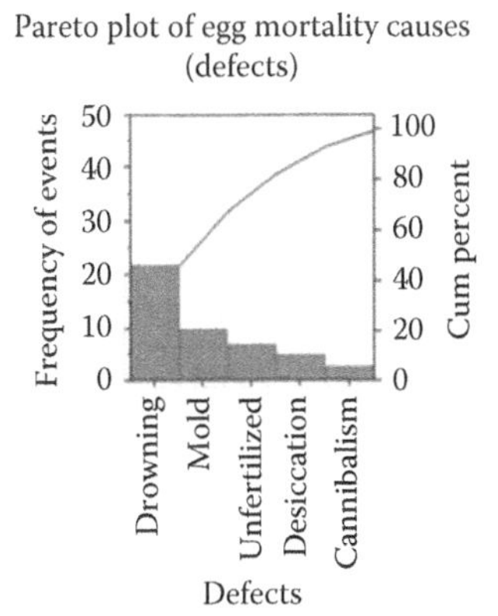

Sources of failure observed	Numbers of failures (or defects)/50	Symptom
Mold	10	Black growth
Desiccation	7	Shriveled egg outside of gel
Drowning	22	Developing embryo in gel
Cannibalism	3	Empty egg shell
Unfertilized	8	Full of yolk, no development

Figure 11.16 Components used in Pareto analysis of causes of egg mortality in *Lygus* rearing.

repeatedly gives readings higher than the true values), insectary workers who regularly omit items from the prescribed diet, or a thermostat that is improperly measuring and improperly controlling the set point temperature for the rearing room. In contrast, random deviations are nonsystematic. These deviations from process capability characteristically conform to a bell-shaped (Gaussian) curve. A process that is in control but that has a high degree of variation in any or all components can be said to be incapable. Conversely, a process can have small amounts of random variation but be out of control. In a well-run, well-controlled insectary, it is crucial to be able to distinguish between deviations that are random (capability error) and those that are nonrandom (process or systematic error). The practical reason for needing to make this distinction is that each type of deviation requires a different approach to eliminate or reduce the source of error.

In general, the upper and LCLs are established by using the mean for a collection of measurements and using the mean plus three times the standard deviation (the "3 Sigma rule") for the upper limit and the mean minus three times the standard deviation for the lower limit. For example, once the water content of a diet is determined in 10 batches of diet, formulated on 10 different occasions, the mean of these 10 samples (e.g., 71%) is the base point ±3x the standard deviation (e.g., 3 x 0.5% = 1.5%) to establish the upper control limit and LCL (i.e., 72.5% and 69.9%, respectively). In subsequent preparations of batches of diet, measurements of water contents that are above 72.5% or below 69.5% are taken to indicate that the diet *process* that is violating these range restrictions must be out of control. These process assessments are automatically assessed and converted in process analysis graphs by programs such as StatView (SAS 1999), making the QC decision process easier for rearing and diet specialists.

Process analysis for QC is further simplified by application of the Pareto analysis system (SAS 1999). Pareto analysis can be a guide to determine the sources of insectary problems including ones that are diet specific as exemplified by the discussion of the information in Table 11.3 (Cohen unpublished data). The parameters in this table were selected, because they were found to be strong indicators of problems (robustness) and because they were easily measured by insectary staff (convenience). In preliminary tests, it was found that continuous measurements provided this information about 10 batches of diet: (1) microbial counts (92 + 9.3, which equals 3x the standard deviation), (2) water content (71% ± 1.56 = 3x SD), (3) FRAP (153 µg/g diet ± 21 = 3x SD), (4) gel strength (350 g/cm^2 ± 18 = 3x SD), (5) pupal weights (142 mg ± 11.2 = 3x SD), (6) eggs/female (97 ± 15.3 = 3x SD), (7) larval duration (9.2 days ± 0.8 = 3x SD).

Table 11.3 Combined Assessment of Weekly Measurements of Three Diet Parameters and Three Insect Parameters over an 11-Week Rearing Period

Date	Microbial	% Water	Gel Strength (g/cm^2)	Pupal wt.	Egg Counts	Mean Larval Duration (Days)
January 1	93	70	355	153	105	8.3
January 7	108[a]	71	353	158	101	8
January 15	93	73	359	149	121	8.1
January 22	91	72	259[a]	122[a]	115	8.4
January 28	91	64[a]	349	128	89[a]	9.1[a]
February 4	93	66[a]	352	132	85[a]	9.3[a]
February 11	95	66[a]	355	125	81[a]	9.3[a]
February 18	89	65[a]	353	122	82[a]	9.2[a]
February 25	119[a]	64[a]	349	129	80[a]	9.5[a]
March 4	86	63[a]	360	131	79[a]	9.4[a]
March 11	96	65[a]	361	134	84[a]	9.4[a]

[a] Rejection of insect or diet batch, because upper or lower limits for parameter were exceeded. Each value for microbial assay includes 5 RODAC™ (Replicate Organism Detection and Counting) plate counts, for % water and gel strength, 8 measurements were made; for pupal weights, 10 measurements were made; for egg counts, 5 pairs of insects were held for 1 week, and for larval duration, 10 larvae were evaluated.

Tabular data such as Table 11.3 can be difficult to visualize for correlation of problem causes. The graph represented by Figure 11.3 shows the type of relationship that can better show connectedness of problems. It is clear from Table 11.3 that the water content in the diet dropped during the fifth week of the rearing cycle. Along with the decrease in water content, there was also a decrease in pupal weight, a decrease in egg counts per female, and an increase in the duration of the larval period. The evident correlation between the three biological parameters and the decrease in the diet's water content presents a possible explanation for the apparent decline in biological quality. However, there was a precipitous decline in biological quality during the fourth week of the cycle represented here, and this decline does seem to be tied to a decrease in water content. However, there was a change in one of the other diet parameters in week 4, a dramatic decrease in the gel strength measured during this time period. The QC team at this insectary would be well advised to consider sudden decrease in gel strength and a longer-term decrease in water content as possible causes for the noted biological decline.

This may explain the week 4 decline in biological quality, but it does not explain the week 5 through week 11 decline (albeit, a less dramatic decline than that seen in week 4). Adding to the complexity of this mystery is the coincidental increase in the antioxidant capacity (FRAP) values of the diet during the same period as the decrease in water content and the decline in insect quality.

With scrutiny of records, the QC team discovered that the diet from week 5 was prepared by a new insectary worker (Cohen unpublished data). Working with this person in an unthreatening way, they learned that the new worker had not been taught the proper leveling and zeroing of the balance used to weigh dry materials. When the components were reweighed on a properly calibrated, leveled balance, it was found that each component weighed several percent more than what was specified in the diet formulation. The water was measured volumetrically (by volume), rather than gravimetrically (by weight), so there was no water error as there was a dry material error. The weighing error also accounted for the increase in FRAP values because of the inadvertent addition of extra dry matter, including the vitamin mixture, which was the main source of antioxidants.

The kinds of charts that can be developed for QC systems are exemplified in Figure 11.4a and b. In Figure 11.4 a typical output of a Statview (1999) program is presented, showing the arrangement of data collected from 10 weeks of water content measurement. Figure 11.4a is an I chart (an individual measurement chart) with the upper control limit and LCL around the mean. In themselves, Figures 11.3a,b and 11.4a give no indication that the process is out of control, but Figure 11.4b, an MR (moving range) chart, shows a plot of the differences in the deviation associated with each data point. This type of analysis calls attention to the deviation noted for weeks prior to 2/11. The indication is an out-of-control process that is evident from the calculated LCL for this period. Provided that the base of data that was used to develop the models depicted here is reliable and representative of the process when it was in control, the departure seen here is a very sensitive indicator of a problem in the system and a process that is out of control and that needs attention.

Finally, the QC charts and the general database can be used to provide information about the problem components of system processes. Information such as the number of defects during certain life stages, the number of rejections of the products associated with subprocesses, or any other body of data on deviation from desired quality can be used to develop a Pareto chart (Figure 11.5). Pareto charts should be used as tools of continuous improvement and a continual source of guidance for decisions about which parts of the overall process (or subprocesses) should receive immediate attention. Pareto charts can be configured to demonstrate the most significant sources of departure from quality or profitability. They can be used to direct the QC and production teams' attention to where improvements are most immediately needed. The example illustrated in Figure 11.5 is an assessment of reasons for rejecting diets based on analysis of subsamples, with tests for the various parameters listed. Clearly from this Pareto analysis, the most common problem was unacceptable water content, and next to this were problems with pH. According to the cum percent column in the Pareto table, the correction of these two problems would eliminate nearly 54% of all rejections of diets. Used in this way, the QC team can contribute to the overall quality of insects by helping with the diagnosis of problems. This diagnosis can become the basis for the process of continuous improvement, which should be the goal of any production facility (or any organization).

Long-chained fatty acids, especially unsaturated ones, are very susceptible to a process of degradation known as lipid peroxidation (discussed in Chapters 3 and 5 and Chapter 17). One of the best ways to tell if a diet or its components have been abused (in terms of improper storage, improper mixing, exposure to pro-oxidants, or microbial damage) is by measurement of peroxidation. When lipids are broken down, at least three destructive processes take place, all subversive to the quality of the diet. All of the fatty acid content that is lost to peroxidation is removed from the nutrient pool, rendering the diet less nutritious. Second, along with the destruction of nutritious fatty acids comes the production of toxic by-products of peroxidation that are both distasteful and toxic to the target insects. Third, the peroxidation process involves the induction of a chain reaction of destruction known as the Haber–Weiss reactions (Chapter 5), a process of free radical formation and free radical attack—a force that further destroys diet quality.

Several tests can be used to determine the extent of peroxidation that has taken place, including gas chromatography, peroxide value, and anisidine values. The most reliable and simple test is the thiobarbituric acid (TBA) test. In this test, the number of micrograms per gram of diet of malondialdehyde (a primary breakdown product of peroxidation reactions) is measured by addition of the TBA reagent along with side-by-side tests with standards incubated and measured in parallel with the diet or diet ingredients. This process is described by Miller (1998). Although lipid peroxidation tests and tests of antioxidant capacity of diets or diet components reveal different chemical events, both tests are good indicators of the same types of deterioration of complete diets or diet components. Therefore, in streamlining a QC framework, an insectary may choose between these two procedures for a routine, relatively simple, robust test of diet quality (or loss of that quality). Both antioxidant tests and a peroxidation tests can determine total diet or diet component *freshness*. Recently, our laboratory has replaced the FRAP test with the DPPH test described in Chapter 17.

The next most vulnerable component and the one that is probably most frequently a problem in the most commonly used insect diets is the gel material (usually agar or carrageenan). The most simple and telling tests of these materials are the gel temperature test and the gel strength test. The gel temperature test is performed by placing a sensitive thermometer in a container such as a Petri dish, then pouring a standard mixture of recently heat-activated and molten material into the container, and then observing the temperature reading at the point that the gel is setting up. Variations in the quality of the gel that may occur from batch to batch can be detected with this method. The other test requires a specialized piece of equipment that is essentially a balance, which can measure the pressure (in grams) required to penetrate the surface of a formed gel. Gel strength testers that are commercially available come equipped with a drive that delivers even force, and such testers also register and memorize the amount of force that was required to penetrate the surface. The readings of such a device are reported in g/cm^2. It is crucial in performing such tests to keep conditions identical from batch to batch. The same percentage of gelling agent and water should be used. It is advisable to perform the test of the gelling agent with water and the gelling agent only, rather than with other diet components. This measure inherently isolates the variations as stemming from the gelling agent and no other components.

It is useful to test every batch of diet for at least one QC parameter. Other useful tests include water or dry matter content, antioxidants (by the FRAP test), microbes counts, a visual (macroscopic and microscopic) inspection to determine homogeneity of particles, and a pH test. The test of water content is simple to perform either on a specialized moisture determination balance or with a standard balance, with diet subsamples that are preweighed and postweighed repeatedly after storage in a drying oven until they reach a steady (equilibrium) weight.

11.11 SOME CONCLUDING COMMENTS: TIPPING POINTS, INCREMENTAL VERSUS SALTATORY FACTORS, AND PARETO THINKING—A CASE STUDY FROM A CACTUS MOTH DIET DEVELOPMENT PROJECT

In recent popular literature, the concept of "tipping points" has become a focus of understanding social and economic system phenomena (Gladwell 2000). It is well understood that certain factors or changes in a rearing system can have very large impact on rearing outcomes (saltatory factors from the Latin *saltare* or "to leap"), while other factors may have little impact (incremental factors). In Figure 11.17, a Pareto analysis shows the result of a set of trials of conditions in the development of an artificial diet for *Cactoblastis cactorum*, the South American cactus moth. The causal factors for failures in rearing cells are depicted in the Pareto analysis and are illustrated in Figure 11.18a through i; these include (1) the overall experimental framework in 32-cell trays, (2) healthy larvae, (3) poor vs. good frass management, (4) pathogen infection, (5) syneresis (not shown), (6) overcrowding, (7) stale cactus flour or other faulty diet components (not shown), (8) improper humidity (excess or deficient), and (9) mold infestation. Clearly, the Pareto chart shows that improper frass management and mold outbreaks were the leading factors in failure of cells. These two factors, which account for more than 50% of all failures in early stages of the rearing system development, were the first problems to be addressed, and correcting them led to a dramatic decline in failures of individual cells. It turned out that managing frass by allowing the larvae to follow their instincts of packing the frass away from feeding sites on diet (Figure 11.18e and f) reduced the mold problem by allowing the frass to dry out below mold thresholds of moisture, and even when mold did start to form, it did not spread rapidly to

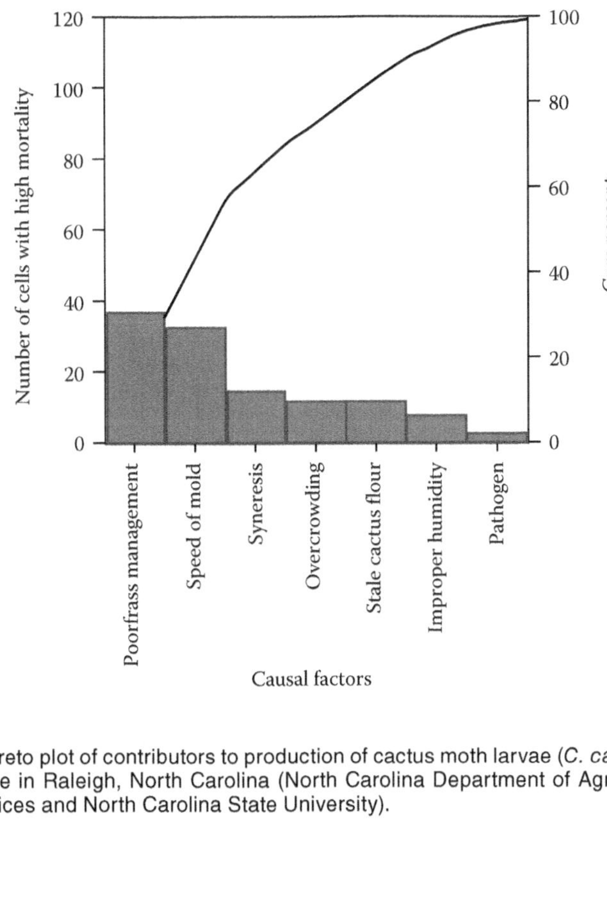

Figure 11.17 A Pareto plot of contributors to production of cactus moth larvae (*C. cactorum*) reared in quarantine in Raleigh, North Carolina (North Carolina Department of Agriculture and Consumer Services and North Carolina State University).

Figure 11.18 **(See color insert.)** The overall experimental setup in 32-cell and 8-cell rearing containers (a), healthy larvae (b), improper humidity (c), larva infected by pathogens (d), properly managed frass (containers and diet permitted larvae to either remove frass from container or to pack it away from diet) (e and f), improper humidity and larval density (g and h), and mold outbreak (i).

the diet. Several other measures were taken to permit *C. cactorum* larvae follow their instincts to deposit frass outside of the cactus pads, or at least to pack the frass tightly and as far from food sources as possible.

As indicated in the discussion of the codling moth earlier in this chapter, the factors that had to be worked out to develop a mass-rearing system for cactus moth were certainly an optimization process. Generally, each component of the system logically must have an optimum that conforms to a normal (or Gaussian) distribution with the continuous variable (e.g., concentration or percentage) on the *x* axis and performance of the insect on the *y*-axis. This is illustrated for diet components in Figure 11.19. Clearly, this is a complex mixture problem (described further in the discussion of why diets succeed or fail (Chapter 4) and diet development (Chapter 10), but for the current discussion, it is useful to view the components in this model as representative of all other continuous variables in a rearing context. The optimization model showing a composite of

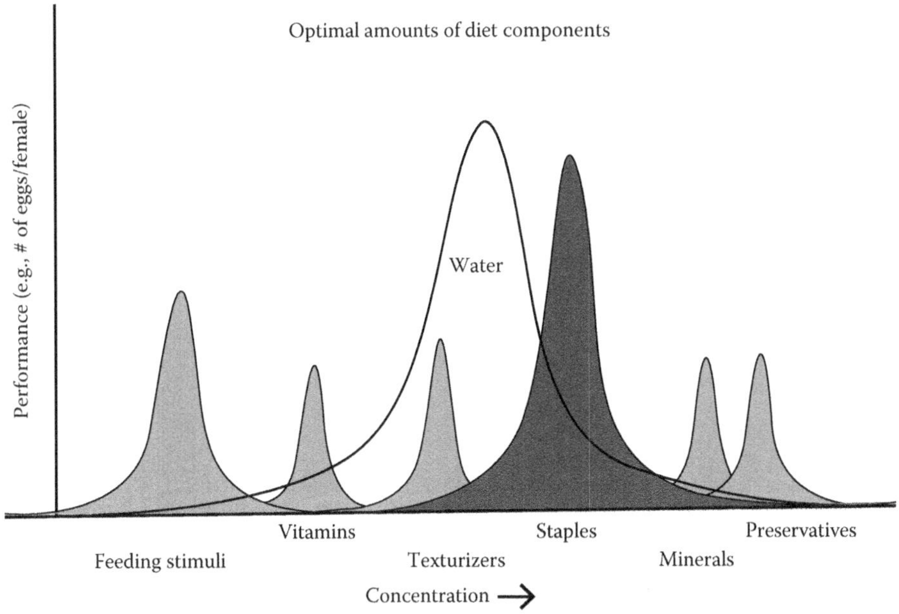

Figure 11.19 Model of optimization in an insect diet such as the cactus moth diet.

continuous variables in percentages of diet components in a typical diet (such as the Yamamoto 1969 medium) illustrates the complex potential of the nutritional, phagostimulatory, and functional outcomes that are possible in diet mixtures.

In this context, continuous variable discussed in the cactus moth rearing system would include humidity inside the containers, air space in rearing cells vs. amount of diet, and space for frass placement vs. excessive open space (where finding diet and drying out of diet becomes a problem). But the major point of this discussion is that for every rearing (including diet) parameter where continuous variation can take place, there will be not only an optimum, but there will be a tipping point where even the slightest change will push the rearing system over the edge and into failure.

Undoubtedly, the most complex feature of rearing systems is that the interplay between numerous components in our rearing systems influences possibly countless other components. As an example of this kind of complexity, a given insect has a given capacity to maintain water balance. Lower humidity puts a fairly predictable stress on water balance, and elevated temperature compounds this stress. Part of that water balance depends on control of spiracles (and gas exchange), but excess CO_2 due to poor ventilation can cause spiracles to remain open, increasing water loss. A slight rise in temperature will cause an increase in metabolic rate with concomitant increases in water loss from increased O_2 uptake and CO_2 output, further increasing water loss. What can add further to this water stress syndrome could be an increase in protein concentration in the diet, which could increase the amount of nitrogen that must be excreted, further increasing water stress by increasing water output required to void excess nitrogenous waste. This complex scenario can be reiterated in terms of microbial factor and environmental factors conspiring to increase microbial stress on colonized insects. In all these cases, there are hidden factors where individual rearing system components are close to their tipping points wherein a seemingly small change in some seemingly benign parameter can cause the system to tip into failure.

CHAPTER 12

Equipment Used for Processing Insect Diets

Small-, Medium-, and Large-Scale Applications

12.1 INTRODUCTION

Scores of papers about insect rearing include the expression "mass rearing"; however, many of these papers deal with production of only a few hundred to several thousand insects per week. "Mass rearing" includes only programs where the number of progeny produced per day equals one million times the number of offspring that can be produced by a single female per day (Mackauer 1972). A similar number is stated by Chambers (1977) who set the boundaries of mass rearing as 10,000–1,000,000 times the reproductive potential of a female from a native population. Dyck (2010) also offers a useful discussion of the scope and complexities of rearing that can be truly considered "mass rearing." As a rule, most sources agree that large-scale rearing operations are characterized by employment of artificial diets and higher degrees of automation than that in smaller-scale operations (Nordlund et al. 2001).

Even after a successful artificial diet is developed, application of large-scale production of that diet is by no means automatic or straightforward. Simply speaking, when we go from experimental formulations, typically less than 0.5 kg, to production quantities, frequently more than 200 kg, we cannot simply multiply all components by 400 and get a diet that is exactly equal to the one prepared in the smaller batch size. The explanation for this phenomenon begins in the geometry of scale.

To illustrate the point, a simple 3D shape, a cube-shaped box with 1 cm sides has a total volume of 1 cm × 1 cm × 1 cm or 1 cubic centimeter (1 cm^3), and the surface area is 6 (sides) × 1 cm^2/side = 6 cm^2. If we wished to double the volume, we could use a box that has double the height so that the volume is now 2 cm^3, the surface area is 2 cm × 1 cm × 4 = 8 cm^2 for the sides = 1 cm^2 for the top and 1 cm^2 for the bottom = 10 cm^2 for the entire box's surface. This shows that for a fixed geometric shape, as volume increases, the ratio of surface area to volume also increases, but not linearly. *In fact, as volume increases, the ratio of surface area to volume decreases.* This generalization has ramifications throughout the physical and biological world, and it is important in diet applications. Volume and mass can be freely substituted for one another when we are dealing with materials, such as insect diets, that generally have very similar compositions (mainly water). Roughly and for purposes of simplicity, 1000 cc (i.e., 1000 mL) of diet equals a kilogram of material. Figure 12.1 shows the differential in rates of cooling of a smaller volume of water compared to that of a larger volume.

Further, Table 12.1 is presented based on empirical measurements of three beakers (approximate cylinders) of 100, 500, and 2000 cc capacity. The shape of the container influences the surface area; cylinders have greater surface areas than do cubes of comparable volume. Next, the ratio of surface to volume (S/V) decreases in the cylinder as well as the cube as volume increases. From these simple geometric concepts, there emerges an important pair of axioms. First, the ratio of

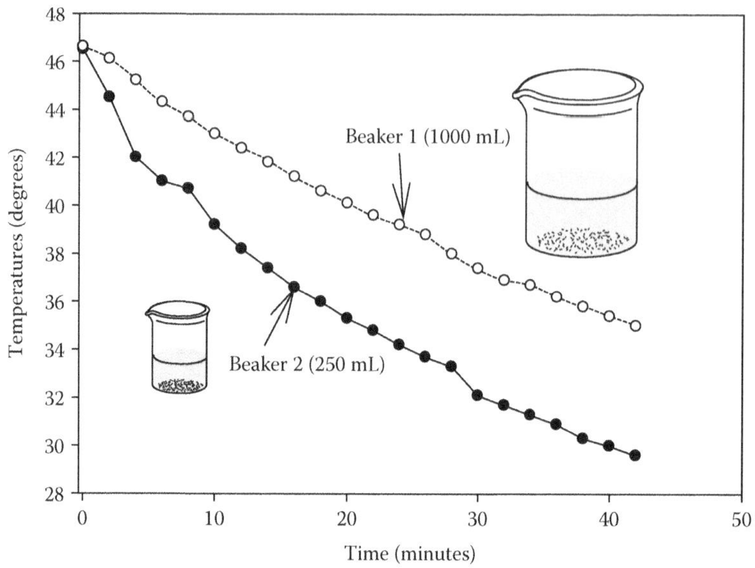

Figure 12.1 The core temperature difference between the beaker with 250 mL and the beaker with 1000 mL of water.

Table 12.1 Dimensions of Beakers (Cylinders) and Cubes of 100, 500, and 2000 cc Volumes

Volume (cm^3)	Cube Sides (cm)	Cylinder Diameter (cm)	Cylinder Height (cm)	Cube Surface (cm^2)	Cylinder Surface (cm^2)	Cube Surface/Volume Ratio	Cylinder Surface/Volume Ratio
100	~4.7	5.2	6.5	130	162	1.3	1.6
500	~7.9	9.0	8.5	377	396	0.7	0.8
2000	~12.6	14.0	15.8	953	1110	0.5	0.6

surface area to volume decreases with increasing size of a given shaped object. Second, the shape of the object influences the ratio profoundly. For example, a 100 cc graduated cylinder whose height is 18 cm (compared to the 100 cc beaker whose height is 6.5 cm) has a surface area of 198 cm^2 and a surface/volume ratio of 1.98, a great increase in that of the beaker (1.6). Some simple calculations can be made of the effects on surface/volume ratios of increasing the length of a cylinder and decreasing the diameter while keeping the volume constant. This thought process will be useful in the discussion of types of diet processing involving heating and cooling of diets. Table 12.1 provides comparisons to demonstrate the effects of geometry on the surface to volume (which equates to surface to mass) ratios in diet containers.

One further introductory point must be made: in every rearing facility where mass rearing is accomplished, there has been incorporation of food science and technology background. Furthermore, of all operations that take place in large-scale rearing facilities, the equipment part has borrowed the most from the food science/food technology community. This is one of the central points of this chapter and indeed of this entire book. Yet, it is suggested here that as a whole, the rearing community has not taken adequate advantage of what the food engineering community has to offer. This statement becomes increasingly clear to one who peruses literature from food engineering journals and texts. For example, in paging through Singh and Heldman's (2014) text, *Introduction to Food Engineering* (Academic Press), it becomes apparent how much rearing specialists use techniques such as freezing, heating, size reduction, and mixing, yet fail to grasp the complex physics and mathematics that make up the food engineering discipline. The graphs and equations in a book like this (and in the hundreds of papers and chapters cited in this book) are indicative of the utility in understanding even seemingly simple processes such as freezing food materials. As a matter of course, rearing specialists who wish to freeze materials to be used in their diet simply place the substances in a conventional, commercial freezer (household freezers or sometimes −80°C freezers widely used in molecular labs) and then give no further thought to the dynamics of freezing: how long it takes, the physics of heat (cold) transfer from the surface into the core, and the consequences of overly slow freezing to the quality of the product.

The discipline of food engineering holds the tools to saltatory improvements in insect diet and rearing technology, not only in the models and approaches to understanding food processing at all levels of theoretical and practical issues, but food engineers have also developed an arsenal of equipment that would enhance insect rearing. The freezing problems mentioned in the previous paragraph have been solved by development of direct and indirect contact freezing systems where food is pressed in plates of ultracold thermal conveyors or bathed in forced air chillers that quickly penetrate packaging materials and outer layers of food masses so that freezing is done with ultrarapid timing. The industrial systems are designed for handling the kinds of large portions that mass rearing facilities must deal with, so the food engineering equipment has ready-made scale that is not within the lexicon of entomologists or the other biologists who are often consulted for rearing advancements.

What has been said about sophisticated and efficient cooling systems applies to preservation, heating, grinding, spray drying, freeze-drying, mixing, and packaging (to mention a few areas of insect diet applications). Chapter 16 discusses future prospects for advances in insect diet science and technology, and a portion of this chapter suggests study of some recent innovations in food science. The remaining parts of Chapter 16 review the principles and practical applications that have been utilized in insect-rearing systems, but the reader is urged to reach out to the literature on food engineering and possibly to consult with food engineers to get input into the insect diet and rearing specialists' problems and concerns.

12.2 APPLICATIONS OF THE GEOMETRY OF SCALE: HEAT EXCHANGE IN DIET PROCESSING

To apply this discussion to diet preparation issues, it will be helpful to consider use of different shaped containers for heat processing diets. For example, the dynamics is compared for two different batch sizes, 0.25 and 1.0 L of a diet that has 90% water. Measured and mixed ingredients in aliquots of 0.25 L are added to a 0.5 L beaker and 1.0 L to a 2 L beaker such as those described in the previous section. It is customary to autoclave liquids in containers with higher volumes than those of the liquids to prevent spillovers. The 0.25 L diet aliquot would have a surface area of about 280 cm^2 and an S/V ratio of 1.12, while the 1 L volume of diet would have an approximate 750 cm^2 surface area and a 0.75 S/V ratio. The heating is to take place in an autoclave set for 20 min at 121°C. The atmosphere in the autoclave reaches the set point within about 5–10 min, the dynamics of heating the two diet aliquots can be considered. Heat is transferred generally by conduction, convection, and radiation. In this setting, a nonstirred viscous medium such as diet, convection should be minimal, leaving radiation and conduction most likely with conduction as the predominant means.

This raises the question about the thermal conductivity of the container and the diet. Commonly, foods have thermal conductivities of less than 1 W/m/K (defined as the coefficient of heat transfer), and glass has a thermal conductivity of 0.52 W/m/K and stainless steel 21 (Fellows 2000). Therefore, the heat must penetrate the walls of the container (which will take place about 40 times as fast in stainless steel than it will in glass) and then penetrate the diet itself. Simplifying the situation to considering only the penetration from the sides of the container to the core of the diet, the heat must penetrate the distance of the radius of each container to reach the core.

What does all this mean to the diet that we are processing? Clearly, the size and shape of the container will influence the cooking rate and other heat-induced and heat-related characteristics or changes in the diet. Furthermore, the size, shape, and composition of the container will influence the degree of heterogeneity in the diet, with the outer regions of the diet (those closer to the initial site of heat exchange) cooking for longer than the core. If the diet was originally developed in batches of 250 mL, it may have very different characteristics if the cooking process were scaled up to 2000 mL, for example. Finally, autoclaving is an inefficient process for diet making, especially large quantities of diets. It is clear from this why the food-processing industry uses cooking methods such as flash sterilizers that maximize the surface of the food to allow more efficient heat transfer. Typically, this means flattening of the food being processed or forcing it to pass through narrow bore tubing or narrow cylinders.

12.3 GENERAL SMALL-SCALE PROCESSING

Typically, newly developed diets are formulated in batches less than 0.5–1.0 kg. This means that when the diet is sufficiently promising to be adopted for mass rearing, a major effort is required to scale up diet production. The standard tools for small-scale diet synthesis are blenders, food processors, magnetic stirring hot plates, and tissue homogenizers. For heating, hotplates, or more often autoclaves, are used, especially if sterilization is required. Pressure cookers are low cost ($30–$50) but can reach temperatures and pressures that approach those of autoclaves at about 1/20 to 1/100 the cost. Currently, microwave ovens are used in many labs for heating diet components, especially to expedite the dissolution of diet components. Small steam kettles are used where higher than ambient pressures are not required.

For making salt or vitamin mixtures, small ball mills, hammer mills, roller mills, or similar devices are sometimes used in smaller laboratories; however, these are generally used in larger facilities and are discussed later in that context. Often, diet ingredients require grinding or milling, processes generally included in the category designated as size reduction (discussed in more detail in Sections 12.9, 12.10, and 12.11). This is accomplished in small laboratories with coffee grinders, or again with some of the more versatile food-processing equipment such as large-scale ball mills, roller mills, and hammer mills.

12.4 MEDIUM- TO LARGE-SCALE DIET PROCESSING

In larger-scale systems, mixers such as blenders and kitchen scale food processors are replaced by vat mixers, and autoclaves are replaced by flash sterilizers. These and other changes involve larger equipment with different processing characteristics. Dealing systematically with the differences in processing characteristics is a requisite for a successful scale-up. A second and equally important difference between small-scale and large-scale diet production is inherent in the process itself, so the special issues of scale are covered here in considerable detail. The take-home message of this section is that it is not a simple matter of going from small batches to large by simply multiplying the proportions of each ingredient and the times of each processing step by the amount that the final diet product is being increased.

An example of the problems in scale became evident in efforts to increase rearing scale using an entomophage diet for *Chrysoperla* spp. (Cohen and Smith 1998, Cohen 1998c). Increasing scale is demanded by economics, especially in terms of labor. Once it was determined that the diet had the potential to be used commercially, it became apparent that the 0.5 kg batches of the diet are too labor intensive and inconvenient. The initial target was to produce a single 50 kg of diet to last for 1 week, rather than repeating production of one hundred 0.5 kg batches. Measuring the ingredients for each batch required about 2 h, so repeating preparation of 100 batches would require 200 h of labor to produce a 50 kg batch. Assuming a salary of $10/h for the insectary worker who prepares the diet, it would take 2 h to produce the standard 0.5 kg batch. This amounts to a cost of \$20/0.5 kg or \$40/kg of diet, strictly for labor. However, making the larger batch all at one time, the 50 kg batch labor cost would be only about \$0.40/kg, assuming that each step of handling the smaller amount of the ingredients would take the same amount of time as it would for the larger amounts. So, for example, it was found that weighing 100 g of ground beef took no longer than weighing 10 kg of the beef. So in one step, 100 times as much beef could be weighed.

The blenders and food processors were replaced with a commercial scale cutter mixer with a 40 L (~40 kg) capacity. The stirring hot plate was replaced with a steam kettle, and the small balances were replaced with industrial scales. The simple multiplication of amounts resulted in poor-quality diets due to several factors. First, the larger mixer had a lower rotational speed. This demanded an increase in the mixing time from about 1.5 min to about 5 min. The increased mixing time also resulted from the physical difficulty of mixing larger amounts of materials. Second, the proportionate amount of water had to be reduced from the 0.5 kg formulation, because the lesser amount of water used in the egg-cooking process boiled away more readily than did the water in the larger batch. This is a simple result of differences in surface to mass ratios in smaller versus larger volumes of material. As the geometry discussion explained, for a given shape, the larger the volume, the lesser is the surface to mass ratio. In the scale-up, using larger volumes of diet altered the surface area exposed to heating and mixing. To solve this problem, several incrementally larger batches of diet had to be prepared with successively lesser amounts of water being added until the final product had a viscosity that was equal to that of the 0.5 kg batches of diet, which had been demonstrated as successful for rearing lacewing larvae.

Beyond the scale of laboratories where diets are developed or where a few hundred to a few thousand of insects are produced per week, rearing of thousands to millions of insects per day must be considered medium to large scale. In this book, "small-scale batch" implies batches of 1 kg or less; "medium-scale batches" are greater than 1 kg but less than 50 kg, and "mass-rearing scale batches" range from 50 to hundreds of kg. These numbers are not completely arbitrary, but rather they are based on the scale of equipment that is required to produce such batch sizes.

Most laboratories have general use equipment such as blenders, volumetric devices, weighing apparatus, and autoclaves, all of which handle no more than 2–5 kg. Large-scale rearing facilities, however, require specialized equipment. Such large-scale equipment may cost thousands to tens of thousands of dollars and generally requires specially trained personnel to keep these implements running. The major emphasis is on what has been used successfully in insect diets, but there is also discussion of food industry equipment that shows promise for seamless adoption to insect diet processing. Even where large process equipment has been adopted for insect diets, this technology often is not documented in the literature for two reasons: (1) large-scale diet technology has often been considered so specialized that it is not of mainstream interest to the entomology community (Cohen 2001) and (2) many insectaries regard their breakthrough technology advancements as trade secrets that they do not wish to share with the rearing community. Therefore, much of the documentation of the practices or technology covered here is drawn from the literature on food science and technology.

12.5 WATER PURIFICATION AND WATER QUALITY

Water is generally the most abundant component in artificial diets for insects, ranging from lower values of 70% of the weight of the diet to highs of about 95% (Singh 1977). This being the case and the fact that water participates in virtually all actions and reactions that take place in diets, before and after they are consumed by target insects, it is very important to have an appreciation for the nature of the effects of water and the importance of water in diet processing. The nature of water's mechanisms in diets is covered in several chapters, especially the one on diet chemistry.

While water purification is central to the success of an artificial diet, formal studies of water quality in diets have received little attention. Water is the predominant ingredient in most diets and can be one of the greatest sources of contamination or other erosion of dietary quality (Sikorowski 1984a, Jay et al. 2005). However, decisions about water quality assurance are often made unsystematically or with little explanation of rationale (Chapters 4, 5, and 10). In small-scale diet preparations, the water that is used is often fairly high quality either distilled or deionized and somehow sterilized—often in the 18 MΩ range of resistivity. However, if such pure water is used in large-scale operations, the cost of preparing or purchasing this water can add greatly to total diet costs. If less purified water could be used with equally high-quality results, it could be prohibitively wasteful to use the more pure water simply out of what is no more than superstition rather than a truly scientific basis. For example, in my laboratory where we rear predatory insects, we began using HPLC grade water that cost us about $4/L, and over the past 20 years, we have evolved into usage of tap water with completely satisfactory results. This simple change has saved us tens of thousands of dollars over the past decade or so. Water can be sterilized by (1) simple heating of either tap water or deionized (or distilled) water by (2) heat/pressure methods such as autoclaves or pressure cookers. The removal of particulate and chemical contaminants is achieved with filtration systems, which can be designed to remove virtually all dissolved materials, organic components, and microbial contaminants. Also water can be purified by distillation. A small-scale ion-exchange/filtration system used for laboratory-scale water purification is shown in Figure 12.2.

Distillation has been a long-standing method of choice for water purification. It involves the heating of water to boiling and capturing the water vapor in a collector whose cooling function returns the water to its liquid state, but now free of nonvolatile impurities. Because some impurities may have remained with the vaporizing water or have entered the collector or storage device, a second stage of distillation is sometimes used to assure attainment of a higher quality of water. The resistivity of the water is the most common way to designate the purity, and a resistivity of 18 MΩ is usually held as a high standard of purity. The resistivity (or the resistance of the water to movement of an electrical current) is inversely related to the amount of dissolved ions that act as electrical conductors. The downside of distillation is that (1) the initial capital outlay is high, even for small capacity systems (which may be priced in the low thousands of dollars), (2) the equipment is labor intensive to maintain, with the heating units developing scale from the minerals that were left behind the distillate, (3) there are moderate to high costs of the utilities for heating the water, and (4) the organic compounds trapped in the water may not be effectively removed by distillation, as they are volatile and may coevaporate with the water vapor.

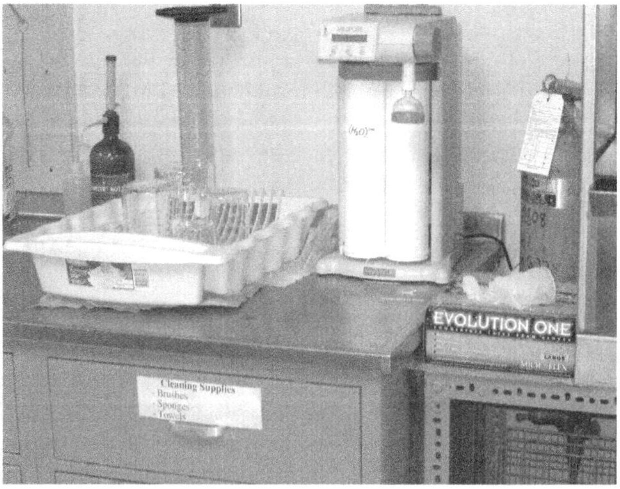

Figure 12.2 Small-scale ion-exchange/filtration system used for laboratory-scale water purification.

Another means of water purification is reverse osmosis (RO), a process that employs pressure to force water molecules through a membrane that is impermeable to most minerals that are common contaminants of the water. RO systems and system maintenance are less expensive than distillation, but the degree of water purification is less than that achieved with distillation, unless RO is coupled with another system such as ion-exchange/filtration.

Filtration coupled with ion exchange is becoming one of the most widely used systems of purification. Filtration/deionization systems are often linked in series with devices for mechanical removal of particulates and organic compounds. Depending on volume and purity requirements, such systems may be a simple, one-stage cartridge or a multiple stage series of cartridges that can equal (or even exceed) the water purity of doubly distilled water, along with the removal of organic compounds. Such systems can produce ultrapure water that rivals that of commercial HPLC grade water. However, such water is expensive to produce, the more elaborate systems costing thousands of dollars and requiring replacement of cartridges according to amount of use. It must be noted that such levels of purity are most likely only required for the most fastidious tests of defined (holidic) diets. It would be wasteful to use such pure water in diets where stray minerals and other contaminants are routinely introduced via nonpure diet ingredients or diet handling implements.

Highly pure water readily scavenges soluble materials such as minerals that contaminate tubing, holding vessels, glassware, and plastic containers. Also, laboratory air can carry many contaminants such as ammonia and carbon dioxide, both of which readily dissolve in clean water, changing its pH and compromising its purity and possibly reacting with diet components or with other adulterants. Therefore, it is recommended that highly purified water be stored in tightly sealed, ultraclean vessels and be treated generally as a perishable ingredient.

Such highly pure water is suitable for smaller-scale production systems, but it would be expensive to use such quality water for the meridic or oligidic diets that are used in large-scale rearing programs. However, depending on the quality of the tap water in a mass-rearing facility, there may be a need for some purification steps for tap water. For example, tap water may have chlorine for microbe reduction and fluorine for protection of the human population's teeth. These elements, as they occur in a water supply, can possibly have a harmful effect on insects as may the inherent microbial load or various minerals that characterize local water supplies. There are radical differences between water quality from region to region and sometimes even within the same water system. Also, it is not uncommon for sources of water to change, for example, from well water to river or lake water as water supplies from an established source diminish.

Another complexity of water as it enters a rearing facility or after some processing is its dissolved oxygen content (Chapter 5). This oxygen can be an aggressive pro-oxidant that can expedite the reactions that destroy sensitive food components such as ascorbic acid and polyunsaturated fatty acids (Chapter 8). Although degassing is routine in biochemistry and analytical chemistry laboratories, it is not deliberately practiced in diet making. Water is degassed by application of a vacuum, often with stirring and heating. It is unfortunate that little research information exists on the efficacy of degassing, considering the fact that such a simple and inexpensive treatment could possibly reduce loss or destruction of components that are sensitive to oxidative destruction. The efficacy of using degassing techniques can be determined on a per-case basis.

12.6 STORAGE OF INGREDIENTS AND COMPLETED DIETS

The chemistry of storage is discussed in Chapters 5, 6, 8, and 13, but this section deals specifically with the equipment involved in storage. Generally, low temperature is preferable to higher temperatures for storage of diet ingredients and completed diets. This is because at lower temperatures, there are lower growth rates of microbes and lower rates of degrading chemical reactions. Storage is further stabilized by use of low water activity, whenever possible. Because of the reactivity of oxygen and its destruction of nutrients, storage in the absence of this element is also desirable. Finally, exposure to light can also be destructive to many substances. Therefore, the most desirable conditions for storage of most ingredients are a cold, dry, dark place that is oxygen-poor. It is also prudent to heed the manufacturer's or distributor's storage instructions, and if no instructions are sent with the product, it would be useful to contact the vendor to get storage instructions.

12.6.1 Storage at Temperatures above Freezing

Refrigeration is almost universally present in even the most humble rearing facilities. There is a tremendous range of size and performance in refrigeration systems, from the minitable top refrigerator to huge walk-ins found in large facilities. The most common temperature range for above freezing units is 2°C–10°C, 2°C–4°C being preferred over the higher temperatures for storage of most materials. Having established that 2°C–4°C is the desired temperature, it is recommended that the personnel responsible for the well-being of the diet materials regularly check the temperature of their refrigeration facilities. This can be done with standard thermometers or more conveniently with the recording units such as the mechanical recorders or the more compact computer-associated units that can help the insectary personnel keep extensive, fastidious records of the history of the components, once they came into the facility. "Events," such as power outages, excessive entry into the refrigeration unit, or fluctuations in the refrigeration unit itself, are nicely chronicled by such devices as the Hobo™ recorders (Chapter 11). Once it is noted that there is a deviation in temperature from the desired range, decisions are made about corrective actions (repair of equipment that may be malfunctioning, transfer of stored material to another location that is more suitable, testing and/or replacement of material that has been subjected to inappropriate temperature).

12.6.2 Completed Diet Storage at Temperatures above Freezing

The storage of individual ingredients of diets at higher than freezing conditions is acceptable for a broad range of components. However, once the diet has been fully synthesized, diets become less stable. Two reasons for this are that (1) the completed diet has much higher water activity than that of most components, thus predisposing each component to be more reactive than it is in its dry state and (2) the completed diet gives the ingredients access to one another. The higher water activity makes the complete diet highly susceptible to microbial attack, to chemical interactions of components (enzymatic and nonenzymatic), to oxidation, and to desiccation. Generally, completed, fully hydrated diets have a shelf-life of no more than 1 week under refrigerator temperatures. Although the shelf-life must be established on a case-by-case basis, Table 12.2 shows the destruction of antioxidants after only 2 days in a refrigerator or 2 days at rearing room conditions.

This point becomes clear with reflection about how a freshly made tuna salad sandwich changes over time in a refrigerator. Most of us would be reluctant to eat the sandwich, even after only 4 or 5 days of storage. The changes that inevitably take place in the stored foods include seepage of fluids from one compartment to another (syneresis), microbial degradation, oxidation, enzymatic and other spontaneous

Table 12.2 Antioxidant Content of Three Diets Stored at Rearing Room, Refrigerator, and Freezer Temperatures

	FRAP			Ascorbate		
Diet	**Rearing Room**	**Refrigerator**	**Freezer**	**Rearing Room**	**Refrigerator**	**Freezer**
Lepidopteran	0.820	1.12	1.30	0.04	0.39	0.61
NI	0.07	0.17	0.21	0.04	0.06	0.07
Entomophage	0.01	0.02	0.02	0	0	0

Source: Data from Cohen, A.C. and Crittenden, P., *J. Econ. Entomol.*, 97, 265, 2004.
Quantities are expressed in ferric-reducing antioxidant potential absorbance units/g of diet on a dry weight basis.

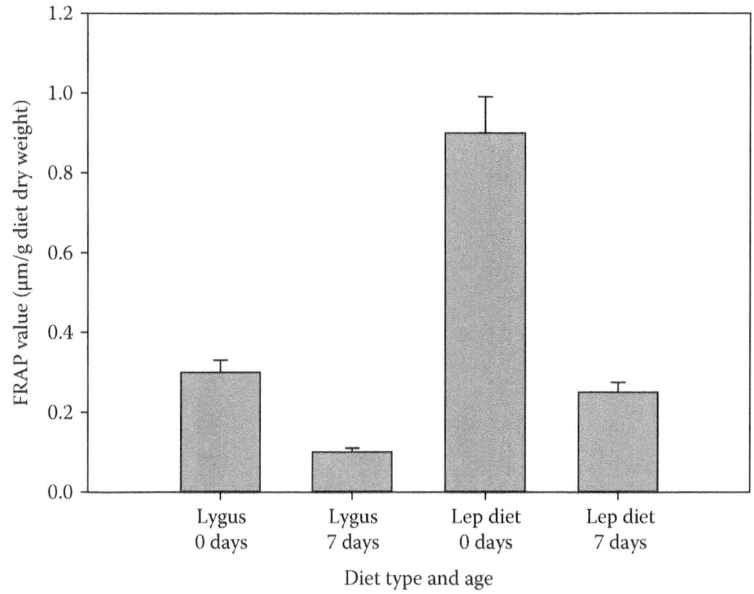

Figure 12.3 The decrease in FRAP value (mainly ascorbic acid content) in a tarnished plant bug diet and a lepidopteran diet after 7 days of refrigeration.

reactions, and desiccation. The changes in appearance, aromas, mouth-feel, and taste are all indicators of changes in the nutritional quality of the stored food, and it is no accident that nature has given us sensory aversion to spoiled food. Insects fed diets that had been stored for weeks in a refrigerator are equipped with a natural sense of what is potentially harmful to them. What is further emphasized is that the insects that are the subjects of our rearing efforts have no choice under the circumstances to which they are typically subjected. If the ascorbic acid is oxidized and the fatty acids have turned rancid, the insects in the cages or rearing units cannot go out and seek other foods. The effects of storage on ascorbic acid content and fatty acid content of a plant bug diet, a lepidopteran diet (Figure 12.3), and an entomophage diet (Figure 12.4) are clearly manifested as losses in these two nutrient classes.

12.6.3 Storage at Temperatures below Freezing

For ingredients stored below 0°C, there are several consequences that result from choices of storage equipment. The range of temperatures of freezers is extensive and requires matching of the products to be stored with storage equipment. Research on the stability of food components for insect diets and the effects of storage is lacking, and the information presented here represents a combination of review of the literature on food science and technology and the author's experience.

In poorly equipped laboratories, it is common for materials to be stored frozen to be kept in the freezer section of conventional household-type refrigerators. The temperatures in these freezers vary from barely below 0°C to –20°C. For most dry ingredients (whose a_w is <0.500 with moisture contents of <10%), storage at <0°C should preserve their nutritional value and palatability for months. For frozen storage of complete diets, for ingredients with higher water contents, and aqueous solutions, freezing does not indefinitely preserve perishable components, because physical and chemical processes continue to take place, even under cold conditions. For example, water does evaporate from its frozen state—a process known as sublimation. A familiar example is dry ice (solid CO_2) moving to the gaseous state without passing through a liquid phase. Likewise, depending on the temperature and the ambient atmosphere, water can go directly from ice to water vapor

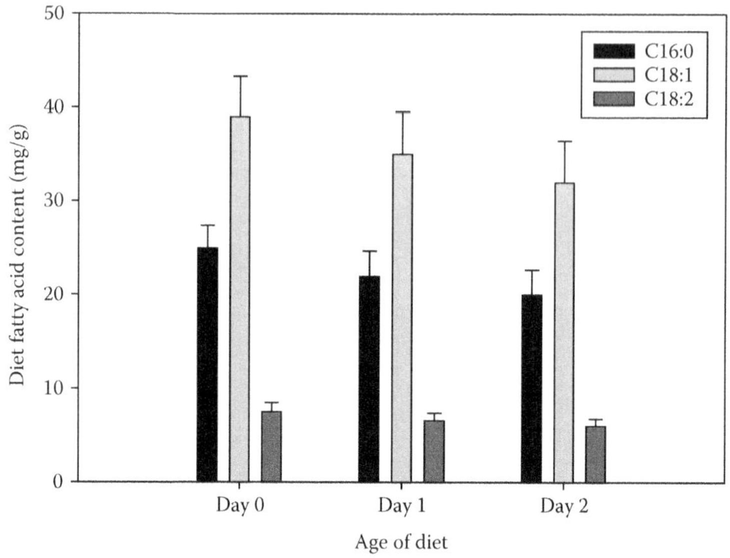

Figure 12.4 Changes in contents of three kinds of fatty acids in an entomophage diet after a week's storage in a refrigerator.

with no liquid transition. This happens rapidly when the frozen water matrix is held under low pressure—a process we know as freeze-drying or lyophilization.

However, when sublimation occurs in diets and ingredients where we expect to maintain a certain water concentration, this is an undesirable change that we could avoid by taking certain precautions. First, storage of frozen materials in a tightly packaged, waterproof container is an obvious option. Second avoidance of prolonged storage of diets and components is advisable. What we mean by "prolonged" depends on the material being stored. In general, plant materials store more poorly than meats (because of the presence of cell walls in plants, as well as a series of autocatalytic enzymes that are released after their compartments are disrupted by crystallization that takes place with the freezing).

Repeated thawing and refreezing also degrades diets and ingredients. This is especially troublesome in frost-free freezers. Frost-free freezers work by cycling through brief warming (thaw) periods interspersed among longer freeze periods. This cycling is highly destructive, especially to ingredients with high water contents. The process of freezing in itself imposes stress on water-containing ingredients by causing separation of the water from solutes or suspended components. This separation causes local changes in pH, concentrations, and access of enzymes to substrates. Freezing also causes the formation of ice crystals, which are disruptive to the integrity of the naturally protective compartments that characterize diet components, especially ones that contain cells or cellular organelles. Such disruption of compartmentalization is a major factor for the lower freezer shelf-life of cellular substances such as plant materials, meats, and some micro-organisms. The logical extension of the concept of the harshness of freezing is that repeated freezing and thawing (or partial thawing) is very threatening to diet and ingredient stability (Fellows 2000).

Slow freezing causes larger ice crystals to form than does rapid freezing (achieved by many commercial grade freezing facilities), and the larger the crystal, the more disruptive it is to compartment (such as cell wall) integrity. However, freezers that are not frost-free introduce another problem. Water vapor from humid air that enters the cold storage unit condenses on the cold surfaces and accumulates to form insulation between the cooling surfaces and the product being kept frozen. This insulation may become several centimeters thick and greatly reduce the cooling efficiency of the freezer, and this results in an elevated temperature, keeping the diet ingredients less protected than would be desired.

The best way of dealing with this situation is to take these precautions: (1) Use freezers that are not frost-free, (2) defrost the freezer frequently as part of a routine maintenance program, (3) transfer the stored ingredients to another suitable freezer to prevent their becoming thawed, (4) keep careful records as part of a plan of action for quality control in storage activity, (5) develop a timetable of usage and removal of ingredients according to ingredient manufacturer's or product authority recommendations, and (6) monitor temperatures in the freezers, preferably with a constantly recording thermometer, taking actions to keep the controlled atmosphere within desired boundaries. Conversely, hydrated diets or diet components can lose water from the surface causing a damage known as freezer burn.

12.6.4 Freeze-Drying

Freeze-drying (lyophilization) is a special case of diet and ingredient storage and processing. In this process, the water is removed from a frozen material by sublimation. Freeze driers consist of an apparatus that will allow frozen products to be held under a very strong vacuum (4.5 Torr or ~600 Pa) while the water is sublimated and collected on cooling coils in a reservoir where the water is retained as ice, away from the product that is being desiccated. The frozen water in the product being dried gains heat from its surroundings and is energized to depart from the product without the product being thawed. Freeze driers range from sizes that will accommodate milligram quantities to those commercial units that will handle hundreds to thousands of kilograms. The water activities of freeze-dried products are less than 0.500 and are therefore resistant to microbial, enzymatic, and oxidative deterioration. To prevent the product from regaining moisture from its environment, it is common practice to store freeze-dried materials in sealed containers, preferably under a vacuum. Depending on the method of storage and the nature of the product, freeze-dried products may have shelf-lives of months. Because this process has not been widely used or researched, it is not known what the upper end of safe storage may be for freeze-dried diets. Therefore, application of this technology is on a case-by-case basis. The discussion of water activity in Chapter 5 explains peculiarities in the isotherms of desorption and adsorption of water in freeze-dried materials. That discussion includes an explanation of the dangers inherent in rehydrating freeze-dried materials.

Drawing from the literature on food science and technology, it appears that freeze-drying is an efficient means of preservation in terms of retaining the nutritional and sensory qualities of the food. Studies with diets for predators and plant bugs indicated that the nutritive qualities and palatability are retained after freeze-drying and rehydration of certain insect diets (Cohen 1998c). Such diets have been used in the author's laboratory after 1 year of storage at room temperature, sealed and under vacuum (Cohen unpublished data).

12.7 STANDARDS OF ACCEPTABLE QUALITY

The food science/food technology industry has well-developed standards for virtually every aspect of food processing. For example, the concepts of PSL or *practical storage life* is defined as "the time the product can be stored and still be acceptable to the consumer," and HQL or *high-quality life* is defined as "the time that a statistically significant difference ($P < 0.01$) in quality can be established by taste panelists" (Evans and James 1993). Examples of PSL include ground beef, which has PSLs of 6 months at −12°C and 10 months at −18°C and HQLs of 3 and 10 months at −12°C and −18°C, respectively, for peas (Fellows 2000). Several features of these standards recommend them as applicable to insect diets and diet ingredients. First, the differences between the PSL and HQL concepts suggest an important disparity between what passes for what is acceptable to the general public and what is a more subtle loss of quality to a panel of taste experts. The chemical and physical changes that cause declines in PSLs and HQLs reflect degradation not only to flavors but also to nutrients. These declines, when they take place in insect diets or diet components, may be below levels that would cause serious problems in humans or domestic stock, but in insects, because the aliquot of diet is all that the insect has to eat may be fatal. Also, there is a sharp difference between shelf-lives of foods that are stored at −12°C and −18°C, a point relevant to developing quality standards for insect diet ingredients and completed diets.

12.8 ULTRALOW TEMPERATURE STORAGE

Ultracold freezers, which keep temperatures of −75°C to −80°C, are becoming increasingly common. Diet specialists may wonder whether storage of heat-labile ingredients would be improved by storage at such ultralow temperatures. Unfortunately, there are no reports of studies directed at testing the hypothesis that ultracold storage of diet ingredients improves insect diets. Certainly, the information on differences in degradation of nutrients and flavor components at various below freezing temperatures provides a theoretical basis for the efficacy of such storage. However, the expense of the ultracold freezers and their maintenance for storage of common ingredients are probably overkill. Application of the storage protocols described in previous sections should be adequate for protection of most diet ingredients. Furthermore, a liability of ultracold storage of some ingredients such as ground meats and cooked eggs may be associated with increases in syneresis (separation of liquids from solid matrices upon thawing; Cohen unpublished data).

12.9 SIZE REDUCTION OF INGREDIENTS

In food processing, size reduction is carried out to increase the ease of mixing and heat transfer during the overall process. For example, in grain products such as pancakes and bread, the whole grains are reduced to flours or meals, which allow them to be mixed efficiently with other solids (also of reduced size) such as salt, sugar, baking powder, and liquids such as raw eggs, milk, oils, and water.

Similarly, in insect diets, size reduction of nearly every component is carried out somewhere in the process of diet production. And similar to the reasons for size reduction in insect diets, mixing and heating efficiency are a major reason for using this process strategy. However, there are other reasons for the size reduction process in insect diets. Insects are confined to a limited access to their diet aliquot, and this can be a source of stress on individuals that fail to get their share of all required nutrients. Each insect must have an appropriate portion of each nutrient within the reach of its mouthparts. This demands that the portions of the diet containing these components are within close enough proximity to one another that they are all within reach. This requirement demands size reduction and thorough mixing of components to fit the scale of the insect's mouthparts. Even if the insect could undertake nutrient self-selection or cafeteria style feeding (Chapter 10), the components must be within a narrow enough size range that they can all be present within the feeding scope of the insect.

However, there is a risk in the size reduction process: the destruction of the nutritional quality of the food or its palatability. One of the chief lessons of this book is that in natural foods (Chapters 4 and 10), nutrients and flavor components are highly compartmentalized. The naturally occurring destructive components, enzymes and oxidation-inducing chemicals, are present in virtually all foods, but they are held in check within the highly organized and protective compartments. However, as soon as we undertake a size reduction step, the protection barriers are broken down, and hydrolysis and oxidation processes can go awry in their destruction of proteins, lipids, complex carbohydrates, vitamins, and flavor-giving components. It is incumbent upon the personnel responsible for diet quality to take measures against these destructive forces. Such measures include timing the size reduction to reduce the timeframe that is available for the destructive processes to take place and using intermediate process steps designed to eliminate or reduce activities of enzymes, microbes, pro-oxidants, or any other factors that are adverse to high-quality diets.

Sometimes simple measures can be effective and protective to diet quality. For example, not mixing egg yolks with whites until the last minute can protect the biotin from irreversible binding by the avidin found in egg whites (Cohen and Smith 1998, Cohen 2000b). Also, using toasted flours and meals can greatly reduce the destructive actions of enzymes such as lipo-oxygenases, proteases, and phenol oxidases. Use of judicious packaging and refrigeration can reduce loss of quality due to contact with air and other pro-oxidation sources. Fellows (2000) points out several examples of the potential destructiveness inherent in size reduction. Among the most impressive are values for the loss of ascorbic acid in maize (from 12 mg/100 g to 0 mg/100 g) and niacin, which is lost in the size reduction process in rice. Besides the enzymatic and oxidative losses of nutrients, size reduction also enhances the losses of nutrients by leaching and discarding of nutritious parts.

Although many diet components are purchased ready for use in diets, some components require some type of size reduction such as cutting, chopping, grinding, or sifting. Some size reducing implements also serve to mix ingredients, and these will be discussed in Section 12.12. A distinction is made between the separation of components of diet ingredients and the actual active process of size reduction. For example, in some rearing facilities, seed meals are sifted so that particles larger than a given size are discarded, and the smaller particles are retained for use in the diet. Such separation is needed in order to protect pumps and coils from damage by particles that are too large to be moved safely through the system. The sifting process is practiced regularly in the Gast facility, for example, to protect the flash sterilizer from damage by the larger particles found in wheat germ (Figure 12.5). This point will be further discussed in Section 12.13.2.

Size reducing implements have been applied mainly to seeds, grain, and to a lesser extent to other plant parts. In my experience, most insect-rearing facilities that use seeds and grains obtain these ingredients in a preprocessed form such as flour (soy, wheat, and rice flours) or meals (corn, wheat germ meals), but when materials such as lima beans, chick peas, and pinto beans are used (such as in the Shorey and Hale 1965 diet or the Patana 1969 diet), the lack of availability of these meals or flours demands on-site preparation. Also, custom (local) preparation may be desirable if the preprocessed materials do not fulfill the requirements as determined by the expert developing or modifying the diet. For example, the soy flours that are commonly marketed have been defatted (some being refatted), but I have concluded that many insects would profit from the natural lipids that are abundant in soy beans before they have been defatted. In my laboratory, we are currently

Figure 12.5 Sifting equipment.

experimenting with changes in the processing of soy flours and meals to make them more suitable for our target insects. This means that we must do custom-milling and heat processing.

There are many kinds of size reduction equipment used in large-scale food-processing plants including slicers, dicers, shredders, bowel choppers, precrushers, hammer mills, fine impact mills, air jet mills, ball mills, disk mills, roller mills, and pulpers (Fellows 2000). Figure 12.6a through c shows a medium-sized hammer mill that can process tens of kg per day. The hammer mill shown here is equipped with three sizes of gratings that can be selected to tailor size reduction of various plant products to a desired particle size. Fellows (2000) explains that the size reduction mechanism of all theses implements can be summarized as a result of the forces of compression, impact, and shearing and that while each implement uses all three forces, generally one force is predominant in each implement. As has been discussed in several previous chapters, the size and shape of components can be very important as factors determining the quality of the diet, so the method of size reduction used to process each ingredient may be of paramount importance in overall diet quality. Therefore, the size reduction equipment is a key factor in high-quality diets. *A case study will illustrate this point.*

After I had developed a simplified working diet for the plant bugs, *Lygus* spp. (Miridae: Heteroptera), we successfully produced 10 generations of *L. hesperus* prior to my being relocated. In my new laboratory, we could not keep our colony alive on what we considered the same diet formulation. Careful scrutiny of all procedures at the new location revealed that the only difference that could be detected was that the lima bean meal was more finely ground in the new facility than

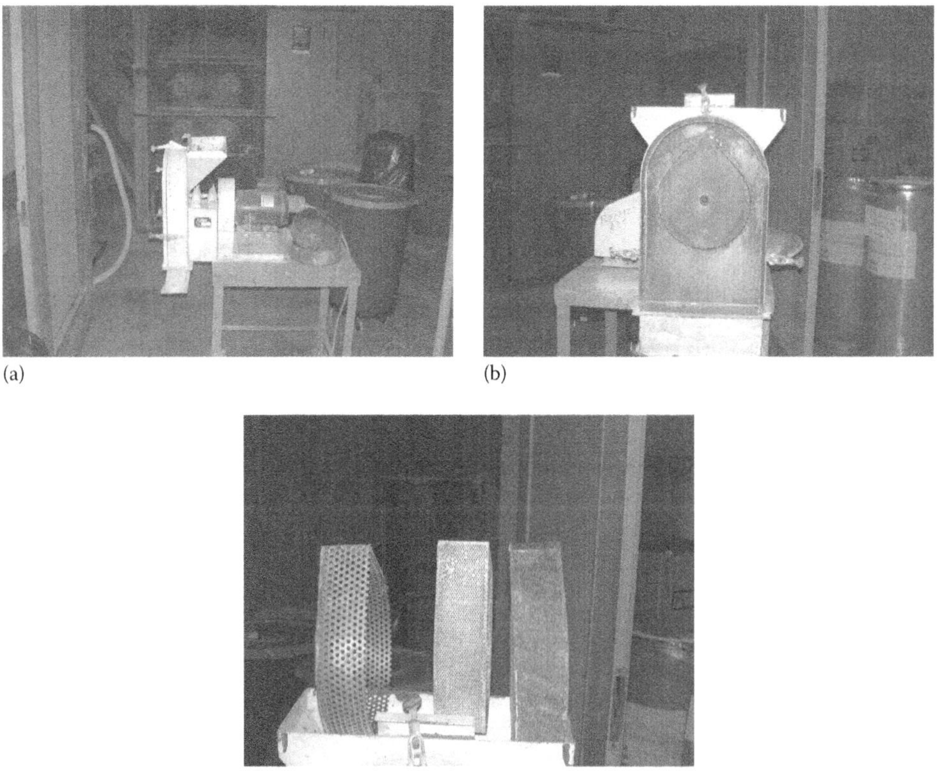

(a) (b) (c)

Figure 12.6 Medium-sized hammer mill that can process tens of kilograms per day. (a) A front view of the hammer mill, (b) the "hammer" apparatus which is exposed for observation by removal of back plate, and (c) three filters that allow particles of 2 mm and less to pass through (left), particles less than 0.5 mm to pass through (center) and only particles less than 0.1 mm to pass through (right).

it had been in my previous location. The finer particles in the newer diet absorbed far more moisture, making the diet more of an immobile paste, rather than a pourable, loose oatmeal consistency that characterized the older version of the diet. We could not determine if the change in diet texture and reduction in moisture availability reduced palatability of the diet or if some other factor such as leaching of nutrients or autolysis from enzymes released from the more finely ground lima bean was at fault. However, in the final analysis, although the water content was identical in diets with each type of ground lima bean meal, the diet with the more course grind was more loose and friable and was a far superior diet for *Lygus hesperus*. This illustrates the extreme importance of using the correct size reduction equipment for processing flours, meals, and any other ingredients.

12.10 SIZE REDUCTION OF MEAT PRODUCTS AND EGGS

Until the invention of diets made of meat pastes for predatory insects (Cohen 1985a, 1999, Cohen and Smith 1998), meat-processing equipment had not found its way into insect-rearing facilities. Recently, the need to make a meat paste of larger quantities of beef liver, ground beef, and other meat products induced the introduction of equipment with a greater capacity than household blenders and food processors. Industrial scale meat grinders, choppers, and cutters are used in large-scale production of various products such as sausage and meat pastes such as liverwurst. These size reduction implements have been found useful in processing moderate sized (~40 kg) batches of entomophage diet. The use of the eggs of domestic fowl has been sparse in insect diets, but a few researchers have made use of small quantities of chicken eggs (Hagen and Tassan 1970, Debolt 1982, Cohen and Smith 1998, Wu et al. 2000). Prior to the invention of the entomophage diet for rearing lacewings and other predators (Cohen 1998c), there had been no need for large-scale egg processing or large-scale meat processing. However, since the patenting and licensing of this diet to several companies, there has been a demand for equipment that can break and separate large quantities of chicken eggs and instruments that allow meats to be reduced to pastes and blended with yeast, preservatives, eggs, and other ingredients. The Adsi egg breaker and mixer is used for shell removal and extraction of large quantities of eggs (hundreds to thousands per hour). For very large-scale egg breaking and separation of yolks from whites, products are available from Ovobel Ltd. Ovobel can include an ultraviolet light to reduce the microbial contamination in the egg product. The Hobart 450 mixer and cutter (Figure 12.7) has been used to replace the smaller blenders and

Figure 12.7 Medium-scale Hobart mixer that cuts and blends up to 40 kg quantities of diet.

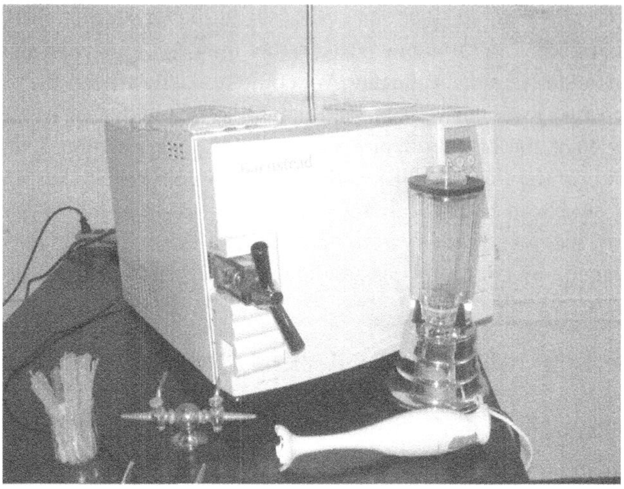

Figure 12.8 The blenders and autoclave used to make small-scale versions of an entomophage diet and a *Lygus* diet.

kitchen food processors (see Figure 12.8) in the scale-up of the entomophage diet (Cohen and Smith 1998, Cohen 1998c, see also Nordlund et al. 2001). Other mixer and size reducers are covered in the next section. Other larger-scale equipment for reduction of meat particle size is available and is known as horizontal food blenders, dual ribbon blenders, and disintegrators (from APV, American Ingredients Co., Corenco, and Breddo). Possibly one of the best potential mixers for high-viscosity liquids and pastes (such as the Cohen 1998c entomophage diet) is the Z-blade mixer (such as that of Winkworth Engineering, Ltd.), which has not been tested in insect diet production.

12.11 SIZE REDUCTION IN PLANT MATERIALS

Where off-the-shelf plant products cannot be used for diets that are required in larger quantities, custom cleaning, sorting, and size reduction processing must be accomplished at a scale that greatly exceeds the capacity of typical laboratory equipment. Examples of typical circumstances where the high capacity and aggressive processors would be useful are the situations where natural plant products are to be included in diets for insects such as the weed-consuming insects whose production is intended to serve as biological control of introduced noxious plants (Spencer 2000). In such cases, stems, leaves, fruits, or roots may be the source of the plant components of diets that consist of a synthesis of conventional food ingredients (flours, protein extracts, salt mixtures, vitamins, etc.) and the "token" ingredients from host plants of the specialists that are the intended targets of this diet technology. Examples are the leaf beetles and weevils for control of purple loosestrife (Blossey et al. 2000); these diets call for the integration of plant parts (purple loosestrife roots) into a mixture with casein, agar, yeast extract, salts, etc. When rearing scale is amplified to the point where tens to hundreds of kilograms of the plant components are required, it will become useful to employ the large capacity plant part reduction capabilities of the equipment mentioned in the above paragraph, along with pulpers, and harvesting/sorting equipment that is used to preprocess sugar cane, sugar beets, and similar plant materials.

One final note regarding size reduction: Matrices and dispersions govern the nutritive, palatability, and stability characteristics that result from the interaction between size reduction of various components and mixing and other processing factors. This relationship is discussed in detail in the chapter on why some diets work while others fail.

12.12 MIXING

Often, the same equipment that is used for heat processing or size reduction is also used for mixing. Small-scale implements such as kitchen blenders and food processors are examples as are larger scale cutters and mixers. In the examples below regarding heat processing, the flash sterilizer has a built-in mixer to keep the diet stirred and suspended prior to the heating process itself, and the twin-screw extruders have ingredient mixing hoppers and introduction jets that allow mixing of new ingredients during the process cycle. The mixing dynamics will be further discussed in Sections 12.13.1 through 12.13.3.

12.13 HEAT PROCESSING

A review of the recent literature on artificial diets reveals that the majority of heat processing of the complete diet (rather than components such as flours and meals, prior to mixing) is some type of steam kettle (Cohen unpublished data). This section will deal with the steam kettle and its benefits and drawbacks, and it will compare this method of heat processing with autoclaving, flash sterilization, and extrusion. Again, as mentioned in the early paragraphs of this chapter, the treatment in the current text cannot be as thorough and penetrating as the explanations in a more specialized work such as that of Singh and Heldman (2014).

12.13.1 Steam Kettles

The steam kettle is probably the most common heat-processing apparatus in medium- to large-scale diet production facilities. It is also the least expensive heating apparatus and the most simple to maintain and use. The principle of the kettle is to have a jacketed area outside the container that holds the diet, and under a pressure, the temperature of the steam can rise well above the boiling point of water. In fact, it is common to use steam kettles at 121°C–150°C. These high temperatures can quickly and efficiently heat the diet in the kettle. Also, it is important to remember that the steam does not directly contact the diet, nor is it vented into the diet preparation facility.

The drawbacks of the steam kettle are similar to those of the autoclave in terms of efficiency of heating large batches of diet. The tendency is to heat the portions of the diet in contact with the container, while the core is unheated. Stirring with an efficient stirrer can help to reduce this problem, but it must be understood that because the diet is not under pressure (despite the fact that the steam *within* the jacket *is* under pressure), it can never be superheated (i.e., heated to more than 100°C). As was specified in Chapter 5, superheating is advantageous for modifying the diet, generally in a favorable way. It will also be emphasized in Chapter 13 that some microbes cannot be completely killed at temperatures below those regarded as superheating. In trying to visualize the characteristics of the uneven, "layered" heating effects, the reader is encouraged to visualize a loaf of bread with the more highly cooked crust on the outside and the less cooked "core" within. Clearly, the crust is chemically, physically, and nutritionally different from the remaining bread.

12.13.2 Flash Sterilizers

The use of flash sterilizers for insect diets was reported by Griffin et al. (1974). The coupling of a flash sterilizer and a modified form-fill seal machine (Sparks and Harrell 1976) supported the production of six million corn earworm pupae in less than 2 years (Tillman et al. 1997). Generally, the flash sterilizer consists of a mixing kettle where ingredients are blended and held until they are pumped to the heating coil where they are subjected to temperatures in excess of 121°C, in some cases up to 140°C. The heating coil is nested within a steam jacket that allows high temperatures to be reached by the compression of the steam that surrounds the coils. Because this is a closed system, the pressure within the diet prevents boiling as would occur when diets in open environments such as steam kettles or hot plates are heated above the boiling point. These higher temperatures and pressures cook the diet quickly, altering the characteristics of various proteins, lipoprotein, and glycoprotein complexes (including denaturation of enzymes) while causing a minimal amount of damage to nutrients such as vitamins, amino acids, or lipids (Fennema 1996). Because the residence time of the diet in the heating coil is short, the process is called flash sterilization, and because of this short residence time, there is minimal damage to the diet, yet efficient destruction of microbial contaminants. Figure 12.9a through e shows a flash sterilizer coupled with a form-fill seal machine and configured for introducing diet and inoculating with eggs.

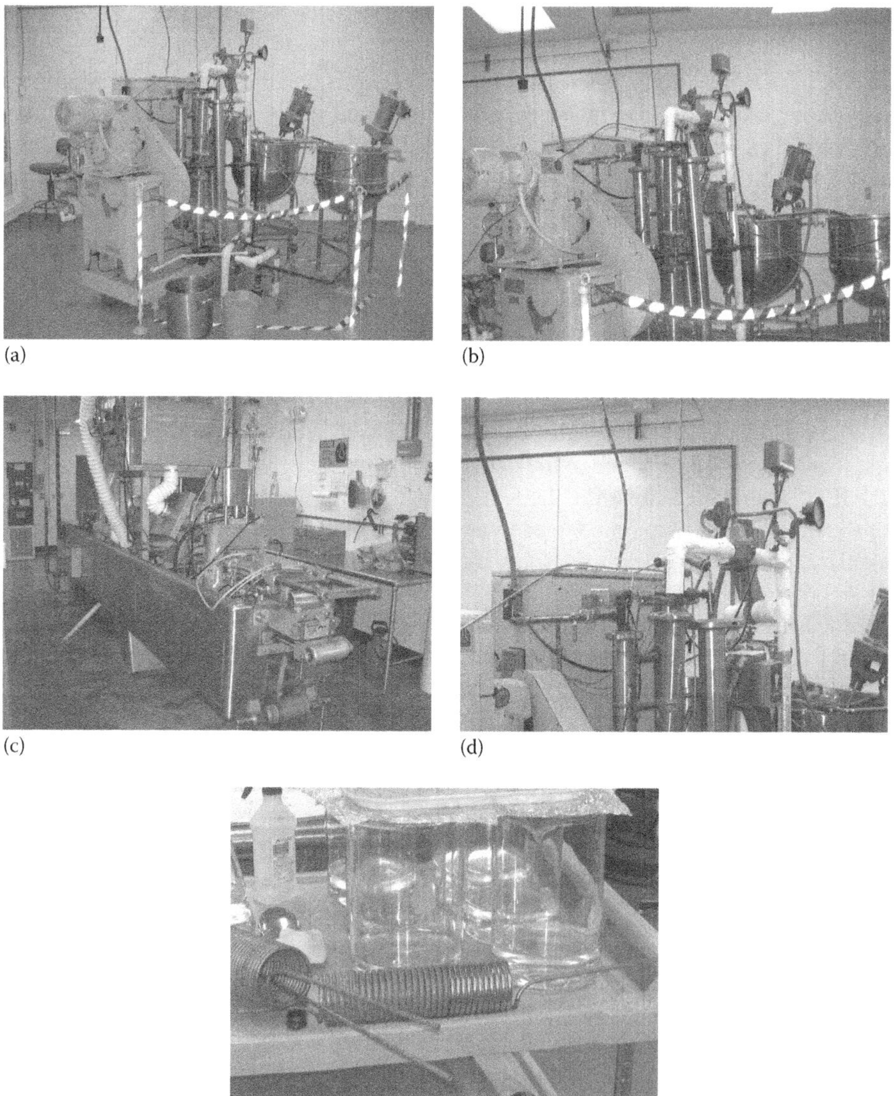

(a) (b) (c) (d) (e)

Figure 12.9 USDA, ARS Gast facility's flash sterilizer coupled with a form-fill seal machine configured for introducing diet and inoculating with insect eggs.

In Figure 12.9e, a comparison is made of a several 4 L beakers and the heating coil that was excised from the flash sterilizer to show the very small bore of the sterilizer tubing, allowing rapid heating and therefore flash sterilization, rather than prolonged heating that comes from trying to heat larger volumes. Thermal death time of even the most heat-resistant spores is less than 1 min at the temperatures employed by the flash sterilizer described here (discussed further in Chapter 13). Furthermore, the residence time of the diet in the heating coil is about 1 min, so this method is effective against most contaminants, according to the *D* values described by Jay et al. (2005) for most potential contaminants, including most spore formers. It should be noted that the coils, with their narrow diameters, are efficient heat exchangers that force the diet into a configuration of very narrow diameter so that heat penetration is virtually instantaneous. The heat exchange efficiency is also enhanced by the stainless steel composition of the coil's tubing.

Once the diet has passed through the heating coils where it has been exposed to temperatures of 137°C–140°C and a pressure of 70 kg/cm^2, it is put through cooling coils that reduce the temperature to about 40°C and is carried via stainless steel tubing to the tray-forming unit where it is delivered in a molten form to fill the newly formed trays and where it will be further cooled to receive the eggs. The steam used to heat the flash sterilizer is supplied by a steam generator. Prior to delivery of the diet, the tray-forming machine heats strips of polyvinyl chloride plastic (15.0 mL gauge, 15 cm wide) to about 150°C and impresses it with a mold, forming a 32-cell tray. The tray is filled with the diet described earlier, cooled with a blast of refrigerated air and moved to the egg-placing area of the tray-former, where a mixture of moth eggs and corn cob grits is sprinkled onto the diet surface, after which the Mylar lidding is seated and sealed to close the tray/cage unit (Tillman et al. 1997). The Mylar lidding (Davis et al. 1990) had been punched with an adequate number of holes to allow the exchange of gases from the cells to the ambient air. The trays are then placed in rearing racks and held in a rearing room throughout the development to the pupal stage when they are harvested and allowed to complete their life cycle in oviposition cages.

The system described here can produce 400 cuts (2–32-cell trays per cut) per hour or over 25,000 inoculated cells per hour or over 200,000 insects per 8 h day. This translates to more than 0.5 million potential *Heliothis virescens* or *Helicoverpa zea*, which are reared as individuals and about 2 million *Spodoptera exigua* per day, provided that the machine is run for 24 h/day. The flash sterilizer requires a cleaning, requiring ½ h per 4 h of run time. The cleaning step is performed to remove the hardened diet that inevitably builds on the inside surface of the heating coils. Cleaning the interior surfaces with an acid wash and fresh water as well as the periodic lubrication of the pump motor are essential routine maintenance, which is otherwise minimal on the flash sterilizing unit. One final note, the machinery described here applies to the lepidopteran rearing program at the Gast facility, but very similar equipment was also used in this facility to produce over a billion boll weevils over the years prior to boll weevil eradication. Much of this rearing and the improvements in these programs were overseen by the late Mr. John Roberson.

12.13.3 Extruders

The food technology industry has employed extruders extensively for the processing of an amazingly broad range of foods since the 1950s (Fellows 2000). Twin-screw extruders have been used in growing numbers since the late 1970s and more prominently in the early 1980s (Harper 1979, 1981). In general, extruders are useful for a wide range of different food types and textures. They have proved to be of unsurpassed benefit in the processing of dairy products, pastas, snack foods, bakery goods, confections, and are even widely used in dry and moist pet foods! They are very efficient in their use of energy, and they are most highly regarded for their versatility. They lend themselves to a variety of methods that they offer to the processing of foods, including mixing, automated addition of ingredients at various stages of the processing, cooking, texturizing, and shaping the final product. The following statement by Harper (1979, 1981) best summarizes the characteristics of extrusion:

> Now, the food extruder is considered a high-temperature short-time bioreactor that transforms a variety of raw ingredients into modified intermediate and finished products. The impetus for these developments has been the following requirements of food processing: (1) continuous high-throughput processing, (2) energy efficiency, (3) processing of relatively dry viscous materials, (4) improved textural and flavor characteristics of foods, (5) control of the thermal changes of food constituents, and (6) use of unconventional ingredients.

An underlying principle of extruders is that a turning screw can produce a great force to propel the incipient product while kneading and heating it. The great leverage that results from the turning screw generates considerable energy that will mix and churn materials with high viscosity and overcome inertia against mixing. The screw can be modified to have different configurations along its length to program different types of processing actions along the path of the food from source to sink. Extruders further lend themselves to modifications along the incipient product's path to allow heating, cooling, pressure changes (adding forces of compression, impact, and shearing), and addition of ingredients at key junctures in the process. The addition of a second screw (i.e., a twin-screw design) provides a tremendous increase in the processing power and versatility of this implement (Harper 1979, 1981, Fellows 2000).

A simplified diagram of a twin-screw extruder is shown in Figure 12.10 (A through G). The raw materials reservoir (A) contains the components that are to be mixed and preconditioned (a process that often involves mixing with water or oils and/or precooking with steam) in a preconditioner (B). The incipient product is fed into the screw drive and is propelled by the feed section of the turning screws. Heat can be added at any point (or points) along the process pathway, and additives can be introduced (vitamins, antibiotics, or other heat-labile components). Along the process pathway, features of the screws may be predesigned to shear, mix, knead, or otherwise modify the incipient product. Finally, the product is forced through the discharge dies, which shape the product into a desired configuration.

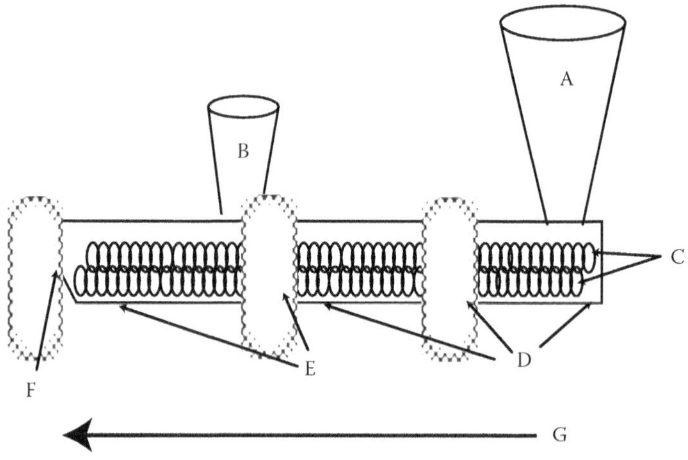

Figure 12.10 Diagram of twin-screw extruder. (A) Raw materials reservoir, (B) water reservoir, (C) co-rotating or counter-rotating screws), (D) mixing blocks and mixing region of screws, (E) restriction discs and kneading region of screws where pressure is greatly increased by restriction discs, (F) die for configuring diet to desired shape, and (G) direction of material (product) flow.

The range of size of the extruders varies from small table top models with capacities of a few kg/h to large machines with capacities of hundreds to several thousands of kg/h. Twin-screw extruders can have corotational or counterrotational screws, which can be intermeshed in a variety of ways that alter the properties of the processing, such as mixing characteristics. Figure 12.10B shows twin screws that have three different configurations along their length. Part of the versatility of the extruder is that these screws and their configurations can be changed so that the same piece of equipment can provide very different processing profiles.

12.13.4 Case Study: The Twin-Screw Extruder in Insect Diets

In the mid-1990s, the Pink Bollworm Facility (USDA, APHIS) converted from steam kettle production to twin-screw extruder (TSE) production of diet for the pink bollworm, *Pectinophora gossypiella* (Edwards et al. 1996). The conversion allowed that facility to move from an impressive production capacity of over 1 million "pinkies" per day to 8–10 million per day. This transition is a historical first demonstration that the very high-throughput technology of extrusion can be successfully applied to insect diets. It further represents advancement in the hybridization of modern food technology and entomology. The TSE has been used since the breakthrough with pink bollworms (PBW) to produce diet for several species of *Lepidoptera*, with considerable success. The paper where Edwards et al. explained the development of the insect diet-making process with the TSE provides excellent information about several questions that rearing specialists would ask of a new method for making diets. They provide the specific information on the various experimental designs for testing: (1) effects of different temperature profiles, (2) effects of different moisture profiles, (3) pressure effects, (4) rate of movement of the screws and the product being extruded, (5) the modifications that had to be made in restriction disks, (6) the rotational angles of the screws, (7) the types of screw subassemblies used for moving diets and kneading them, and finally (8) the exit dies and the related diet-shaping strategies. Most of these points are summarized in Figure 12.11. The specific details of the TSE are as follows:

> The twin screw extruder used in these studies was a Continua 37 (Werner and Pfleiderer Corp., Ramsey, N.J.) previously described (Edwards et al., 1994). The machine had screw diameters of 37.4 mm, an overall screw length of 1480 mm, and 9 zones (barrel sections) with independent temperature controls. The length to diameter ratio (L/D) was 39.7. Diet was fed by a twin screw gravimetric feeder. Feedwater, vitamin slurry, and corn oil were added to liquid ports in zone 1 by variable stroke piston pumps. Additional water was normally added in zone 7. Water mass flow rates were measured with Coriolis type flow meters.

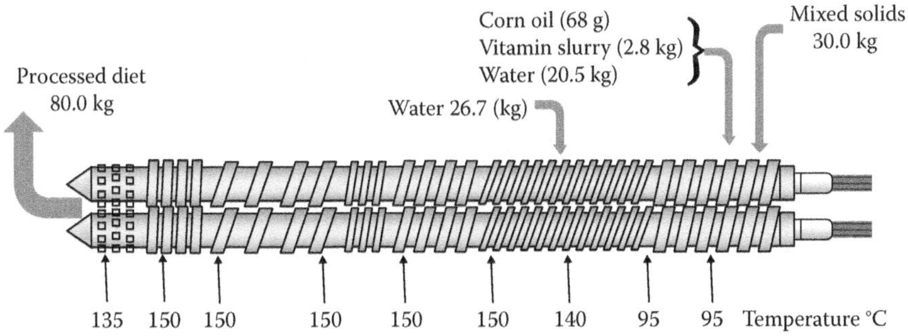

Figure 12.11 The screws from the Continua 37 (Werner and Pfleiderer Corp., Ramsey, NJ) twin-screw extruder used by Edwards et al. (1996) to produce artificial diet for pink bollworms.

The paper provided exhaustive documentation of the bioassay results that validated the diet parameters compared to alternative parameters (such as water content and temperature of processing conditions). Also the study included microbial contamination tests under various process conditions, and furthermore, the authors provided documentation that thiamine, which is probably the most heat-labile component in the PBW diet, was nearly 100% conserved in the extrusion process.

Finally, Edwards et al. (1996) expressed the advantages of the TSE process as follows:

> The twin screw extrusion process for production of sterile to near sterile pink bollworm diet has the following advantages: 1) produced by a continuous process; 2) has excellent process control characteristics; 3) has a short residence time in the extruder; 4) is adaptable to automation; 5) gives reduced labor and operating costs; 6)is scalable to large capacities; 7) will eliminate the need for formaldehyde; 8) produces suitable product over a wide range of feed compositions and moisture contents; 9) results in excellent, repeatable product quality; 10) is capable of producing widely different product textures, and 11) should be adaptable to other insect diets.
>
> The other high temperature continuous process currently used for diet sterilization is a flash sterilization process using a scraped surface heat exchanger for processing the boll weevil diet (Griffin et al., 1974; Roberson and Wright, 1984). Compared to twin screw extrusion, flash sterilization has the following disadvantages: 1) diet must be pumpable, thus limiting allowable moisture levels; 2) cleaning more difficult; 3) it is difficult to add ingredients to the process stream; 4) large size sterilizers are not available; and 5) fewer diets could be amenable to processing."

All of the advantages listed by Edwards et al. (1996) have proved to be true. Extruded diet has been used for several species of insects other than the PBW, including several species of *Lepidoptera* (Hannah Nadel, personal communication), and a dry extruded diet has been used as food for crickets as a convenient food used during shipping (Cohen unpublished). It is also a crucial factor that the versatility of the extruder process allows diet production with materials that are difficult or impossible to pump. This means that insects that require large particle–sized materials can be provided with extruded diets. The versatility and power of the twin-screw extruder in insect diet processing would recommend it as a revolutionary improvement.

However, extruders still (nearly 20 years after their introduction to the world of insect diets) have not been adopted as widely as might be expected, given their very positive features. The possible reasons for this are (1) expense: a small—"research"—model costs upward of $100,000 for a basic unit; (2) technical difficulty: extruders require a fairly substantial training; and (3) entomologist and other members of the rearing community tend to be conservative about innovations with the thinking, "If it's not broken, don't fix it." And "We have been using steam kettles for years, why change now?" Hopefully, this final reason for failure to explore the vast possibilities of twin-screw extrusion and other novel techniques from the food industry will change with education in diet and rearing science and technology (Table 12.3).

Table 12.3 Profiles of Autoclave, Steam Kettle, Flash Sterilizer, and Twin-Screw Extruder

Equipment Type	Throughput (kg/h)	Microbial Kill Capacity	Cooking Homogeneity	Destruction of Ascorbic Acid	Relative Cost	Versatility
Autoclave	1–5	High	Low	High	Low to moderate	Low
Steam kettle	5–25	Not spore formers	Low	Moderate	Low to moderate	Low
Flash sterilizer	50–100	High	High[a]	Low	Moderate to high	Moderate
Twin-screw extruder	50–200	High	High	Low	High	Very high

[a] Mixing homogeneity is moderate.

12.14 PACKAGING AND CONTAINERIZATION

The topic of containerization (or the technology of cages, enclosures, or other means of holding insects captive for manipulation) in insect rearing is complex and generally not within the domain of this book. However, in several rearing settings, the diet container and the cage are one in the same and are, therefore, discussed here. The main part of this section, however, will be addressed to the characteristics of diet packaging per se.

In cases where the diet container is also the insect's cage such as in the lepidopteran and weevil rearing described earlier, the diet is generally delivered into the container in a molten form, which, upon cooling, fills a considerable portion of the enclosure. Such cages/diet containers are usually made of plastic, in some settings, preformed units that are filled either manually or by machines. In some larger-scale facilities, diet containers are made on site from tray formers or cup formers that press the desired shape out of a flat plastic fed into the machine from a roll (such as what is used in the Gast facility).

An important consideration in this aspect of diet container/insect cage technology is that the amount of diet being added to the container is a very important part of the economics and the ecology of rearing. First, in the type of rearing considered here, the diet serves as the environment for the target insects, which bore into the diet, are surrounded by it, are reliant upon it as a source of free water and as a source of moisture that modifies the microclimate within the container. Second, the amount of diet must be adequate to provide all of the food that the insect can consume, but it should not be so much beyond the consumption requirements that it is wasteful of material that typically costs from $1.00 to $5.00 per kg for the complete diet (including materials and labor but excluding initial capital outlay and overhead). An example of the magnitude of the potential waste is seen in the following breakdown. In a facility that produces about 0.5 million Heliothis virescens per week, the cost of diet (labor and materials) is about $2.5/kg, or about $0.02/cell. If there is twice as much diet in each cell, this amounts to about $0.01 per cell wasted times 0.5 million cells (assuming an unrealistically high yield), and the total wastage per week would be $5,000 per week.

12.15 FUTURE PROSPECTS

One of the most exciting and promising subjects in insect rearing is the application of modern food technology principles and practices to the formulation, production, preservation, and use of insect diets. The technology described by Fellows (2000) has potential bearing on diet processes that include various kinds of heat treatments, sterilization, and improvement of nutritional quality by value-added processing, extension of the range of diet materials, upscale of the capacity, automation, packaging, and storage of diets. For example, an in-depth consideration of what has been accomplished with twin-screw extruders in the food industry should encourage insect diet experts to devise ways that the extruder's versatility can be employed to increase our ability to mix and heat process ingredients that were previously unsuitable as diet components for various insects. Soy flours and soy protein extracts have shown enormous potential in human foods and nutrition and an incredible versatility, but before the potential of soy could have been realized, researchers had to find novel ways of manipulating it to tap into its potential. Likewise, soy products have been (in this author's opinion) underused in insect diets, especially when one considers the high nutritional value of these products.

The field of food-processing technology as articulated by Fellows (2000) holds a blueprint, that, if read properly, offers the potential directions that insect diet science and technology should follow. Fellows covers all the major and up-to-date aspects of food processing, including the general theory of process control, food component separation, heat and cold processing of foods (for alteration of nutritional quality, flavor, and preservation), size reduction, mixing and forming, preservation, and packaging technologies. Insect diet experts and entomologists, in general, have virtually ignored (or have been ignorant of) this well-established technology. Having to "reinvent the wheel" countless times or never having known about inventions that are pertinent to our field retards progress in our endeavors to use artificial diet technology.

What will be needed to bring to fruition the full potential of the food technology knowledge base is the thorough education of entomologists (or insect diet specialists) in all aspects of food science and technology—most especially the field of food-processing equipment. The reciprocal—recruitment of food scientists into entomology—would greatly enhance and advance insect diet applications, especially in mass-rearing domains. This would further require a more than casual education in entomology and rearing for professionals in the food science and technology domain. The integration of food science and entomology and the adoption of the methods and advancements in food technology in research and applied programs in insect diets will accelerate the advancement of mass rearing beyond the current boundaries and carry the field fully into the twenty-first century.

CHAPTER 13

Microbes in the Diet Setting

13.1 OVERVIEW OF MICROBE/INSECT INTERACTIONS IN THE REARING SETTING

There are three major ways that microbes play a role in insect rearing: (1) they may be symbionts that live in or on the bodies of insects in a neutral relationship (sometimes known as commensalism) or in a way that is mutually beneficial to themselves and their insect hosts (a partnership relationship known as mutualism), (2) they may occur as parasites or pathogens of the insects that we are trying to rear, and (3) they may occur as contaminants of the diets or other rearing materials in our cultures. In the first type, the symbionts may be essential to the well-being of the insects, most often in a nutritionally beneficial relationship where the microbe guest repays its host's hospitality by synthesizing an otherwise unavailable or poorly available nutrient. Termites and their microbes are excellent examples of such a mutualism as are cockroaches and their symbionts and most homopterans and their microbes. In using artificial diets for insects that rely on such relationships, the danger is not in contamination but rather in the disruption of the relationship by inadvertent removal of the microbe or of its means of gaining entry in or access to the host. Even with the so-called commensals, it is possible that there is a cryptic benefit (i.e., a hidden mutualism) that we may disrupt by our rearing efforts. These relationships are further discussed in Chapter 7.

The second type of relationship is the subject of the field of insect pathology and is not within the scope of this book to do more than briefly characterize this large field. This subject, pioneered by Steinhaus (1940), has become a major subdiscipline of entomology and is important far beyond the scope of rearing facilities. In nature, many insect populations are profoundly affected by their pathogens, and in a growing number of cases, insect pathogens have been exploited as biological control agents. However, in insectaries, whose purpose is to colonize robust, healthy insects, the entry of a pathogenic microbe into the colony can be a disaster. The nature of insect colonies is close contact between individuals, often under stresses such as nutritional problems, overcrowding-induced concentration of waste products, agonistic behavior between individuals (fighting or cannibalism), and other factors, which help spread disease. These factors often result in clinical levels of disease that threaten the colony.

The third type of relationship is one where there is an accidental introduction of microbes in the diet or other aspects of handling (cages, oviposition materials, or laboratory air). The conditions that are intended as optimal for the insects are also ideal for many opportunistic microbes. The metabolic action of the microbes living within the diet may deplete the diet of key nutrients as well as introducing distasteful or toxic substances to the medium that is intended to nurture the target insects. In general, accidental microbial invasions are at best neutral and at worst harmful to rearing efforts.

However, not all instances of microbial contamination are necessarily destructive. It is possible that microbial contaminants (which are accidental or facultative diet associates) can alter the diet directly or modify conditions in ways that render the diet more palatable or nutritious to target insects. In this circumstance, the accidental introduction of a microbe improves the diet, but because the introduction was accidental, the reason behind the benefit is not understood or controlled by the diet specialist. Such microbe-induced changes have long been used to improve human foods in fermentation such as the processes used in making yogurt, cheese, wine, beer, pickles, soy sauce, and many other processed foods. One of the most promising areas of potential new developments in insect rearing and insect diet research is in the novel domain of *probiotics*. Probiotics are discussed in the Section 13.12. However, it should be emphasized that the subject of benefits of insect diet-borne microbes has not been well researched.

13.2 *SYMBIOSIS*, MUTUALISM, AND COMMENSALISM: MICROBES THAT HAVE BENEFICIAL OR NEUTRAL RELATIONS WITH INSECTS

Tanada and Kaya (1993) define *symbiosis* as an interaction where two species of organisms live together in a close association. Tanada and Kaya explain that while both members of the association are technically symbionts, conventionally they are divided into separate, distinguishing categories: host and symbiont (sometimes written as "symbiote"). They explain that there are several possibilities that these close associations offer the host (the larger organism) and the symbiont (the smaller organism that lives in or on the host). The association may be neither harmful nor beneficial to either associate, an interaction called *inquilinism* (sometimes called neutralism). The association may be beneficial to the symbiont and neutral to the host, a phenomenon called *commensalism*. If the association is beneficial to the symbiont and harmful to the host, it is called *parasitism*. Finally, if the process is beneficial to both species, it is called *mutualism*. It is the two latter categories that are of greatest concern to insect diet specialists.

It should be noted that within the vast array of insect species and insect/microbe mutualistic interactions, there are cases where the microbes are outside of the insect (ectosymbionts) and other associations where the mutualistic microbes are inside the insect's body (endosymbionts). In both types of associations, the host insect has special structural/physiological and often behavioral adaptations that enhance the association. Some entomologists (e.g., Farrell et al. 2001) have described the external relationships (such as leaf-cutting ants and their "fungal gardens" and ambrosia beetles with their "fungal agriculture") as farming or cultivation activities. Whether the interactions are external or internal with respect to the insect host, the associations are of tremendous importance, and if we are attempting to rear the insects that are adapted to these associations, it is important to understand them so that we can encourage them in our efforts to rear healthy insects.

Parasitism by microbes will be covered in a later section of this chapter, and mutualism is discussed here. It should be emphasized that terms such as mutualism and commensalism are really approximations of the interactions, and evolution is constantly taking place in the biological world, with respect to microbiome/host interactions. Also, it needs further emphasis that our views of these interactive processes are based on incomplete capacity to observe the dynamics; so much of our body of interpretation is based at least partially on inference. And finally, there are microenvironmental contexts where interactions can change in such a way that what may have started off as an instance of commensalism between the host/microbe interaction can turn into either parasitism (pathology) or mutualism.

Reviews of mutualism (Campbell 1990, Tanada and Kaya 1993, Douglas 1998) in insects explain mutualism as being mainly of nutritional benefit to the host. There is a lesser amount of information about the symbionts contributing to other functions of the host, including production of digestive enzymes and pheromones. One of the most fascinating and complex symbiotic relationships is one where viruses (called polydnaviruses) associate with the parasitic wasps causing the insect hosts of the wasps to change their physiology in several ways that benefit the wasps that inject the viruses into their lepidopteran hosts (Stoltz and Vinson 1979). The viruses and the wasps mutually benefit from the association at the expense of the host, whose defenses are suppressed by the viruses. The viruses also modify the metabolism of the host to the benefit of the parasitic wasp.

Most other microbe–insect associations are less complex but no less beneficial to the symbiont–host team. According to Dadd (1985), most of these associations are found in insects that habitually consume diets that are of marginal nutritional value such as plant saps (xylem sap, phloem sap, and nectars—"true saps," not predigested, concentrated nutrients as discussed in Chapter 7). However, despite the fact that many of these associations have been recognized for nearly a century and the fact that they have been intensely studied, there remains a great deal of uncertainty about many of the essential details of these associations. For example, the role of symbionts in aphids, whiteflies, and other homopteran species remains unclear as to the exact nutrients that are provided by microbes. Although earlier work implicated that the symbionts synthesized sterols that were essential to their hosts (Dadd 1972, Mittler 1972), more recent work (Campbell and Nes 1983, Douglas 1998) reports that the microbial symbionts that inhabit the Homoptera are incapable of synthesizing these important lipids. This conclusion leaves students of insect nutrition with the uncertainty as to how much sterol is required by these insects and where the sterols are coming from. A perplexing question remains as to the source of sterols where aphids have been fed diets that are supposed to be sterol free and where the aphids had been rendered aposymbiotic (symbiont free) as a result of treatment with antibiotics such as chlortetracycline. In a review of the relationship between aphids and their chief bacterial symbiont, *Buchnera*, Douglas (1998) concludes that while the bacteria, which may make up as much as 10% of the aphid's biomass, do not provide sterols or fatty acids to the insect, they do somehow contribute a sterol-sparing effect that remains to be explained. Douglas also concludes that the bacteria do contribute to the amino acid metabolism and nutrition of their hosts but not their vitamin nutrition.

Campbell (1990) and Tanada and Kaya (1993) examine some of the other interactions between symbionts and their hosts. Their consensus is that in many instances, in addition to their contribution of nutrients, the symbionts often contribute digestive enzymes to their hosts. Among the best documented relationships is that of termites and their symbiotic Protozoa and bacteria. Symbionts that reside in the hindgut of termites may make up as much as one-third of the host insect's biomass and produce several digestive enzymes that help the host with its nutritionally poor diet, especially with the production of cellulase, which allows the termites to utilize the otherwise indigestible polysaccharide, cellulose (Tanada and Kaya 1993). An interesting topic that has recently emerged is the role of the rickettsial agent, *Wolbachia*, in determining reproductive compatibility of its host. Increasingly, insects of various taxa are being found to have their reproductive activities influenced by one or another of the various species of the symbiont (Werren 1997). Werren explained that these rickettsia are obligatory symbionts that are cytoplasmically inherited and transmitted horizontally between arthropod species. Occurring in the reproductive tissues, *Wolbachia* have been shown to have several effects on their hosts, including alteration of sex ratios by feminizing genetic males, causing reproductive incompatibility, and inducing parthenogenesis. These effects can strongly impact the quality of insect colonies.

Earlier work by Brooks (1963) and later reviews by Campbell (1990), Tanada and Kaya (1993), and Dillon and Dillon (2003) describe several other examples of insect/microbe symbiotic interactions, including many cockroaches, long-horned beetles, hematophagous flies, fleas, and many other insects that eat wood, blood, plant saps, or other nutritionally poor or incomplete diets. These authors also describe the regions of the insects' body where the symbionts reside, including the hindgut of termites, the midgut (either in diverticula or in specialized cells), or in the hemocoel, either as forms that are free of insect cells or within groups of cells known as mycetocytes, or in gut diverticula (discussed further in Chapter 7), or finally in organs known as mycetomes.

13.2.1 Associations of Insects and Their Microbiome Bacteria

An extremely important aspect of insect homeostasis, including healthy steady-state conditions in insect-rearing domains, is to be found in the insect gut. The high numbers and impressive biomass of bacteria in insect guts are compelling factors in the argument that microbial populations must play major roles in the health of insects and other higher organisms. For example, Dillon and Dillon (2003) mention (citing Savage 1977), "The consortium of microbes inhabiting the human gut (10^{14}) is estimated to outnumber the somatic and germ cells of the body (10^{13})." Dillon and Dillon point out that the relative numbers in insect digestive systems are neither less impressive nor any less important than they are in human gastrointestinal systems. However, there are still significant gaps in our knowledge of the taxonomic profiles and the ecological roles of the bacteria in most insects' guts, with the possible exception of termites and some cockroaches that have been studied more aggressively than other insects.

In earlier studies of insect/microbe symbiotic relationships (especially mutualism and commensalism), techniques, which depended on ability to culture microbes, limited the ability to diagnose relationships with bacteria, fungi, and protozoans. The microbes that were too fastidious to be cultured were missed in screening techniques, and even when microbial associates were visualized by light or electron microscopy, identification by visual means was problematic. However, now with the ability to use molecular techniques, conventional culture techniques can be complemented by genetic analysis, and new associations are being discovered daily. Furthermore, with proteomic and metabolomic technology, we can fill in many of the missing pieces about the nature of the associations in terms of metabolic products or enzymes provided by or enhanced by the microbial associates.

Dillon and Dillon (2003) have summarized several of the techniques that are being used to elucidate the complex relationships in insect digestive systems, including 16S rRNA gene analysis, denaturing gradient gel electrophoresis (DGGE), in situ hybridization with species-specific rRNA probes, can add a great deal of resolution to the characterization of gut microbiota. Also Dillon and Dillon mention the value of using the analysis (fractionation) of cytosine to guanine (C–G) ratios to assess microbiota community diversity within insect guts: a procedure that would be useful in assessing the kinds of changes that take place in insects as they go through the process of domestication or adaptation to artificial diets. Other techniques that have been used to clarify the insect/microbiota relationships include labeling microbes with molecular-based markers such as green fluorescent protein to track associations of microbes in specific gut locations (Hurst and Jackson 2002).

Two recent examples of the progress in clarifying the microbiome/insect interactions in silkworm larvae are presented here. Anand et al. (2010) used conventional microbial culture techniques coupled with enzyme analysis to help demonstrate the profiles of microbes in silkworm larvae and their microbiome. They found

> Gram positive *Bacillus circulans* and Gram negative *Proteus vulgaris*, *Klebsiella pneumoniae*, *Escherichia coli*, *Citrobacter freundii*, *Serratia liquefaciens*, *Enterobacter* sp., *Pseudomonas fluorescens*, *P. aeruginosa*, *Aeromonas* sp., and *Erwinia* sp. Three of these isolates, *P. vulgaris, K. pneumoniae, C. freundii*, were cellulolytic and xylanolytic, *P. fluorescens* and *Erwinia* sp., were pectinolytic and *K. pneumoniae* degraded starch. *Aeromonas sp.* was able to utilize the CMcellulose (a substrate for analysis of cellulose) and xylan. *S. liquefaciens* was able to utilize three polysaccharides including CMcellulose, xylan and pectin. *B. circulans* was able to utilize all four polysaccharides with different efficacy. The gut of *B. mori* has an alkaline pH and all of the isolated bacterial strains were found to grow and degrade polysaccharides at alkaline pH. The number of cellulolytic bacteria increases with each instar."

In a study of the effects of different diets on microbial profiles in silkworms, Liang et al. (2014) used molecular techniques for identifying the genera of microbes under the rearing conditions with conventional mulberry leaves vs. novel use of lettuce leaves as food. They used culture-dependent, clone library analysis of 16S rRNA gene and DGGE approaches to evaluate the changes in bacterial profiles (shown in Figure 13.1).

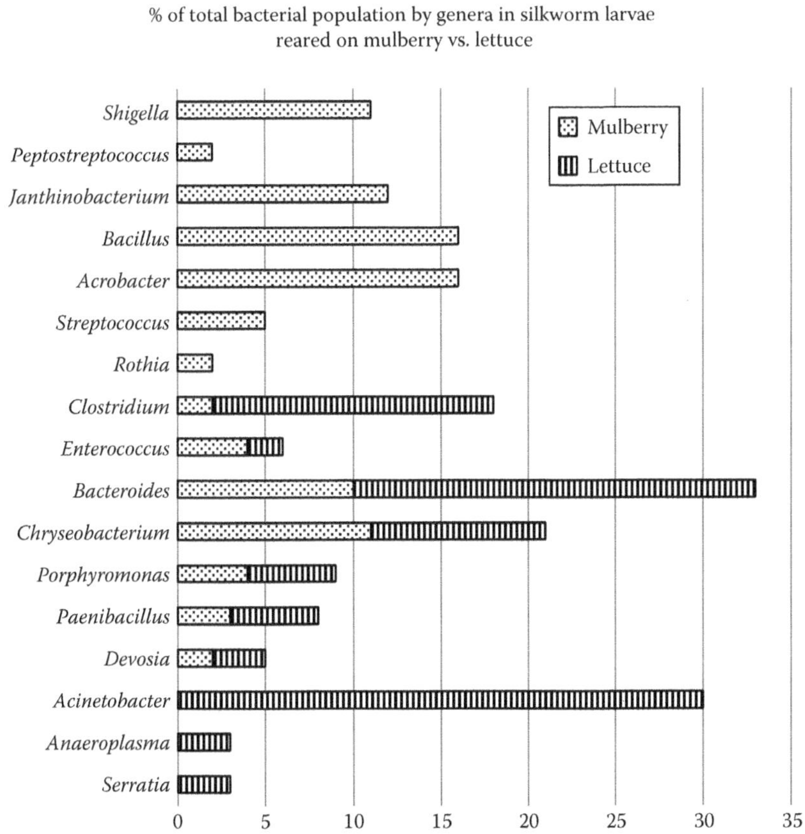

Figure 13.1 Numerical profile of gut bacteria profiles of larval guts of silkworms reared on diets of mulberry vs. lettuce. (Redrawn from Liang, X. et al., *Appl. Microbiol. Biotechnol.*, 98, 3769, 2014.)

Clearly, the original bacterial profiles in mulberry-fed larvae were dramatically replaced, with *Shigella*, *Peptostreptococcus*, *Janthinobacterium*, *Bacillus*, *Acrobacter*, *Rothia*, and *Streptococcus* being completely displaced by *Serratia*, *Anaeroplasma*, and *Acinetobacter.* Also, there were shifts in the relative numbers of *Clostridium*, *Enterococcus*, *Bacteroides*, *Chryseobacterium*, *Porphyromonas*, *Paenibacillus*, and *Devosia* populations (Figure 13.1). Liang et al. (2014) reported further that the larvae on the lettuce diet or markedly less fit than the mulberry-fed larvae, and while it is unclear whether the changes in the fitness of the larvae is attributable to the differences in the food quality of lettuce compared to mulberry or the changes in microbiome profiles, this study points to the need for further work along this line where diet modifications are evaluated in terms of nutritional quality vs. microbial profiles. Such experiments will require careful planning to assure that nutritional quality is held constant while microbial profiles vary and vice versa.

Another important question, especially in the rearing setting, is how the symbionts reach these sites in their hosts. A related question is how the microbes are transmitted from one individual to another. This is important, because if the target insects are mutualistic (of benefit to the host insect that we are trying to rear), it is crucial to provide rearing conditions that allow the normal maintenance of the interaction. Any inadvertent disruption of the host inoculation process (sterilizing eggs, removal of frass before an inoculation can take place, or use of an antimicrobial that kills the microbial partner as described by Costa et al. 1997) can strongly and negatively impact rearing systems.

In some symbionts, the transfer of microbes takes place as part of the reproductive cycle where the symbiont enters a forming egg in a process called transovarian infection. This is the case with aphids, whiteflies, and other homopteran species. In other cases, Tanada and Kaya point out the symbionts are placed on or in the eggs by specialized structures in the female's reproductive system and enter the egg via the micropyle from which the neonate ingests the microbes (as in the case of the olive fly). Aside from such elaborate means of infection, there are more simple mechanisms such as those of termites whose well-developed social behavior facilitates symbiont transfer to neonates and recently molted individuals via feeding on anal secretions from infected colony members.

On several occasions, disruptions of symbiont transfer have been observed to be caused by ill-advised use of antimicrobial chemicals and occasionally by simple problems with malnutrition (Dadd 1985, Tanada and Kaya 1993, Dillon and Dillon 2003).

13.2.2 Yeast Dynamics in Insect Diets: From Live Yeasts to Commercial Yeast Products

In nature and in artificial rearing systems, yeasts and yeast products are of tremendous importance. Yeasts and yeast products can be divided into several categories that help us understand the nature of these microorganisms in insect feeding. First, yeast consumption in natural vs. artificial systems can be discussed. Consumption of live yeasts and other fungi is important in nature, and for some insects this feeding habit applies to artificial diets as well. In this vein, drosophilids and tephritids, which are saprophagous and especially attuned to yeast consumption, are considered. Also, however, there are other trophic groups such as wood-feeders (e.g., beetles, termites, and ants) that depend extensively on mainly non-yeast fungi. One of the best-known and most charismatic relationships is between the fungus cultivating ants and the fungi that they cultivate—where both the fungi and the ants derive mutualistic benefits (Schultz and Brady 2008, Hölldobler and Wilson 2009). Other instances of improvement of substrates that are poorly nutritious to insects include the vast array of ambrosia beetles (Hulcr and Cognato 2010), and gall-forming flies (Heath and Stireman 2010).

In many of these cases, the insect has highly evolved regions known as mycangia for carrying the fungi. Tri-trophic insect/microbe/substrate systems involve insect species that utilize microbes that are external to the insects' bodies. Generally, the microbes are fungi (filamentous or yeast) or bacteria that are trophically considered as saphrophagous, but in some cases, the host organism (substrate) is alive. Plant galls (Heath and Stireman 2010) are examples of tri-trophic relationships where the substrate (= host plant) is alive. The insect/microbe relationship may be facultative or obligatory. Although there is a fairly large body of literature on these interactions, there remain some important relationships that remain unclear—especially those pertaining to the exact nutritional nature of the interactions (Kobayashi et al. 2008).

Unquestionably, from a global perspective, the most widely cultured insect species is the *Drosophila* species complex, especially *D. melanogaster*, and although this species is reared on hundreds of different diet formulations, it is a primarily yeast-eating insect. Furthermore, although the primary yeast species used to culture *D. melanogaster* is *Saccharomyces cerevisiae*, its more natural foods include several species of yeast, most of them outside of the genus *Saccharomyces*. In an excellent study by Stamps et al. (2012), the authors showed that in mixed cultures of about 20 yeast species, the presence of *D. melanogaster* had a very strong influence on the species distribution (Figure 13.2). The precedence for this type of influence by drosophilid flies had been established by a considerable body of ecological literature by research groups from the University of Arizona and University of California Davis (e.g., Starmer et al. 1977, 1986, 1990). In these studies and others from these research groups, it becomes evident that there is an intricate association between drosophilid flies and yeast that are associated with a wide variety of plant species, but this association is especially well documented in cacti in the Sonoran Desert, where the plants are well isolated and present an island biogeography where the yeast/fly associations are in an intense evolutionary flux. Other authors such as Anagnostou et al. (2010) have advanced the understanding of the relationship from a nutritional ecology perspective.

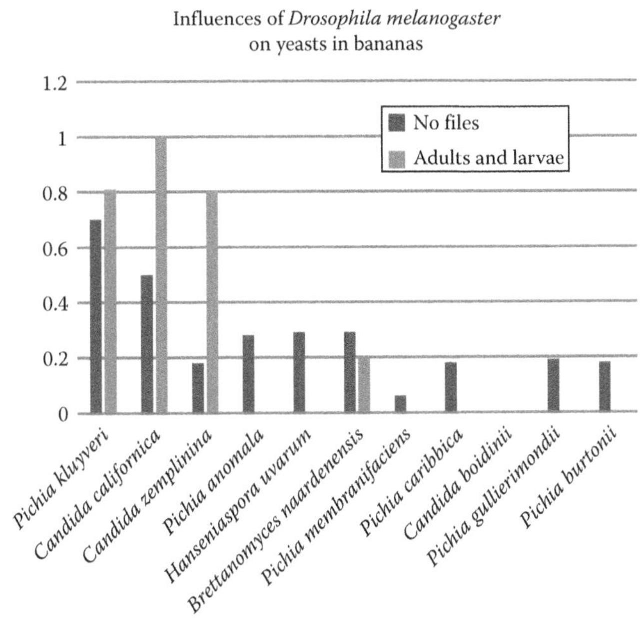

Figure 13.2 Shows the tri-trophic interaction common in feeding systems where external microbes prevail. (From Stamps, J.A. et al., *PLoS ONE*, 7(7), e42238, 2012.)

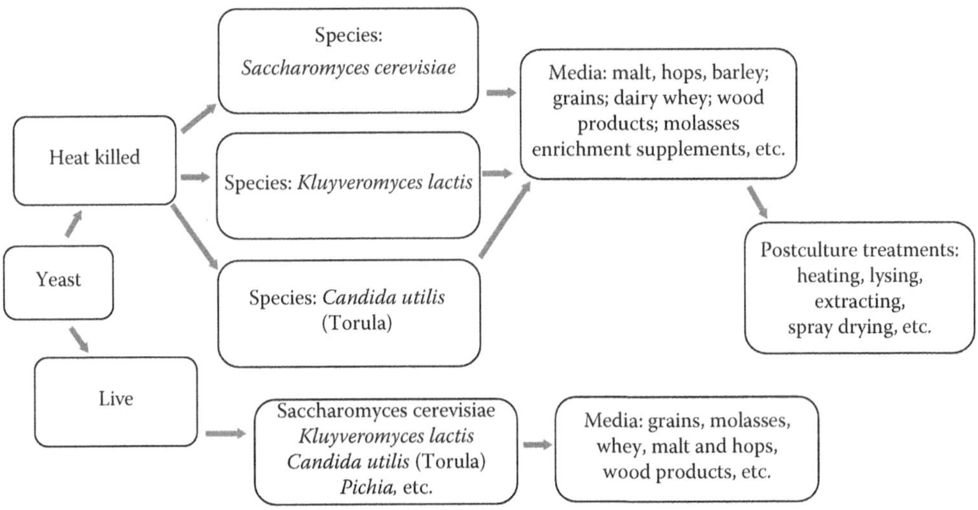

Figure 13.3 Processing scheme for yeast showing main steps in the production of live and heat-killed yeasts as products.

However, for the focus of this book, the use of microbes as foods is confined to artificial diets, and this excludes many of the more trophically complex relationships mentioned earlier. The deliberate use of live microbes in rearing situations is fairly limited and takes place mainly in rearing *Drosophila* spp. (Cooper 1960). It is common for people rearing *Drosophila* spp. to add live yeast to colonies, though several publications describe axenic defined diets for *D. melanogaster*, indicating that this species does not require live yeast to meet its basic requirements (Figure 13.3).

Yeasts that are used in insect diets are mainly heat-killed bakers'/brewer's yeast (*S. cerevisiae*) or Torula yeast (*Candida utilis*, formerly *Torulopsis utilis*). Occasionally, other species of yeast, mainly in the genus *Saccharomyces*, for example, *Saccharomyces fragilis*, are used in diets (Singh 1977). At least 20 species of yeast have been shown to be present in association with insects in the field, especially drosophilids and tephritids (e.g., Vacek et al. 1979, Starmer et al. 1976, 1977, Stamps et al. 2012), but there has been little attention to processing these yeasts and including them in diets in the same way that *S. cerevisiae* and *C. utilis* have been used. It is especially interesting that despite repeated and extensive demonstrations of the importance of interactions between a wide variety of yeast species with drosophilids and tephritids, most insect diet practitioners have confined their choices of yeasts to these two species (*S. cerevisiae and C. utilis*).

The one important exception to this is the use of food Wheast as described by Hagen and Tassan (1970) for lacewings and other predatory insects and by Standifer (1980) for honeybees. Wheast is a product of the dairy industry that was available three to four decades ago when dairy whey was considered a problematic waste product of the cheese industry. Dairy yeasts such as *S. fragilis*, which can digest lactose sugar, were used to remove the milk sugar from the product. Because lactose made up 70% of the dry weight of sweet dairy whey and because it was nondigestible by conventional sewage decomposers, whey liquid caused disposal problems. The engineering solution to the disposal problem was, quite cleverly, to use dairy-derived microorganisms to ameliorate the lactose problem. In this disposal process, it was discovered serendipitously that the combined product, *S. fragilis* yeast, and the remaining whey materials proved to be a highly nutritious source of proteins and other nutrients (e.g., B vitamins and minerals). The processing facilities soon became factories for converting whey into a novel product known as Wheast, sometimes called "Food Wheast." When new uses for dairy whey came into practice, Wheast was no longer produced by the dairy industry, and an excellent and most promising diet source for insects became unavailable. Currently, some companies sell a product sometimes called Wheast, but this material is not what can be considered as "true Wheast" unless it is a product of fermentation or biomanufacturing interaction between the dairy substrate, sweet dairy whey, and a specialized yeast that is capable of hydrolyzing lactose. It is noteworthy that the standard bakers' yeast *S. cerevisiae* cannot break down lactose, as this yeast lacks the enzyme β-galactosidase. Therefore, if one who is trying to produce Wheast simply mixes live *S. cerevisiae* with dairy whey or if they mix heat-killed brewers' yeast with sweet dairy whey, they may have a fairly nutritional product, but it will not be the same thing as "true Wheast." The nature of yeast products as they might affect target insects (e.g., flavors and nutritional content) results from a complex dynamic between the yeast and its substrate. This is illustrated in a simplified diagram in Figure 13.4.

13.2.3 Some Other Cases of Fungal/Insect Relationships That Are Beneficial or Essential to Some Insects

A subject that has received considerable attention in the ecological and evolutionary communities but little attention in the rearing community is the relationship between many insects and microbes that live outside the insects' bodies. This observation is surprising in light of the importance of many of the insect species that depend upon the relationship with external microbes—species that are of economic or extensive importance as research subject, as well as ecologically key insects such as bark beetles that can reshape a forest within a few years. There are also associations of external microbes that are involved in gall formation in plants, relationships that include a fairly wide variety of insect families in Hymenoptera and Diptera. Less well-characterized relationships include a huge array of saprophagous insects that use a vast array of foods (rotting plant tissues, rotting animal tissues) in a nearly universal range of habitats. Even one of the most commonly encountered insects, the common housefly (*Musca domestica*, Diptera: Muscidae) depends heavily upon bacteria to *improve* the nutritional quality of its cosmopolitan foods.

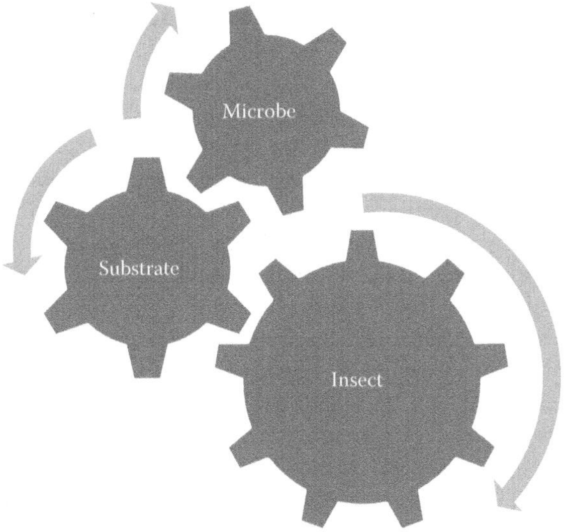

Figure 13.4 Model of insect/microbe/substrate complex.

The dependence on fungi (and sometimes bacteria) in several species of wood-eating (xylophagous) insects is underscored by the presence of specialized microbe-carrying structures called mycangia. Mycangia are adaptations (*fungal pockets*) of the exoskeleton where glandular cuticular pits in mandibles, on thoracic regions, or even, in the case of wood wasps, the ovipositor, are present. These cuticular pockets are adapted in concert with behavioral modifications for fungal uptake and fungal inoculation to efficiently infect the host tree to the insect's benefit. In some cases, such as with many ambrosia beetles, the sole source of food is the fungus (Francke-Grossmann 1967), and in other species, the fungus contributes to the beetles' nutrition by modifying, enzymatically, the digestibility or nutrient content of otherwise intractable woody materials (Paine et al. 1997).

In light of all the complexities of the relationships between various xylophagous beetles and their microbial (mainly fungal) symbionts, it might be expected that fully successful artificial diets that support mass rearing have proved evasive goals. There are a few dozen publications on increments of progress both in developing a practical rearing system and in unraveling the nutritional and dietary relationships between the beetles (mainly ambrosia beetles) and their fungal mutualists (these works being represented initially by Saunders and Knoke 1967, then Norris and Chu 1985, Klepzig et al. 2001, and recently by Castrillo et al. 2012). The diets for xylophagous beetles have generally contained sawdust or ground wood (often either powdered phloem tissue or ground bark from the species of trees that are hosts to the insects in question), and in addition, the diets have standard components such as agar, starch, sugars, and other functional diet components. Some diets that were designed to define nutritional relationships contain purified cellulose powder along with other chemically defined nutrients, but these wood-free diets have not succeeded in supporting sustained beetle rearing, even when the fungus was present (e.g., Norris and Baker 1968). One more point that is key to the problem of artificial diets for xylophagous beetles is that the texture of the diet must meet the physical needs of the insects, which are adapted to tunneling, ovipositing in special portions of the tunnels, depositing frass in specific locations, and other aspects of the biology of an organism that is highly specialized. Future success in the advancement of diets and rearing systems for xylophagous insects will depend on our understanding much more fully the trophic relationships and the textural matrix where the insects must go through their complex life cycles.

13.2.4 Other Side of the Coin: Microbes That Cause Disease—Pathogens

The previous section dealt with microbes that are beneficial to our target insects. This section deals with the disease-causing agents, pathogens. Tanada and Kaya (1993) describe disease as an abnormality that occurs because of "physical or physiological derangements." They point out that disease may be caused by living or nonliving agents, including nutritional or mechanical factors. However, for the present discussion, only "derangements" caused by living agents, specifically microbial-caused diseases, will be considered. The major microbial groups that are known to cause diseases in insects include viruses, bacteria, protozoa, and fungi. All of these groups are represented in insect colonies, and all can be connected in some way with artificial diets. A useful distinction about pathogens is that some are *obligatory* (i.e., they must live parasitically at the expense of a host) and others are *facultative* (i.e., they can take opportunistic advantage of a living host).

By their nature, viruses are obligatory pathogens. Another characteristic of viruses is that they are generally fastidiously specific to a given taxon, though some viruses do cross taxonomic lines (such as the West Nile virus). The other groups (bacteria, fungi, and protozoans) have representatives from both categories: obligatory and facultative pathogens. For example, the bacterial species *Bacillus cereus* and *B. thuringiensis* are commonly found in soils and can be free living, but both can also be virulent pathogens of insects. Similarly, *Enterobacter*, *Serratia*, *Pseudomonas*, and *Proteus* can all move from free-living forms to becoming facultative pathogens of a wide variety of insects (Tanada and Kaya 1993). Similarly, *Aspergillus* and *Mucor* (both common fungi found in a variety of environments) can become serious pathogens and can readily be transferred to insects via their diet (Sikorowski 1984a,b). Because they are common environmental microbes, all of these species can become contaminants of artificial diets and can therefore become pathogens that broadly affect colonies. The spread of these pathogens is facilitated by communal feeding such as is prescribed for the mass rearing of *Lygus hesperus* (Debolt and Patana 1985, Cohen 2000c) and for other insects such as various armyworm species and *Heliothis* spp. (Patana and McAda 1973). Also, pathogens may gain entry into a potential host via feeding contact with frass (fecal matter) from infected individuals or from consumption of infected cadavers (Tanada and Kaya 1993). The microsporidian protozoans have a spore portion of their life cycle where they are very resistant to environmental stresses such as high temperature, desiccation, or harsh chemicals. The spore stage is also very infectious and can readily be transferred into diets in any kind of communal feeding situation, or they can be transferred through contact with cage materials such as those found in many mass-rearing situations.

Clearly, a dilemma of insect diet specialists is that however effectively potential pathogens (and diet-destroying contaminants) are removed from diets or killed, there is always the possibility of post-sterilization infection of the diet. This possibility is the basis of strategies that are aimed at making the diets inhospitable to microbes, including the use of prophylactic chemicals that are intended to endure for the required life of the diet. These chemicals fall into the general category of preservatives, but they can be more specifically grouped as antimicrobial agents (antifungal, antiviral, antibacterial, or antiprotozoan chemicals), pH adjustors, and humectants (which adjust water activity). Many microbial problems with insect diets are shared by the food industry, and many remaining problems faced by diet specialists have already been researched extensively by the food microbiology community.

13.2.5 Damaging Effects of Contaminants That Are Not Pathogens

There is substantial information on adverse effects of microbial contaminants on boll weevils (Sikorowski and Lawrence 1994). These adverse effects are from contaminants that can survive internally in weevils but are not pathogens. Sikorowski and Lawrence describe the adverse effects of several bacterial and fungal contaminants, including reduction in pheromone production, a reduction in fatty acid composition of the insects' bodies, reduction in free amino acids in the hemolymph, damage to the gut, and adverse changes in the overall metabolic rate. The exact mechanism of the deleterious effects is not clear, but possibly these effects are from microbial toxin-induced damage to the gut of severely infected insects. This damage may prevent the digestion or absorption of nutrients (Sikorowski and Lawrence 1994). It is also possible that the microbes depleted the nutrients in the diet in support of their own accumulation of biomass. The deterioration of diet and the direct negative effects of microbial contaminants make insects generally more susceptible to other environmental stresses, including infection from true pathogens (Tanada and Kaya 1993). Factors that deplete nutrients add to the susceptibility of otherwise healthy insects, making them susceptible to a variety of environmental stresses.

13.3 CONTAMINATION: MICROBIOLOGY OF FOODS AND INSECT DIETS

As mentioned earlier, there is a useful literature on the microbiology of foods (Jay et al. 2005), and this body of knowledge is highly applicable to insect diets. Microbes are everywhere that other living organisms exist—in the air, in the water, in the soil, on and in plants, on and in insects, and on and in the people who take care of insects. The diet ingredients used to nurture insects, the containers where the insects are held, the implements such as paper towels, sponges, and all other materials associated with rearing contain microbes, often in very large numbers, such as on paper towels, unless rigorous efforts are made to sterilize these components (Sikorowski and Lawrence 1994).

13.3.1 How Microbial Contaminants Enter Diets

The major sources of diet contamination (Figure 13.5) are (1) the insectary workers, (2) the diet ingredients, (3) the processing equipment and containers, (4) the environment, and (5) the insects themselves (Sikorowski and Lawrence 1994). These authors explain how humans are the most important source of microbial contamination in insectaries. We carry various microbes on our skin, in our hair, our eyes, nasal passages, mouth, digestive tract, urogenital system, and on virtually everything we wear. This extends to our possessions—including decorations in the rearing facility. Things that we might surround ourselves with in a typical office setting can contaminate the facilities that we

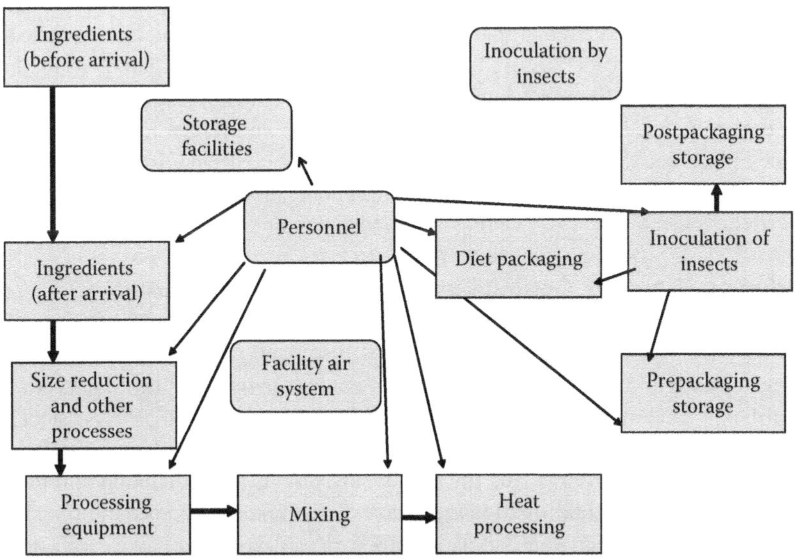

Figure 13.5 Possible sources of microbial contamination in rearing setting.

make efforts to keep microbe free. It is unsound laboratory practice to use expensive sterilization equipment and antimicrobial additives to reduce contamination risks and then circumvent these measures by introducing new sources of contaminants. This is a touchy subject and one that is difficult for insectary managers to deal with because enforcement of standard operating procedures (SOPs) regarding hygiene can make the enforcers appear to be insensitive to emotionally charged facets of their employee's lives. However, the guidelines and tenets that I offer here are at the core of the major means of protecting insect colonies from excessive risk of contamination—*basic sanitation and reasonable hygiene*. The issues of management/personnel interactions and the concept that the precautions that protect the insects also protect the workers are discussed in Chapter 14.

The most effective means of reducing contamination problems is through helping insectary workers to gain a clear understanding of basic sanitation and a plan of sanitation methodology. The only way sanitation SOPs work is when there is total cooperation of the insectary workers. This means that the first line of defense against microbial contamination is *education*. Information on managing contamination is already well established in the food processing community. The standards used by reputable food processing plants are a good model for insectary sanitation protocols. However, the standards for insectaries should be higher than those for human foods, because contamination problems in insect diets are amplified by several factors: (1) the diets are generally nutritionally complete both for the insects and inadvertently for the microbes, (2) the diets are to be held at room temperatures for extended periods of time—a circumstance not expected of human foods, and (3) because of the confinement inherent in rearing situations, the diet which we present to our captive insects affords no choice of alternative foods that may be present in nature. In contrast to human foods, the aliquot or ratio of insect diet that we offer is all that they have to utilize.

13.3.2 Insectary Workers as Sources of Contamination

The first and most delicate issue is the cleanliness of the workers that handle diets. The cleanliness of workers' and visitors' clothes, bodies, and especially their hands can influence the number of microbes introduced into the diet. Washing hands after restroom visits, trips outside of the clean area, and handling food (or cigarettes and makeup) are crucial. Because even clean clothes carry numerous microbes, protective clothing is strongly advisable, including lab coats, face masks, head wear (hats or hairnets), gloves, and even footwear (such as clean booties that fit over shoes). These outer garments must be provided as freshly laundered or disposable materials. While such protective garments will not be sterile, they are bound to be less contaminated than clothing that was worn in less controlled conditions than those of the insectary. Along with these precautions, it is essential that an action plan be established regarding departure from and return to clean rooms where diet handling takes place (NASA 1969, Sikorowski and Lawrence 1994). In summation, Sikorowski (1984b) observed, "After investigating many factors, I concluded that workers are the major source of microbes in the rearing facility as well as an important source of diet contamination."

13.3.3 Reducing Microbial Contaminants from Non-Diet Sources

The diet-based sources of contamination will be discussed elsewhere in this chapter, but here the environment-based and worker-based sources of contamination will be addressed. Sanitation is the most significant force in reducing contamination (Sikorowski 1984a,b). First, access to the key places in the insectary *must* be restricted to essential personnel. The key areas are the diet preparation room, the insect-holding rooms, and the sites where insects are inoculated into their containers and diets. Specific plans for contamination reduction should take into consideration any other place, specific to each rearing setting, where diets and their insect targets are processed or held. Except for obvious emergencies, only trained personnel should be allowed to enter sensitive rearing areas, and even visiting dignitaries should be either excluded or properly trained and attired as described later. The practice of allowing visitors that are not properly prepared (in terms of protective gear and sanitation practices) can become a serious breach of sanitation efforts.

Good sanitation practices call for thorough hand washing with soap and warm water before entering or reentering the insectary (after visits to restrooms, break areas, or other areas that are not held as clean areas). Good practices also include use of protective clothing, including hair covering, face masks, gloves, and coveralls, aprons, or lab coats—all of which are freshly laundered. Because insectary workers are being asked to wear protective clothing for protracted periods of time, insectary managers provide comfortable, as well as protective, equipment—including surgical face masks that fit the workers both protectively and comfortably. Work areas are scrubbed with appropriate antimicrobial cleaning agents, including ones that contain chorine, iodophors (iodine plus carriers), or quaternary ammonium compounds (Sikorowski and Lawrence 1994). Work areas should be as nonporous as possible so that cleaning agents can gain access to the cryptic microbes. Stainless steel and like materials that are widely used in food preparation facilities are ideal, especially when they are formed into configurations that are smooth and easily reached with cleaning agents.

This issue is also discussed in Chapter 14, which deals with the safety of personnel in concert with the well-being of the insects. It will be emphasized there, but it bears repeating that no amount of administrative and managerial effort will succeed in achieving the high levels of sanitation that are needed to overcome and prevent microbial contamination problems if the insect-rearing specialists are not educated in matters of how and why to take measures of quality sanitation practice and if those specialists are not committed to carrying out their duties in a way that is compliant with good laboratory practices. A positive and motivating force is that the basic sanitation procedures are at once not only beneficial to the insects that are being reared, but also congruent and completely compatible with safety practices that help keep the rearing specialists from biological or chemical hazards.

13.3.4 Diet Ingredients as Sources of Microbial Contamination

Most raw materials for diets, especially unprocessed foods, contain live microbes. In the dry form of these ingredients, the water activity is too low to support the rapid growth of their resident microbes. However, if these materials are incorporated into the diet without proper sterilization or other microbe-reducing procedures, they can serve as a major source of contamination once the diet is fully synthesized. Once most diets are synthesized, they are nutritionally complete with high enough water activity to make them ideal media for many contaminants. With some diets, it can be effective to filter sterilize such ingredients as vitamin solutions that are added after the other components are heat sterilized (e.g., Cohen 2000c). Liquid diets that are solutions (including most Homoptera diets) may be filtered through membranes with pore sizes of 0.45 or 0.23 μm. Such filtration removes all but the smallest viruses or prions, neither of which appears to be significant diet contaminants. However, diets with high sugar contents or with high percentages of dissolved nutrients will have high viscosity, and such diets require strong vacuum and sometimes multiple-step filtration (a progression through larger pore filters prior to ultrafiltration). After the removal of microbial contaminants, sterile techniques must be used in the completion of diet handling, because the reintroduction of contaminants defeats the purpose of sterilization.

However, most artificial diets for insects cannot be filter sterilized. This leaves such limited choices as chemical treatments or high-energy physical treatments. The chemical strategies are discussed elsewhere in the chapter, but they present several disadvantages that make the physical methods more appealing. The main physical method is heating. Heating to temperatures below 100°C are generally considered: either blanching or pasteurization. Neither of these methods kills all microbial contaminants; at best, they reduce the populations to manageable levels and often remove the most virulent pathogens. The purpose of using these relatively low-temperature methods is to reduce microbial populations without destroying either or both the nutrient value or the sensory qualities of delicate ingredients. For example, the addition of chicken eggs, sugar solution, and acetic acid as a heated mixture to the meat paste (Cohen and Smith 1998) is intended as a method of blanching the meat products to lower their microbial counts, without rigorous cooking of the meats. Previous studies showed that cooked meat products were unacceptable to the predaceous insects that were targets of mass-rearing programs (including *Chrysoperla rufilabris*, *Geocoris punctipes*, and several other predator species).

The steam kettle is a common method of microbe reduction in insect diets. Steam kettles employ high-temperature and high-pressure steam in the jacketed part of the kettle, and they operate by conduction of heat through stainless steel sides and into the diet. The heating process is inherently uneven with the diet closest to the sides being subject to very high temperatures and the inner parts of the diet being insulated from the heat by the diet itself (aqueous media being notoriously poor conductors of heat). To compensate for the poor heat transfer, steam kettles usually employ vigorous stirring to distribute the heat more evenly, but with viscous diets, such stirring still does not distribute the heat evenly and leaves pockets of poorly heated diet. Also, because the heating is done openly (i.e., not under pressures above one atmosphere), the higher-than-boiling temperatures needed to provide complete sterilization are not reached in steam kettles (in contrast to autoclaves, flash sterilizers, and extruders—all of which operate at pressures elevated above ambient).

Although flash sterilizers were first used for insect diets more than 25 years ago (Griffin et al. 1974), this technology has not been as widely used as the steam kettle, probably because of higher costs and maintenance expenses of flash sterilizers. However, coupling of the flash sterilizer with packaging equipment allows a superior basis for mass production of many kinds of insects (Griffin et al. 1974, Tillman et al. 1997). Also, flash sterilization allows production of a completely sterile, well-mixed diet whose components are heated to a point that is likely to detoxify undesirable components such as anti-digestive soy proteins and secondary compounds such as saponins (Fukushima 1991). Likewise, the use of twin screw extruders in the production of insect diets, though introduced more than 5 years ago (Edwards et al. 1996), has not yet caught on in many other rearing facilities; at least, use of this implement has not reached the literature as survey by Cohen (2001). This is unfortunate in light of the fact the extruder technology has been demonstrated over the past quarter century to be an excellent way of producing large-scale amounts of foods that are free of live microbes and yet whose sensory and nutritional qualities are superior to foods processed in other ways (Fellows 2000). This topic will be discussed in greater detail in Chapter 12, where the basics of extruders and other diet-processing equipment are discussed, and also in Chapter 16, where new directions in diet processing are recommended in conjunction to the food engineering background.

This knowledge base from food microbiology is applicable to insect diet technology for several reasons: (1) insect diets and foods of humans and livestock have many components in common; (2) the methods are applicable because many microbial taxa are common to both fields; (3) both groups face the problem of preserving nutritional quality and palatability while ameliorating contaminants; (4) food safety, food preservation, and microbe-based processing have been thoroughly researched because of their impact on human welfare; and (5) the knowledge base is very well regimented and highly reliable because of its long history (having gotten its start by the most renowned pioneer in food science, Louis Pasteur in 1870). Pasteur must be heralded as one of the most famous pioneers in insect pathology with his early work on pébrine, a protozoan disease of silk worms—research that saved the silk industry.

13.4 USING A MIXTURE OF TWO OR MORE KINDS OF PREVENTATIVE ACTIONS TO REDUCE MICROBIAL CONTAMINATION

The environmental and dietary conditions that contribute to contaminant proliferation are nutrient profiles, temperature, pH, reduction/oxidation potential, and the types and amounts of various antimicrobial compounds inherent in insect diets. For example, Hedin et al. (1978) pointed out that

Table 13.1 Effect of pH on *D* Values for Spores of *Clostridium botulinum* Heated to 115.6°C

pH	*D* Value for *C. botulinum* in Tomato Sauce
4.0	0.128
5.0	0.260
6.0	0.491
7.0	0.515

D values are presented in minutes.

many of the naturally occurring compounds, now known as plant secondary compounds, conferred natural antimicrobial qualities to the cotton plant parts eaten by a variety of insects. These compounds included terpenes, tannins, caryophyllene, and gossypol (Hedin et al. 1978).

Because of the multiple factors that make insect diets targets of microbial proliferation, it is reasonable to apply multiple strategies or an "integrative approach" to quality preservation of diets as has been shown to be useful in food preservation (Taoukis and Labuza 1996). Using extremes of pH, temperature, low water activity, high concentrations of antimicrobial substances, or oxidation–reduction potentials not only deter microbial growth, but could also degrade the palatability of the diet for the target insect. However, if a combination of milder conditions were applied, it is possible that a set of conditions that were unacceptable for the microbes but tolerable or even preferable for the insects could be found. This type of integrative approach has not been applied deliberately in any published studies of insect diets, but such an approach shows promise and may be a way of reducing reliance on chemical prophylactics (Cohen unpublished data). The integration of rational sanitation technology along with diet manipulation to reduce hospitality to microbes could revolutionize rearing practices by reducing costs of diets and greatly increasing their quality. An example of how pH and temperature interplay in microbial mortality is seen in Table 13.1.

13.5 COMMON CONTAMINANTS IN INSECTS, INSECT DIETS, AND REARING SETTINGS

A wide range of major microbial taxa can be contaminants in insect diets: viruses, bacteria, protozoa, and fungi. For example, some of the most common contaminants in boll weevil diets are shown in Table 13.2, and common microbes in a meat- and egg-based entomophage diet are shown in Table 13.3. Most commonly, environmental bacteria and fungi are the contaminants in insect diets. They opportunistically exploit the nutrient-rich diets and conditions which favor or at least allow their growth. Several kinds of bacteria are widely found in diets, including *Bacillus subtilis* and other species of *Bacillus*, *E. coli*, *Staphylococcus* spp., *Salmonella* spp., *Streptococcus* spp., *Lactobacillus* spp., *Leuconostoc* spp., and *Micrococcus* spp. (Sikorowski et al. 1980, Sikorowski 1984a). While some of these bacteria live only in what is called a vegetative phase, some such as the various species of *Bacillus* are spore-formers. This means that when using heat treatment to kill the bacteria (or other treatments as well), the vegetative bacterial cells will die readily, but the spores will remain dormant, but alive, until conditions are suitable for their reemergence. The implications are that mild heat treatment may be effective for the removal of some bacteria, but spore-formers require more extreme treatments. Several opportunistic human pathogens are found in insect diets (Sikorowski 1984a). For example, *Aspergillus* and *Mucor*, both molds, and the bacteria, *Pseudomonas* and *Streptococcus*, were commonly present in boll weevil diet (Sikorowski 1984b).

Like spore-forming bacteria, the fungi use spores as a means of reproduction as well as vegetative propagation. The vegetative phases of fungi are relatively susceptible to heat treatment, but the spores, according to Jay et al. (2005), are resistant to high temperatures and a variety of other environmentally harsh treatments (e.g., extreme pHs, salt concentrations, and desiccation).

Table 13.2 Bacteria and Fungi Commonly Found in Diet-Reared Boll Weevils and Their Diets

Microbe	Taxon
Micrococcus luteus	Bacteria
Staphylococcus aureus	Bacteria
Streptococcus spp.	Bacteria
Lactobacillus plantarum	Bacteria
Leuconostoc mesenteroides	Bacteria
Bacillus sphaericus	Bacteria
Pseudomonas aeruginosa	Bacteria
Enterobacter aerogenes	Bacteria
Corynebacterium humiferum	Bacteria
Serratia marcescens	Bacteria
Aspergillus niger	Fungi
A. flavus	Fungi
Rhizopus nigricans	Fungi
Cladosporium sp.	Fungi
Fusarium sp.	Fungi
Several unidentified yeast species	Fungi

Source: Sikorowski, P.P., Pathogens and microbial contaminants: Their occurrence and control, In *Boll Weevil Mass Rearing Technology* (P.P. Sikorowski, J.G. Griffin, J. Roberson, and O.H. Lindig, Eds.), pp. 115–169, University Press of Mississippi, Jackson, MS, 1984b.

In a study of the sources of microbial contamination and the effectiveness of several antimicrobial agents, it was shown that diets lacking antibacterial and antifungal agents supported a growing population of bacteria and fungi that reached an asymptote by 24 h after the onset of incubation at 27°C (Inglis and Cohen 2004). These authors showed that principal spoilage bacteria were species from the genera, *Carnobacterium* and *Lactobacillus* (Table 13.3). They also showed that the pH of the diet decreased progressively due to the activities of the bacteria and that the antimicrobial chemicals added to the diet in the original diet formulation were very effective in reducing the growth of the microbial populations associated with liver, ground beef, and eggs, the main source of contamination. Although the three major diet ingredients had substantial and diverse microbial flora, only a limited number of a few

Table 13.3 Bacteria and Fungi Found in a Meat- and Egg-Based Diet for Entomophages

Microbe	Taxon
Carnobacterium pisicola	Bacteria
C. divergens	Bacteria
Lactobacillus curvatus	Bacteria
L. sakei	Bacteria (eggs)
L. mesenteroides	Bacteria (eggs)
Enterococcus spp.	Bacteria (eggs)
Pseudomonas putida	Bacteria (ground beef)
E. aerogenes	Bacteria
C. humiferum	Bacteria
S. marcescens	Bacteria
Candida zeylanoides	Fungi (yeast)
Torulaspora globosa	Fungi (yeast)
Yarrowia lipolytica	Fungi (yeast)
Cladosporium sp.	Fungi
Fusarium sp.	Fungi (eggs)

Sources: Inglis, G.D. and Cohen, A.C., *J. Econ. Entomol.*, 97, 235, 2004; Cohen, A.C. and Smith, L., *Biol. Control*, 13, 49, 1998.

species consistently thrived in the completed diet. This type of accounting of the microbial ecology of diets deserves further attention in future studies of microbe/diet interactions, and it would be valuable to further include the insects in this complex scheme of factors that affect diets and diet quality.

13.6 OTHER TECHNIQUES USED TO REMOVE, REDUCE, OR AMELIORATE MICROBIAL CONTAMINANTS

13.6.1 Filtration

Diets or diet components that are solutions can be sterilized by filtering through a 0.45 or a 0.22 μm filter. Such filtration will remove most viruses and all larger microbes such as bacteria, fungi, and protozoa. The benefit of filtration is that heat is not required, so heat-labile components are protected. By filtering a diet, microbes are removed without recourse to chemical agents, which could be destructive to the insect. For small-scale diet production, filtration can be efficient and relatively inexpensive. Finally, filtration can be accomplished with relatively inexpensive equipment that requires little space. For example, a 2–4 L side-arm flask and its accompanying filtration system (minus the vacuum pump) have a footprint of less than 0.33 m × 0.33 m (Figure 13.6).

However, there are several limitations to filtration. It is useful only for diets that are presented as solutions. Such diets are generally limited to strictly liquid feeders such as xylem sap or phloem sap feeders (including many members of Homoptera such as cicadas, certain leafhoppers, aphids, and whiteflies). Because blood cells would not pass through a filter that is fine enough to remove microbes, it would not be applicable to suspension feeders such as blood consumers, including biting flies, bed bugs, and hematophagous reduviids. A further limitation of filtration is that very viscous liquids such as diets with over 5% sugar would create problems in generating an adequate suction to pull the diets through the filter. Once the diet is sterilized by filtration, it will be subject to recontamination by the insects that feed on the diet, as well as by environmental contaminants.

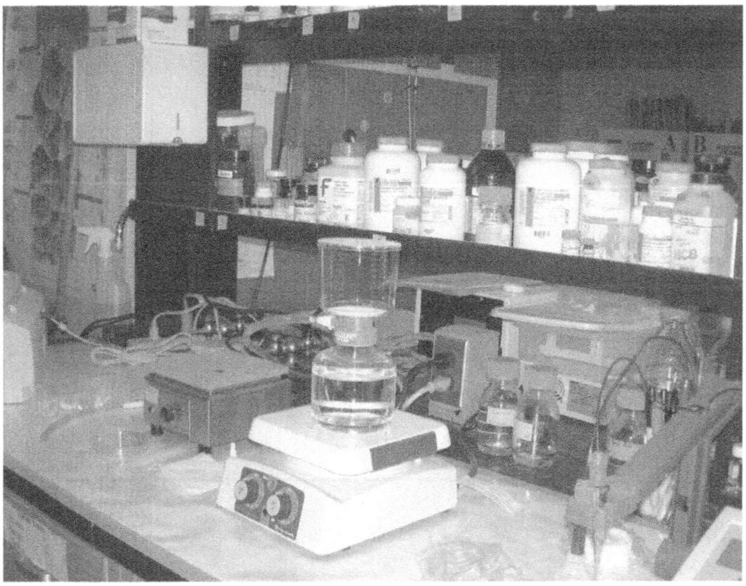

Figure 13.6 A 1 L disposable filtration system equipped with a 22 μm filter. The system requires either a vacuum pump or a "house vacuum" line.

The requirements of a filtration system are a reservoir source, receptacle, a filter holding component that is leak free, the filter itself, and a vacuum source. The receptacle vessel must be sterile, or the purpose of the filtration is defeated. This usually means autoclaving or some other means of sterilization. For larger levels of production, there are filtration systems designed to remove particulates from juices, beer, and wine. Such systems may be applicable to insect diets, especially if smaller contaminants such as viruses are not the major problem. The larger systems are discussed thoroughly by Fellows (2000).

13.6.2 Heating

The most widely used means of elimination or alleviation of microbial contamination in insect diets is heat processing. Heating is a means of preservation of the diet. Heating is also used to preserve diet components such as toasting of soy flour or wheat germ before storage. The effects of heat on diet components and processing, in terms of nutritional and biochemical characteristics, are discussed in Chapters 5 and 12, on diet chemistry and diet *processing* equipment, respectively. The range of elevated temperatures varies from mild blanching (50°C–100°C) to reduce mainly surface contaminants to temperatures that exceed the boiling point of water, often with pressure to increase the temperature further. The mildest heat treatment used with insect diets is blanching, and pasteurization is slightly more aggressive. Blanching was used in a diet for entomophagous insects—a mixture of blended eggs, sugar solution, and acetic acid heated to boiling to blanch a paste made of beef liver, ground beef, yeast, and preservatives (Cohen and Smith 1998, Cohen 1998c). The mixture reached 65°C–70°C upon first being mixed. It should be noted that this temperature range is within the range of temperature that is considered pasteurization (Jay et al. 2005), provided the temperature is held at this level for an appropriate period of time. For example, Jay et al. explain that if a food such as raw milk is held at 63°C for 30 min (considered LTLT conditions or "low temperature: low time"), at least 10^6 colony-forming units are killed. This is equivalent to holding the food material at 72°C for 15 s or at 100°C for 0.01 s. Although it is somewhat reassuring that heating can cause such impressive destruction of microbial contaminants, it must further be noted that this killing capacity of heat treatments does not include thermophilic organisms or spores. This is evident from Figure 13.7.

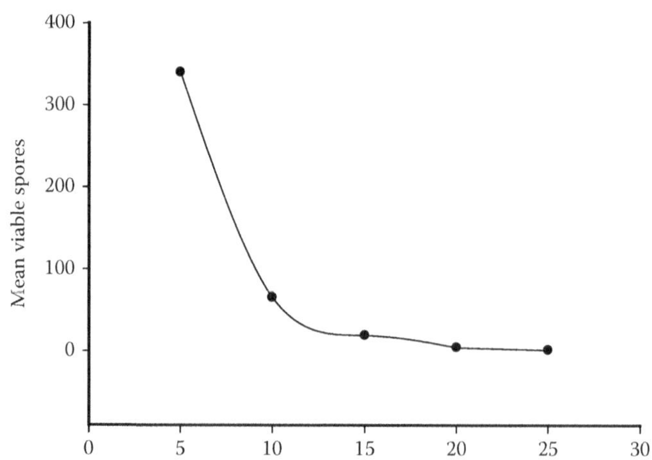

Figure 13.7 The viability of flat sour (fungal) spores in canned pea brine held for various time intervals at 116°C. (Data from Gillespy, T.G., *Recent Adv. Food Sci.*, 2, 93, 1962.)

The objective of this procedure was both to combine the lipoproteins in the eggs with the lipids of the meat products and to reduce the microbial count. It was discovered earlier (Cohen unpublished data) that the predators for which the diet was developed did not thrive on cooked meat products as well as they did on uncooked meats. Therefore, the blanching was considered a compromise to reduce microbe counts, though it would not sterilize the diet. Reductions of as much as 99% of the vegetative cells of bacteria have been recorded, but probably typical values are lower than this (Jay et al. 2005). The other advantages of blanching, other than microbial reduction, are discussed in the chapters on diet chemistry and diet processing.

13.6.3 Thermal Death Time and *D* Values

Two important concepts in communicating susceptibility of microorganisms to heat-caused destruction are the thermal death time (TDT) and the *D* value (Figure 13.7). The TDT is defined as the time (usually in minutes) required to kill a given number of organisms (Jay et al. 2005), and *D* is the amount of time required to kill 90% of the organisms present (Jay et al. 2005). Often *D* is determined for given organisms at 121°C, which is also the most typical temperature used in autoclaves. For food or diet processing, reaching temperatures in excess of 100°C generally requires an elevation of pressure above that of standard atmospheric pressure. Otherwise, the water boils away quickly from the substance being heated, and superheating temperatures cannot be reached. The superheating is essential to kill most of the spores in spore-forming bacteria (such as the bacilli and clostridia) and the various fungi. Some of the more heat-tolerant bacteria and fungi (thermophiles), even in their vegetative (active growing) stages, are resistant to high temperatures. Jay et al. (2005) list several species of bacteria that continue to grow at 73°C (~163°F). Many of the common contaminants may not grow at such high temperatures, but they can survive at 100°C only to grow prolifically at rearing room temperatures. This topic is covered further in Section 13.11, and in excellent detail in Singh and Heldman (2014).

13.6.4 Factors That Affect Thermal Tolerance (*D* and TDT Values)

A wide array of factors have influence on the thermal tolerance of microbes, including pH, fat content, protein content, water activity, salt content, oxidation–reduction potential, and a whole host of other factors relating to diet composition (Jay et al. 2005). These factors are important because they can subvert efforts to use heat to pasteurize or sterilize a diet and render ineffective what should be a reliable means of treating potential microbial contaminants or pathogens. For example, the thermal death as indicated by *D* value of the anaerobic bacterium that causes botulism (*C. botulinum*) was shown by Xezones and Hutchings (1965) to be strongly influenced by the pH of the food medium. This is shown in Table 13.1 from which it is evident that the amount of time required to kill the bacterial spores was about four times as great at pH 7.0 than it was at 4.0.

Similarly, the higher the fat or protein content, the higher the tolerance that microbes have to thermal death. Jay et al. (2005) point out that thermal tolerances are affected differently by different salts, with some salts affording protection of microbes and others reducing the thermal tolerances. As water activities or moisture contents are reduced, thermal tolerances increase. Also, increasing carbohydrate contents of foods (or diets) increases resistance to thermal death of most microbes (Jay et al. 2005). Because insect diets generally contain proteins, fats, and carbohydrates, they may inherently protect microbial contaminants from thermal destruction. This must be considered as yet another chaotic outcome because it is not intuitively expected that such protection from thermal death would be afforded by these diet components. Only the empirical experience such as that reported in food science literature brings this potential threat to our attention.

13.7 COLD TECHNIQUES

Cold techniques (refrigeration, freezing, and freeze-drying) are largely confined to preservation of diets or components to reduce microbial growth rates rather than to kill microbial contaminants. It is essential to realize that cold does not effectively kill microbes; it merely slows their growth or metabolic rate. In fact, microbes are generally so resistant to cold that freezing and freeze-drying are routinely used to store and preserve cultures. Even such extreme treatments as lyophilization to total dryness, storage in a −80°C freezer or holding in liquid nitrogen actually act to *preserve*, rather than destroy many microbes.

However, cold temperatures are used widely as preservative techniques for diets and diet ingredients (and even for the insects themselves, under special circumstances such as in storing and shipping). If prolonged storage is desired, it should be carefully planned and monitored because a number of microbe species are well adapted to thrive in even extreme cold conditions. Such organisms, called psychrophiles and psychrotrophs, can grow at temperatures between 0°C and 7°C—typical refrigerator values. There are a few species that can grow on foods in freezers at −18°C (Jay et al. 2005). However, the vast majority of microbes are prevented from growing at low temperature as is dictated by the relationship described as Q_{10}, as indicated by the following equation:

$$Q_{10} = \frac{\text{velocity of chemical reactions at a given temperature } \{T\} \pm 10°\text{C}}{\text{velocity of reaction at } T}$$

The Q_{10} of the metabolic reactions of most organisms is between 1.5 and 2.5. This means that moving diets from room temperature of approximately 22°C into a 2°C refrigerator will decrease the metabolic rate to less than one-fourth the rate characteristic of the higher temperature. The other side of the Q_{10} coin is that changes in diet storage (such as moving diets from 2°C–4°C refrigerator storage to 24°C–28°C rearing rooms) can cause increases in microbial metabolism by several fold, thus promoting microbial deterioration of diets and setting up opportunities for facultative pathogens to take over the host insects that we are trying to rear.

13.8 CHEMOTHERAPY AND CHEMICAL-BASED PROPHYLAXIS

Table 13.4 explains the terminology relating to concentrations of diet additives, and Table 13.5 lists the major antimicrobial agents that have been used in insect diets. Table 13.6 lists the levels of antimicrobial compounds determined to be safe for at least one species of insects, and Table 13.7 presents information on the effective concentrations of antimicrobial compounds. These lists indicate that there have been several antifungal agents used and several antibacterial agents but fewer antiprotozoan agents. There are several antimicrobial compounds that have been reported under several different names that all represent the same chemical. For example, methyl *p*-hydroxybenzoate has been called methyl paraben, Nipagin, methyl paracept, Tegosept M, and methyl chemosept. These ambiguities can cause confusion, but they can be dealt with by use of a good reference such

Table 13.4 Terminology Related to Concentrations

Verbal Expression	Equivalent Number of Parts
Parts per hundred (=1%)	1
Parts per thousand	10
Parts per ten thousand	100
Parts per hundred thousand	1,000
Parts per million	10,000

Table 13.5 Antimicrobial Agents Used in Insect Diets, Their Effective Concentrations, Effective pH Range, Target Microbes, and Reference

Antimicrobial Agent	Concentration Range (ppm)	Effective pH Range	Target Microbes	References	Comments
Formalin (=37% solution of formaldehyde gas in water)	~1000	Broad	General	Debolt (1982)	Toxic to humans
Benzoic acid (also in the forms sodium benzoate and potassium benzoate)	~2000	Acid	Fungi	Funke (1983)	Low human toxicity
Methyl paraben (Nipagin, methyl paracept, Tegosept M, methyl chemosept, methyl *p*-hydroxybenzoate)	~1000	Acid	Fungi	Funke (1983)	One of the earliest uses of this antimicrobial
Propyl paraben	~300	Acid	Fungi	Funke (1983)	Low human toxicity
Sorbic acid (also used in forms such as potassium sorbate and sodium sorbate)	~800	Acid	Fungi	Funke (1983)	Low human toxicity
Propionic acid	~800	Acid	Fungi	Funke (1983)	Low human toxicity
Fumadil B, Fumidil = Fumagillin (Merck)		Soluble in bases and organic solvents	Protozoa (*Nosema* spp.) + general	Budavari (1996)	Moderate mammalian toxicity
Nisin		Acid	Bacteria	Jay et al. (2005)	Low human toxicity
Streptomycin sulfate	~100	Acid	Bacteria	Hsiao and Hsiao (1974)	Used as antibiotic in humans
Chlortetracycline (aureomycin)	~200	Acid	Bacteria	Chang and Jensen (1972)	Used as antibiotic in humans
Benlate (Benomyl fungicide from duPont)	~50	Acid	Fungi	Hsiao and Hsiao (1974)	Moderate mammalian toxicity
Oxytetracycline (Oxacycline, Terraject, Terramycin)	~200	Acid	Bacteria	Chang and Jensen (1972)	Used as antibiotic in humans
o-Phenylphenol	~25	Acid	General antimicrobial	Singh (1977)	Toxicity similar to phenol
Kanamycin sulfate	~500	Acid	Bacteria	Singh and Howe (1971)	Used as antibiotic in humans

as the Merck Index, as indicated later. Unfortunately, several antibacterial and antifungal agents that show great promise in the food industry have not yet been introduced into the insect diet domain. These compounds have the very positive attributes of having low metazoan (multicellular animal) toxicity. These compounds are discussed by Jay et al. (2005) and include a group of antimicrobial agents called bacteriocins, which are produced by *Bacillus* spp. and some strains of *Lactococcus lactis*. These compounds include subtilin, tylosin, and nisin.

Of these compounds, nisin has gained the widest use as a general antimicrobial agent and may offer a great amount of promise toward antimicrobial therapy for insects. This generalization is made

Table 13.6 *Safe* Levels of Antimicrobial Compounds

Antimicrobial Compound (Merck Name)	Safe Level (ppm)	Toxic Level (ppm)	Antimicrobial Target
Aerosporin (polymyxin)	1,600	3,000	Bacteria
Albamycin (novobiocin)	400	1,000	Bacteria
Aureomycin (chlortetracycline)	600	1,500	Bacteria
Bacitracin	8,000	50,000	Bacteria
Bradosol (domiphen bromide)	2,000	4,500	Anti-infective—general antimicrobial
Chloromycetin (chloramphenicol)	600	2,500	Antibacterial and antirickettsial
Erythrocin (erythromycin)	800	4,000	Antibacterial
Ethanol	15,000	40,000	General antimicrobial
Formalin	4.32	15	General antimicrobial
Gantrisin (sulfisoxazole)	400	2,000	Antibacterial
Kantrex (kanamycin)	3,200	10,000	Antibacterial
Methyl-*p*-hydroxybenzoate (methyl paraben)	100	1,000	Antifungal (slightly antibacterial)
Mycifradin sulfate (neomycin)	1,000	5,000	Antibacterial
Penicillin G potassium	2,000	30,000	Antibacterial
Potassium sorbate	2,000	15,000	Antifungal (slightly antibacterial)
Sodium benzoate	1,000	10,000	Antifungal
Sodium propionate	4,000	15,000	Antifungal
Streptomycin sulfate	2,000	20,000	Antibacterial
Terramycin (oxytetracycline)	400	2,000	Antibacterial
Tetracyn (tetracycline)	400	3,000	Antibacterial
Vanocin (vanomycin)	4,000	8,000	Antibacterial

Sources: Adapted from Singh, P. and House, H.L., *J. Insect Physiol.* 16, 1769, 1970a; M. Windholz (Ed.), *Merck Index*, Merck and Co., Inc., Rahway, NJ, 1983.

Table 13.7 Effective Concentrations of Antifungal Agents

Antifungal Agent	Effective Concentration (ppm) at pH 3.0	Effective Concentration (ppm) at pH 5.0 {"Safe" Levels: "Toxic" Levels for *Agria affinis*}	Effective Concentration (ppm) at pH 7.0
Methyl paraben	800	1,000 {100: 1,000}	1500
Propyl paraben	200	300	500
Benzoic acid	400	2,000 {1,000: 10,000}	Not effective
Sorbic acid	400	800 {2,000: 15,000}	Not effective
Propionic acid	800	800 {4,000: 15,000}	Not effective

Sources: Adapted from Funke, B.R., *Bull. Entomol. Soc. Am.*, 29, 41, 1983; Singh, P. and House, H.L. , *Agria affinis. J. Insect Physiol.* 16, 1769, 1970a.

in light of the wide use of nisin in food preservation (Jay et al. 2005), making it probable that it will be safe for use with insects both in terms of low mammalian toxicity (for protection of insectary workers) and in terms of insect toxicity (for protection of the target insects). Other possible antimicrobial agents that have not gained entry into the insect diet regimen include monensin, which has been used in cattle as a probiotic, and natamycin, which is promising as an anti-yeast and general antifungal compound. With all these compounds, several caveats apply: (1) their recommended use is as an occasional therapeutic rather than as long-termed prophylactics (i.e., prolonged, resistance-inducing uses are to be avoided); (2) they should be tested to establish their nontoxic nature to each species of target insect on a case-by-case basis; (3) the economics of their use should be established; and (4) their effectiveness against the specific microbial organisms should be worked out, again, on a case-by-case basis.

It is evident from Table 13.5 that the effective pH range of most antifungal agents is in the acid range. In fact, Funke (1983) explained that it takes at least twice as much benzoic acid or sorbic acid to provide the same level of suppression of *Aspergillus niger* at pH 5 than it does at pH 3. At pH 7, both of these antifungal agents become completely ineffective. There are two reasons behind the effectiveness of these antifungal agents at low pH. First, the mechanism of the agents dictates that for most of the antimicrobial agents, the molecules must be present in their nondissociated form (which occurs only in the acid state), and second, the lower pH range is in itself unfavorable to most species of fungi (including *A. niger*) and bacteria (Jay et al. 2005).

Also, most of the antimicrobial compounds listed in Table 13.5 are insoluble in aqueous diets. Therefore, special precautions must be taken such as dissolving the antimicrobial agent first in a solvent such as ethanol and mixing the ethanol/antimicrobial solution with the remaining diet ingredients. Usually, procedures that call for this measure include a step (such as heating or bubbling an inert gas through the medium) to eliminate the solvent. As an example of the problems of solubility, if the acid form of sorbate is used in a diet (sorbic acid per se), less than 0.25 g will dissolve in 100 mL of water. However, if the potassium form of sorbate is used (potassium sorbate), as much as 58 g can be dissolved in 100 mL of water. Although one would never use a 58% solution of sorbate in a diet, it is possible that more than a 0.25% solution (~2000 ppm) may be desired. For example, as Funke (1983) points out, the working range of sorbic acid is greatly diminished as the pH increases, and if the target insects will not tolerate lower pH, it may become necessary to use a higher amount of sorbate to provide any degree of antifungal activity.

A review of the literature on diets reveals that the predominant antimicrobial agents are sorbic acid (or one of its salt forms such as potassium sorbate), methyl *p*-hydroxybenzoate (methyl paraben), streptomycin, and Aureomycin (chlortetracycline). In a random survey of 100 artificial diets, 58% incorporated methyl *p*-hydroxybenzoate, 43% incorporated sorbate compounds (sorbic acid, potassium sorbate, or sodium sorbate), 20% included streptomycin sulfate, and 24% included Aureomycin (chlortetracycline). An unfortunate and confusing problem that permeates diet literature is the multiple names for the same antibiotic compound. Consultation of the Merck Index can clarify some of the ambiguities. Some guidance as to how to use the Merck Index is provided here.

In an early study on rearing southern corn rootworms (Guss and Krysan 1973), it was reported that achromycin was to be used in the diet to prevent microbial growth. Although the discussions of compounds covered by the Merck Index are presented in alphabetical order, achromycin is not found in the general discussions under that heading. However, it is found in the index where there is reference to the compound number under which the primary name for the compound is listed (9021 in the Merck Index that I am using). Looking under the numerical listing, 9021, the primary listing "tetracycline" appears. Immediately after the name "tetracycline," is the formal chemical name, 4-(Dimethylamino)-1,4a,5,5a,6,11,12a-pentahydroxy-6-methyl-1,11-dioxo-2-naphthacenecarboxamide. Besides this huge formal name, there are 22 other names listed for this compound (including achromycin). The Merck Index listings also give the structure, empirical formula, references as to how it was prepared, patent information, and physical and chemical descriptions of its crystalline structure, boiling point and several other very useful pieces of information, such as its mammalian toxicity (LD_{50} of 807 mg/kg in rats). The solubility in various solvents is also provided. For example, 1.7 mg of tetracycline (achromysin) will dissolve in 1 ml of water at 28°C. Something else that is very useful to diet specialists is the explanation of uses of the listed substances. It is explained that tetracycline has used as an antiamebic, antibacterial, and antirickettsial agent. In light of the fact that most authors of diet papers do not explain the purpose of any of the ingredients in their diet formulations, it is helpful to have a source of information that gives a hint at the purpose of a given diet component.

Acetic acid, lactic acid, propionic acid, and sorbic acid all have antimicrobial properties both because they lower the pH of their surrounds to levels below optima (or tolerance) of most microbes and because of their direct effects on the metabolism of microbes (Jay et al. 2005). Therefore, adding any of these acids sets up conditions that are inhospitable to microbes and further improves the antimicrobial qualities of virtually all antimicrobial agents (Table 13.5). Because of their effectiveness as antimicrobial agents and as adjuncts to other antimicrobial efforts, these acids should be further considered for improving insect diets. It is a further advantage that these organic acids, especially acetic and lactic acids, are economical, and they add to desirable qualities of other food components such as detoxification of soy (and other) components (Fukushima 1991).

13.8.1 Quantity Equivalencies

Table 13.4 shows the equivalency of a 1% solution or mixture. For example, if a formulation calls for a 1% solution of potassium sorbate, here is the number of parts of the material in question in relationship to all other components. Using Table 13.4, one can quickly convert to deduce that a 10,000 ppm solution is equivalent to a 1% solution. Likewise, if one wishes to convert the 800 ppm that is the approximate working value for sorbic acid (as indicated by Table 13.7), it is equivalent to a 0.08% concentration of sorbic acid.

13.9 PHYSICAL/RADIATION TECHNIQUES

A method of preservation of foods against microbial contaminants that is gaining increasing use is that of employing various kinds of radiation (Fellows 2000, Jay et al. 2005). The use of radiation has not gained much attention in the insect diet community, but there are some promising aspects of these techniques that merit consideration. The types of electromagnetic energy that have been used in the preservation of foods are ultraviolet rays, beta rays (streams of emitted electrons such as those in cathode ray tubes), gamma rays emitted from the excited nucleus of elements such as cobalt 60 and cesium 137 (^{60}Co and ^{137}Cs), x-rays, and microwaves. Ultraviolet radiation is known to be an excellent antimicrobial agent, if it reaches the target microbes. However, this is a very nonpenetrating type of radiation and is useful only for surface sterilization. The other forms of radiation mentioned here are far more penetrating, a factor that poses a greater risk to personnel as well as conferring a great microbe-killing potential. The most frequently used types of radiation in food processing are gamma rays and microwaves. The latest inventions and improvements in the preservation and prevention of microbial contamination of foods are presented in excellent details in Singh and Heldman (2014). Although some of these innovations are presented in Chapters 12 and 16, there is far more valuable information in this source on food engineering than can be covered in the present text. However, it should be mention here that Singh and Heldman present information about the latest technology that prevents microbial contamination without adding harsh chemicals or treating foods (and potentially insect diets) with amounts and types of energy that destroy nutritional or sensory qualities of the foods under treatment. This concept is important in the next topic decontamination and deterioration of food (diet) quality.

13.10 DECONTAMINATION PROCEDURES CAN DETERIORATE DIET QUALITY

Scientifically based studies of the tolerance of various species of insects to a wide variety of antimicrobial agents are essential to the development of tools for insect-safe reduction of contaminants (Singh and House 1970a, Sikorowski et al. 1980, Alverson and Cohen 2002). The choice of antimicrobial chemicals can profoundly influence biological parameters that indicate insect quality. For example, the degree of damage that can result from using incompatible antimicrobial agents is dramatically illustrated in Figures 13.8 and 13.9. It is evident from all four parameters of the bioassays that methyl paraben was very destructive to *L. hesperus*. Next, it is also important to note that the formalin treatment was also less suitable than the other antifungal agents and that all the antifungal compounds were harsher to the *L. hesperus* population than the control, which contained no antifungal agent. It is noteworthy that methyl paraben is one of the most commonly used antifungal

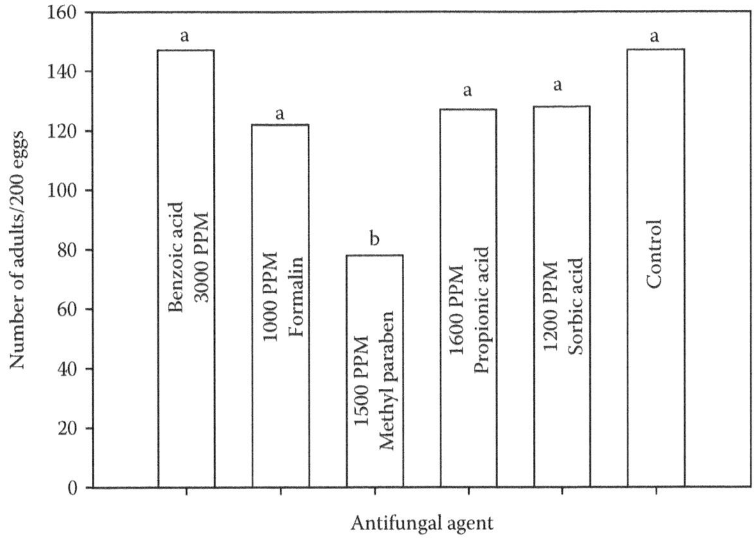

Figure 13.8 Mortality effects of adding several antifungal agents to the diet of *Lygus hesperus*. (Adapted from Alverson, J. and Cohen, A.C., *J. Econ. Entomol.*, 95, 256, 2002.)

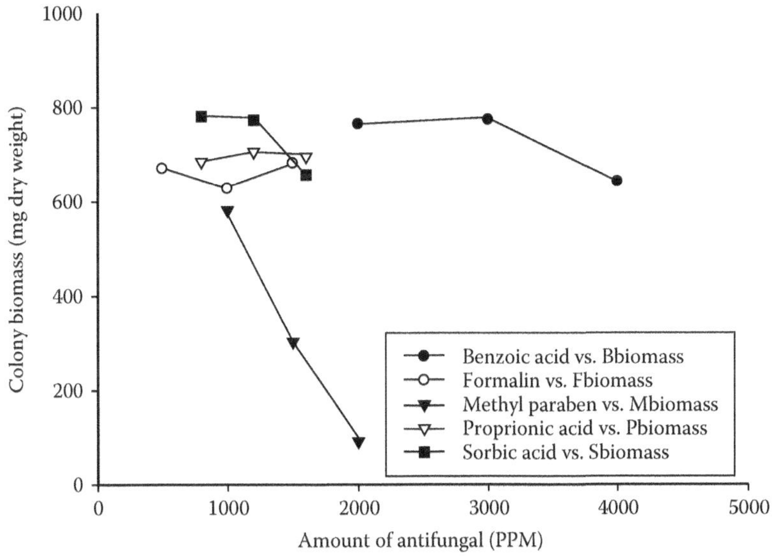

Figure 13.9 Dose–response curves of several kinds of antifungal agents in relation to colony biomass. (Adapted from Alverson, J. and Cohen, A.C., *J. Econ. Entomol.*, 95, 256, 2002.)

agents (as discussed elsewhere in this chapter) and that formalin is also widely used and has been the principal antifungal agent in rearing programs for *Lygus* species (Debolt 1982, Debolt and Patana 1985, Cohen 2000c) and was one of the antimicrobials throughout the history of the colony used as the source of subjects for the Alverson and Cohen (2002) study. It is also important to note that although some of the biological parameters measured (such as the number of survivors and biomass) are related and will inherently be correlated, other parameters are independent of one another statistically, and a correlation between them reinforces the interpretation that certain antifungal agents (methyl paraben and formalin) are physiologically very harsh to the subjects compared to other agents that are relatively benign (benzoic acid and sorbic acid). Figure 13.9 shows a dose–response curve of tarnished plant bugs fed diets with various concentrations of antifungal agents.

Another point that has not received adequate experimental analysis is the metabolic and physiological effects on our target insects of the various antimicrobial compounds routinely used in rearing systems. Zha and Cohen (2014) demonstrated that several aspects of what is conventionally called "nutritional ecology" (Chapter 9) can be strongly affected by some of the most commonly used antimicrobial agents, including methyl paraben, potassium sorbate, and sodium propionate. Using painted lady butterflies (*Vanessa cardui*) and tobacco budworms (*Heliothis virescens*) showed that 1,000 ppm had little adverse effect on nutritional profiles of both species, and 5,000 ppm had slight impact, but efforts to use 10,000 ppm of each antimicrobial agent had strongly negative impact on both species, especially the painted lady butterflies (Figure 13.10). These results of that study indicate that further attention is needed as to the effects of antimicrobial agents on several metabolic and nutritional ecology parameters, especially with experiments on the effects of these compounds on different species of insects. It is also worth further attention to determine the interactions between the insects' defensive systems and the chronic use of antimicrobial chemicals.

13.11 FINDING A SAFE MIDDLE GROUND: OPTIMIZING AND BALANCING MICROBIAL CONTAMINANT TREATMENTS WITH INSECT WELL-BEING

Sikorowski et al. (1980) tested 14 antibiotics against 6 species of bacteria commonly found in the boll weevils then reared at the Robert T. Gast Boll Weevil Rearing Laboratory (USDA, ARS) and found that medication of the substrate used to cover the diet (sand or corn cob grits) greatly reduced bacterial contamination of the weevils. They also found that the insects were not adversely affected by any of the antibiotics as determined by bioassays of pheromone production, number of adults to emerge, weight of adults, number of eggs per female, and egg hatch rate. They discussed the treatment of the surrounding non-diet medium as a shielding of the boll weevils rather than an inclusion of the antimicrobial agents within the diet. This procedure is certainly a more gentle way of medicating an infection than the direct application of the antibiotic into the diet. However, the authors warned that antibiotics should only be a temporary remedy under "...unusual circumstances when sanitary measures are temporarily interrupted." They warn against the prolonged use of antibiotics as a measure that ensures that only resistant cells will survive and ultimately become the dominant population of microbes. This is a warning that Sikorowski has issued over his several decades of contributions to insect-rearing science and technology, and careful attention to these caveats would improve most rearing systems. In virtually all of his works, Sikorowski emphasized the need for meticulous sanitation measures and careful education of insectary workers to give them the tools and motivation to carry out these measures. Such sanitation is a much more reasonable approach than continuous recourse to antimicrobial chemicals.

One of the most comprehensive experimental treatments of this issue was reported by Singh and House (1970a,b), who tested 21 antimicrobial compounds, using the sarcophagid fly, *Agria affinis*, as a target species. They divided the concentrations that they tested into four categories or ranges including "safe, primary inhibition, secondary inhibition, and toxic." The authors took into account the rate of growth and the mortality of *A. affinis* larvae reared on a chemically defined diet with five concentrations of each antimicrobial compound added to diets that were otherwise free of antimicrobial substances. The results of the study, reported in Table 13.6, show that within a given species of insect, there is a tremendous range of response to different kinds of antibiotics. This is similar to the results of the experiments on responses of *L. hesperus* to antifungal agents (Alverson and Cohen 2002), but interestingly, the responses of the two species are dramatically different with respect to the degree of sensitivity to methyl paraben.

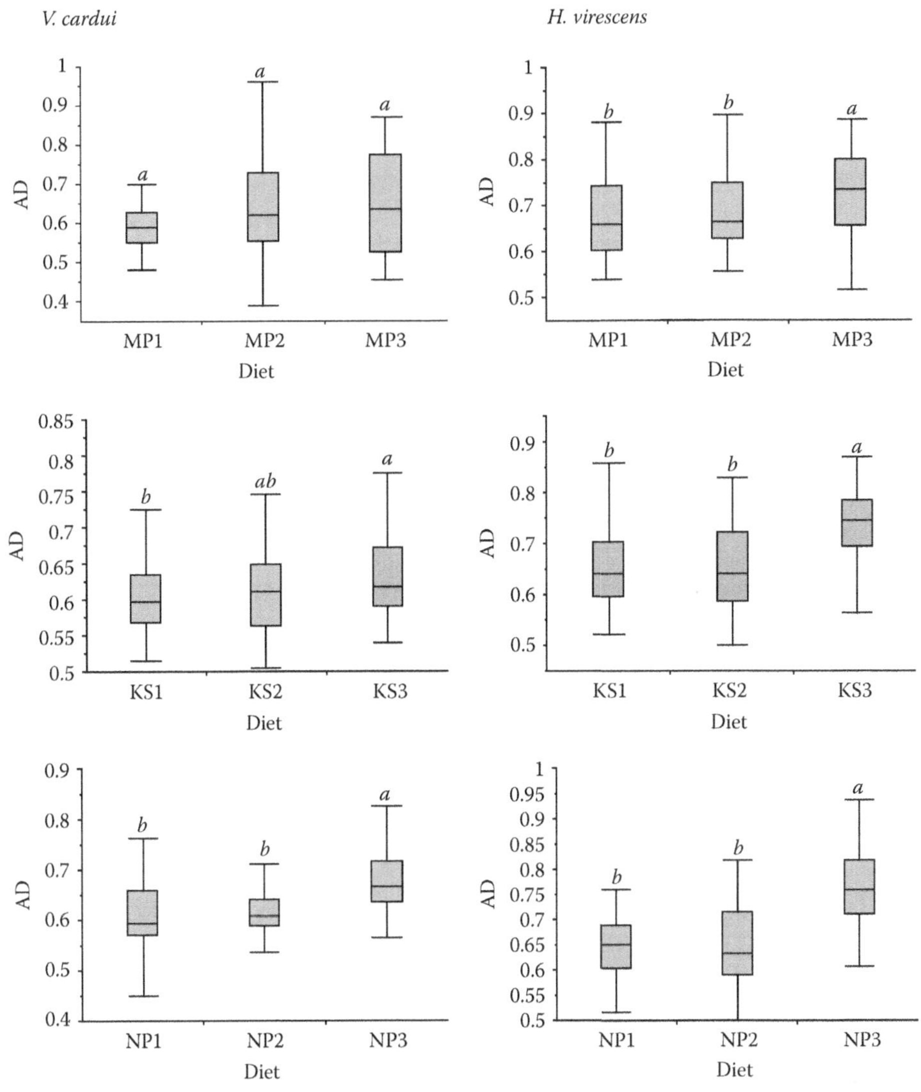

AD (approximate digestibility) = $\frac{E - F}{E}$

where

E is the amount of food ingested

F is the amount of frass produced, with both referring to dry weights

Figure 13.10 Effects of different concentrations of three antifungal agents (MP = methyl paraben, KS = potassium sorbate, NP = sodium propionate; 1, 2, 3 = 1,000, 5,000, 10,000 ppm) on approximate digestibility (AD) of *V. cardui* and *H. virescens*. Different letters indicate significant differences across antifungal doses (Tukey's test, $P < 0.05$). (From Zha, C. and Cohen, A.C., *Entomol. Ornithol. Herpetol.*, 3, 120, 2013, doi: 10.4172/2161-0983.1000120.)

Table 13.7 shows the levels of antifungal agents that are effective against one of the most common fungal contaminants, *Aspergillus niger*. The effective level for control of *Asp. niger*, according to Funke (1983), is 1000 ppm at pH 5.0, but the "toxic" level of this antifungal agent is 1000 for *A. affinis* reared at pH 5.8. Therefore, extrapolating that the effective level at a diet pH of 5.8 would be something higher than 1000 and lower than 1500 ppm; in order to gain control of *Asp. niger*, the concentration of methyl paraben, one of the most commonly used antifungal agents, would have to be above the toxic threshold of *A. affinis*. It is also evident from Table 13.6 that benzoic acid, sorbic acid, and propionic acid are effective against *Asp. niger* at "safe" levels for the representative species, *A. affinis*; these antifungal agents start losing their effectiveness above pHs of more than 5.0, and they become completely ineffective at pH 7.0 (Funke 1983).

This discussion leads to the balancing act that insect diet specialists must perform in order to keep microbial populations minimized while providing nourishing diets and hospitable conditions (temperatures above room temperature, high humidity) for the insects. A governing rule of living systems is that conditions that are optimum for one group of organisms are frequently optimal for other groups. If a diet is nutritionally complete for the insects' needs, it is inherently complete enough for microbes' needs. Many of the "environmental" microbes that are commonly found in foods, people's bodies, the air, containers, and water are less fastidious than the target insects, and they will grow readily on the diets (or on and in the insects themselves). Likewise, the microbes generally do not do well with low pH, low water activity, and low temperature, but neither do the insects. Those who design insect diets are charged with finding a balance of conditions such as low enough pH to decrease microbial activity but high enough to meet the behavioral or physiological needs of the insects. Likewise, if a target insect is found to accept a water activity lower than 0.950, many of the opportunistic microbes will be ruled out, and a water activity of 0.900 would rule out most microbes as potential contaminants. If the insects are found to tolerate a pH of less than 5.0 and a water activity of less than 0.900, this combination would tremendously increase the stability and microbial resistance of the diet. A further opportunity to reduce microbial contamination in a way that is safe to insects is to take advantage of the many "natural" antimicrobial substances such as caryophyllene, gallic acid, gossypol, and tannins that are potentially very useful in reducing bacterial or fungal populations in diets (Sikorowski 1984a).

13.12 FUTURE PROSPECTS IN THE MICROBIOLOGY OF INSECT DIETS: PROBIOTICS, PREBIOTICS, AND NOVEL ANTIMICROBIALS

One of the most exciting and innovative fields to emerge in the past few years is a discipline that deals with chemicals called probiotics. The probiotics concept has roots both in modern microbiology and in folk medicine and other ancient practices that emerged prior to our having a thorough understanding of microbes and their beneficial and destructive potential. Probiotics deals with means of using beneficial microbes or chemicals that encourage beneficial microbes (called *prebiotics*) to perform their ecological roles in a manner that is beneficial to people and their domestic animals and plants. Beale (2002) points out that current attention by researchers in probiotics is in developing treatments for ear, intestinal, urinary tract infections, and reduction of blood cholesterol, cancer risks, and even tooth decay—through use of "friendly" microbes. The general strategy of probiotics is to introduce beneficial microbes that outcompete their harmful counterparts or to modify the target environment so that existing beneficial species are favored to increase their numbers and their beneficial effects in localized regions of the target organism's body. So, for example, members of the bacterial group *Lactobacillus* and its relatives (the group involved in milk fermentation that results in yogurt production and various other dairy products) lower the pH of their environment, making that setting unsuitable for other microbes, including pathogens. It has long been understood that foods are protected by such processes, and to a great extent, the microbial-based fermentation industry is based on this phenomenon as exemplified by the low spoilage potential for foods such as pickles, sauerkraut, and other strongly acidified foods (which share the qualities of pleasing taste and resistance to colonization of spoilage microbes). Researchers in the field of probiotics point out that in humans and their domestic vertebrates, there is an intricate interplay and a scarcely understood relationship between the hundreds of species of symbiotic microbes and their vertebrate hosts. As the intricacies of these interactions become better understood, it becomes growingly clear that the healthy, homeostatic condition of the host vertebrate is profoundly affected by the microbial "guests."

With a few notable exceptions such as some aphids, termites, and cockroaches, even less is understood about the normal healthy relationships between insect hosts and their diet-borne symbiotic microbes. This topic is discussed in depth elsewhere in this chapter in the section titled, "Symbiotic Interactions between Insects and Microbes in the Rearing Setting," but it should be pointed out here that there has been far too little reported research on the potential beneficial interactions between insects and the microbes via the diet, that is, the probiotic associations that can benefit target insects and become economically sound environmentally friendly improvements in diet technology. Research in the area of diet improvements via probiotic and prebiotic strategies seems to be one of the most promising potential innovations in insect diet science and technology for the near future.

13.13 STUDIES OF BIOFILMS

Another "hot topic" in the microbiology community and in food science microbiology, specifically, is the subject of biofilms and how they pertain to microbial contamination (Jay et al. 2005). Biofilms are aggregates of microbes that occur in a huge variety of settings from dental plaque and gastrointestinal tract microbes to slime layers on rocks in streams and on surfaces of food (and insect diet processing plants). Generally, biofilms are communities of mixed species of bacteria, fungi, and sometimes algae. These communities form complexes that secrete a highly resistant extracellular matrix around themselves and the communally secreted digestive enzymes that break down foods whose nutrients are absorbed by the members of the community. Because of the resistance of the extracellular matrix, mechanical or chemical disruption of the films is difficult, and they are difficult to kill. The degree to which microbial contaminants in insect diets exist as biofilms has yet to be evaluated, but observations of diets or diet ingredients that have been stored for any length of time would suggest that the hard-to-kill microbes exist in biofilms and that is a factor in the failures of antimicrobial agents to prevent diet contamination. Future studies of this subject may provide profitable information to help develop a more scientific basis for the control of microbes as diet contaminants.

13.14 INTEGRATION OF FOOD INDUSTRY SANITATION WITH INSECT DIET PRODUCTION

A vast improvement in insect diet quality and offsetting of diet deterioration would come from wholesale adoption of the standards and practices of the food science and food technology community. The standards of personnel hygiene, attire, protective equipment, and conduct (restrictions from eating, smoking, applying makeup, permitting unauthorized visitors into restricted areas) are well defined and well established in the food handling community. Once insect diet specialists come to terms with the realization that the standards of the food community should be held as *minimal* for the insect diet community, there will be a vast improvement in sanitation and in the diet conditions that had previously suffered from lax sanitation conditions.

It is not suggested that the well-being of a target insect is more important than the well-being of a person! The point that has been emphasized and reemphasized in this book is that the expectations that are held for insect diets are far more exacting than those of an aliquot of human food. The typical lifetime aliquot of insect diet for a given target insect is about equal to one or two bites or mouthfuls of human food (about 50–100 g). If a person ingests a few thousand nonpathogenic microbes or a food whose nutritional content has been deteriorated by microbes, the person in question will probably never miss the nutrients or know the difference because he/she will have the opportunity to eat many, many other such mouthfuls of food, some of which will have nutrients that make up for the missing ones from the meal in question. However, if the diet aliquot for our target insect is contaminated, and if the diet is held for several weeks at rearing room temperatures, there will be ample opportunity for the microbes to proliferate and completely overcome the diet with growth, removing nutrients, introducing toxins, and causing a complete breakdown of palatability and nutritional quality. The insect in question under these circumstances does not stand a chance! The insect is being held captive, and it is a victim of its limited circumstances. The same issues apply to the other forms of diet deterioration discussed in other chapters, desiccation, oxidation, enzymatic deterioration, light-driven vitamin breakdown, and all other processes that erode diet quality. All of these forces can conspire to ruin the chances of our target insect's developing into a healthy product that will do the job that it is the purpose of the rearing facility.

CHAPTER 14

Safety and Good Insectary Practices

14.1 INTRODUCTION: SAFETY AND GOOD INSECTARY PRACTICES ARE COMPLETELY CONGRUENT

There are two sides to safety and good practices in insectaries: the protection of the personnel and the protection of the insects whose high quality and economic production are the main purposes of the insectary's mission. It is the responsibility of the insectary's management staff to assure that its workers are protected from threats to their health and well-being in their workplace, and it is further the responsibility of the workers themselves to protect themselves and their fellow workers in the execution of their duties. It is, further, the responsibility of all insectary personnel to carry out their duties to the best of their abilities to strive to produce a healthy population of insects that are of the highest possible quality. Fortunately, these two charges are mutually inclusive. What emerges as a felicitous bottom line is that the precautions that are good for human well-being in insectaries is also good for the insects. The simple truth behind this issue is that good understanding of the whys and wherefores of their jobs (i.e., education) makes insectary workers more efficient, capable, and safer employees, and their products profit from their more enlightened approach to doing their jobs.

Another fortunate reality about insectary safety is that the production and handling of artificial diets are not among the most hazardous practices. Most of the ingredients in diets are nutritional chemicals or complexes that are inherently nontoxic, and most equipment routinely used in insect diet production have been built with users' safe use in mind. This does not mean that there are no hazards in insect diets or other insectary operations, one of the most notable examples being the tragic death of boll weevil diet expert Robert T. Gast. The untimely death of Dr. Gast came as a result of an accident that occurred in conjunction with the use of an autoclave and organic solvent-based cotton boll extracts. In retrospect, it is now clear that the use of a highly flammable solvent and the high temperatures of an autoclave are not compatible. Even today, acetone, diethyl ether, ethanol, hexane, and other organic solvents are routinely used as solvents to introduce lipids and lipophilic extracts into diet mixtures (as discussed in Chapters 3 and 5), and generally, the solvents are removed by bubbling an inert gas through the diet mixture or by allowing the solvents to evaporate from diets exposed to air currents. Removal of organic vapors is strictly to be confined to a properly functioning fume hood. The other hazards found in insectaries fit these categories: chemical, mechanical, microbial (or generally biological), and electrical.

14.2 CHEMICAL HAZARDS

A responsibly managed rearing facility (or laboratory) will keep on hand a complete list of chemicals and their material safety data sheets. These sheets, now widely found in computer formats, contain the type of risk involved with various chemicals, the recommended protective clothing and protective equipment that should be used in conjunction with the potentially dangerous materials, and some guidelines on how to dispose of them. In larger rearing facilities and laboratories, regular safety orientation programs are given through a professional safety officer, but in smaller programs, safety matters become the responsibility of a limited circle of personnel, and some books, videos, and judicious use of websites can be informative and a good step toward responsible building of a chemical safety program.

The chemical hazards, in general, can be toxins, allergens, or dust hazards. Most substances used in diets are used also in human foods, so they are inherently nontoxic or only slightly toxic. One of the most toxic substances still in fairly common use in insect diets is formalin. Formalin (used to prevent microbial contamination) is volatile, a contact irritant, and a poison (with a toxicity indicated by its LD_{50} of 0.80 g/kg in rats, according to Merck). An important reason for using high levels of caution with formalin (or formaldehyde) is that it is now known to be a carcinogen. The precautions that are recommended for use with formalin can be applied to other volatile or contact toxins. First, the protective wear that should be used when handling formalin and other potentially toxic substances include chemical-proof gloves, coveralls or lab coats, safety eyewear, and chemical-proof aprons. The formalin should be measured and added to diets within a certified fume hood so that vapors are not breathed. Chemical spill kits should be kept handy with personnel trained to use the kits in the event of a spill. Excess formalin should be discarded in appropriate waste containers, and the glassware used to handle this toxin should be rinsed properly under the hood so that danger is avoided from fumes or contact in the open diet preparation area (outside of the fume hood). It should be noted that once the formalin has been added to the diet and the diet has been processed and provided to the insects, it still contains the traces of this toxic material (intended to be retained as a means of reducing microbial growth). The general thinking about this situation is that the evaporation of such small volumes of formalin over long periods of time and with dilution into large volumes of air in the insectary render the risks of inhalation as minimal. Every laboratory or mass-rearing facility that incorporates the hazardous materials described here should have posted protocols for safe handling and disposal, as well as clearly described instructions as to how to treat accidents or spills. Such instructions should be not only made available on a passive basis, but the protocols should be part of an ongoing teaching process that is periodically revisited by all staff members who may use or be exposed to the chemicals. Education, training, and motivation are the most important tools for achieving a safe working environment.

As with formalin, other volatile or dust hazards should be used in a certified fume hood. It must be emphasized that a fume hood is not the same thing and does not function the same way as does a laminar flow hood. Fume hoods draw air from outside the hood and force the air out of the laboratory or diet facility, venting the contaminated air into the atmosphere outside of the building. Fume hoods create a negative pressure directed so that fumes and dusts cannot return into the laboratory or diet room where workers can be exposed to airborne hazards. In contrast, laminar flow hoods operate with a positive pressure, with layered (laminated) air that is pre-filtered and is then passed over the work surface in the hood so that the work area is bathed in clean air that returns to the laboratory. A helpful way of thinking about these two kinds of hoods is that the fume hood is strictly for protection of the personnel to prevent toxic substances from entering the laboratory or diet room, and laminar flow hoods protect the materials on the work surfaces from airborne contaminants. A fume hood and a laminar flow hood are shown in Figures 14.1 and 14.2.

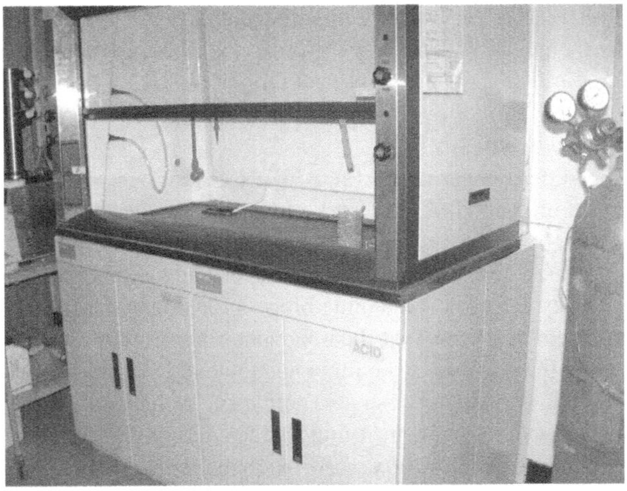

Figure 14.1 Fume hood. Note that the fume hood must have its sash (window) lowered to a position where it is both protective to workers and airflow is optimal.

Figure 14.2 Laminar flow hood. The laminar flow cabinet is designed to keep work sites scrubbed by filtered air, but it does not afford safety from fumes or dust.

It should be noted that the fume hood (Figure 14.1) has a sash that must be lowered to a prescribed point so that its draw of air is adequate to remove the fumes from the hood's work area and to further afford mechanical protection to the user's face and upper body in case of an explosion or fire. Most fume hoods are equipped with electrical outlets that allow mixers, hot plates, balances, and other diet preparation equipment to be used within the confines and protection of the hood's work surface. However, it should be recognized that any electrical implement that may cause an arc or sparks should not be used in conjunction with flammable solvents. It should be understood that working in the hood does not eliminate the very high explosion or fire risk associated with mixing electricity or high temperatures and flammable solvents. In contrast with the fume hood, the laminar flow hood (Figure 14.2) has no sash and instead has a completely open face where the positive flow of air (opposite of the negative flow in the working area of the fume hood) can keep the work area scrubbed with filtered air, preventing contaminated air from outside the hood from reaching the work surface and the materials that are to be protected from contamination. *It is an absolutely imperative rule of thumb that the fume hood is to protect personnel from airborne contaminants, and the laminar flow hood is to protect the materials (diets and diet components) from contamination, mainly microbial contaminants.*

It should be realized that proper use of fume hoods includes careful planning and compliance with several protocols. Fume hoods do not work properly if the sash (window that can be elevated or lowered to close the hood) is not in a proper position at about half-mast. Also, if the hood is cluttered with bottles, hot plates, and assorted other materials, there will not be proper air flow to safely remove the toxins or dusts. Part of the planning process for safe and efficient use of the hood is to have measuring equipment nearby so that the toxic materials are not carried for long distances in the laboratory, defeating the purpose of the fume hood. Good laboratory practices also include the installation of a safety shower and eyewash near to the site of hazardous materials usage. Finally, with the use of the hoods and other safe measures, the laboratory or diet room workers should wear the proper protective clothing. An insectary worker with recommended protective clothing is shown in Figure 14.3.

Besides formalin, other commonly used chemicals used in insect diets include phosphoric acid, formic acid, hydrochloric acid, acetic acid, sodium hydroxide, potassium hydroxide, and various organic solvents, all of which should be used with precautions to prevent volatile fumes from reaching laboratory personnel (i.e., in a fume hood or with some other kind of suitable venting system). Besides volatile chemical hazards, there are other substances that can be breathing hazards or otherwise potential contaminants of the laboratory or rearing facility's air. The substances are the fine powders that can be energized to become airborne where they can be inhaled, ingested, or eye irritants. Not only are primary potential hazards such as powdered antibiotics to be treated with proper care, but otherwise nontoxic materials such as flours and fine meals can become dust hazards. It is wise to use prudent measures of dealing with dusts, including use of proper ventilation, use of dust masks, proper eyewear, gloves, and garments such as aprons, lab coats, or coveralls.

The protective clothing described earlier is part of the precautions that good laboratory practices dictate. These protective devices should be kept clean, preferably with disposable gloves and dust masks (or other appropriate situation-suited breathing protection) and with freshly laundered

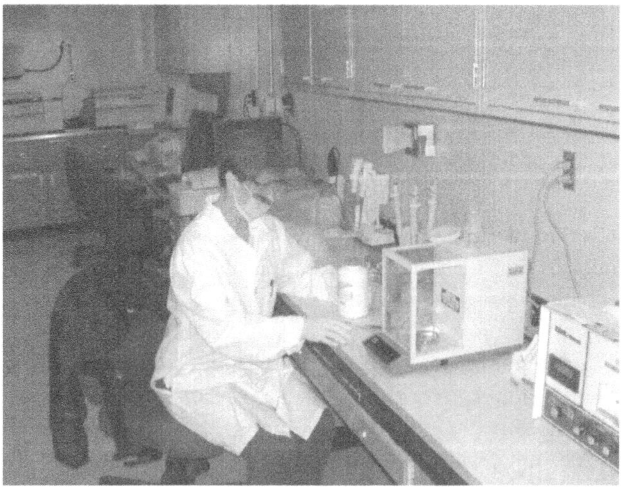

Figure 14.3 An insectary worker with recommended protective clothing. Note that the worker is wearing a lab coat, gloves, goggles, and a surgical-style dust mask that is fitted properly and contains a partial eye guard.

or disposable garments. There is a reciprocal benefit to the use of the proper protective equipment. The people are protected from the hazards, and the insects are protected from the contamination introduced by the personnel. This subject is covered in more detail in Chapter 13.

14.3 PROPER STORAGE AND DISPOSAL OF POTENTIALLY HAZARDOUS CHEMICALS

Although most materials used in diet-making processes are foods and are therefore inherently nonhazardous and do not require special hazard-based precautions, there are several materials that require special care. The classes of hazardous compounds commonly used in diets or diet-producing functions include the following: organic solvents, acids, bases, disinfectants, and antibiotics.

Alcohols are the most common organic solvents used in or with diets, mainly ethanol (ethyl alcohol) and to a lesser extent isopropanol (isopropyl alcohol). These alcohols are often stored in large volumes; therefore, their flammability makes them a considerable fire hazard, demanding the use of tightly sealable containers preferably metal drums, if more than 2–4 L are to be stored. The alcohols should be stored in cool, well-ventilated facilities, and it would be prudent to have a special storage room for these substances. Such a room should be secure and accessible only to authorized personnel who are educated in safe handling of hazardous materials. The storage shelves on which these and other solvents should be stored are to be located in lower portions of the facility so that risks of containers falling from excessive heights are bypassed. Other solvents that are sometimes used are hexanes, pentanes, and ethers. Similar precautions to those suggested for the alcohols should be incorporated for these solvents. Ethers are extremely volatile, especially diethyl ether, which is notorious for forming explosive peroxides after prolonged storage. Ethers and plant (or other diet material)-based extracts made from ethers should be avoided, but if they must be used in a facility, they should be carefully dated, stored in a cool place, preferably in a metal can designed for diethyl ether, and the material should be disposed of within a tightly monitored rotation schedule. All that is said about virgin (unused) materials pertains to the precautions that should be taken with wastes containing these solvents. Diligent and proper labeling of parent or virgin substances and wastes that are to be stored and disposed of are prudent practices.

Acids (e.g., propionic acid, lactic acid, formic acid, hydrochloric acid, acetic acid, phosphoric acid) and bases (e.g., sodium hydroxide, potassium hydroxide) can be corrosive and can promote a rapid and complete deterioration of their container. They should therefore be stored in the containers in which they were shipped, and the container should be monitored at regular (at least once per 3 months) for signs of deterioration. Acids and bases should be stored in appropriate metal cabinets that are marked as to their contents. Because acids and bases can react with one another in sometimes violent chemical reactions, they should not be stored together so that spills of acids and bases will not lead to a disastrous mixing. Example of the proper kinds of chemical storage facilities and cabinets for acids and bases are shown in Figure 14.4.

Antimicrobial agents, including bactericidal chemical and fungicides, should be stored in a place where they will not be contacted by insectary workers. This includes prevention of skin contact, inhalation, or contamination of foods. In this vein, antimicrobial agents should be treated with respect as potential toxins and also with their potential to promote resistance to these agents by microbial species. One of the most important risks connected to antibiotics is the generation or enhancement of development of resistance by microbes that are or can become human pathogens. Also Levine (1973), Sikorowski et al. (1980), and Sikorowski and Lawrence (1994) point out that exposure to antibiotic compounds, even in the low concentrations characteristic of those in insect diets, is not without risk. Therefore, any stream of prolonged exposure of antibiotics to microbes should be avoided. This means that workers should take special care to wear face masks, gloves, and outer garments to protect their skin and clothing when handling these agents. Cleaning agents such as sodium hypochlorite

Figure 14.4 Example of the proper kinds of chemical storage cabinets for acids and bases.

(household bleach) and other agents that contain ammonium compounds, soaps, detergents, and other disinfectants should be stored where they will not be exposed to excess heat or mix with diet ingredients or other laboratory chemicals. Sodium hypochlorite is a strong oxidizing agent, and while it is an excellent disinfectant and cleaning agent, it should be handled and stored with care.

14.4 MICROBIAL HAZARDS AND OTHER BIOLOGICAL HAZARDS

The biological hazards in rearing facilities tend to be minor in most facilities, but there are some risks that require attention and a respect for prudent practice. First, it should be realized that the antibiotics discussed in Chapter 13 are not very toxic in themselves but can have long-term risks if used improperly in situations where they can contribute to development of resistance in common microbes. The same kind of cautions that the medical and veterinary communities advocate are reasonably applied to insect rearing. If antibiotics are used without careful discrimination, caution, and planning, they can help in the inadvertent selection of resistant microbes that can later become environmental and medical hazards.

On a more short-term basis, the most immediate risks to insectary workers are microorganisms that are inadvertently colonized on media intended for insects, especially bacteria and fungi. As was discussed in Chapter 13, there are several species of microbes that are common contaminants in diets, insects, and other rearing facility locations. The most direct and potent defenses against these microbes are good sanitation and using protective tools in a well-planned, systematic, and diligent way. Cleanliness in the workplace and personal hygiene are among the most important protections against microbial risks.

Within the domain of personal hygiene are frequent and systematic hand washing with warm water and soap after removal of safety equipment. Hands should be washed whenever the work area is left, before breaks, lunch, and leaving work. It is also beneficial to the diet preparation if the same care in hand washing is carried out when starting work, returning from breaks, and returning from lunch.

14.5 SPECIAL ISSUE OF SMOKING IN CONJUNCTION WITH REARING

In this vein, it must be emphasized that under no conditions should smoking, eating, putting on makeup, or other personal activities be allowed in the laboratory, insectary, or diet facility. This raises a difficult issue for insectary managers. Use of tobacco can be a potential risk to the insects that we are trying to rear. Nicotine and its fellow toxins in tobacco are known to be potent insecticides, and it is of questionable wisdom to use tobacco products in the same time frame such as within minutes to hours before handling diet materials, insect cages, or other materials that will come in contact with the target insects. This imposes a special problem upon insect diet and rearing specialists and insectary managers. Management of smoking in the workplace has become an increasingly well-recognized issue, in light of recent attention to the human health effects of second-hand tobacco smoke. In light of these concerns for risks to human health, smoking is being banned in increasingly larger numbers of public buildings. However, the workers that return from smoking breaks with clothing, hair, and skin that emanate residual smoke components are exposing the insects to these substances, which may be harmful to those insects. Although no studies are available that scientifically demonstrate such harmful effects, it is a possibility that the "halo" of smoke components may be counterproductive to the rearing accomplishments that these workers are trying to achieve.

14.6 MECHANICAL AND THERMAL HAZARDS

The nature of diet production and the equipment involved in this activity are sources of several kinds of mechanical injury, potential burns, and some fire potential. Blenders, chopping equipment, stirrers, cutters, mixers, and various other kinds of high-energy equipment are characterized by parts that can dismember a worker who has a single lapse of attention. Almost all such equipment has (or should have) a cover or other type of protection that when closed protects personnel from reaching in where moving parts can be contacted. Another hazard associated with the equipment with forcefully moving parts is a potential for diet materials to be thrown from the equipment with sufficient energy to cause injury, especially eye injury. Again, strictly followed (and enforced) protocols should mandate that equipment be used only with covers, lids, or other protective equipment in place.

The next line of defense is the use of protective wear, which must include protective eyewear and gloves. Also important are body coverings and protective footwear. The eyewear is one of the most frequently misused protections in laboratories and rearing facilities. Where it is possible for harsh chemicals or mechanical projectiles to become airborne, simple eyeglasses or eyeglass-like goggles are *not* adequate. The hazardous materials can get under or around such simple eyewear. Where severe hazards are present, full face shields should be worn, for example, when strong acids such as concentrated sulfuric or hydrochloric acids are used. The gloves, like eyewear, should be suited to the specific hazard and degree of risk. It is always a difficult matter to resolve the question of finding a middle ground between the protection gained from bulky or sometimes sticky gloves and the desirable dexterity that is lost by using excessively bulky gloves. The issue of the competition of safety/cleanliness vs. dexterity/comfort must be resolved on a case-by-case basis, hopefully with the collective wisdom and cooperation of the management and workers.

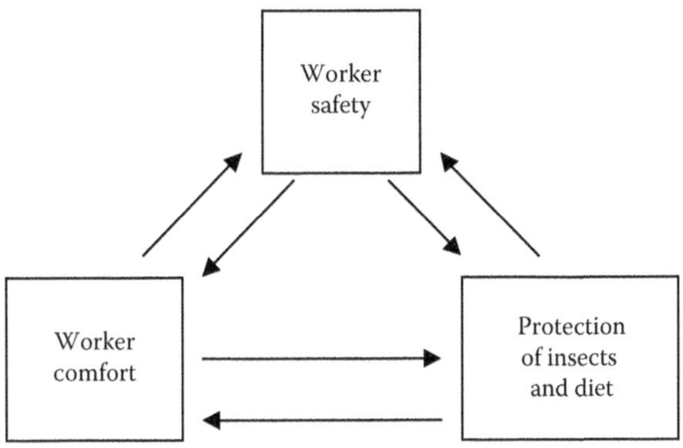

The various kinds of mechanical hazards are very diverse and depend upon the nature of the specific equipment in the diet facility or laboratory. Often, the mechanical hazards are associated with size reduction and mixing equipment. Blenders and food processors have rapidly rotating blades that have a powerful potential to cut, pulverize, or otherwise mechanically damage body parts of personnel. Rapidly rotating blades can throw shards out of the apparatus and cause severe eye damage, so such implements should be operated with protective covers or doors in place. It is tempting for an intense and diligent diet specialist to keep the top off of a blender so that he/she can introduce an additive while the blender is operating. Some blenders even have a removable portion of the lid to permit such a practice, and even this should be discouraged. Another mechanical hazard that is potentially dangerous is the long-shafted propellers in stirring implements such as that shown in Figure 14.5. Loose garments (sleeves, apron, or lab coat bottoms), long hair, or appendages can get caught in such implements. What is also evident in Figure 14.5 is a protective rail and warning colors to remind workers to be cautious of the hazards in the specified area. The areas where these mechanical hazards exist should be well marked; unauthorized personnel should be restricted from such areas, and personnel who use this equipment should be thoroughly educated and frequently reeducated about the hazards.

The diet facility is inherently a place where thermal hazards abound. Hot plates are capable of inflicting severe burns, and there is often no indication that the hot plate is heated. The lights on many hot plates are illuminated only when the element is conducting current, but the plate remains hot, even when the element is not illuminated. It is all too easy for a hot plate that had been recently used and is still hot to be touched by another worker who did not use the plate but is using the workbench after the hot plate user has departed from that workplace. A good idea is for such hot plates to be marked with a prominent sign that reminds everyone nearby that the plate is still hot. Care should be taken to remove the sign after the plate has cooled to avoid the "cry wolf" effect in this and other safety issues. Other hot places include autoclaves, ovens, steam pipes, flash sterilizers, and other diet heating implements should be marked as to their thermal hazards and preferably fenced off from unauthorized personnel.

Another kind of mechanical hazard in insectaries is the threat of hearing damage. As is the case in nearly all industrial settings, machinery operation often generates loud noise. Industrial-scale size reduction equipment, blenders and mixers, and packaging equipment often create sounds that are in excess of 70–80 dB. Many such machines generate high-pitched sounds that are penetrating

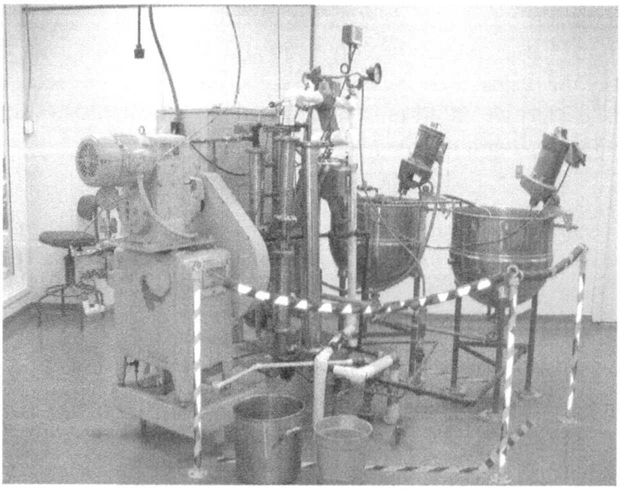

Figure 14.5 An example of a potentially dangerous mechanical hazard: the long-shafted propellers used in stirring implements.

and potentially damaging, and the insidious factor is that the damage from sound is often noticed after a delay of months or years after exposure. Adequate hearing protection is easy to provide and easy to use. Good earplugs can be used to prevent damage from lower-intensity sound, and the larger, more protective "earmuffs" afford a greater amount of protection against more intense sound. As is the case with other protective equipment, for the sake of sanitation, devices used for hearing protection (especially earplugs) should not be shared with other workers.

14.7 ELECTRICAL HAZARDS

Potential electrical injuries are no more common in insect-rearing facilities than they are in any other laboratory or industry setting. All electrical equipment should be grounded properly and wiring kept within codes and local, state, and federal standards. Electrical outlets should be ground fault protected, again, according to local codes. Because diet production involves wet materials, spills in conjunction with electrical wiring are a continual source of potential electrical shocks. This hazard could be averted by planning strategies that keep cords in places where spills will not wet plugs or cord/instrument interfaces. The use of proper protective clothing appropriate to handling wet materials (rubber boots and rubber gloves) will afford protection from electrical hazard and spill hazard events. Electrical cords should never be strung across the floor or surfaces where traffic flows. Electrical cords can be tripping and slipping hazards as well as agents of electric shock potential.

14.8 IN GENERAL

As a general rule, safety programs and all of their components should be developed by or in conjunction with well-trained professionals in compliance with state and/or federal standards such as those prescribed by OSHA (OSHA 2000). Often, the compliance with a rigorous safety program requires a change in institutional culture. This change is likely to be what many managers and workers consider cumbersome, especially if it involves modification of behavior patterns that have been established, sometimes over decades, and such patterns have become habits that are firmly ingrained. However, when it is realized that the changes that must be made are for a composite of good reasons based on worker safety (OSHA issues) and insect well-being, most workers will comply with the new standards and regulations. A good program of education in how and why such regulations are being implemented is the basis for general acceptance of the principles of good laboratory and insectary practices.

CHAPTER 15

Introduction to Texture, Consistency, and Rheology in Insect Diets

Chris Daubert, Chris Pernell, and Allen Carson Cohen

15.1 INTRODUCTORY COMMENTS: AN OVERVIEW OF TEXTURE IN DIETS

In this chapter, we present a summary of established concepts about the texture or consistency of insect foods, including artificial diets. We characterize these concepts in the context of feeding mechanics in insect feeding behavior and within the constraints of anatomy and physiology discussed in Chapter 7. To meet the texture/consistency requirements of insects, it must first be recognized that there is a tremendous range of feeding specializations. These specializations include strict fluid ingestion, extra-oral digestion (of solid food materials, including tissue/cellular structures in plants and animals), ingestion of soft solid materials such as fat body, ingestion of particulate materials such as flour and meal, and ingestion of hard materials such as wood (including woody plant tissues and bark). The difficulty in dealing with the incredibly broad range of insect foods that are used by the millions of species of insects is part of the challenge of grasping concepts of texture in insect foods and the fact that many insect foods (like human foods) are combinations of the broad categories or hybrids of the texture types that we are trying to characterize.

We extend these characterizations into the objective and potentially quantitative treatment of insect foods by introducing the concepts of rheology (the study of stress and strain in materials). We adopt the practices of the food scientists (food rheologists) with concepts and techniques that can both help us better understand the feeding/diet dynamics in insects and help us better evaluate the artificial diets that we use in our rearing systems.

15.1.1 Overview of Texture and Consistency and Their Role in Diets

"Texture" is used frequently in literature on artificial diets for insects to mean the consistency of a given diet (solid, liquid, slurry, or gel), for example, by Singh (1977). Throughout this book, "texture" or "consistency" is used more or less interchangeably to refer to the overall physical nature of a diet such as in distinguishing diets for tarnished plant bugs as slow-running (moderate viscosity) slurry vs. aphid diets as being runny (low-viscosity) solutions or suspensions. The importance of texture or consistency has been discussed formally in only a few publications such as Singh (1977) and Moore and Singh (1985). In many other publications, these physical factors are given only slight attention such as the discussion of casein in a diet for pink bollworms, Vanderzant and Reiser (1956) wrote:

> The physical properties of the medium, such as hardness and texture, may also have contributed to the improved results. Agar, which has no nutritive value, produced a firm gel in the casein medium whereas the coagulable property of the protein alone thickened the albumen medium. However, it is unlikely that the physical properties of the medium could have produced such a marked effect upon the growth of the larvae. In the experiments performed in developing the albumen medium, wide variations in the water, fat, sugar, and albumen content, which changed the physical consistency markedly, failed to change the growth rate of the larvae.

It is very interesting and representative of the pervasive thinking among insect diet and rearing specialists that the "physical properties of the medium, such as hardness and texture" could have produced marked effect on growth. The underlying assumption in this case and fairly much throughout insect diet literature has been that nutritional values are by far the most important determinant of diets that work well. The assumption is further clarified here where Vanderzant and Reiser explained that in albumen-based diets, changes in the diets' consistency failed to improve the growth rate of pink bollworm larvae, and because the various ingredients changed the physical consistency markedly, they concluded that the casein alone must have been responsible for the improvements in the agar and casein diet. Unfortunately, these authors did not test casein with and without agar, so the conclusion that texture is not important in comparison with nutritional value was concluded and remained a standing concept in diet studies.

A notable exception to this can be found in the history of diets for one of the most successful cases of mass rearing for sterile insect technique (SIT) efforts with the very serious fruit pests: codling moths. A highly informative history of efforts to optimize diet texture is to be found in the sequence of diet modifications for the codling moth, *Cydia pomonella* (Linnaeus) (Lepidoptera: Tortricidae).

Dyck (2010) has reviewed this history in depth, but briefly here is a summary of the evolution of texture control in diets for this important pest, which has been managed effectively by a highly effective program of SITs in several laboratories in the Pacific Northwest such as the Okanagan-Kootenay Sterile Insect Release Program facility in Osoyoos, Canada, and USDA, ARS laboratory in Yakima, Washington, DC. In these and other codling moth mass-rearing facilities, it was recognized that small-scale rearing techniques with conventional texturizing materials such as agar did not translate to larger-scale production needed for SIT. Researchers noticed that diets either dried out too soon or retained too much moisture causing microbial outbreaks and prompted difficulties in harvesting late larvae or pupae that were needed for sterilization (Brinton et al. 1969, Howell 1970, 1972, Hathaway et al. 1971). These problems prompted researchers such as Brinton et al. (1969) to replace agar with saw dust and wood shavings. These changes made huge improvements in the diet's texture in terms of larval development and recovery of prepupae and pupae. The sawdust and wood chips also made a significant impact on reducing the cost of the mass-produced diet whose most expensive component was agar.

15.1.2 Texture and Consistency in the Context of Insects' Feeding Habits: Solids, Liquids, and Slurries

Within the context of the vast diversity of insects' feeding habits, it is always an oversimplification to characterize feeding mechanisms and the kinds of physical characteristics of foods. Clearly, it is an oversimplification if we classify insects with chewing mouthparts or piercing and sucking mouthparts and then expect that an insect such as wood-eating Asian long-horned beetles to have identical feeding issues with leaf-eating insects such as silkworms or insects with piercing and sucking mouthparts such as whiteflies, which feed on phloem sap, to have very similar feeding issues to tarnished plant bugs, which use extra-oral feeding mechanisms to attack cellular or solid components. In the context of recognition of these complexities and subtle differences between foods, even ones that are broadly classified within certain groups, there is a compelling argument for using tools well established in food science to clarify the nature and causes of these differences in insect diets.

To give further sense of the scope of diet materials' textures or consistencies, intact wood must be considered solid, and dry grain must also be considered as matter in a solid state. Clearly and intuitively, students of wood-eating insects such as the many species of termites, bark beetles, long-horned beetles, and carpenter ants would distinguish between the ability of these insects that attack relatively solid wood and grain-eating insects that feed on powdery wheat, rice, and other grains that have been milled into a fairly fine powder (flour) or meal. Furthermore, the consistency of the woody foods and grain meals is different from leaves, stems, and roots that are eaten by many kinds of plant-eating specialists. Yet, most of us consider the leaves, fruit, and other fresh plants to be solids (despite the presence of liquids or gels within the cells of these plant materials).

In contrast with these clearly solid consistencies, saps and nectars are runny fluids, and the insects that consume these materials have mouthparts and digestive systems adapted to handle liquids (e.g., piercing and sucking mouthparts discussed in Chapter 7). Vertebrate blood and insect hemolymph are examples of fluids with sometimes high concentrations of cellular components (blood cells). To consume these complex liquids, biting flies and other hematophagous arthropods employ special adaptations such as anticoagulants to reduce viscosity, but the nature of these liquids' texture is clearly a free-flowing state of matter. It was emphasized in Chapter 7 that many insects used modifications of their feeding process known as extra-oral digestion to render into a freely ingestible liquid solid or semisolid materials. And finally, in rearing facilities, we very commonly use gelled diets to meet the needs of insects with a fairly wide range of textures or consistencies of their natural foods.

These points raise questions about texture differences between formulations of diets within these rather broad groups. So, for example, in an aphid or whitefly diet intended to simulate phloem sap from a cucurbit plant, what kinds of differences are there between a diet with 20% vs. 40% sucrose? Clearly, there are differences in taste, osmotic potential, and nutrient reward (in terms of energy and materials) between these two solutions, but what about differences in ease or difficulty in feeding on these materials? The major determinant of relative difficulty of ingesting these fluids is viscosity, and as can be seen from Table 15.1, the viscosity (expressed as kg m^{-1} s^{-1}) is three times as great for 40% sucrose as it is for 20%. As explained by Kingsolver and Daniels (1995) (and by Mittler and Dadd 1963a and later discussed by Oten et al. 2014), the physical constraints on the ingestion of fluids by insects with tubular food canals in their piercing and sucking mouthparts are determined by the Hagen–Poiseuille relationship:

$$Q = \pi R^4 \Delta P / 8 \eta L$$

where

- Q is the volumetric flow rate
- R is the radius of the tube (between about 0.5 μM for whiteflies and about 20 μM for tarnished plant bugs)
- L is the length of the stylets (500 μM in whiteflies to 2000 μM in tarnished plant bugs)
- ΔP is the pressure differential
- η is the apparent fluid viscosity

This relationship suggests that feeding dynamics of fluid feeders with feeding tubes that are relatively long and narrow can be expected to be very much impacted by the viscosity of their foods. Also, understanding this relationship is of value for researchers who are either developing or

Table 15.1 Density and Viscosity of Selected Liquids Encountered by Insects That Feed on Fluids

Solution (% Solute)	Temperature (°C)	Density (as kg m^{-1})	Viscosity (as kg m^{-1} s^{-1})
0	10	999.7	0.0013
0	20	998.2	0.001
0	30	995.7	0.0008
0	40	992.2	0.0006
Sucrose 20	15	1082	0.0025
Sucrose 20	20	1081	0.0019
Sucrose 20	25	1079	0.0017
Sucrose 40	15	1178	0.0075
Sucrose 40	20	1176	0.0062
Sucrose 40	25	1174	0.0051
Sucrose 60	15	1289	0.0795
Sucrose 60	20	1286	0.0585
Sucrose 60	25	1284	0.044
Xylem sap	20	1000	0.001
Phloem sap	20	1010	0.002
Vertebrate blood (0.4 hematocrit)	37	1025	~0.002

Source: Derived from Kingsolver, J.G. and Daniel, T.L., Mechanics of food handling by fluid-feeding insects, in R.F. Chapman and G. deBoer, eds., *Regulatory Mechanisms in Insect Feeding*, Chapman & Hall. New York, 1995, pp. 32–73.

modifying diets (Chapter 10). In this context, the discussion of other aspects of fluid feeding by Kingsolver and Daniels (1995) is of tremendous value to researchers and diet specialists who are charged with better understanding of the basics and improving our practical issues in fluid diets for blood feeders, xylem sap consumers, and phloem sap feeders.

As for solid foods, or modifications of solid foods found commonly in artificial diets such as agar, carrageenan, or other hydrocolloid-containing diets, the rheological characteristics have received relatively little attention. An exception to this generalization is a paper by Davis et al. (1995) where it is reported that testing leaf toughness revealed that maize leaf resistance to feeding by lepidopteran pests was related to the relative force required to sheer leaf whorls by a Mullen tester, which applies hydraulic force to cutting surfaces and provides a readout in Pascals of force that cut the tissue (known as a TAPPI test of bursting strength, TAPPI Test Methods 1991). Other publications on rheological measurements of artificial diets or foods of insects are virtually nonexistent. There has been some effort in mass-rearing facilities to use gel strength (the amount of force required to penetrate a gel, i.e., penetrometer tests) to evaluate agar. In the USDA, ARS Pink Bollworm Facility in Phoenix, AZ, and in the USDA, ARS Biological Control and Mass Rearing Research Laboratory at Mississippi State University (this laboratory is no longer in service), gel strength testers were used to measure the quality of new shipments of agar. However, texture, consistency, and rheological properties have remained a much neglected aspect of insect diet science and technology.

All of the foods discussed here have their own special issues regarding palatability, nutritional content, bioavailability, and the special adaptations the insects using these foods must possess in order to "make a living" using these materials. However, one special issue that has received too little attention in the diet literature is the texture or consistency. In Figures 15.1 and 15.2, several of the various food consistencies eaten by insects are shown, including fresh plant leaves or broccoli florets, semisolid artificial diet that simulates the consistency of insects' internal organs, flour (a stored grain material), phloem sap, and wood. Ingestion of free liquids including open sources of water, dew, nectar (floral or extra-floral), or hemolymph that has exuded from wounds made by predators or parasitoids involves a simple lapping or sucking process, while feeding on cloistered liquids such as vertebrate blood, xylem, or phloem sap requires special probing mechanisms, and in such feeding, the viscosity of the liquid

Figure 15.1 **(See color insert.)** (a through e) Liquid and slurry feeders. (a) Whiteflies (*Bemisia tabaci*, Aleurodidae: Hemiptera) feeding on hibiscus phloem sap. (b) Hemlock woolly adelgids (*Adelges tsugae*, Adelgidae: Hemiptera) feeding on cell contents of xylem ray parenchyma. (c) *Euthyrhynchus floridanus* L. (Hemiptera: Pentatomidae) feeding collectively, via extra-oral digestion, on a mealworm pupa (*Tenebrio molitor* L. Coleoptera: Tenebrionidae). (d) Tarnished plant bug *Lygus lineolaris* (Miridae: Hemiptera) feeding on extra-oral digestion products from broccoli florets. (e) *Sasajiscymnus tsugae* (Coccinellidae: Coleoptera) feeding on slurry-type artificial diet. (f) *Geocoris punctipes* (Hemiptera: Geocoridae) using extra-oral digestion to liquefy and consume prey contents.

can significantly influence rate and quantity of ingestion. Examples of feeding on cloistered or compartmentalized liquids include aphids that consume phloem sap, leafhoppers that consume xylem sap, and mosquitoes and other biting flies that consume vertebrate blood. This range of food consistency varies from dilute liquids to nutritious internal contents of prey insects or from nutrient-rich seeds to intractable materials such as wool and wood. The topic of viscosity in liquid diets is discussed later.

A more derived form of feeding on liquids or slurries is extra-oral digestion, discussed in Chapter 7, at length. Extra-oral digestion is a feeding strategy that has evolved in at least some representatives of most orders of insects (Cohen 1995), and it is a way that insects can deal with foods (natural or artificial) that are not in an appropriate form when the insect encounters them. While extra-oral digestion can be considered as an adaptation related to chemical/enzymatic processing and selection of foods, it is also important to understand it in the context of textural modification. Extra-oral digestion, as it is conventionally conceived, involves production of enzymes by the insect, but it can also be accomplished by symbiotic microbes (Chapter 13). In either case, as a result of the application of secretions and/or microbial products, structures of foods (including macromolecules) are attacked, suspended in digestive fluids, and ingested, usually as a slurry (Cohen 1995).

In brief, the class Insecta, with its vast variety of species and adaptations to an extensive array of foods, displays adaptations to every kind of food texture or consistency possible. It is an underdiscussed subject that, besides meeting nutritional requirements, insect diets must also conform to textural constraints that are poorly known. Elsewhere in this book, issues such as concentration of nutrients in plant saps are discussed as is the special character of the insects' feeding apparatus for using liquids that require extreme measures to concentrate the nutrients to a useful level. However, the physical texture of the range of insect foods has not be treated sufficiently, and it is the purpose of this chapter to bring to light special aspects of texture that insects must find suitable with regard to all foods to which they are adapted to utilize.

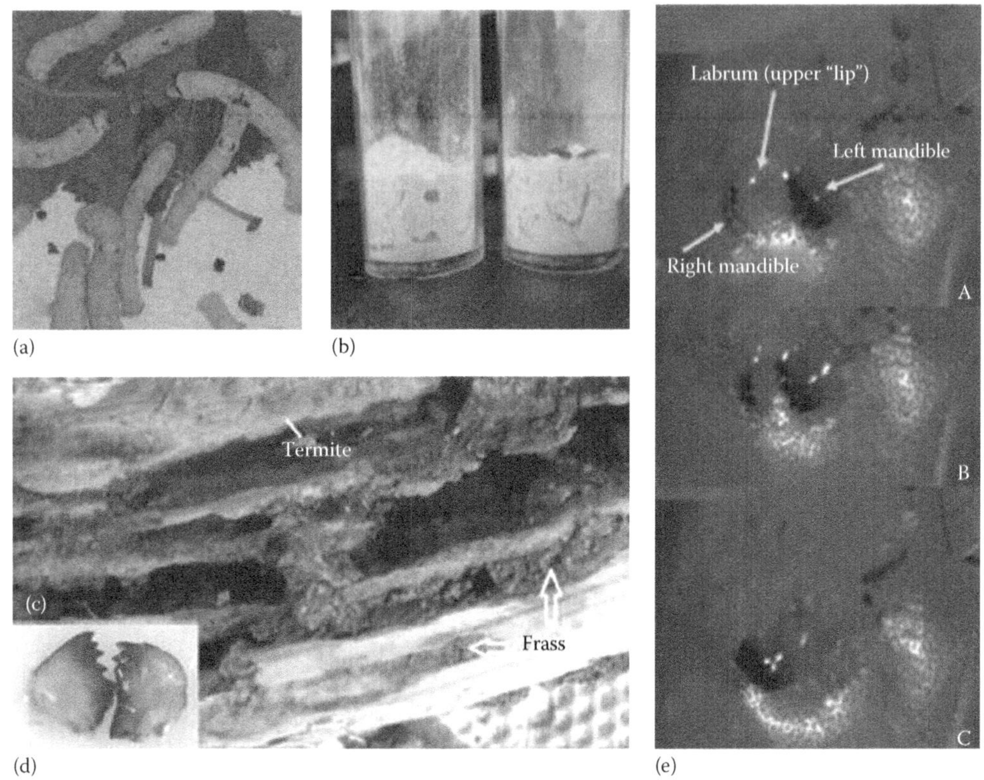

Figure 15.2 **(See color insert.)** (a through e) Showing insects that feed on solid foods. (a) Silkworm larvae, *Bombyx mori* L. (Bombycidae: Lepidoptera) feeding on mulberry leaves. (b) Confused flour beetles *Tribolium confusum* (Tenebrionidae: Coleoptera) feeding on wheat flour and wheat bran (not tunnels). (c and d) Termite colony feeding on wood, showing galleries and frass produced by termites (c) and mandibles of termite, capable of breaking down wood. (e) Stop action images showing movements of mandibles of a typical chewing insect (*Manduca sexta*, Lepidoptera: Sphingidae).

15.1.3 Bulking Agents and Texture

A topic that has received less attention than it deserves within the domain of texture or consistency is the nature and role of bulking agents. Throughout literature on insect diets, there is a sprinkling (sometimes a flow) of reference to the addition of so-called inert components to give "bulk" or texture to diets. Most often, such material is a highly purified cellulose such as microcrystalline cellulose. Early on, researchers working with plant-feeding insects with chewing mouthparts recognized that the nutrients such as casein, yeast, sugar, and vitamins would not provide the physical characteristics that provided a plant-like texture, even when agar was used to gel the diet. Singh (1977) discussed this, and he provided formulations for dozens of diets where cellulose was added.

An example of research that is at once informative and practical is found in the report by Chaudhury et al. (2011), where they successfully replaced a gelling agent (sodium polyacrylate polymer) with a cellulose medium. This report is useful, also, because it updates the considerable thinking over the past five or six decades behind modifying texture of the screwworm larval diet to help increase the efficiency, productivity, and quality of these insects. The importance of such discoveries and innovations is difficult to overemphasize in programs where seemingly small changes in a diet component can leverage huge economic advantages, and where a significant portion of a program's budget is dedicated to an expensive hydrocolloid such as agar or polyacrylate polymer, and tens of thousands of dollars per year can be saved. If the texture modification also effects a biological advantage, it can make a previously unfeasible program successful.

15.1.4 Physical States of Various Insect Artificial Diets

Because of the extremely broad range of foods used by insects, artificial diets may be formulated as simple liquids, slurry diets, gelled diets (true gels vs. "pseudo gels"), solid particulate diets, or solid continuous-phase diets. However, many natural diets and artificial diets are complex combinations of these diet states or hybridizations of two or more states. For example, the diet seen in Figure 15.1e (for *S. tsugae*) is a gelled slurry made from coagulated egg yolk particles suspended in an alginate gel and consisting further of dissolved nutrients such as fructose and water-soluble Vanderzant vitamins.

Simple liquids: In nature, nectars, xylem sap, phloem sap (see whiteflies, *B. tabaci* in Figure 15.1a and cell sap in Figure 15.1b), and apoplastic sap: in artificial diets, solutions of sugars, proteins or amino acids, minerals, vitamins, and sometimes traces of lipids such as sterols. Such diets have been used for aphids, whiteflies, leafhoppers, and other insects with feeding apparatus and digestive systems adapted to ingestion and concentration of dilute nutrients and removal of excess water.

Slurry diets: Several artificial diets for predators (Cohen 1985a, Cohen and Smith 1998), plant bugs (Debolt 1982, Cohen 2000c), and other insect species are semisolid (or semiliquid) complexes of aqueous solutions suspending particles or entanglements of solids. Depending on the viscosity, a slurry may run fast, slowly, or not at all. Familiar examples of these three semi-states of slurry behavior can be found in hot cereals such as oatmeal or cream of wheat.

Solid diets—conventional solids: Insects that eat wood, feathers, and wool (to name a few) can be thought of as using truly solid food, and insects that feed on milled stored grain products are also consuming what can readily be viewed as a true solid, albeit already past the need to bite into a fairly hard material such as intact wood. All of these solid foods require some kind of mouthpart and feeding behavior specializations to accommodate ingestion or penetration of hardwood or flour. Insects that eat plant tissues such as leaves or fruit and insects that eat "solid" animal tissues (such as carrion beetles and flesh flies) can be considered well within the range of solid food consumers, though the complex structures of plant cells with vacuoles, cell saps that are gels, and amalgamations of liquids bound or trapped by entanglements of proteins and carbohydrates make the concept of "solid food" a bit of an oversimplification.

Solid diets—gels: In insect diets, true gels are results of gelling agents (generally hydrocolloids) and water that has been bound by and stabilized by the gelling agent (Chapters 3 and 5 also treat this topic). True gels are prevalent among insect diets, with agar and carrageenan being the pervasive gelling material—both being heat-activated. There are also some diets made with alginate salts, and these are cold-gelling. The gelled diets are very complex because they consist of mixtures of dissolved materials in water (sugars, amino acids, antifungal agents, etc.), and the solid materials such as yeast suspensions, soy flour, wheat germ, plant cellular debris, and other nonsoluble materials surrounded by the gelled liquids. Proteins that are often added to gelled diets such as caseins, whey proteins, soy proteins, albumens, and sometimes mammalian serum proteins may form complex entanglements with the particles and the surrounding gels. These physical relationships in insect diets are not well understood because they have had little attention paid to them.

15.2 RELATIONSHIP BETWEEN CONSISTENCY (TEXTURE) AND RHEOLOGY

As treated in several places in this text (Chapters 3, 5, 7, and 10), studies of the roles of chemically based acceptance of foods by insects (feeding choices and phagostimulation by specific chemicals) have occupied a huge amount of the research efforts of scores of entomologists for well over a century. One of the most impressive examples of this approach is that of Vincent Gaston Dethier (1915–1993). He worked for over five decades on how flies and caterpillars recognized their food and what went into their choices based on chemo-sensory inputs and processing. He used elegant techniques to determine the abilities of insect to discriminate between different kinds of sugars, amino acids, plant secondary compounds, salts, and other factors that drove the tastes and choices of foods. And like Dethier, sometimes with his methods, sometimes with derivations of his labial eversion reflex, or sometimes with the determination of neurons firing, other researchers have treated chemical basis of choices in most compelling and informative experiments.

Chemical sensing as part of a food choice system has even been used to help develop artificial diets or to guide researchers to proportions of diet components by behavioral tests based on nutrient self-selection and geometric feeding models (discussed in Chapter 10). However, there is an almost total disregard for textural or physical qualities of insect foods, especially artificial diets. There are a few instances when physical/textural features are mentioned or considered in papers of insect feeding systems and diet technology, but these comments are perfunctory and never the central topic of a study of feeding biology. In contrast, in studies of human responses to foods, texture is an important part of the evaluation of why certain foods possess or lack sensory appeal.

It is hoped here that calling attention to the qualities of insect foods, especially artificial diets, will prompt further attention to textural/consistency aspects of insect foods. We suggest here that there are two possible positive outcomes of such attention: (1) using textural (rheological) analysis can be used as an important process/quality control tool for improving the production of artificial diets and (2) coupling textural analysis of changes that accompany modifications of diet components or processes of diet synthesis with properly designed bioassays of the insects' responses to such changes, which can help us better understand the basic issues in insect feeding biology.

15.3 SOME FOOD SCIENCE BACKGROUND

Because this section draws much of its teaching from food science, it is appropriate to begin the discussion with background on the physical basis of texture analysis of human foods. It will also be useful to use some examples from human foods because readers share experiences of everyday human foods that can serve as a basis for establishing terminology. It should be noted that terminologies used in food science such as in *texture profile analysis* (TPA) for concepts such as *hardness*, *chewiness*, *gumminess*, *stringiness*, *springiness*, *cohesiveness*, and *adhesiveness* (to name a few) are initially subjective sensory qualities. Food scientists try to use these terms that are factors in the physical palatability of foods in an objective framework that can be physically and mathematically described with the help of appropriate equipment and models. It is also important to point out that food scientists have human subjects (e.g., food tasting panels) who can express their impressions of and responses to foods that have been prepared in some way that modifies texture. So food scientists have both subjective human tools and objective instrumental tools to interpret the nature—especially the sensory qualities—of food texture.

An important discipline in food science is called *rheology*. Rheology is the study of the effects of stresses and strains on materials, in the current case—foods. It encompasses the use of special instruments to help us understand how defined forces change or impact materials in terms of viscosity, viscoelasticity, deformation, and other qualities that can be described subjectively as mouthfeel (Daubert and Foegeding 1998). A major driving force behind food rheology is to better understand the nature of foods' sensory qualities beyond the chemical factors such as sweetness, sourness, bitterness, and aroma. Rheology is a powerful way to explore why certain chocolates are more appealing because of qualities such as creaminess or sausages have a greater or lesser appeal because of their relative springiness. While cream cheese would not have an acceptable mouthfeel if it were springy, a piece of celery would not be deemed acceptable to most people if it were creamy. Food rheologists give us objective descriptions of how fat substitutes in puddings or ice cream can change viscoelasticity (for the better or the worse) or gumminess, adhesiveness, and other nuances of mouthfeel and texture.

As a foundation for application of texture analysis (rheology) to insect diets, it would be helpful to consider a typical approach that food scientists might take in dealing with a problem such as reduction of fat content in yogurt. Given the rationale for fat reduction that a lower fat yogurt would be healthier but less palatable, food scientists might culture a lower-fat milk and use some kind of texturizing agent such as the hydrocolloids xanthan gum, guar gum, and/or carrageenan to replace or simulate the "body" that the higher fat content imparts in the yogurt—giving the product a smooth and viscous mouthfeel (possibly called a creamy mouthfeel). It should be noted that the aforementioned hydrocolloids do not add to the calorie count in the product, as they are not digestible by humans.

Because the changes in the product affect mouthfeel, it is appropriate to assemble a food tasting panel to judge the palatability of the incipient product. Such a panel is generally a trained group of tasters, and the products are provided to panelists in a blind experimental design. However, because it is apparent that judgments about mouthfeel are inherently subjective, it is also useful to subject the modified food products to rheological analysis with instruments designed to simulate stresses and strains on foods that would take place in the human mouth. Therefore, the yogurt products would be further tested with rheological instruments to provide much more objective evaluations of the modifications of texture. Ideally, the subjective human tasting assays would be in complete agreement with the objective tests. With a well-derived grasp of treatments of foods, scientists and food technologists have developed systematic and highly predictable tools to engineer improvements and establish process control and quality control capabilities. These capabilities translate into improved food quality and economics of food product supply. Likewise, application of rheological principles coupled with good bioassay techniques can lead to great improvements in the management of insect diets.

The physical characteristics of insect diets can be thought of heuristically in terms of these questions: (1) how does consistency meet the sensory needs and feeding mechanisms of the target insects, (2) how does the consistency relate to the nature of diet components and the interaction of these components, and (3) how can we use the measurement of consistency of the incipient or completed diet as part of a process control and quality control system?

15.4 APPLICATIONS OF RHEOLOGY TO INSECT DIETS

Once we understand that all insect diets (natural or artificial) have rheological properties (textural or consistency characteristics) that can be measured objectively, and with considerable precision, we can use studies of stress and strain (deformation) in ways to get basic (fundamental) knowledge and applied knowledge of the foods that we provide to our insects. In fact, knowledge of rheological properties of natural foods can help us better interpret what insects are doing in nature when they select certain foods in preference to others. Just as we have asked the question repeatedly, "How much is a mustard family specialist responding to secondary compounds peculiar to that family?" we can also ask about how much some rheological property affects an insect's choice of food.

But for the purpose of this discussion in an insect diet book, the focus is on two kinds of rheological inquiry: (1) how can we use rheological tools to help us understand the feeding biology of our target insects and (2) how can we use rheological tools as a process and quality control portion of our rearing systems. The first question is one that will require very careful thinking and experimental design to help tease out where a factor is of rheological character vs.

where it is affecting a chemo-sensory mechanism of food acceptance. The second question (the PC/QC assessment) is much closer to being useful immediately, especially in the mass-rearing systems where the risk of losing tens of thousands of dollars to hundreds of thousands is always imminent. A third possibility for rheological studies with insect diets is the application of theory and equipment that helps us better understand fundamental physical properties of diets. In food science, rheology has been used to understand the nature of gels, molecular entanglements, and interactions of complex components in food mixtures. Insect diets are generally at least as complex as food mixtures such as bread or pudding, and rheology could help researchers understand the nature of their materials used in diets and the processing steps involved in synthesizing them. For example, when the transition was made for making pink bollworm diet with a twin screw extruder instead of a steam kettle, there were differences introduced due to the processing differences (e.g., higher temperatures and higher pressures), and rheological testing helped to point out the differences in the products (Edwards et al. 1996). These differences seemed to have affected the insects positively, in terms of improved feeding and biological performance related to the new diet processing (Edwards et al. 1996). Similarly, palpable differences (e.g., differences in the feel of diets when they are squeezed between the fingers) indicate that changes in gelling agents or lipid components result in textural differences. Using the appropriate rheological test such as large strain or small strain analysis, viscometry, shear analysis, or various other tests of diets' structural/rheological properties can provide sensitive, objective, and repeatable tests of diets. Such tests can be useful in development of reliable process control systems.

Clearly, there is a need to have a means of objectively measuring features of the various consistencies of diets to help us better understand insects' behavioral responses to texture qualities. Such tests would not only be useful in studying or predicting feeding behavior on both natural and artificial foods, but the tests would also be valuable in process control in diet synthesis. The relationship between texture analysis and both palatability and process control has been well established in the human food science and technology community (Steffe 1996, Daubert and Foegeding 1998). Daubert and Foegeding (1998) point out that (1) "Rheology is the study of the deformation and flow of matter" and (2) rheology deals with *stress* (force per unit area) and *strain* (deformation per length).

15.4.1 Using Rheological Testing as Part of the Process Control and Quality Control System

As we improve our understanding of the rheological properties of diets so that we become familiar with the types of responses that diets show under controlled test circumstances, we find that various components, various processes, and interactions between components can change the fundamental properties of the diets in question. This realization can form the basis of a process control system and a quality control system for the diets in our rearing system. Once we realize that our diets have a set of characteristics and that these characteristics are diagnostic of the physical nature of the diet, we can project that realization into a system of questions about the properties of diets from batch to batch and properties that are susceptible to change with deliberate or inadvertent changes in our diet-making process. With rheological testing instruments and procedures, we have a reliable, objective, and fast way of measuring deviations in our diet-making processes. And once we establish which rheological qualities (characteristics) of our diets are most telling about important deviations that detract most from quality and capability of our diets, we can use these measurements routinely and in a timely way.

An example of how this evaluation process can be started is offered here for clarity. We made several versions of the Yamamoto (1969) diet for tobacco hornworms, and we subjected samples of these diets to a practical kind of rheological evaluation known as texture profile analysis. So, for example, we made Yamamoto diet with agar (high gel strength agar from Sigma) and another formulation with the gel-forming hydrocolloid, carrageenan (Gelcarin 812 from FMC). Yamamoto diet is a wheat germ/casein/torula yeast diet with various additives for nutritional supplementation and for antimicrobial effects (Table 4.6). We used an Instron Universal Testing Machine, where we applied procedures of compression analysis to cylindrical diet plugs, with a conventional downward pressure, an upward phase, a second downward phase, and a final lifting of the pressure plate (Figure 15.3a through d). The forces (stresses) and the responding deformity of the diet plugs (strain) were constantly measured by the instrument and reported to a computer that was programmed to chart (Figure 15.3d) and report (in Newtons of force) the responses of the material. These responses are presented in a summarizing plot in Figure 15.4.

The forces and responses plotted in Figure 15.4 are translated into human sensory terms (hardness, gumminess, resilience, etc.), which may or may not have direct application to the insects themselves, but it must be noted that there is an objective reality to the measurements of stress and strain in insect diets, and these measurements are telling about the physical nature of the diets, and if repeated properly, these measurements can tell a tremendous amount about the nature of the materials and the processing of the diets. In this example with the agar-based and carrageenan-based Yamamoto diets, the hardness of the carrageenan diet was nearly twice that of the agar diet (Figure 15.5). The gumminess and chewiness were about twice as great in the carrageenan diet than they were in the agar diet, but resilience was greater in the agar diet than in the carrageenan formulation. The springiness was about equal for the two diets. We did similar tests with diets formulated with other components such as agar diets with oil and with soy protein substituted for casein.

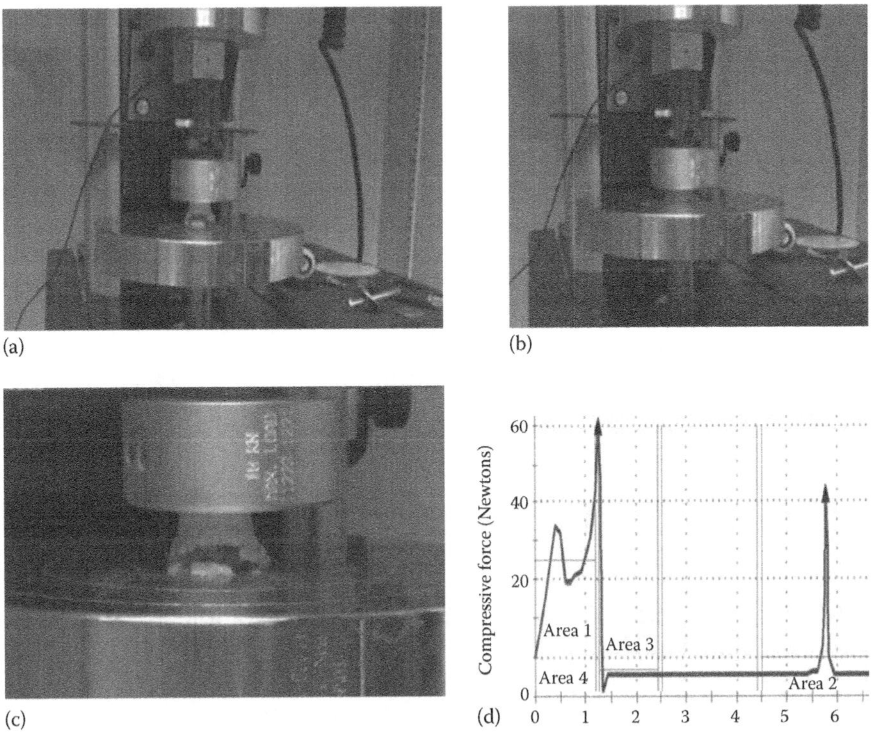

Figure 15.3 (a through c) show an Instron Universal Testing Machine in a test of the Yamamoto (1969) diet made with carrageenan for comparison with Yamamoto diet made with agar. These figures show the test cylinder (compression plate) in its raised position (a), in the full down position (b), and a close-up of the diet and analyzer after the first cycle (c). (d) Is the diagram of the two-cycle process. The components of this test (TPA) are shown in Figure 15.4.

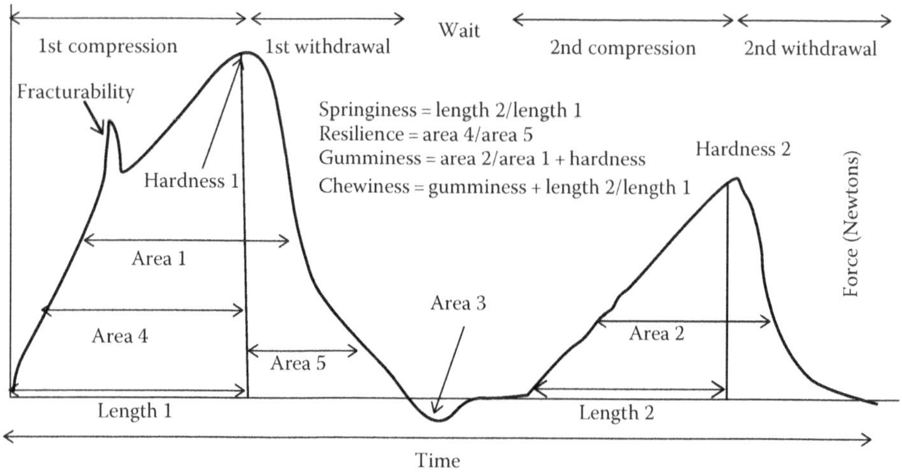

Figure 15.4 Defining terms expressed in a representative texture profile analysis (TPA).

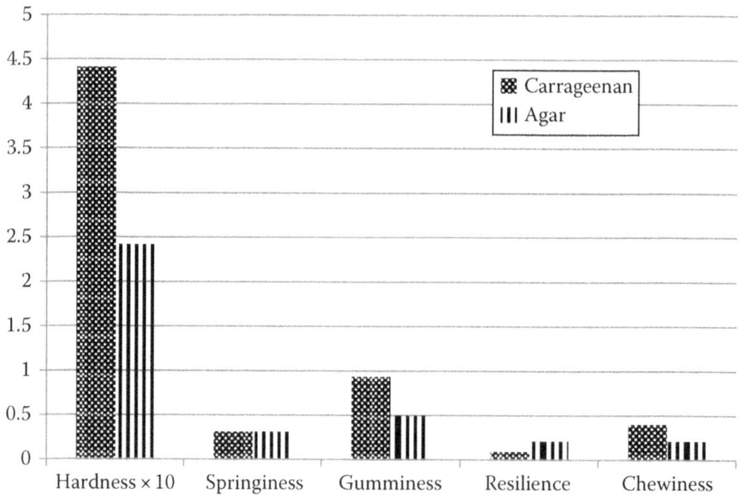

Figure 15.5 Results of texture profile analysis of 10 samples of Yamamoto diet made with agar vs. the same diet with a carrageenan replacement for agar as gelling agent.

15.4.2 Using Rheological Testing for Establishing a Rheological Basis of Palatability

The bioassay techniques used widely in diet development can be coupled with objective rheological techniques to parallel the food tasting panel coupled with texture analysis with human foods. Once the rheological characteristics of insect diets are established in adequate detail, bioassays can affirm the trends and help provide a base of information about the importance and scope of relevant rheological properties. Experimental design will be challenging where we would need to tease chemical sensory responses from textural sensory responses. So if an insect showed preference for a diet with carrageenan over agar (given that all other components were the same in both diets), is the preference based on a taste difference or a textural difference (greater hardness, lesser resilience, etc.) between the two gelling agents?

15.4.3 Factors That Are Known to or Seem Likely to Influence Rheological Characteristics in Insect Diets

The kinds of interactions between diet components discussed in Chapters 3 and 4 include synergism and antagonism. Both synergism and antagonism can influence things like cross-linking of hydrocolloids that are involved in the rheological characteristics. It is well established, for example, that calcium and potassium enhance cross-linking in carrageenan molecules and form stronger gels but may increase syneresis (fluid "leakage"). Proteins that may cross-link with agar can form stronger gels than ones where the same portions of agar and proteins were present, but conditions of cross-linking could not take place. We know also that pH can affect gelling actions, and chelating agents can effectively remove cations (such as calcium, magnesium, or potassium), preventing the cross-linking that would otherwise strengthen and stabilize a gel. As much as caseins are used in insect diets, it is unfortunate that more attention has not been paid to the rheological behavior of caseins in relationship to other diet components. The rheological outcomes of various dietary additives such as plant parts in powdered form, oils, lecithin, types of mineral supplements, and various protein sources all need further study.

CHAPTER 16

Future Prospects for Insect Diets

Introduction

There are several recent developments that point to excellent potential advancements in diet science and technology. These recent developments are, of course, founded upon the historical breakthroughs such as the landmark studies that proved that insect can be reared on nonnatural foods, the discovery that insects, like all other animals thus far studied, require basic nutrients such as vitamins, certain amino acids, and sterols, and that insect diets can be benefited by the inclusion of inexpensive, readily available complex foods such as wheat germ, other plant-derived flours, meat products, and poultry eggs. In fact, in nearly every case where a major diet or rearing breakthrough took place, the researchers had used interdisciplinary outreach into seemingly disparate fields such as genetics, biotechnology, biochemistry, food science, engineering, or microbiology.

Other future prospects for insect diet science and technology reside in the applications of insect diets to programs aimed at improving biologically based pest management, including amplification in the ability of rearing facilities to produce huge biomasses of insects that could be used for programs in biological control, sterile release programs, genetic pest management, production of pathogens, and even in the production of biological products that result from a combination of efficient rearing technology and genetic modification of insects or insect pathogens. The latter category includes the exciting prospect of using insects as a basis for producing high-quality, high-yield pharmaceuticals. A further innovative use of insects that has not yet been explored sufficiently is the use of artificial diets as delivery systems for substances (potential or putative toxins, feeding deterrents, nutrients, and antinutrients) to be tested against otherwise healthy insects. Control of weeds and noxious invasive plants by insects is in its embryonic stage and will become more promising when scale-up technology is applied to true mass production of weed-consuming species. Insect rearing and diet professionals can make huge contributions to the conservation and preservation of rare and endangered species, and they can make great progress in the efforts to enhance biodiversity by application of moderate- to large-scale technology and the science behind this technology to apply to threatened insect species.

Currently, insect mass-rearing technology is being used to produce insects as foods for other organisms, beyond their use as prey or hosts for biological control programs. Many conservation efforts in zoos, aquaria, or in private and public sector facilities depend upon high nutritional quality insects that must be readily available to feed captive animals that are endangered or whose habitats are being threatened. The range of animals in such programs is impressive, with captive mammals such as rare primates, birds, reptiles, amphibians, and fish as targets of conservation efforts. This use of insects is similar to the growing industry of insects as foods for pets. A surprising number of people worldwide keep insectivorous pets such as lizards and frogs, and these exotic pets require insects that not only will nourish them but also are safe from toxins such as plant secondary compounds that could poison a chameleon or gecko that may have cost its owner hundreds of dollars (or more for special varieties)! And, in fact, there are many issues that challenge diet scientists to produce insects of highest possible nutritional quality: for example, vertebrates require fairly high amounts of calcium in their diets, but insects tend to have low calcium concentrations. Strategies are sought to boost calcium in captive-reared insects and to maintain or boost profiles of amino acids, antioxidants, and other nutrients.

In a related area that is of emerging interest worldwide is the potential for using captive-reared insects as food for people. While it has been a practice dating back to prehistoric times for people to eat insects that had become available periodically, the deliberate production of insects for large-scale human food is a recently emerged topic. Several papers have appeared recently, many of them citing a book or website (http://www.food-insects.com/) by Dr. Gene DeFoliart (see DeFoliart 1999). This topic will be more fully explored in a later section of this chapter.

Such achievements and advancements will come in a reasonable amount of time only if insectary workers in general and insect diet professionals in particular are provided a formalized and thorough education and with the respect and standing deserved by specialists in such a highly technical, scientifically based field.

16.1 APPLICATION OF FOOD SCIENCE AND FOOD TECHNOLOGY PRINCIPLES

Insect diet development, improvement, delivery, and quality control have profited from the application of established and well-developed food science and food technology procedures and principles. However, such applications have been few and far between. The somewhat arbitrary application of food science and food technology basics results from the fact that the entomologists and their cooperators and support personnel have generally not been educated in the rudiments of food science. The vast industry, the extensive equipment, and the considerable body of knowledge generated by the food science community have been virtually untapped by insect diet professionals with some notable exceptions highlighted in this book.

If the education of current and future insect diet professionals were to include courses in food microbiology, food chemistry and physics, and food processing technology, there would certainly be a revolution in the insect diet domain. There are several kinds of improvements that can be made in insect diets by use of special processes derived from food science and technology, such as fermentation technology, extrusion applications, modern flash sterilizer technology, size reduction, and drying of key plant materials. Application of the highly sophisticated flavor technology from food science can be used to great advantage in developing token stimuli techniques, and current food industry applications of using modified atmospheres, can enhance diet storage and stability technology. Also, further development of freeze-drying and spray-drying applications can greatly improve diet component and synthesized diet shelf-life. Finally, the state of the art use of antimicrobial and antioxidant technology would be welcome improvements in the insect diet and rearing community.

16.2 PROGRESS IN EQUIPMENT APPLICATIONS

In order to apply the procedures described earlier, insect diet professionals will need to become well acquainted with the current food processing equipment, and researchers in the domain of diet development and diet improvement will have access to such equipment. For example, the incredible versatility of the twin screw extruder offers tremendous opportunities for diet improvements based on various programming regimes that can be applied with ease to extruders to alter mixing rates, temperatures, and pressures. Food science and food technology researchers have done wonders to vary the textures, moisture content, nutrient and antinutrient contents, and sensory characteristics of a huge variety of foods, using various programs in their extruders. The various kinds of textured soy products are excellent examples of this kind of food engineering that can be used as a model for insect diet modification. The potential fully exists for rendering relatively inexpensive diet components such as flours from various plant species into nutritionally rich diets. Along with the nutritional improvements from such components, it is probable that the accompanying texture modifications would eliminate or reduce the need for extraneous and expensive gelling agents. Further simultaneous benefit of the extruder application is the elimination of antimicrobial agents. However, for this type of multiple benefits to be realized, there must be an application of a base of knowledge that comes from a hybridization of the food science community and the insect biology community.

16.3 FOOD MATRIX ANALYSIS

This author is convinced that the principal reason for the success of most or all of the time-proven diets is that the interplay between diet components as they are arranged or ordered into a dispersive or matrix state predisposes the nutrients and other diet components to be (1) palatable to the insect, (2) available to the insect in terms of its mouthpart size and configuration, (3) of a chemical and a physical form that lends itself to the bioavailability in the insect's gut, and (4) preserved from various sources of degradation. Further study of the interplay of components and how this interplay serves to either stabilize or destabilize components will be fruitful in explaining how successful diets work; such studies will undoubtedly remove much of the stochastic (or "blind luck") elements of planning and designing diets and diet-synthesizing processes. For example, once we have learned that immobilizing iron and keeping it sequestered from lipids will reduce lipid peroxidation, we can apply stabilization techniques such as the use of gelling agents that are especially adept at surrounding iron with a result of gaining stability of our diet. In this context, it would be a further improvement of the diet's nutrition and stability if we could apply techniques for compartmentalizing ascorbic acid, iron, and lipids so that the newly recognized effects of negative synergy between iron and ascorbic acid could be bypassed.

16.4 DEVELOPMENT OF SYMPTOMOLOGY OF NUTRITIONAL DEFICIENCIES

An area of insect nutrition that is lagging far behind its counterpart in mammalian nutrition is the development of a list that ties nutritional deficiencies with specific disorders or clear-cut symptoms. Very general disorders such as wing deformities, slow growth, asynchronous development of similar-aged members of a colony, diminished size, or behavioral abnormalities have been linked to numerous causes, including many non-nutritional ones. Future efforts to link specific nutritional causes and effects and to provide unambiguous symptoms with a nutritional deficiency syndrome will be welcome and will advance the field of rearing insects on artificial diets. Although the inherent difficulty in developing such a symptomology is compounded by the large number of species that are reared on diets, it would be hoped that carryover from one insect species to another will broaden the base of usefulness of a systematic linkage. Also, the application of the microanalytical procedures such as capillary electrophoresis, small-volume chromatography, and what has recently been called nano-spectroscopy should help researchers who are seeking physiological and biochemical correlations with nutrient deficiency. For example, low-volume spectroscopy has great promise in the detection of abnormally low (or high) levels of antioxidants in the hemolymph or other compartments of insects, even those of relatively small size.

16.5 DEVELOPMENT OF HIGHLY REFINED BIOASSAYS

Throughout the past three or four decades, entomologists and a variety of plant specialists have used artificial diets as delivery systems for various ingredients that were to be tested either for nutritional efficacy (including feeding stimuli and true nutrients) or as toxins, feeding deterrents, or other anti-feeding qualities. Recently, researchers have tried to use artificial diets to deliver and test products from genetically modified organisms (GMOs), reasoning that the artificial diets offer a simplified way of presenting a single, isolated chemical to test its potential toxicity. For example, Forcada et al. (1999) used artificial diet to deliver Cry 1 Ac toxin to *Heliothis virescens* via an artificial diet, by applying a solution of the toxin to the surface of a gelled diet that was poured into wells in a feeding arena. This is in contrast with the presentation of a plant that has been genetically modified to over-express some protein or product of a protein-regulated pathway. Whole plant systems are understood to be complex in terms of their physical and chemical compositions, so isolation or interpretation of the role of the toxin becomes complicated. In contrast, the thinking about artificial diets has been that if the diet itself (inherently more simple than a living plant) is nontoxic, then the addition of the toxin to the diet should permit detection of toxicity of the putative toxin. Also, naturally occurring plant secondary compounds have been tested by several researchers using artificial diet as a delivery system for the putative toxin (Cohen and Urias 1988, Timmermann et al. 1999).

However, there are complexities that should be addressed prior to assuming that the artificial diets are nutritionally or chemically inert and completely passive backdrops for putative toxins. This point is especially pertinent to situations where negative results are obtained, which may be interpreted to mean that the putative toxin had little or no effect on the target insect. Depending upon how the putative toxin was added to the diet, the toxin may be avoided behaviorally by the target insect, or the toxin may be detoxified by the processing (such as heating) or the other diet components. In the study by Cohen and Urias, for example, raw liver and ground beef were chief components of the diet—both ingredients containing a wide range of enzymes, including ones that could have altered the possible toxicity properties of the materials being tested. For putative toxins that may work entirely or in part by mechanisms that affect nutritional physiology, the high level of nutrition of the artificial diet may mask an otherwise adverse effect of the substance being tested. Likewise, putative toxins that work by creating an oxidative stress may be entirely or partially detoxified by the various antioxidants that are both deliberately and inadvertently added to artificial diets. The converse of this generalization also is possible, where added components that are nutrients in their own right, such as iron and ascorbic acid, may interact in a destructive way to induce lipid peroxidation (as discussed in Chapters 5 and 8).

The point here is that many artificial diets are highly complex entities, and the application of these diets as toxin delivery systems must be approached with due caution and respect for the complexity of the diet and the diet/insect interaction.

16.6 DEVELOPMENT OF PROCESS CONTROL SYSTEMS

Although the topic of statistical process control (SPC) is discussed throughout this book, it merits a special mention in this chapter on new directions that promise to advance insect diet science and technology. Given the importance of SPC in production systems that range from human foods to automobiles, it is amazing that the topic has gotten so little attention in entomology, especially in insect-rearing systems. It has been nearly a quarter century since Lorraine and Bruzzone (1992) wrote a paper titled "Process Control: An Untapped Resource for Managers of Sterile Insect Programs." That paper pointed out clearly and authoritatively that process control is of enormous potential in the economics and especially the quality of insect-rearing system; however, as of August 2014, that paper has had zero citations according to the Web of Science! Key word searches of various terms pertaining to process control ("control charts," "Pareto diagrams," and any related terms) reveal the inattention to SPC in the entomological community.

16.7 APPLICATION OF FERMENTATION AND GMO TECHNOLOGY

The use of fermentation technology is at once one of the oldest and most useful means of processing human foods and also one of the most active areas of current research and technology improvement. The use of various microbes from almost limitless species and strains has allowed the improvement of utilization of the nutritional value of many foods, the preservation strategies, and the enhancement of flavors, often turning a low-value food into a highly nutritious, high-value food—a generalization possibly best illustrated by the soy processing industry. In light of this splendid potential for applying fermentation technology and other aspects of processing such as gene manipulation, in conjunction with fermentation settings, it is disappointing to report that the literature remains empty of examples of applications of these technologies to insect diets and diet components. It is an exciting prospect that should be encouraged to have researchers with a competence with insect diet work learn the fermentation and genetic modification techniques or join forces with scientists and technologists who already have the expertise to apply these prospective tools to improvement of diets. The value of soy flour and other soy products has already been demonstrated in the domain of insect diets, and such techniques as heat processing has been shown to detoxify and otherwise improve the soy proteins (Cohen et al. 2005). However, much more potential exists for applying microbes such as *Aspergillus* spp. as agents of further improvement of soy. The model of food improvement by fermentation has been applied for human foods, with Asian practices of soy fermentation going back for thousands of years. The same kinds of benefits of selective fermentation practices should apply at least as well to making soy and other plant products better ingredients in insect diets. In regard to GMOs, researchers who would modify microbes that are used commonly in fermentation to over-express insect yolk proteins such as vitellins could revolutionize the in vitro rearing of many parasitoids and predators, which have been shown to thrive on these yolk proteins (Cohen and Patana 1985). Fermentation processing is reviewed in detail by Fellows (2000). In the section that follows, a broadly modeled blueprint is presented for using inexpensive and readily available raw materials to bioengineer components designed specifically for insect diets.

16.7.1 Improved Diet Components from Fermentation Technology Engineering

The proof of principle that fermented or microbially processed foods can be highly efficacious in insect diets has been established with the use of yeast products in the past century. A survey of more than 400 diets in the Singh (1977) *Artificial Diets for Insects, Mites, and Spiders* text revealed that more than 50% of all successful diets contained some kind of yeast product, generally brewer's yeast, torula yeast, or some kind of yeast extract or lysate/autolysate. The highly promising results from using Wheast™ (which is a product made from sweet dairy whey fermented by specialized dairy yeast) suggest the potential of using what we now call "biomanufacturing" technology—basically, a food engineering process that can be highly controlled to turn lower-value starting materials into high-value nutrients.

The establishment of a "minimal hypothesis" that fermented or otherwise bioprocessed foods can be very beneficial to insects has been accomplished countless times over the history of rearing on artificial insect diets. For more than a century, yeast products such as brewer's yeast, torula yeast, "Wheast," yeast hydrolyzates, and yeast extracts have been used in insect diets, extensively, with great success. Also, though to a lesser extent, bacterial modifications of insect foods have been recognized as highly nutritious to members of Diptera ranging from houseflies to mosquito larvae, and bacterial bioproducts have been used in rearing various species of aquatic insects such as mayflies, stoneflies, and some Hemiptera.

However, the fermented or bioprocessed foods were ones that were already available for other uses such as vertebrate nutritional supplements, health foods for humans, by-products from the brewing and dairy industry, or microbial culture materials. The concept of designing or engineering bioprocessed foods directly applicable to insect diets has been a neglected domain. So, for example, brewer's yeast—though it is one of the most widely used insect diet components—is a by-product of beer making. So-called nutritional yeasts that are produced as "primary yeasts," which are designed to be used as primary foods (Bekatorou et al. 2006, Stone 2006), have potentially greater nutritional value than yeast products that are by-products, requiring added processes such as debittering, which can detract from palatability and nutritional value. The main feature of designed or engineered bioproducts is that they can be modified by the use of specially selected media components (such as soy meal, molasses, and other low-cost ingredients—Solaiman et al. 2007) to have higher mineral content, desirable lipid content (including a high sterol content, which would be desirable for insects), or specialized protein components. Most importantly, the Asian food model of fermented soy products teaches us that use of fungi and/or bacteria to modify soy beans can both add to the nutrition of the product and remove antinutrients such as protease inhibitors, lectins, phytic acid, and other factors that detract from soy beans' use as a food for humans or livestock. The emerging appreciation of the value of microbial actions on insect foods in nature makes the concept of deliberate or designed food engineering to improve insect diet components most appealing and promising. A bioprocessing strategy being used in the Insect Rearing Education and Research Program at North Carolina State University is shown in Figures 16.1 through 16.3.

Some very available raw materials that have great nutritional potential for insects are soy meal (or flour), sweet dairy whey, and molasses. Soy meal has a high protein content with all the insect-essential amino acids (discussed further in Chapter 4), and it also contains valuable lipids, including unsaturated fatty acids and phytosterols. It also contains some vitamins and most of the minerals required by insects (again, further discussion in Chapter 4 offers more details). However, soy is known to contain several antinutrients (protease inhibitors, lectins, and phytic acid, to mention a few). Sweet whey protein contains a high concentration of high-quality proteins with an excellent essential amino acid profile, but it contains large amounts of lactose (70% on a dry weight basis), so it presents problems for use directly in insect diets. However, these two components have been researched for food engineering improvements that could inexpensively improve these materials (Hirabayashi et al. 1998, Karr-Lilienthal et al. 2005, Song and Chang 2005, Jang et al. 2006, Park and Bae 2006,

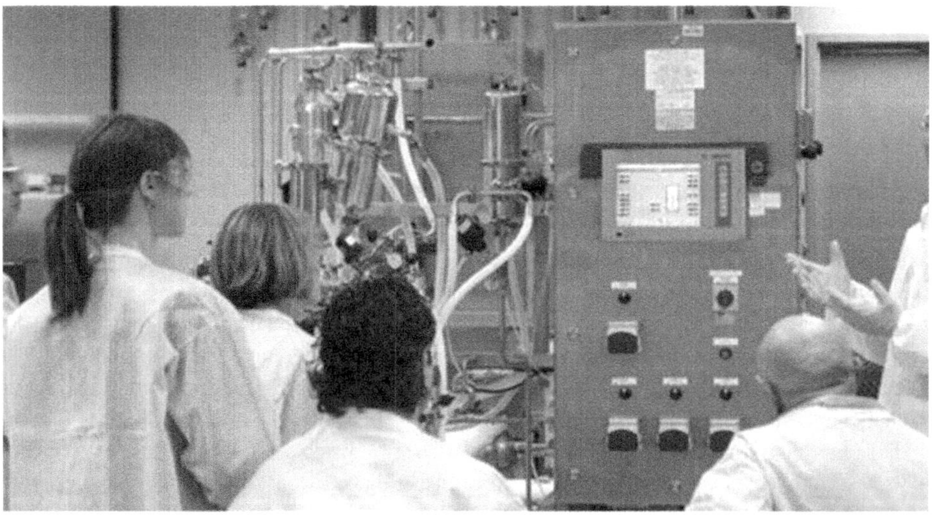

Figure 16.1 Instructor explaining biomanufacturing process in a bioreactor engineering class. The class is doing hands-on operation of a 35 L bioreactor at the BTEC facility (Bio-manufacturing Training and Education Center at North Carolina State University). Bioreactors are used in industrial-scale systems for the production of conventional and bioengineered food and pharmaceutical products.

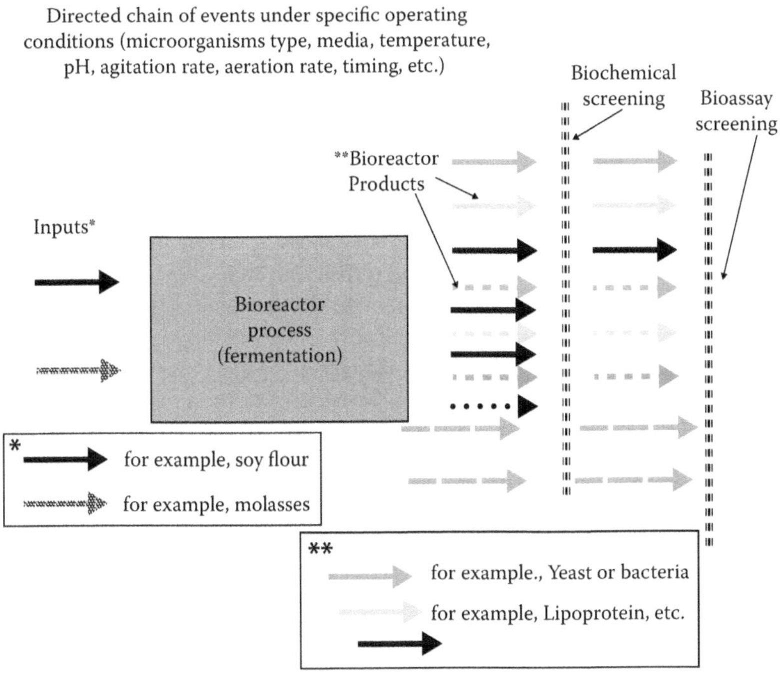

Figure 16.2 Generalized diagram of bioprocessing or biomanufacturing foods for insects, showing inputs (soy flour and molasses), microorganisms and supplementary raw materials, and generalized screening scheme.

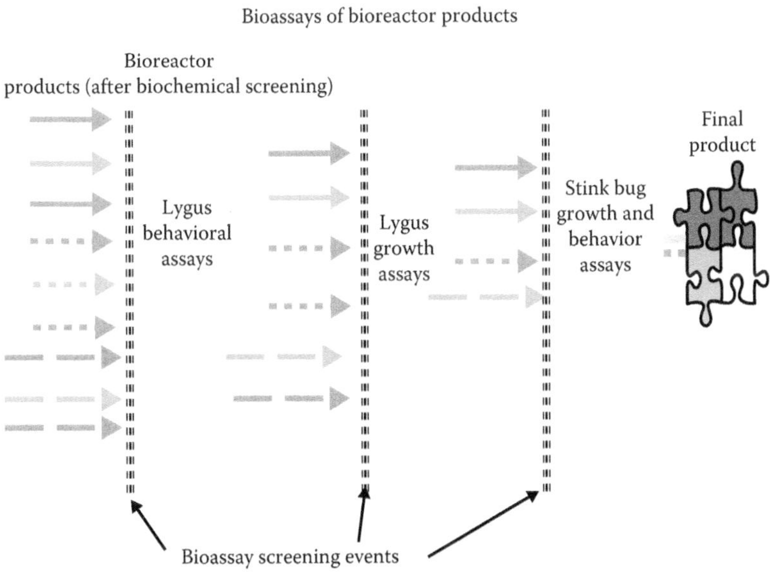

Figure 16.3 Diagram of screening scheme for biomanufactured products engineered to meet feeding requirements of target insects such as Lygus bugs (plant bugs) or stink bugs.

van Kempen 2006, Feng et al. 2007, He and Guo 2007, Huang et al. 2007, Solaiman et al. 2007, Berger et al. 2008, Donkor and Shah 2008, Frias et al. 2008, Pham and Shah 2008a,b). In all of these reports, processing soy or related legumes by enzymatic techniques or by both conventional and unconventional industrial processes is presented. For example, Solaiman et al. (2007) show how a yeast that has not been used conventionally can be used in a soy–molasses medium to produce a biosurfactant (a glycol-lipid). Although a biosurfactant is not what we would want to produce in diet development engineering, it exemplifies the kinds of novel and potentially useful nutrients or functional materials that can be produced by engineering common starting materials such as soy, molasses, and whey. The point of this discussion is that there are many very promising but unexplored technologies in existence that have great potential in the development of designed or engineered diets and diet components.

An important and promising concept is the engineering of diet components through bioprocessing with microbes (essentially fermentation) to make existing diet components more nutritious, more palatable, and possibly less expensive. The idea here is to develop "designer" (actually designed) diets that incorporate the best of multiple components such as molasses, soy flour, and other inexpensive raw materials. It must be emphasized here that the expression "engineering" is not meant in the sense of *bioengineering* by genetic means, but rather *conventional* food processing engineering. The possibilities are great with a raw material like dairy whey, soy meal (or soy flour), which is very nutritious and available in bulk and at economically feasible costs. However, as discussed later, there are drawbacks with soy meal and dairy whey—mainly their characteristics of having antinutrients or indigestible (or non-bioavailable) components that must be ameliorated.

Once the raw materials and the microbial components are decided upon, the application of modern food and pharmaceutical equipment such as bioreactors (Figure 16.1) can be used with optimization techniques such as those illustrated in Figure 16.2 to optimize the production of novel and most valuable diet components. The biomanufacturing process described here could result in diet components that contain all or nearly all the nutritional and even functional components that now must be added to diets separately and mixed. The most exciting prospect, however, is to be able to biomanufacture insect foods that are tailored to the insects' needs. As shown in Figures 16.1 through 16.3, products from designed or bioengineered foods can take advantage of relatively inexpensive foods and utilize components with low to moderate nutritional content (or which contain antinutrients such as those found in unfermented soy products) and convert the low-quality foods into highly palatable microbial products. By selecting the appropriate microbe such as specialized yeast that are highly attractive and palatable to many insects and amplifying the biomass of the microbial component, a uniquely nutritious product can be produced. For example, it is well documented that minerals that are nutritionally valuable can be made more bioavailable by adding the inorganic salts of the minerals to a yeast-growing medium and allowing the yeast to convert the salts into organic compounds.

A quotation from Stone (2006) explains the complexities and the extensive range of products possible:

> Fermentation Metabolites: Production of fermentative metabolites by the yeast cells, sometimes referred to as "nutrilites," is the principle behind the yeast culture fermentation. A specific culture media is inoculated with live yeast cells, allowed to ferment under a specific set of conditions, and then the entire fermented media is dried … it contains both the residual live yeast cells used in the fermentation, as well as the metabolites or metabolic by-products the yeast produced. Yeast Culture is a very complex product and contains both the "intracellular" yeast cell nutrients and the "extracellular" metabolites of fermentation. Simply feeding yeast cells will not duplicate the total "nutrient" range found in yeast culture.
>
> When yeast cells ferment sugars, they secrete metabolic by-products … alcohol is a good example of an extracellular metabolite resulting from yeast fermentation under brewing conditions. Variations in metabolite profile are why different fermented products (e.g., breads, beers, wines, whiskies) all taste and smell differently from each other. There are many different metabolites secreted, depending on the composition of the media fermented and the conditions of the fermentation process, including peptides,

> alcohols, esters and organic acids. In the case of beer fermentation, these extracellular metabolites are washed away from the yeast cells and are contained in the beer and are not part of the brewers yeast. Just remember that Yeast Culture is unique among all other yeast products and must be evaluated as a fermented product containing "undefined nutrients" and not based on its "known nutrient" profile like protein, vitamin, and amino acid profile.

In this discussion, it is apparent that the fermentation/biomanufacturing process is an example of a "whole is greater than the sum of its parts" phenomenon. Yeast (or other microbes) makes use of the raw materials of its medium, making inorganic minerals, simple sugars, and amino acids into often more palatable and nutritious compounds. The most simple example of this is that the inorganic salt sodium selenite and amino acids in yeast culture are converted into selenomethionine, a highly bioavailable form of selenium, which has powerful antioxidant properties (Stone 2006). Similarly, inorganic salts of calcium, iron, and zinc are converted by yeasts such as *Saccharomyces cerevisiae* to organic species of these divalent cations that are much more bioavailable than their inorganic counterparts. The feasibility of this process has been unequivocally demonstrated for human health foods (such as high selenium yeast), but it has not been deliberately applied to insect diet components and remains a fertile ground for future improvements of insect diets.

16.8 ADVANCED TECHNOLOGIES FOR DETECTING AND HANDLING MICROBIAL CONTAMINANTS

The many advancements in food safety science and technology include new methods of detecting microbial contaminants as well as reducing populations of such contaminants. The diagnostic techniques include DNA techniques, especially those that use PCR and blotting methods. Application of these techniques as diagnostic tools for detecting microbial contaminants in insect diets and diet components will help diet professionals pinpoint the species of contaminants with which they routinely deal, and this specific information will be a strong basis for chemically based treatment or alleviation strategies. The technology available now for treating foods includes various radiation applications and other relatively new techniques such as ohmic heating, ultrasound, pulsed light, magnetic field, pulsed electric fields, and high-pressure systems (Fellows 2000), all of which have excellent potential to rid diet materials (and diets) of live microbes while having only minimal negative influence on sensory and nutritive properties of these materials. However, these techniques need to be tested in an orderly scientifically based procedure. Although there are some very promising new antimicrobial substances that have been recently identified or developed, the use of a chemotherapy is always clouded by the potential of the chemical agent as being harmful to the target insect, and the use of chemical antimicrobial agents is also tainted with concerns over the expense of these chemicals and the potential for microbes to develop resistance to such agents, especially when they are overapplied or are used chronically.

16.9 ADVANCEMENTS IN TECHNIQUES TO CHARACTERIZE THE SPECIES AND NATURE OF SYMBIONTS

Many of the microbes that are true mutualists with insects are so fastidious in their nutritional and environmental requirements that they defy identification by the conventional growth media techniques that had been used in past efforts to characterize normal and beneficial microflora (Douglas 1998). The better that the symbionts are understood, the greater are the chances of developing techniques that allow us to rear the target insects but simultaneously respect the integrity of the microflora. For example, the enhanced understanding of *Wolbachia* spp. in various insects and other arthropods (Werren 1997) makes it clear that these microbes have a profound effect on the functioning of their hosts. Likewise, associations of *Buchnera* spp. have dramatic effects on the well-being of their mutualistic partners, the aphids (Douglas 1998). Yet, the use of conventional antimicrobial substances to control undesirable contaminants has been one of the major tools despite the fact that it has long been a concern that these chemical agents, especially when used chronically, may damage or offset the normal insect/flora relationships. Hopefully, along with the increased understanding of the nature of these relationships and with better tools to combat contaminants, insect diet professionals will be able to deal with the contaminants and the desirable symbionts with tools that more finely pinpoint their targets. These issues are further discussed in Chapter 13.

16.10 APPLICATION OF ADVANCED NANO-ANALYSIS TECHNIQUES FOR NUTRIENT EVALUATIONS ON AN ULTRASMALL SCALE

Currently, there is a revolution in the realm of micro- and nanoscale analytical chemistry with the introduction of such small sample techniques as capillary electrophoresis, microliter-scale spectrophotometry, and ultra-small volume HPLC (especially when these techniques are coupled with micromanipulation processes). These techniques will finally allow the analysis of samples of such small size that they can pinpoint the biochemical nature of foods that are selected by insects. It was emphasized in several previous chapters how important it was in the development of successful artificial diets to determine exactly what the target insect was ingesting or selecting from its food source. However, in the not too distant past, it was impossible to analyze the insect "bite-sized" samples. The ultrafine analytical techniques can be used to determine accurately exactly what the target insect is ingesting by measuring the restricted area within the food source or within the gut contents.

The areas that will probably profit most immediately from the nanoscale analytical chemistry are in the detection and evaluation of lipids that are ingested by insects, the chemical characterization of the various proteins, and the chemical interplay of the various components involved in antioxidant chemistry. The lipid chemistry of the past three to four decades had been largely dependent upon the thin-layer and gas–liquid chromatography (TLC and GLC or GC) techniques that required milligram to gram samples, and for sterol analysis, even larger samples were required. The replacement of packed GC columns by capillary columns and the incorporation of ultrathin TLC plates reduced the sample sizes required but were still not sensitive enough for the analysis of insect "*bite-sized*" samples. Refinements in HPLC analysis, including ultrasensitive detectors and small-bore columns, have reduced the sample sizes required and have made possible the resolution of species of molecules that are similar to one another, including the various kinds of phytosterols—a domain that is practically a blank page in insect nutritional requirements for most target species.

Proteins have long been recognized as key components in insect nutrition and more recently as important antinutrients (such as lectins and enzyme inhibitors). However, without the technology to fully characterize the proteins ingested or bypassed by insects in their small scale of feeding, it has been difficult to unravel the Gordian knot of protein chemistry in the nutritional domain. Now, with the ability to perform high-resolution analyses of protein structure, including the details of glycoprotein and lipoprotein characterization, and protein function, as well as protein digestion characteristics, proteins can be better understood within their digestive/nutritional context. The reality is becoming better realized that protein functions are much more complex and elaborate than just their role as repositories of amino acids, and it is becoming better understood that the nutritional value of proteins varies from one type of protein to another, not merely according to the amino acid composition. There are many potential applications of this newfound knowledge of protein functionality in the feeding systems of insects, including the circumventing of antinutrient qualities and the improvement of bioavailability of proteins that were previously considered of low quality or even as toxic. Application of a thorough understanding of proteins in the context of the target insects' feeding systems can be effective in increasing the economics and scale of rearing species that were previously not within the reach of the rearing community.

The roles of various components in the processes of oxidation and reduction and especially the factors in oxidative stress will hopefully be better understood in the insect feeding systems where these processes play a prominent role. The biological applications of the cascade of metal (especially iron) toxicity in the Haber–Weiss (or Fenton–Haber–Weiss) reactions can be better understood as these reactions pertain to insects if the microscale analyses of these processes can be performed. Pioneering studies such as that of Paes et al. (2001) have outlined the importance of these reactions in the midgut of certain insect species, setting the stage for our need for and approaches to understanding the cascade of reactions (Fenton or Haber–Weiss reactions) involved in oxidative stress in insect guts. But further studies are needed, especially those that use the nano-analysis techniques that are becoming increasingly available. One of the new technologies available for the study of metabolic and microbial interactions is the nanoSIMS (secondary ion mass spectroscopy) procedures described by Carpenter et al. (2013).

16.11 APPLICATION OF RESEARCH TECHNIQUES WITH ADVANCED MICROSCOPY TOOLS

Advanced microscopy tools, including fluorescence and confocal microscopy and several kinds of electron microscopy, can help us better understand insect diets as complex matrices and the elaborate interplay between the gut/food interaction. The integration of the microscopy techniques with the nano-analytical techniques discussed in the previous section can help resolve how various insect foods are treated in the insects' digestive systems. One example of the power of nano-procedures (or ultramicro techniques) is presented in a recent paper by Carpenter et al. (2013) where the identification and functional characterization of termite gut microbes is executed with a procedure called nanoSIMS analysis. The authors correlated application of techniques that would help researchers trace the breakdown of foods, and the uptake of the breakdown products could be of great importance in explaining why and how certain materials have high degrees of bioavailability while others lack this desirable quality. For example, the ability to track various forms of iron (the variety of chelates, non-chelated forms, and organic forms such as heme iron and ferritin-bound iron) could be of tremendous help in explaining why some iron-containing diet ingredients are useful while others are useless or even toxic. The complex story of the role of amorphous vs. crystalline ferric phosphate in the gypsy moth abnormal performance syndrome is a case in point.

16.12 IMPROVED DIET COMPONENTS BY APPLICATION OF ENGINEERED FOODS THROUGH REFINEMENT OF PROCESSING TECHNOLOGY

The human food science and technology community and the animal feeds community have made remarkable advancements in food processing and packaging, preservation, and food stability in the past decade since the first edition of this book was published. The fundamental issues or concepts in food engineering and processing technology remain fairly constant with recurring themes of fluid flow, heat transfer, preservation, cooling, evaporation or condensation, water activity (or generally psychrometrics), mass transfer, membrane separation, and desiccation appearing as central themes in food engineering (Singh and Heldman 2014). However, the research on improvements in our understanding and application of these concepts is growing exponentially as documented by the thousands of publications in this field yearly.

For example, Kumar et al. (2008) developed a continuous flow microwave heating system to treat vegetable purees. They explained the system as follows:

> Continuous flow microwave heating is a promising alternative to conventional heating for aseptic processing of low-acid vegetable purees. However, non-uniform temperature distribution and control of processing parameters are the major hurdles in the implementation of continuous flow microwave heating. This study was undertaken to overcome issues associated with the scale-up of a continuous flow microwave system from pilot plant scale to industrial scale and to conduct extended run times of 8 h based on the procedure developed.

And, they wrote,

> The slow rate of heat transfer from the wall to the center of the can limits the maximum size of the can. This is a major hurdle for more widespread use of vegetable purees as ingredients in the food industry. Excessive thermal treatment of the product also results in degradation of color, flavor, texture, and nutrients.

And, of course, most of the same issues apply to heating insect diets to activate gelling agents, to cook raw materials, and to sterilize the diets, especially large batches as are used in mass-rearing facilities. The kinds of problems solved by the use of flash sterilizers (pasteurizing equipment) as explained by Griffin et al. (1974) are nearly identical to the problems in heat-treating vegetable purees: uneven heating causing differential cooking and overheating product so that vitamins and other heat-labile components are destroyed.

The fast and uniform heating by the continuous flow microwave heating system allowed purees to be heated to about 120°C at a flow rate of 3.78 L/min, and residence time for food in the system was about 8 s, which is comparable to the flash sterilizer treatment of boll weevil and tobacco budworm diet (Griffin et al. 1974). Kumar et al. also mentioned that the product, once heat treated, could be automatically sent to a packaging system, which would be ideal for artificial diet used in insect-rearing systems. The point of this discussion is that the novel heat processing technology developed in the food science and technology domain is one of many technologies that could be available for the insect-rearing community.

16.13 EXPANDING THE SCOPE OF INSECT REARING: NEW USES FOR MASS-REARED INSECTS

Worldwide, there is a very large infrastructure for insect mass rearing where the large numbers of insects produced daily are used for (1) sterile insect techniques to suppress pest populations, (2) entomophagous insects for biological control, (3) feeder insects for captive animals, including pets and zoo/aquarium animals, and (4) education or celebration functions (such as for schools or for release of butterflies at weddings or other celebrations), (5) pollination, and (6) biomanufacturing of pharmaceutical products (Hughes and Wood 1998). Insects produced for research purposes are very numerous considering all the research laboratories in educational institutions and industrial labs where insect testing and studies are conducted, but these are smaller facilities that are widely spread geographically. However, with all these programs, the biomass of insects that they involve would be dwarfed by the production of insects for human food, should this become a widely accepted practice.

Recently, there has been a resurgence of thinking about using insects as food for humans—beyond the traditional insect cuisine as a novelty or seasonal supplement in tribal or cultures in rural settings (Carrington 2011, Klunder et al. 2012, van Huis et al. 2013). There are several recent papers that present information on the nutritional value of insects including three papers by Finke (2002, 2005, 2007) and papers by Mishra et al. (2003), Longvah et al. (2011), Zhou and Han (2006), and

Xia et al. (2012). These papers present information about the nutritional value of selected insects in the context of use as vertebrate foods, especially humans. The paper by Longvah et al. (2011) is an especially good model of how information about insects as human food should be presented because these authors included experiments on net protein utilization of the silkworm prepupae and pupae (Figure 16.5). Although the work was done with rats as model vertebrates, the evaluation of protein utilization is much more meaningful than a simple presentation of protein composition.

This increased attention to entomophagy by humans follows some modern pioneers as advocates of the potential benefits of using insects as food for cultural, economic, and environmental benefits, including Bodenheimer (1951), Taylor (1973), and Defoliart (1975). The monograph by van Huis et al. (2013) and published by FAO treats the subject extensively, including documentation of the nutritional value of a wide range of insects and strategies for encouraging programs that help facilitate the use of insects either as staples or as supplements, including ideas about converting insect biomass into a flour, which would at once enhance the palatability of the insects and possibly disguise the nature of the insects as a food for people whose culture has not traditionally espoused insects as food.

Although there is considerable enthusiasm for the possibility of using insects as a possible way to fill a void that is anticipated with a burgeoning human population, estimated by the UN (United Nations Department of Economic and Social Affairs, World Population to 2300. 2004. Executive Summary, Page 2) to grow from seven billion in 2014 to more than nine billion in 2050 (an increase of almost 30% of the current level in 36 years), there is a desperate need to find new ways to feed people. Many advocates of entomophagy by humans express optimism about the potential of insects to increase their biomass quickly, and they cite analyses of the nutritional composition of insects, especially their relatively high content of protein of quality comparable to some of the best foods currently in wide use (dairy products, chicken eggs, and plants such as soy with high levels of protein). These entomophagy enthusiasts point to the long-standing practices of various peoples throughout the world who eat a large variety of insects (van Huis et al. 2013).

However, most of the insects that are eaten by people come from harvested wild insects, with a few coming from rearing or insect cultivation situations. The sustainability of food sources from wild populations is a major problem because the populations of insects from the wild are sporadic and unreliable. Recognizing the problem of sustainability of foods that humans can depend upon, many human entomophagy advocates turn to mass rearing as a kind of sustainable farming approach. The insects as food advocates point to mass-rearing facilities such as the Pink Bollworm Facility in Phoenix, the Mexican Fruit Fly Facilities in Mission, Texas, and the San Miguel Petapa (SMP) Mexfly Production Facility in Guatemala. Each of these facilities produces about 200,000,000 insects/week for sterile release (SIT) programs (Gersabeck et al. 2010). The cost of producing this number of Mexican fruit flies under the conditions in the facilities in Texas and Guatemala is about $2.4 million/year. A preliminary model for using insects produced under representative mass-rearing conditions as human food is presented in Figure 16.6.

Given a recommendation of 50 g of protein per day as the requirement for humans (from recommendations by WHO 2002), and given a biomass estimate of 0.05 g/pupa (from Churchill-Stanland et al. 1986), the calculations in Figure 16.6 were developed. The cost of rearing Mexican fruit flies was based on a report from Fanson et al. (2014). The idea here is not to suggest that Mexican fruit flies from the Mexican and U.S. (Texas) mass-rearing facilities should be used as food, but instead the numbers are used as realistic costs and other requirements of mass rearing based on actual production numbers. The mass-rearing system that is cited in this estimation is one that has taken at least 20 years to develop and tens to hundreds of millions of dollars to implement at the current production levels. Therefore, the costs of production are based on a system where the infrastructure is already in place and was arrived at through very costly and exhaustive research, which was considered feasible because of the high risk and high costs of infestation of the fruit flies that were targeted. The facilities, equipment, and staffs of these facilities are already paid for, and the costs mentioned by Fanson et al. (2014) are not amortized to repay the building of the infrastructure.

This raises the question about whether the cost of the insects that can be produced for human food, under current levels of knowledge and infrastructure, is feasible. If the production of insects in a kind of farming/mass rearing is going to be used with any hope of feeding the burgeoning human population mentioned earlier, how much money will it cost per daily ration? The calculations from Figure 16.6 are based on a scenario where the insect component of a person's diet will be entirely from insect biomass, which would cost about $1.43–$1.90 (U.S.)/day. This system is used as an example because fruit fly production is at the top of the list of mass-rearing facilities (similar to the screwworm mass-rearing yields), and the efficiency of these facilities is highly refined and therefore an excellent model system. The point of this discussion is to show the constraints that must be estimated and dealt with, and it should not be interpreted that fruit flies are suggested as the optimal target for mass-rearing insects for human food.

The two figures are based on data provided by Dr. Hugh Conway (Mexican Fruit Fly Production Facility in Mission, Texas). The range of adult weights is between 15 mg and 20 mg/fly, depending on rearing conditions. The difference in yield is a very important factor that points out the need for improvement and optimization efforts for a system of food production from insect mass-rearing facilities. If only one-fourth of the daily ration was to come from insects, then the cost for the insect part of the diet would be between $0.36 and $0.48/day. This would be something like the rice dish in Figure 16.4 (a stir-fry recipe from the Raleigh BugFest 2014). It should be noted that if crickets were used instead of Mexican fruit flies at the current prices for crickets indicated by some Internet shopping, the quantity of crickets required to provide 50 g of protein would be about $12.50, or for using crickets to meet one-fourth of the daily human requirement, it would cost about $3.13/day (for a recipe such as the stir-fry in Figure 16.4).

These two estimations (the fruit flies and the crickets) are based on currently existing technology. If a first approximation of a feasible cost of meeting human protein nutrition needs were set at $1.00 or below, then the cricket costs would have to be reduced to less than one-third of the current prices, and the fruit fly costs would have to be reduced to about two-thirds of the current costs (with the caveat that the fruit fly production infrastructure is already established, and no amortization was figured into the cost as discussed here).

A few further comments are in order about potential production on a mass scale of insects for human food: the other nutritional features must also be considered, and the literature cited presents an optimistic view of the nutritional value of insects. They contain a fairly high energy content per unit

Figure 16.4 Cricket and mealworm in stir-fry dish served at BugFest in Raleigh, North Carolina, in 2014. People were lined up by the hundreds and waited for 10–20 min to sample the insect cuisine.

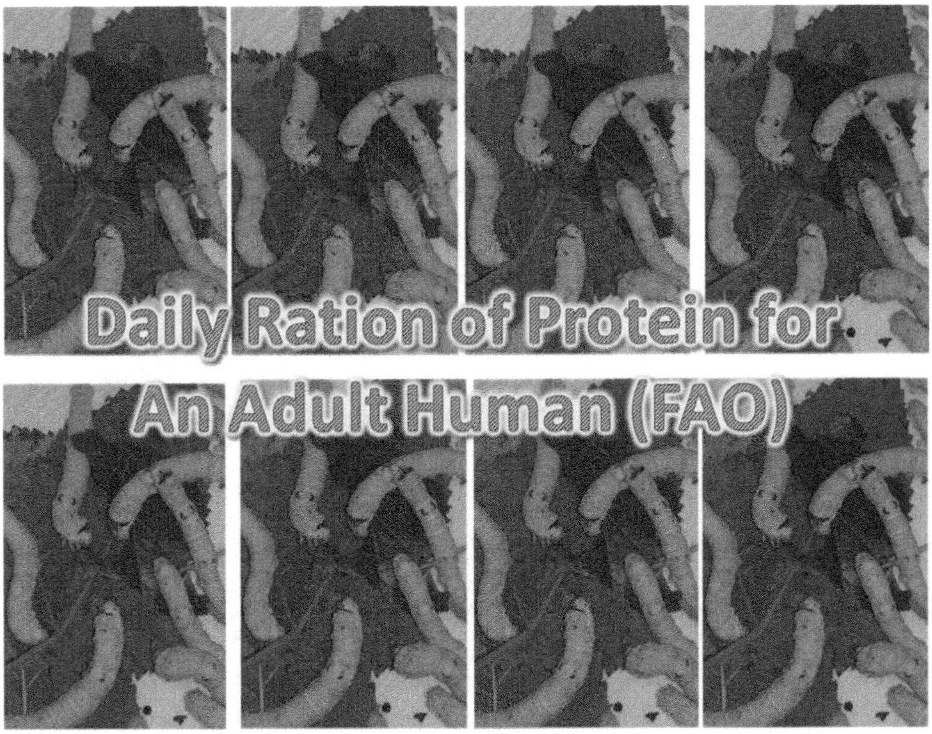

Figure 16.5 Full-sized silkworm larvae (*Bombyx mori*) close to pupation. Larvae are about 5 g fresh weight, withabout10%–20%protein.Therefore,a1-dayrationofproteinforahumanadultwouldequatetoabout50 silkworm larvae, the number seen in the eight frames of this image.

of biomass, a fairly good lipid profile, many of the minerals, and some of the fiber and vitamins that humans require. There is not adequate information about protein (or other nutrient) utilization of insects by humans. As far as mass production is concerned, those of us in the field of insect rearing especially those with experience in scaling up for mass-rearing capabilities know that management of microbial infestation (including pathogens and contaminants), dealing with environmental and genetic exigencies, obtaining and maintaining equipment for automation, and training competent personnel, is not as easy as many people without rearing expertise may think. Another point that is often raised is that there will be a need to bridge the cultural gap to convince people that they could (or should) eat insects. Current practices cited by van Huis et al. (2013) and experiences at the Raleigh BugFest convince me that the social or cultural barriers are not going to be as serious as the technical and scientific barriers that are indicated throughout this book and throughout the literature on insect diets and rearing systems.

Although there is considerable interest in expanding the use of insects as food by developing enhanced production technology, there are some serious drawbacks or potential limitations that must be fully considered. These drawbacks or considerations are the ones (modeled in Figure 16.6) and must be addressed if production of insects as human food is to become a serious and sustainable possibility.

While van Huis et al. (2013) present a compelling case for the need for increasing food sources for an increasing human population over the next century, and while they provide numerous examples of insects as human food and how advantageous expansion of current practices would be, it must be recognized that most of the instances of human entomophagy involve insects that have been harvested, rather than reared. Among the few exceptions—where insects such as mealworms and silkworm larvae and pupae (Figure 16.5) are eaten—insects such as termites, grasshoppers, ants, and other species are gathered and used by very small numbers of humans.

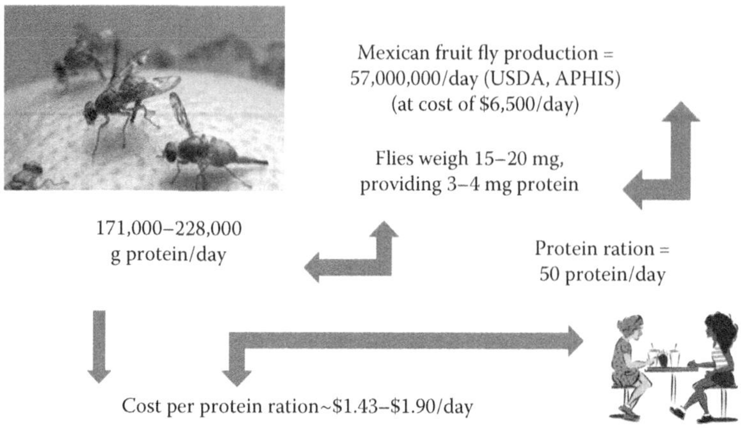

Figure 16.6 A model of the potential for using current mass-rearing technology for the production of mass-reared insects for use as food. The diagram is based on reports from USDA, APHIS mass rearing of Mexican fruit flies in facilities in Guatemala, Mexico, and the United States (Texas) where the daily production of pupae for sterile insect techniques costs about $6,500/day for 57,000,000 flies. Based on weights of flies (15–20 mg/adult) and an estimate of 20% protein, the cost of providing a ration of protein is between $1.43 and $1.90/day also based on WHO/FAO standards of 50 g of protein per adult human per day. The efficacy of this projected use of insects is discussed in the text.

While van Huis et al. (2013) mention impressive success in mass-rearing facilities, it must be considered that a substantial technology still needs to be developed to solve insectary problems including (1) selection of species of insects that meet the palatability and nutritional needs of serving as a staple in human diets, (2) domestication of the insects chosen, (3) provision of a diet that is suitable to sustain a healthy colony of insects that are produced in numbers that allow harvesting for food and continuation of the colony, (4) prevention of microbial contamination and pathologies—both of which being inherent in insect colonization, (5) provision of the environmental conditions that allow year-round insect production systems, (6) development of automation whenever possible to reduce labor costs and especially to improve process control that leads inexorably to improved and controlled quality, and (7) education and training of rearing personnel in all areas of insect mass-rearing technology. It is well documented across many industries that educated personnel who understand how to do their job and why they follow standard operating procedures produce higher-quality products at economical levels of productivity (Figure 16.7).

Figure 16.7 Cost of using crickets as a daily ration of human protein (nutrition) needs.

16.14 APPLICATIONS OF RHEOLOGICAL PRINCIPLES OF TEXTURE ANALYSIS FOR TEXTURE IMPROVEMENT AND PROCESS CONTROL

Throughout the history of insect rearing with artificial diets, there has been occasional but sparse mention of the importance of matching a diet's texture or consistency to the insect's feeding adaptations. Only sporadically have authors commented on the characteristics of a diet in terms of texture-linked sensory responses: this is in contrast to the thousands of references in hundreds of papers that treated chemical sensory responses. There are even fewer cases of application of analysis of some texture feature as a process or quality control measure.

In Chapter 15 of this book, there is a discussion of some of the aspects of texture characterization by rheological analysis, using texture profile analysis, measurement of diets' and diet components' viscosity and elasticity, large strain analysis, small strain analysis, and shear viscometry. Once diet specialists become aware of the rheological properties of diets and when they and their fellow entomologists have the tools to measure the diets objectively in terms of textural nuances, these specialists can start developing a systematic understanding of the sensory qualities of the diet in the context of how textural sensory qualities affect the target insect's feeding behavior. These rheological tools, as is the case with food science applications to human foods, will allow an objective analysis of how modifications in diets affect the diet's mouth feel, penetration resistance, flow characteristics, and many other factors that may be crucial to the acceptance and utilization of diets by target insects. Finally, besides the improvements in the understanding of insects' responses to diets and improvement of the diets' textural qualities, the rheological tools will provide an ideal set of components for process control and ultimately quality control of diets and the insects produced on those diets.

16.15 TWENTY-FIRST-CENTURY INSECT DIET PROFESSIONAL: SUGGESTIONS FOR A NEW CURRICULUM AND EDUCATIONAL PROFILE

The formal education of insect diet professionals has been typically a major in entomology or, less often, various biology majors (including zoology, plant pathology, or botany). The typical curriculum of entomology majors includes general entomology, insect systematics, insect ecology, insect physiology, and anatomy (all core courses), and a smattering of various specialty courses such as toxicology, genetics, resistance management, and a variety of other courses that provide little or no background in the basics of rearing and diets. The insect diet professionals and other rearing specialists who have been educated in other biological science majors have even less formal exposure to rearing and diet science and technology. A survey of curricula across the North America has revealed that insect rearing and diet work get virtually no formal attention (Cohen 2001). Food science curricula typically offer core courses in microbiology, food chemistry and physics, food processing equipment, and a variety of related satellite courses, but as would be expected, they do not prepare students for work in entomology. Therefore, between the two disciplines, entomology and food science, insect rearing and diet science and technology fall through the cracks, and the education that would be most valuable for rearing and diet professionals is completely missed. This fact helps explain the uncertain status of the insect rearing and diet professional and the problems faced by such professionals in terms of various forms of recognition (discussed by Cohen 2001).

What would go a long way toward correcting this problem and toward advancing insect rearing and diet science and technology would be a new curriculum designed deliberately to provide the kind of background needed to make the advancements outlined in this chapter and suggested throughout this book. The new breed of insect diet professionals would have a core background with at least one course each in general food science, food chemistry and physics, food microbiology, general biochemistry, general entomology, insect physiology and anatomy, insect ecology, and insect systematics. These courses are taught at a level where prerequisites are required, so it is implied that the incipient diet professional would have had general and organic chemistry, physics, math courses, statistics, general biology, general ecology, botany, general physiology, genetics (population and molecular), and the other elective courses that enrich biology and food science majors. In essence, what is called for is a hybrid major (a strongly integrated double major) with a strong background in both entomology and food science. The kinds of engineering contributions that are needed could come from people with the background in food technology engineering with a solid background in courses in entomology and biology (including microbiology). I have seen over the years that expectations that people with pure mechanical engineering background have been expected to step into a biologically based domain (an insectary) and solve problems that have intricate and intimate relationships to biological issues, and I have seen the frustration on the part of both the entomologists and the engineers in not being able to understand each other's problems and needs. Rather than approaching this problem with acrimony, it should be dealt with by understanding that the two domains had been brought together with a lack of mutual background education in the two seemingly disparate areas. Education of rearing and diet professionals will bridge this chasm.

In this context, Table 16.1 provides a detailed syllabus of two insect-rearing courses that are being offered online at North Carolina State University. These courses are an outgrowth of the author's experiences in teaching "special topics courses" in insect rearing and several years of teaching 1-week workshops for rearing professionals.

16.16 PUBLISHING ON INSECT DIETS IN THE TWENTY-FIRST CENTURY

The body of publications on insect diets and rearing systems is extensive over the twentieth century and the first 15 years of the twenty-first century, but it is a record with uneven quality and many inconsistencies. As with any academic discipline, there are some landmark papers and many lesser papers, but in rearing papers, especially ones on diets, the standards for what is publishable are not clear. As Cohen expressed in 2001 in an article advocating the formalization of insect rearing, it is often difficult to publish papers on rearing accomplishments, and he expressed the need to develop a set of standards and guidelines for papers written on this topic. But along with standards and guidelines, there is a burden that the journal editors and reviewers must accept—to make available a place where authors who meet the standards of science and technology can expect to have their works published. As the publication situation exists today, there is no dedicated section in any of the journals of the largest entomological society in the world for publishing a rearing paper. The journals have sections on biochemistry and physiology (including molecular biology papers), on ecology, on horticulture, on insect/plant interactions, on biological control, on host plant resistance, on forest insects, and so on; but there is no section on rearing—despite the fact that well over 60% (more than 90% for some subjects) of the insects reported in the various journal sections are products of rearing, usually on artificial diets (Cohen 2001).

Almost everyone in entomology wants the insects available to them to do their studies or to execute their programs, but there is excessive difficulty for those who develop the rearing technology to publish their findings. It is amazing that the subject—insect rearing and diet science and technology—upon which so much depends does not have a reliable and respectable place where its scholarship can be shared in a reliable way. This is not to say that insect rearing and diet papers do not get published: they do, but the process is haphazard and vague, and the quality of rearing and the products of rearing—insects suffer from the lack of standards and guidelines. The spotty quality of the papers in rearing is a testimony to these assertions, and my own experiences over the past 35 years serve as further testimony that there is a need to (1) guide researchers on what constitutes a worthwhile publication, (2) provide a competent set of editors and reviewers to judge works fairly and productively, and (3) have a place where the papers can be published and where they can be found with a simple search, rather than a forensic-type analysis.

Table 16.1 Online Insect-Rearing Courses Taught from NCSU

Course 1A Syllabus

Course 1A consists of 32 short video lectures on a variety of topics related to insect rearing. The following is a list of the topics that you will learn in the course

Overview (Who Rears Insects and Why?)

1. Course Philosophy
2. Who Rears Insects? Why Do They Rear Them? Where Are They Reared?

Insectary Structure

3. Insectary Layouts/Floor Plans
4. Insectaries as Environments
5. An Overview of Rearing Containers
6. Basics of Insectary Sanitation
7. Basics of Insectary Safety

The Biological Characteristics and Requirements of Insects

8. Insect Life Cycles in Rearing Settings
9. Insect Feeding Systems I: Nutrition
10. Insect Feeding Systems II: Mouthparts
11. Insect Feeding Systems III: The Digestive System
12. Gas Exchange
13. Movement and Internal Communication
14. Rearing-Related Components of Insect Development
15. Reproduction
16. Circulation and Transport
17. Social Biology in the Rearing Setting
18. An Overview of "Healthy" Insects

The Domestication Process (What Are the Steps in and Consequences of Domestication?)

19. Insect Genetics: A Brief Overview
20. Domestication: An Overview
21. Domestication: How It Affects Rearing and Insect Quality
22. Understanding Insect Feeding Systems as They Pertain to Rearing

Understanding Insect Diets (What Is in Diets and What So the Components Do?)

23. Types of Diets and Diet Terminology
24. Components of Diets and What They Do
25. More about Components of Diets
26. Interactions of Insect Diet Components
27. Stability of Insect Diets and Components
28. Water in Diets (Water Content and Water Activity)
29. Texture of Insect Diets I
30. Texture of Insect Diets II
31. Handling and Treatment of Diets and Diet Components
32. Important Issues in Diet Presentation

Course 1B Syllabus

Essentials of Insect Reproduction (Mating, Oviposition, and Diapause in Rearing Systems)

33. Mating in Rearing Systems
34. Oviposition in Rearing Systems
35. Diapause in Rearing Systems

(*Continued*)

Table 16.1 (*Continued*) Online Insect-Rearing Courses Taught from NCSU

Microbial Management

36. Kinds of Microbes Relevant to Rearing
37. Kinds of Microbial Interactions (Contaminants, Symbionts, Pathogens, and Commensal Microbes)
38. Managing Microbes I
39. Managing Microbes II
40. Managing Microbes III

The Insectary as Environment

41. Temperature and Its Biological Importance
42. Humidity and Its Relationship with Rearing Issues
43. Light (Photoperiod, Intensity, and Quality)
44. Gas Exchange
45. Noise and Vibrations: the Insectary Soundscape
46. Air Quality
47. Special Cases I (Soils)
48. Special Cases II (Aquatic Systems)

Introduction to Statistical Process Control (SPC) in Rearing

49. What Is Statistical Process Control?
50. What Are the Sources of Variation in the Rearing Setting?
51. Using Fishbone Diagrams
52. What Do You Measure and How Do You Measure It?
53. Using SPC Tools for Problem Solving and Process Improvement
54. Using Pareto Charts and Pareto Thinking
55. Using Fishbone Diagrams
56. Developing and Using Process Control Charts

Managing Processes in Rearing Systems (Instruments and Equipment)

57. Basic Laboratory Instruments for Rearing I: pH and Weighing
58. Basic Laboratory Instruments for Rearing II: Water Activity and Water Content
59. Basic Laboratory Instruments for Rearing III: Volumetric and Spectrophotometric Measurements
60. Basic Laboratory Instruments for Rearing IV: Microscopes
61. Diet Instruments I
62. Diet Instruments II

What Every Rearing Specialist Should Know: Developing Education and Academic Background for Self-Improvement as a Rearing Specialist

63. Further Coursework
64. Reading Books and Journals
65. Networking and Tapping into Other Disciplines
66. Using the Internet as a Resource

Supplementary Modules: Videos on Insectaries

Visiting an Insectary I: The Gast Facility
Visiting an Insectary II: The Pink Bollworm Facility
Visiting an Insectary III: The North Carolina State University Insectary
Visiting an Insectary IV: The Otis Gypsy Moth Facility
Visiting an Insectary V: The NC State Quarantine Facility
Visiting an Insectary VI: The NCSU Insect Rearing Education and Research Rearing Rooms

Note: The self-paced courses are presented with feedback from self-graded quizzes. They allow students of insect rearing to take classes at home or at work and maintain seamless communication with their instructor.

What was stated in the paper on formalization of insect rearing (Cohen 2001) remains valid today and demands even more urgently that several issues need to be corrected. Briefly, these issues include (1) sections in existing journals (or establishment of new journals) where rearing papers are welcomed, (2) establishment of a cadre of competent and willing editors and reviewers who can handle the flow of rearing papers, (3) an improvement in the scientific and/or technological quality of papers being written such as statements of rationale in every paper and statements of hypothesis and proper explanation of experimental design in scientific papers, (4) an acceptance for the publication of technological methods that report new findings or improvements in existing systems, (5) proper citation and acknowledgment of previous works, and (6) a recognition (as is modeled by the publications in food science and technology) that worthwhile and useful accomplishments may be incremental rather than saltatory and major breakthroughs.

With an understanding and adoption of the interdisciplinary components of insect diet and rearing systems, such as those characterized in the current book, there should be a renaissance of inquiry in rearing systems. Integration of the many techniques that are available today can make diet and rearing science and technology an exciting an even more productive field, but there must be a home—an outlet for the scholarship that is the outcome of the process of this inquiry.

16.17 TWENTY-FIRST-CENTURY INSECT DIET AND REARING PROFESSIONAL: AT FORMAL PROFESSIONAL STANDING

Clearly, if the educational gap as described earlier is bridged, there will be an inherent tendency for the status of diet professionals to improve. The process of consciously and deliberately designing a rearing and diet curriculum as the basis of formal education will enhance the ability of the diet professionals to do their jobs with greater scope of understanding. This improvement will undoubtedly boost the stature of diet and rearing professionals as will their newfound roles as well-trained specialists with a highly technical and scientifically based educational base. The formalization of the education process will also be likely to evolve into a better understanding and appreciation for the rearing and diet specialists by the colleagues and customers of the rearing and diet professionals. The formalization of the insect diet and rearing education can take the form of certification for support personnel and advanced degrees such as MAs and PhDs with an emphasis on rearing, diets, and food science for scientists.

CHAPTER 17

Introduction to Analytical Procedures for Statistical Process Control of Diets

The purpose of this chapter is to provide a set of methods that can be adopted by rearing facilities to allow them to test diets at various stages of the diet-making process. All of these methods have been used by the author in his laboratory, so they have been thoroughly vetted. The methods were selected because they were determined to be useful and generally not highly technical. However, some of the methods require some skill in the use of equipment and reagents, and they all require some practice. The assumption for some of the methods is that the users will have had some courses and/or experience with equipment such as spectrophotometers, volumetric equipment (especially pipettes), balances, and a grasp of microbiology (for the microbial tests). There is also a safety issue with methods that require use of solvents such as petroleum ether and acids or bases. The safety procedures discussed in Chapter 14 certainly apply here, and unless the personnel performing tests such as the sulfuric acid–based anthrone or phenol test have appropriate experience and facilities to handle concentrated sulfuric acid properly, these tests should be avoided.

17.1 DEVELOPING STATISTICAL PROCESS CONTROL PROCEDURES

As discussed in Chapter 11, the time to begin developing process control procedures is when things are working well in the rearing process. The purpose of statistical process control (SPC) is to determine when and where the normal working process has gone out of control; therefore, it is essential to have a baseline of normal process characteristics.

17.1.1 Protein Analysis: An Overview

Protein analysis is extremely important because of the roles of proteins in foods and in nearly every aspect of biochemistry. For several reasons, protein analysis is also extremely difficult to do properly. Despite more than 100 years of research on various methods of analysis, the difficulties remain obstacles. An understanding of the nature of proteins helps us understand why these difficulties exist.

Proteins are made of 18–20 different kinds of amino acids (structures and chemical characteristics in Chapter 3), which are joined together by special connections called peptide bonds (Chapter 5). Different kinds of proteins can be made of anywhere between a few dozen amino acids to thousands of amino acids in each protein molecule (giving larger proteins a molecular weight of

400,000–500,000 Da). Proteins have many biochemical functions: as enzymes, as structural units (e.g., making up skin and hair), and as nutrient storage (such as seed and egg proteins). Proteins may be water soluble or insoluble. Their characteristics are strongly influenced by pH, temperature, and mineral environments.

Soluble proteins are more easily analyzed (quantified) than insoluble proteins. Many tests have been established to quantify soluble proteins: absorbance at 282 nm, Bradford, Lowry, etc. In each of these tests, we use a standard, highly purified protein such as bovine serum albumen (BSA), which we compare with our protein of interest by using a *standard curve*, which is a plot of *concentration of protein* (*x-axis*) vs. absorbance at a given *wavelength* (*y-axis*). In each of these tests, the absorption of a light (or ultraviolet) wavelength is related to the amount of a specific amino acid in the protein (such as tyrosine or phenylalanine in the 282 test) or the amount of copper dye that binds with a specific portion of a protein. This means that these tests have an inherent error that comes from differences in the amount of tyrosine or phenylalanine in given proteins. So, for example, a type of protein that has many tyrosine units per protein molecule will give a high reading of absorption at 282 nm, while a potentially much larger protein with few tyrosine or phenylalanine will give a lower absorption value. With this point established, it is evident that two kinds of protein with molecular weights of 20 kDa (kilodaltons), one having one tyrosine, the other having 10 tyrosine units, would give a 10 × difference per unit of protein weight.

Method	Comments
Kjeldahl[a]	Measures total nitrogen from each amino acid, overestimates samples with nonprotein nitrogen (such as urea, ammonia, and alkaloids), is "fooled" by melamine or other nonprotein nitrogen-rich sources. It is an indirect method, and it takes special equipment (for digestion, distillation, and titration of samples after they are digested by aggressive acid and heat treatments). This method has been used for more than 100 years and despite its problems remains the most conventional method of food analysis and evaluation of ecologically important proteins (e.g., to develop a nitrogen budget for ecological communities).
Absorbance at 282 nm	Estimates protein according to the number of tyrosine and phenylalanine molecules present in protein.
Copper dye–binding methods	Used with known standards for comparison to measure only dissolved proteins, can be accurate only if most of your protein is dissolved.
Acid orange 12	Dye-binding method where proteins are bound by orange dye, causing proteins to precipitate. Works with ALL proteins, soluble and insoluble. Binds with basic amino acids in proteins (lysine, arginine, and histidine).
Hydrolysis and amino acid analysis	This is the most accurate and by far the most complex method. It involves the complete digestion of proteins by strong acids, strong bases, or protein-digesting enzymes. Once the proteins are digested into their component (free) amino acids, the amino acids are subjected to chromatographic analysis, which requires equipment that costs roughly $100,000 and demands technical expertise to do reliable analyses. The procedures take days.

[a] Please see Chang 1998 for a thorough explanation of Kjeldahl analysis.

The Cohen laboratory has investigated all of these methods and has determined that the *acid orange 12* method is the most reliable one for the purposes of routine protein determination for process control in insect diet production. The theory behind this method is that the dye called *acid orange 12* binds with the basic amino acids that are present in nearly every protein. Note: there are exceptions to this generalization, including gelatin and collagen (both of which contain no or virtually no basic amino acids). Therefore, if it is desirable to estimate the protein content of gelatin or collagen, there would be virtually no reaction from *acid orange 12*. Comparing the responses of various proteins to *acid orange 12*, it has been determined that on a per-protein weight basis, a protein like BSA will give a higher number than an equal weight of beta-lactoglobulin. It must be understood that every other protein analysis method (except for total amino acid analysis after hydrolysis) offers similar errors.

When *acid orange 12* is added to a protein-containing sample and mixed thoroughly, the dye that gives the solution an orange color *binds* with the proteins' basic amino acids, *and the dye is removed* from the solution, *making the mixture less intensely orange* than it was before the dye–protein interaction. *Therefore, the rule of thumb is* that *the lighter-colored the solution after binding, the more protein is present.* It should be noted that the dye that has bound (reacted with) the protein plus the protein itself form a precipitate and are forced to the bottom of the centrifuge tube during the process of centrifugation.

To use this relationship to make quantitative measurements of protein content, a spectrophotometer is used. The spectrophotometer produces a given light value (color or wavelength) and passes a beam of monochromatic light through the solution and gives a reading of how much light has passed through the test tube (or cuvette, or the well in a plate reader plate). The more a solution in the beam absorbs light, the higher the absorption reading. In the case of *acid orange 12*, it has been established that the solution most intensely absorbs light of the wavelength 482 nm (nanometers). The protocol provided here has been established with a series of *protein standards* with different concentrations of known protein standards (e.g., beta-lactoglobulin or BSA) and with a range of solutions that contain 1, 2, 3, and 4 mg of protein in each mini-test tube (microcentrifuge tube). Then, the *acid orange 12* dye is added in its undiluted form, shaken vigorously in a vortex mixer, then (after a period of incubation) the mixture is centrifuged so that the more dense protein/dye precipitate can fall to the bottom of the tube in a pellet, leaving the upper layer (called a supernatant) that contains the remaining dye that has not reacted with protein. Theoretically, all the protein that was in the solution has reacted with the dye, and again the rule of thumb applies: *the more protein that there was in the mixture, the less color (dye) remains in solution.*

This test is designed so that it can be done with miniaturized conditions to save expenses of materials and to reduce the lab space needed to do these assays. *The test was further designed to give a straight line to represent the relationship between the amount of protein and the amount of absorbance.* This is very important because the accuracy of the analysis depends upon the linearity of the relationship. It has also been determined that there are limits to the linearity of the relationship between sample protein and *acid orange 12. Therefore, it is important to perform these tests within the range specified.* We have found that for wheat flour, given the amount of reactive protein, the optimal amount of wheat flour to test is about 25 ± 5 mg. This amount of wheat flour should give an absorbance for the samples that will be within the range of the standards (1–4 mg of protein).

Another important aspect of the tests is that once the mixture of dye, protein, and water in the micro-centrifuge tube is processed (dye added, shaking, centrifuging), a sample of the supernatant (upper layer of dye) must be diluted with water so that it will be within the absorbance range of the spectrophotometer (please note that the readings of a spectrophotometer are reliable only between the absorbance values of 0.000–2.000). A good working solution is 50 μL of the upper phase (supernatant) with 1500 μL water, giving absorbance readings within a reliable range.

Some other considerations are as follows:

1. The finer the sample is ground, the more the dye has a chance to gain access to the proteins. In the standards, where all the proteins are in solution, the dye has complete access to the basic amino acids. The dye can "find" the target amino acids and react with them. If the particle size is large, and the dye cannot "find" hidden proteins, the dye cannot react with them, and these hidden proteins will not register in the readings. Therefore, it is fairly important to use finely ground samples and to allow an adequate reaction (incubation) time for dye/protein reaction to take place.
2. Because the *acid orange 12* dye reacts with lysine, arginine, and histidine (all very important essential amino acids), this method is also roughly a test of protein quality. Many researchers and livestock professionals regard lysine values as the ultimate measure of protein quality.
3. Because wheat flour and other nonsoluble nutritional products (yeast, milk [which is not entirely soluble], corn meal, bran, etc.) cannot be completely accessed by the dye, there will tend to be an understatement of the amount of protein present. You can compensate for this as the Udy people do

by having your sample tested initially with Kjeldahl testing, then use the Kjeldahl number and the *acid orange 12* dye number to develop a correction factor (a constant), which you would then multiply by your percent protein done by the dye method. This correction factor would hold up as long as you were using wheat flour samples. If you changed to yeast, bran, or milk, you would have to get a new Kjeldahl number for that correction factor. This is only in case you want to know (as closely as possible) the *absolute* amount of protein in your sample. Our suggestion is that you retain a sample of high-quality wheat flour from a batch that you know works well for your insects, then use that standard wheat flour (or yeast, or bran, or milk) whenever you get a new batch of product.

17.1.2 Protein Evaluation Method

This system can be used to measure the relative or absolute amounts of protein in a variety of food materials, including wheat flour, soy flour, or various plant materials. Although accurate quantification of protein is difficult, the dye-binding method (using acid orange 12) is exceptional, both in its reliability and in its simplicity. An important point is that this process (total protein analysis) is given in careful detail, but the same steps can be applied to the analysis of carbohydrates, antioxidants, and any other components where spectrophotometric analysis (or plate reader analysis) is applicable.

Method of protein quantification:

1. Establish and retain a sample of the type of material such as wheat flour or soy flour that has been determined (tested) to be completely suitable for rearing the target insects. Save about 100 g in a vacuum-sealed container, preferably in aliquots (individually sealed subsets of the whole batch of material). Prior to aliquot distribution, make sure that the sample is well mixed (homogeneous) and that each aliquot represents the batch as a whole. Then, separate about 1 g fractions.
2. Measure the sample as follows (please note that it is important to use precise and accurate measurements, so volumes should be measured with good volumetric equipment such as micro-pipettes, and weighing should be done with a balance that reads to at least 0.1 mg [0.0001 g]):
   a. Make up fresh standards. Once a set of protein standards is prepared, they can be refrigerated for about 2 weeks, provided that they are kept sealed and refrigerated. Make standards as follows:
      i. Weigh 0.500 g (500 mg) of beta-lactoglobulin or BSA or a similar pure protein.
      ii. Dissolve the protein in 50 mL of deionized water (= DIW). This results in a *10 mg/mL* concentration, and this amount will be used as a *working solution* to make dilutions for the other standards. Next, take 2 mL of the working solution and add 8 mL of DIW. This is a *2 mg/mL* solution of standard. Next take 4 mL of working solution and add 8 mL of DIW. This is a *4 mg/mL* solution of BSA. Next, take 6 mL of working solution and add 4 mL of DIW. This is a *6 mg/mL* solution of BSA. Finally, take 8 mL of working solution and add 2 mL of DIW, resulting in an *8 mg/mL* solution of BSA.
      iii. Take 0.5 mL (=500 μL) of each standard (1, 2, 3, 4 mg of protein standard), and add each to 2 mL capacity plastic centrifuge tubes. Mark all tubes with a Sharpie pen.
      iv. Next, prepare the samples by carefully weighing 0.035 g (35.0 mg) of the wheat flour being tested. Repeat this with eight more samples of about 35.0 mg each and place each sample in the sample's own 2 mL centrifuge tube. Add 500 μL of DIW to each sample tube and mix sample into water with a vortex mixer for *5 min in a pulsed vortex mixer.*
      v. Add 1.5 mL (1500 μL) of undiluted acid orange 12 dye solution to each of standards and samples. *Cap tubes and vortex all tubes for 30 min in a pulsed vortex mixer at highest possible speed.*
      vi. Allow all standards and samples to incubate for 30 min.
      vii. Turn on your spectrophotometer, setting the wavelength to 482 nm and letting it warm up. After about 30 min of warm-up, with a cuvette containing 1.5 mL DIW, set blank to 0.
      viii. Centrifuge all tubes (standards and samples) for 5 min at high speed.
      ix. Now, remove 50 μL of each tube (dye solution plus standard or sample) and place in cuvette. Add 1.5 mL (1500 μL) of DIW, allowing solution to equilibrate (it is a good idea to mark the cuvette at the top away from the light path with the concentration or sample name ["standard 1," "sample 1," "sample 2," etc.]).

x. Next, read each cuvette's absorption and record data in a data table like this:

Standard or Sample	Absorbance
1	1.481
2	1.167
3	0.853
4	0.535
Sample 1 (31.29 mg)	0.940
Sample 2 (36.05 mg)	0.853
Sample 3 (36.16 mg)	0.841
Sample 4 (35.90 mg)	0.881

3. Graph the standards in Excel (or some other program that will allow establishment of a linear regression).
4. Use the linear regression function in Excel to calculate the regression equation and R value. Note, if the R value is below 0.90, the analysis process should be repeated with greater care in making measurements. The regression equation is calculated as follows: ($y = -0.315x + 1.797$ for the data from standards 1–4).
5. The absorbance reading for each sample is the y value that you must solve for using the regression equation, which in this example is $y = -0.315x + 1.797$. Therefore, for sample 1, the y value is 0.940. So, this equation must be solved: $0.940 = -0.315x + 1.797$. To solve the equation for x (which is the amount of protein in the sample), subtract 1.797 from both sides giving you $-0.857 = -0.315x$. Next, you divide both sides by −0.315, giving you $x = 2.72$. Therefore, the 31.29 mg sample has 2.72 mg of protein in it or 2.72/31.29 × 100, giving 8.7% protein. This calculation is repeated for each y value. Once the average for all samples is calculated, there is a basis for establishing the table of protein concentrations that will be used in the construction of a control chart or other graphic procedure to allow diet personnel to decide on whether any given sample is "in control."
6. Remembering that this amount of protein is subject to the errors discussed in the section of this chapter on Protein Analysis, a comparison of these values can then be made when future shipments of the material in question are received or when other portions of the diet-making process are being tested. It is through this comparison of the ongoing process with past successful applications of the process that decisions can be made as to whether or not the process is in control at each given point where testing is taking place (see Figure 17.1). It is only through this type of comparison that it can be decided that a new sample or process step is acceptable or if it must be modified or replaced.

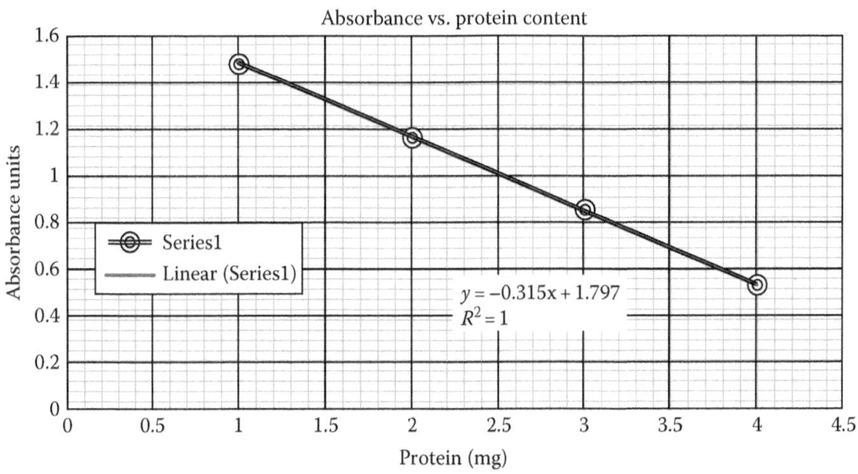

Figure 17.1 Standard curve for acid orange protein analysis.

17.1.3 Lipid Analysis in Process Control of Diets or Diet Components: An Overview

17.1.3.1 Background on Lipids in Diets

Throughout this text, the importance of dietary lipids is emphasized. The fact that all insects have been shown to require a source of dietary sterol (some form of the sterol nucleus) indicates only part of the importance of supplying target insects with appropriate amounts and forms of lipids. The role of lipids as precursors for insect hormones (immune system compounds such as prostaglandin and eicosanoids, as well as membrane lipids from fatty acids and ecdysterones and membrane structures from sterols) adds to the value of lipids in nutrition and the fact that structural and energy roles of fats further reinforce the point about lipids as crucial diet components. Finally, the fact that lipid-soluble vitamins and antioxidants cannot be adequately supplied if the overall lipid concentration is too low adds a conclusive argument that lipids are central in insect diets. This leads to the argument that lipids must be considered a key factor in a scheme of process control in insect diets. However, because lipids are among the most complex chemical groups, biochemically, and because lipid analysis can be extremely complex and specialized, rearing/diet specialists must be judicious in selecting the target lipids and how they should be evaluated and analyzed.

17.1.3.2 Rationale for Lipid Analysis in Diet Process Control Systems

Starting with the premise that the right kinds of lipids and the right amounts are important factors in insect diets, it becomes clear that any deviation from optimal lipid composition can lead to problems in the insects' fitness. Clearly, if specific lipids such as cholesterol or plant sterols are essential, anything that causes a deviation from optimal concentrations or suitable proportions of these nutrients can be problematic. Also, many lipids are very vulnerable to degradation, including enzymatic degradation, oxidative degradation (such as lipid peroxidation), and losses due to separation.

Some of the major lipid groups and functions are presented in Chapters 3 and 5, but for the purposes of devising a scheme of process analysis, it is sufficient to understand the term "lipid" as it is used here to include complex and simple lipids, neutral lipids, and polar lipids. There are many kinds of analytical techniques routinely used for lipid analysis in biochemistry laboratories and in many food science and technology laboratories. These methods include chromatography (gas, liquid, thin layer, paper) and various separation procedures that take advantage of differential solubility of lipids vs. non-lipids, allowing target components to be isolated or purified by partitioning. Often techniques such as iodine number are used to determine the degree of saturation of fatty acids or saponification number characterizes the fatty acid content of triacylglycerols (Min and Steenson 1998, Pike 1998).

Many of the methods for lipid analysis require specialized equipment and techniques that demand considerable skill and training of analysts. For example, gas–liquid chromatography (GLC or GC) requires the use of a gas chromatograph that costs upward of U.S. \$15,000–\$20,000, a supply of gases such as purified helium, hydrogen, and air, gas cylinder regulators that cost upward of \$400 each, special syringes or automatic injectors, data systems, and a supply of reagents for extraction and derivatization. The training for proper operation and interpretation of data demands that the technician who would perform the GLC analysis spend at least 6 months and the operator would need to spend at least 2–3 days per week performing routine tests. This demand of equipment, supplies, and skilled labor is prohibitive for most rearing operations.

Many of these techniques are useful for a detailed analysis of lipids, especially in a research laboratory, and many can be applied to a process control system. However, two techniques are selected here because they are both robust and simple enough to be performed with only a moderate amount of training and equipment that costs less than \$2000–\$3000.

17.1.4 Crude (Proximate Lipid Analysis: Using a Soxhlet Extraction Method)

In this procedure, you will use the Soxhlet apparatus to extract lipids from samples of wheat germ. This method is useful for many kinds of diet components, but wheat germ will be used as a model. The sample should be ground to a fairly fine powder (particle size ~50–100 μM) using a hammer mill, cutting mill, or ball mill (other mills such as Wiley mill, coffee grinder, mortar and pestle, and cyclone mill may also be used). The sample should be dried under a standard drying condition, at 105°C for 20 h, which aggressively drives off virtually all the water (or at 60°C, which drives off most of the water but protects the volatiles from being removed). Before starting the process, be sure to chill the coolant (a mixture of 50:50 antifreeze: water will work well) to under 0°C.

Caution: This work should be performed in a fume hood where care is taken not to let petroleum ether fumes enter the lab and where the petroleum ether is kept from open flames or sparks because of its highly flammable nature. *Also, this test should be performed only by personnel trained in safe handling of solvents.*

1. Place about 2–4 g of ground dry sample into a cellulose thimble, making careful note of the thimble weight alone and thimble weight with sample (this allows you to calculate the sample weight used in your analysis).
2. Place thimble with sample in the extraction chamber.
3. Place some boiling chips (Teflon or mineral chips) in the solvent flask. Pre-weigh the round bottom solvent flask (together with boiling chips), then add about 100 mL of petroleum ether.
4. Assemble the Soxhlet apparatus as shown in Figure 17.2a and lower (using the ring stand clamps) the apparatus into the heating mantle (as shown in Figure 17.2b).

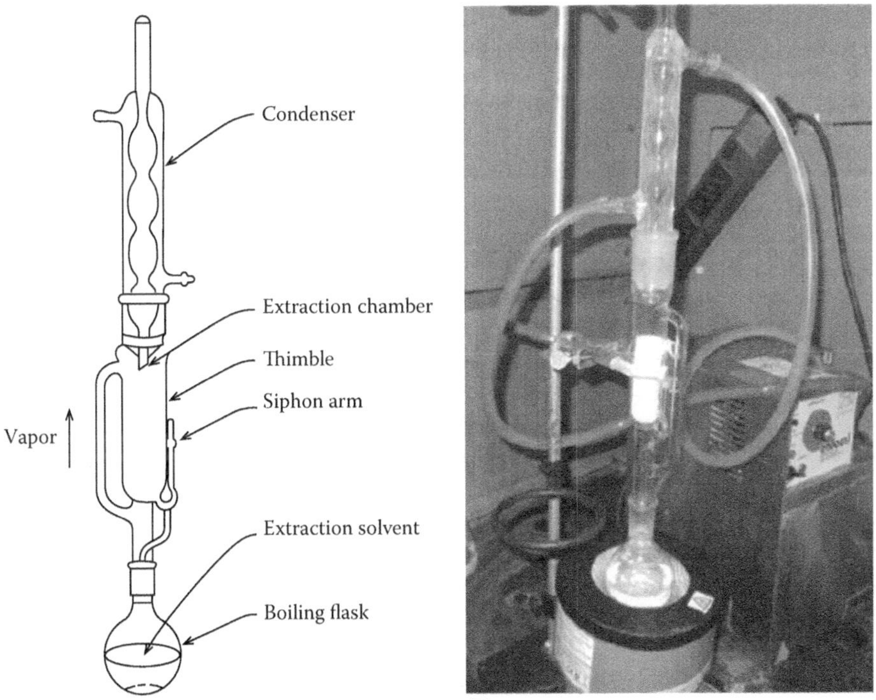

(a) (b)

Figure 17.2 (a and b) Soxhlet apparatus, with boiling flask on bottom (containing solvent such as petroleum ether), extraction chamber containing cellulose or glass thimble, and condenser on top with cold antifreeze and water mixture circulating through fluid space surrounding the rippled glass.

5. Turn on the heating mantle so that the petroleum ether comes to a rapid boil. Allow the evaporated ether to pass around the sample thimble and rise into the condenser. The solvent vapor will condense in the cold finger in the condenser, and it will drop as a liquid directly into the thimble where it will saturate the sample, and the level will rise. When the level has risen to the inverted "U"-shaped siphon arm, it will flush back down into the solvent flask.
6. Each flush is considered a single cycle (flush cycle), and after about 10 flushes, the sample should be fully extracted. Once the process has begun (it's slow at first, but then accelerates), it should take no more than 45–75 min to go through 10 cycles.
7. Shutting down: once the sample has been extracted, the heat should be turned off, and the Soxhlet apparatus should be raised (using the ring stand clamps).
8. Once everything is safely cooled, the solvent flask can be separated from the extraction chamber, which can be removed from the condenser. The solvent flask can be set aside in the fume hood and allowed to evaporate fully (allow about 24 h for total evaporation).
9. Now, the sample lipids (petroleum ether–soluble materials) remain in the flask. The flask can now be weighed again, and the difference between the final flask weight and the original weight is equal to the extractable material (mostly lipids).
10. The flask can now be washed with three flushes of petroleum ether, and the flask and boiling chips can be used for the next analysis.

Some hints: (1) It is a good idea to have at least as many flasks as you plan to sample (if you are doing five tests of subsamples per batch of diet material, then you should have five flasks so that you do not have to wait for 24 h between your samples). (2) Once you have done the extractions a few times, you will determine the variability of this test in your hands, and you should be able to reduce the number of analyses per batch to three instead of five. (3) The extract can be used to do further lipid analysis such as thin layer chromatography or GC to determine lipid groups (by TLC) or fatty acids and sterols (by GC). (4) The procedure can be used with other solvents, and it can be used to make extracts from various materials (such as plants that your insect consumes in nature).

17.1.5 Carbohydrate Analysis: An Overview

Carbohydrates are highly important and very diverse components of insect diets (Chapters 3 and 5 explain this fully). They function as sources of energy and materials for building cells (growth), as feeding stimuli, bulking agents (helping to shape texture), and various other functions as components of diet matrices. Also, many of the materials called "dietary staples" in this book have variable carbohydrate contents. This includes various wheat products (wheat germ, wheat bran, and wheat flour), soy products, other plant materials, including leaves or roots that may be added to diets, as well as eggs, meats, and microbial components such as yeast products. The variation may be in terms of specific sugars—including monosaccharides and oligosaccharides—starch, glycogen glucans, etc. The carbohydrates may be soluble (such as free sugars) or insoluble such as starch and other polysaccharides. There are nearly as many methods of analysis as there are different kinds of carbohydrates, including chromatography (GC, HPLC, TLC, and paper and column chromatography), various enzyme techniques, several physical methods such as differential solubility, and the very widely used spectrophotometric techniques. The technique explained here (the anthrone test) is a spectrophotometric procedure that is used for many kinds of carbohydrates. Many carbohydrate analysts use a similar colorimetric technique, the phenol–sulfuric acid procedure.

17.1.6 Carbohydrate Analysis: Soluble Carbohydrate Analysis

1. Prepare sample extracts by carefully weighing material that is to be extracted (we can use wheat germ as an example) and add water. A good starting place would be 20 mg of wheat germ in 2.0 mL water (about 10 mg/mL); vortex for 15 s, then centrifuge at 10,000 rpm. The supernatant will contain extractable (soluble) sugar. This sample will be used in your analysis, and it should be prepared

freshly just before your tests, or it should be kept refrigerated if it is to be stored for 2–3 days or frozen if it is to be stored for longer than 3 days. *Note: However you treat your sample, you must remain consistent for all the samples that you are working into your process control database!*

2. Prepare anthrone reagent: 150 mg anthrone powder is dissolved in 100 mL of sulfuric acid/water (75% sulfuric acid). Remember to use extreme caution in handling sulfuric acid; wear proper gloves, goggles, lab coat, and an acid-resistant apron. Always add acid to water to prevent explosive splattering.
3. Prepare standards: 100 mg of glucose in 100 mL water (=1 µg/µL). From this standard, make dilutions of 1 mL standard and 4 mL water (=0.2 µg/µL), 2 mL standard and 3 mL water (=0.4 µg/µL), 3 mL standard and 2 mL water (=0.6 µg/µL), and 4 mL standard and 1 mL water (=0.8 µg/µL).
4. Add 50 µL of each standard to Eppendorf centrifuge tube so 50 µL of undiluted standard provides 50 µg of glucose standard. Next, 50 µL of the standard with 0.8 µg/µL = 0.8 × 50 or 40 µg of glucose; the 0.6 µg/µL standard provides 30 µg of glucose; the 0.4 µg/µL standard provides 20 µg of glucose; the 0.2 µg/µL provides 10 µg of glucose in 50 µL; and finally, a standard (blank) with no glucose added (deionized water only) provides the 0 standard.
5. Next, add 1500 µL of the anthrone/sulfuric acid reagent to the standard; cap the tube carefully and swirl the tube to allow reagent to mix with the standard.
6. Using a floating holder for centrifuge tubes, heat tubes at 90°C for 20 min.
7. Place holder and tubes in ice water bath for 2–5 min.
8. Using a disposable transfer pipette, transfer the contents of the tubes (*carefully*) to semi-micro-cuvettes or to plate wells, if a plate reader is to be used.
9. Read absorbance at 625 nm, then construct standard curve as shown earlier. (Please note that the method of solving for Y in terms of mass of sugar equivalents of glucose is the same method as described in the section on protein analysis in 1b.)
10. Solve for Y to estimate the sugar content of the sample (in grams), then divide the number of micrograms of sugar per unit of sample weight. Record this figure for use in the SPC control chart (Figure 17.3).

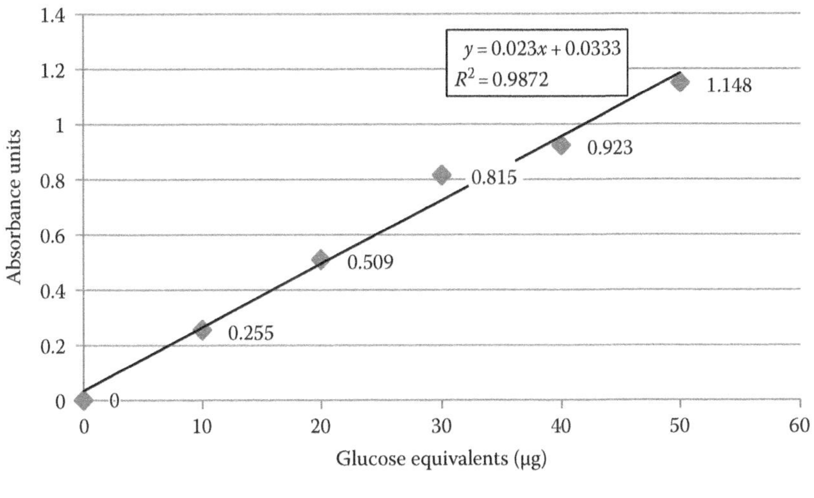

µg Glucose	Absorbance Units
0	0
10	0.255
20	0.509
30	0.815
40	0.923
50	1.148

Figure 17.3 Anthrone carbohydrate test: absorbance of anthrone-reacted glucose measured at 625 nM. See text for details.

17.1.7 Antioxidants and Free Radical Scavengers: An Overview

As discussed in several places in this book (especially Chapters 3 through 5), the role of antioxidants or free radical scavengers is becoming increasingly well documented. It is also well known that antioxidants can be unstable under diet preparation or storage conditions. Therefore, it can be a very important issue to be able to evaluate the overall antioxidant capacity or free radical scavenging capacity of a diet component or the synthesized diet. There are several methods of measuring antioxidant or free radical scavenging capacity, one of the most simple and widely used being the 1,1-diphenyl-2-picryl-hydrazyl (DPPH) antioxidant or free radical scavenger assay (Sharma and Bhat 2009). The free radical DPPH happens to be purple colored when it is intact, but it loses that color as it reacts with free radical scavengers such as ascorbic acid, tocopherol, and various other chemicals that have free radical scavenging (neutralization) potential. In the test described here, you will use ascorbic acid as a standard—a well-documented antioxidant/free radical scavenger. You will see in this test that the free radical scavengers from your diet can take out the purple color (such as seen in Figure 17.4a). You will measure the extent to which the scavenger is "neutralized" from your diet's antioxidants/free radical scavengers. This important test can serve as a measure of freshness of your diet or of the raw materials that you put into your diet. The test can even be modified by linking with the enzyme ascorbate oxidase to help you determine specifically the ascorbate in your diet after various treatments.

It will require an effort to determine the proportions for the samples in your rearing program, but once you establish a set of limits, you can continue to use the proportions and conditions routinely. For example, if you use 25 mg of dry pulverized diet and extract this sample in 1 ml of acetate-buffered methanol (60:40 methanol:acetate), you can remove aliquots of 5, 10, 20, and 50 µL in 1000 µL of DPPH reagent. The amount of extract that gives you readings of absorption that fall within the range of standards shown in Figure 17.4b would be the amount that you would use for all your readings. Once you solve the linear equation to determine the micrograms of ascorbic acid equivalents of free radical scavenging in your sample, you can correct for the weight of the sample and use the final value for the final value for determining the free-radical scavenging capacity in terms of "equivalents of ascorbic acid per g of diet."

17.1.8 Antioxidants and Free Radical Scavengers: Analytical Procedure

1. Prepare extract of diet material (or other substance to be analyzed for free radical scavenging/antioxidant activity) by using an appropriate-sized aliquot (such as 25 mg of wheat germ or powdered leaf material). Place the aliquot in 2.0 mL of the buffered methanol (described later). In preparing the sample for use in SPC measurements, it is important to be consistent in drying the sample thoroughly and grinding the sample to a consistently fine powder and to use exact proportions of the solvent (60 methanol:40 0.1 M acetate buffer at pH 5.5).
2. To prepare buffer, add 28.8 mL of 1 M acetic acid and 273.3 mL of 0.3 M sodium acetate in a volumetric flask and add deionized water to the 1.0 L mark. Mix HPLC grade methanol 60:40 with the buffer and use this to make standards with ascorbic acid and the DPPH reagent, as well as the extraction buffer.
3. To prepare DPPH solution, add 8.0 mg of DPPH (1,1-diphenyl-2-picryl-hydrazyl or its synonym, 2,2-diphenyl-1-picrylhydrazyl available from Sigma Chemical Company, St. Louis, MO) to 100 mL of buffer from Step 2. Dissolve completely and keep stored in aluminum foil covered or amber glass container. This solution is about 200 µM of DPPH.
4. Prepare the ascorbic acid standards fresh each time you do the test. Prepare a stock standard of 1000 µM ascorbic acid by adding 0.176 mg of ascorbic acid to 1.0 mL of buffered methanol (from Step 1). For convenience, you can use larger amounts of ascorbic acid and buffer, which can help you make your measurements more accurately.
5. Into each of five centrifuge tubes, add the following amounts in microliters of deionized water and ascorbic acid solution (from Step 4) in ratios of 20:0, 15:5, 10:10, 5:15, and 0:20 water to ascorbic acid solution. These will give you a total of 20 µL in each tube of 0, 5, 10, 15, and 20 µM of ascorbic acid.
6. Next add 1000 µL of the DPPH/buffer solution, cap, and shake.

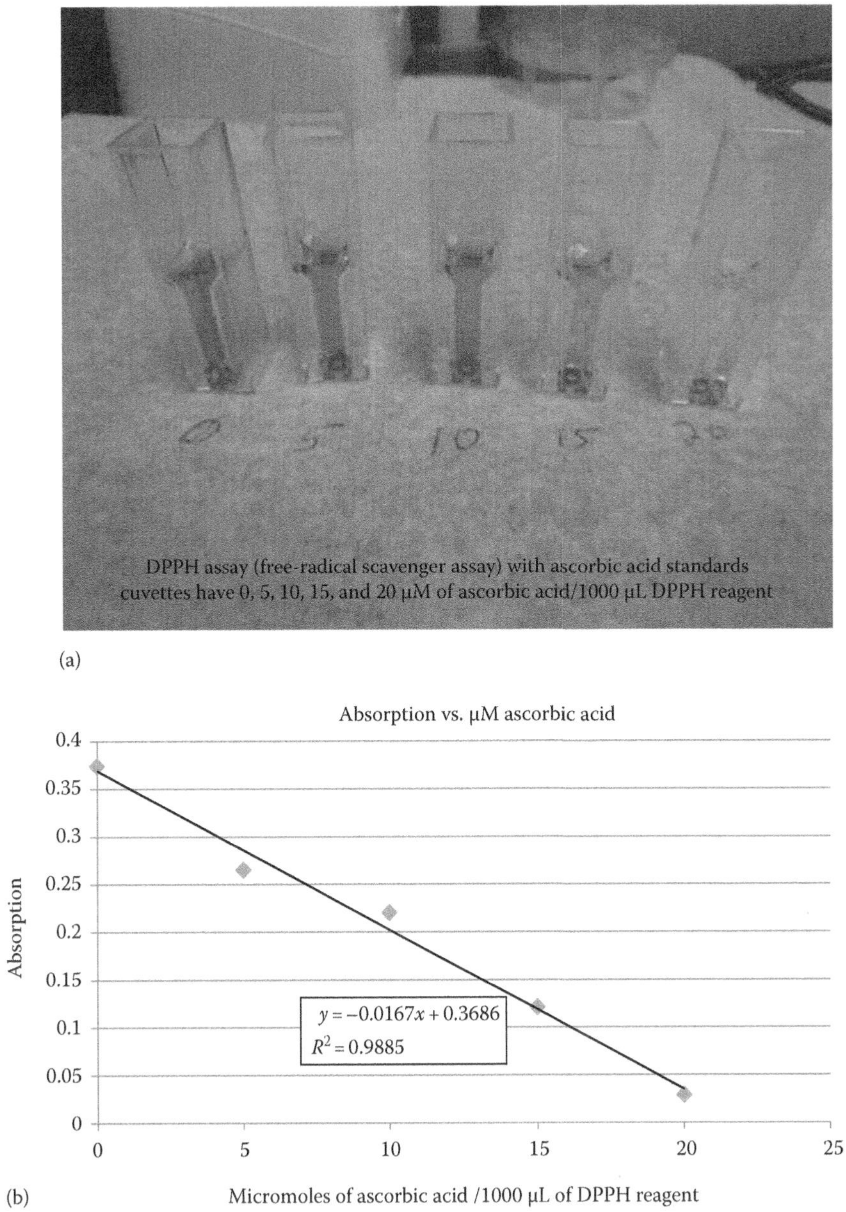

Figure 17.4 (a and b) DPPH assay with ascorbic acid as the standard. The regression equation is presented in the figure and is used to calculate the amount of free radical scavenging capacity present in diet samples (or other samples such as insect hemolymph).

7. Incubate the solutions at room temperature for 30 min in the dark (under a cabinet or in a drawer will do the job).
8. Transfer (with clean pipettes or careful pouring) the solution into cuvettes and read in a pre-warmed spectrophotometer at 517 nM (or a wavelength close to 517 as long as all your measurements are consistently made).
9. Record your absorbance for each standard and sample, then construct a standard curve using an Excel type plot, then calculate the samples' concentrations in terms of "free radical scavenging equivalents to ascorbic acid." This will be the basis of your SPC evaluation of your diets (Figure 17.4).

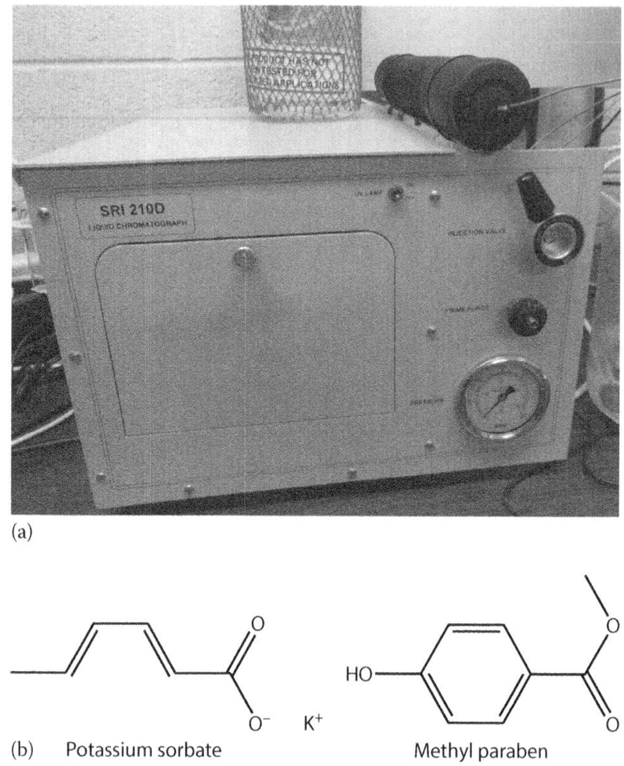

Figure 17.5 (a and b) An HPLC used to determine antifungal agents and two of the commonly used preservatives (antifungal agents), potassium sorbate and methyl paraben.

17.1.9 Antifungal Agents: An Overview

Figure 17.5a and b shows an HPLC, the structure of potassium sorbate and methyl paraben, and an analysis of these two preservatives by isocratic (no gradient) HPLC with an internal standard of phenylthiourea (PTU) (Figure 17.6).

Commonly, we use antimicrobial agents such as methyl paraben (methyl 4-hydroxybenzoate), sorbic acid (or sorbate salts such as potassium sorbate), and propionates (such as propionic acid or sodium propionate) to prevent insect diets from becoming overrun with environmental molds. However, sometimes we find mold growing on some of our diets, even after using chemical prophylaxis. When this happens, we face the question of whether our antimicrobial agents are properly distributed in the diet (i.e., are they mixed thoroughly?), or have they somehow become deactivated?

This exercise demonstrates a technique of extracting and analyzing sub-fractions of a diet to quantify sorbate and methyl paraben to determine the variation detectable in various portions of a single batch of diet. *Note*: in undertaking a quality control/process control project, we use this technique and perform our analyses over several batches of diet. This will allow us to determine (1) variability *within* each batch of diet and (2) *between* batches. It will also produce a set of *baseline data* to compare any possible deviations from the controlled system that we are trying to establish.

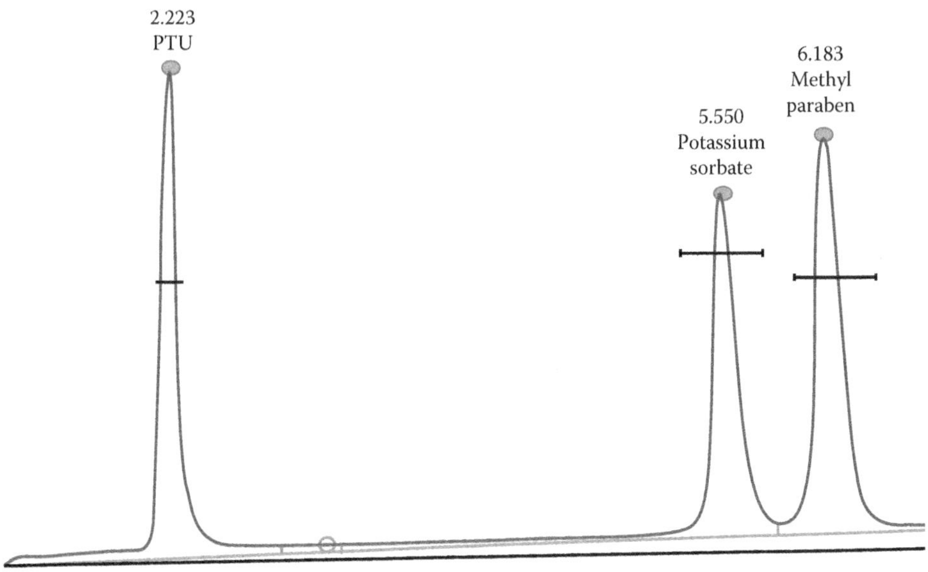

Figure 17.6 Chromatogram of methyl paraben and potassium sorbate analysis with HPLC.

17.1.10 Antifungal Agents: Analysis of Potassium Sorbate and Methyl Paraben Using *HPLC*

We will use a fairly simple HPLC method of analyzing the extracts. But first we will use the first part of this exercise to learn a method of making extracts for this analysis and several other kinds of diet analysis.

Preparation of Extracts

1. Collect five subsamples (aliquots or portions) of diet.
2. Freeze-dry the portions.
3. Grind dry diet using a mortar and pestle or tissue grinder.
4. Weigh out about 1 g (dry weight) portions placing diet powder into 15 mL centrifuge tubes, being careful to record the weight of each sample number.
5. Add 15 mL of methanol to each tube (under a hood and with protective gloves).
6. Vortex tubes for 30 s, then centrifuge at 2000 rpm for 5 min.
7. Filter methanol extract decanted from tube using a syringe 0.45 μm filter.
8. Inject 50 μL into HPLC.*
9. Run samples, noting the quantities of methyl paraben and sorbate in each sample.
10. Divide the amount of each compound by the dry weight of the sample that you introduced into the centrifuge tubes. This amount should be the number of micrograms of each preservative per gram of dry diet.

* HPLC conditions: use an HPLC at a wavelength of 254 nm, flow rate 1.0 mL/min. solvent 55:45 buffer: methanol. Buffer should be 0.05 m potassium dihydrogen phosphate adjusted to pH 4.40 and filtered through 0.22 μm filter system. If your HPLC is equipped with a column heater, set to 35°C. Please note that the system should be preprogrammed using standards of about 500 ppm methyl paraben and potassium sorbate in methanol and an internal standard of 500 ppm PTU. Now, the analysis is ready to be completed. This will require that extracts with internal standards be injected into the HPLC and run for each sample needing to be analyzed.

11. By performing this test on multiple samples, you can determine the mean and standard deviation for the batch of diet in question. By collecting data from several batches of diet, you have the basis of a quality control/process control system for preservatives in your diet.
12. For the analysis method with internal standards, a concentration of PTU about equal to the amount of methyl paraben and potassium sorbate should be present in your standard.

17.1.11 Measuring pH of Gelled Diets: An Overview

As explained in several places in this book, especially Chapters 3, 5, and 13, the pH of the diet has profound influences on palatability, susceptibility to microbial attack, susceptibility to enzymatic and other chemical reactions that can make or break a diet's capabilities, and many other factors that contribute to the diet's bioavailability and stability. In developing a process control system, pH lends itself to simple, inexpensive, and reliable measurements that can help with the diagnosis of raw materials, mixing problems, or other processing issues. Because pH is nondestructive and unobtrusive, measurements can be made at several key points in the diet-making process or at the completion of the process. The main caveat about pH measurements is that they depend upon the diets having reached room temperature (not right out of the refrigerator or heated above ambient). An important point is that diet personnel often are unaware of the existence of flat pH probes that are specially designed to measure the pH of gels. A conventional pH probe is fine for liquids, but to measure a gelled diet, the readings must be made with a specialized flat surfaced probe (Figure 17.7).

17.1.12 Measuring pH of Gelled Diets: Procedure for Developing Statistical Process Control for Gelled Diets

17.1.12.1 Procedure for Developing an SPC Diagram of pH in Gels

1. Cool diet to room temperature (not refrigerated either).
2. Calibrate pH probe and meter with a buffer standard close to the anticipated pH of the diet (use a standard of 7.0 or 4.0 for most diets).
3. Press flat base of probe firmly against the diet surface.
4. At least three measurements should be made for each batch of diet to enable statistical evaluation; five or six measurements will provide more statistical reliability without a major investment of time.
5. Make a table of readings and graph the readings in a sequence that represents the batches that were measured, preferably in an order of time sequence when diets were made or according to some other parameter such as replacement of a staple or supplementary component.
6. Clean probe carefully with bleach solution (5% commercial bleach), rinse in tap water, and store probe in storage solution.

pH of competed diets (using flat probes)

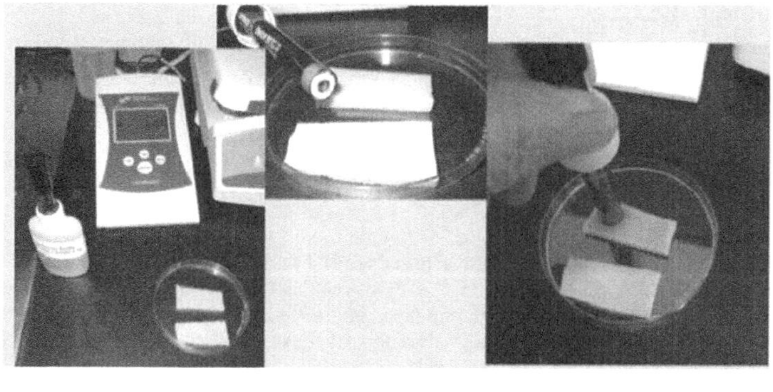

Figure 17.7 Using a pH meter equipped with a flat probe.

17.1.13 Measuring pH of Liquid Diets or of Liquid Components in Diets

1. Cool diet to room temperature.
2. Calibrate pH probe with calibration buffer close to the anticipated pH.
3. Insert the probe into liquid diet and allow to stabilize before recording readings.
4. At least three measurements should be made for each batch of diet to enable statistical evaluation; five or six measurements will provide more statistical reliability without a major investment of time.
5. Make a table of readings and graph the readings in a sequence that represents the batches that were measured, preferably in an order of time sequence when diets were made or according to some other parameter such as replacement of a staple or supplementary component.
6. Clean probe carefully with bleach solution (5% commercial bleach), rinse in tap water, and store probe in storage solution.

17.1.14 Measuring Water Activity: An Overview

Water activity is discussed in several places in this book in terms of its contribution to chemical reactions in diets (Chapters 3 and 5) and in terms of its relationship with microbial growth (Chapter 13). It has also been discussed in terms of the relationship between freshness of dry diet components and product freshness. Simply put, the longer a material is stored, the higher the water activity will be up to some upper limit, which is above the point of staleness or deterioration. Commonly, freshly acquired materials such as soy flour, wheat germ, rice meal, corn meal, and even salt mixtures or sugar will start off with water activity of less than 0.25, and they will increase in water activity, often above 0.45–0.55. These higher activities can support some microbial growth, but more commonly, they will lend themselves to deterioration reactions with free radicals and enzymatic deterioration. The rate of water uptake by a product depends upon these factors: (1) its characteristics in terms of moisture absorption potential (the extent of hygroscopic potential), (2) packaging of the material (sealed with humidity absorbent vs. storage in a relatively open container), (3) environmental conditions (relative humidity and storage temperature), (4) how frequently the package is opened to get access to the material. The recommended conditions for packaging are discussed elsewhere in the book (Chapters 3 and 5). However, for the development of a protocol for measuring and managing water activity, there is a relatively simple procedure that lends itself excellently to process control. With each batch of diet, a small subsample (about 5 g) should be measured with a water activity meter (Figure 17.8). (There are several brands of water activity meters, which range from under U.S. $2,000 to nearly $10,000 for research models. In the Cohen laboratory, we use the less expensive PawKit.)

17.1.15 Measuring Water Activity: Statistical Process Control Procedure

1. As soon as the material (e.g., wheat germ, brewer's yeast, or soy flour) arrives in the rearing facility, it should be divided into aliquots (subsamples) that are slightly greater than the amount that will be used in a batch of diet. This measure is taken so that the original material is not repeatedly warmed to room temperature and exposed to laboratory air with microbes, humidity, oxygen, and possibly other contaminants. The packaging should allow little space for free air, and the packaging should be capable of being sealed. It is strongly recommended that the material be packed with an oxygen and humidity absorbent.
2. Place sample (about 5 g) in measuring cup, being careful to keep the sample level under the rim of the cup (so that it does not touch the sensitive measurement apparatus). Be sure to use the pad that comes with the instrument (for the PawKit) to be sure that complete closure of the measuring chamber takes place.
3. Start the apparatus by pressing the button on the left so that the hash marks on the water activity symbol are actively flashing (a_w).
4. Wait for about 5 min until the "complete" notice beeper sounds.
5. Record sample water activity and proceed to next sample.

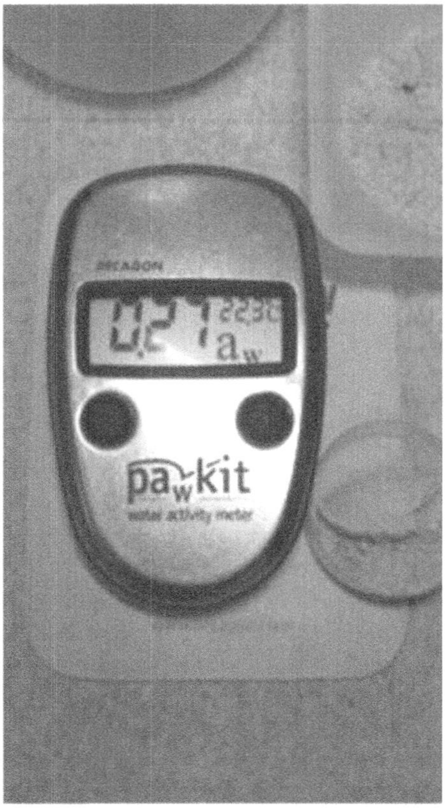

Figure 17.8 A PawKit portable water activity meter from Decagon.

17.1.16 Measuring Water Content in Diet Components for SPC: An Overview

Although water activity and moisture content are related, they must be understood to be different entities. Water activity is a thermodynamic parameter and has to do with the energy of the water in a material (e.g., diet or diet component). The water activity (being thermodynamic) expresses the tendency of water to move from one region to another (in this way, it relates to osmotic pressure, which is an expression of water activity, often in a physiological system). Water activity will predict the net movement of water, which will always move "downhill" or from regions of greater activity to regions of lower activity. Moisture content, however, is simply the total amount of water in a material (percentage of water) (Figure 17.9).

The standard way of measuring moisture content is to drive off the water with heat. A most simple and inexpensive way of measuring water content is by pre-weighing the sample in question, heating it at a specified temperature (in a drying oven) for a prescribed time, then reweighing the sample to determine the weight difference, which is taken to be the water content, which when divided by the original weight tells us the moisture content. Or we can determine moisture content with a moisture balance, which is designed to both weigh a sample continuously and dry it by the application of heat (infrared or via a halogen lamp).

(a) (b)

Figure 17.9 Moisture determination balance with top up (a) and top down (b). Moisture balance (Ohaus M845).

17.1.17 Measuring Water Content in Diet Components for SPC: Procedures

1. Place the aluminum tare pan in the holder and set on the weighing arms on the balance. Make sure that the pan does not bind with any other parts and then press the "Tare" button to zero the balance.
2. Place the sample on the pan and spread out as much as possible to maximize surface area to make drying as efficient as possible.
3. Close the lid of the balance, then press the "Start" button to begin the drying process. At the end of the process, record the final percentage of moisture, making sure that the sample is fully dry.
4. Discard sample and begin the process with a new sample.

17.1.18 Texture/Rheological Measurements: An Overview

Textural and rheological analysis of diets is discussed more completely in Chapter 15, but the explanation here is intended to help diet personnel to use inexpensive equipment to develop a set of data that will lend itself to testing the completed (synthesized and completely processed) diet. The instrument (a Wagner Instruments fruit tester) in Figure 17.10 costs under $800 for the stand and the penetrometer, and an operator can learn in minutes how to make fairly reliable measurements with a less than 3% error. Other kinds of instruments for penetration force measurement or viscosity. As might be expected, the more sophisticated instruments such as some of those discussed in Chapter 15 allow much more interpretation of the structural characteristics of diets and diet components, but a lot can be learned from the simple instruments.

17.1.19 Texture/Rheological Measurements: Procedures

1. Place diets that have cooled to room temperature on a flat, level surface (or warmed to room temperature, if they had been refrigerated).
2. Having selected and fitted the proper-sized probe for your diet samples in place, place the diet under the penetrometer and lower the gage with a steady, even, force.

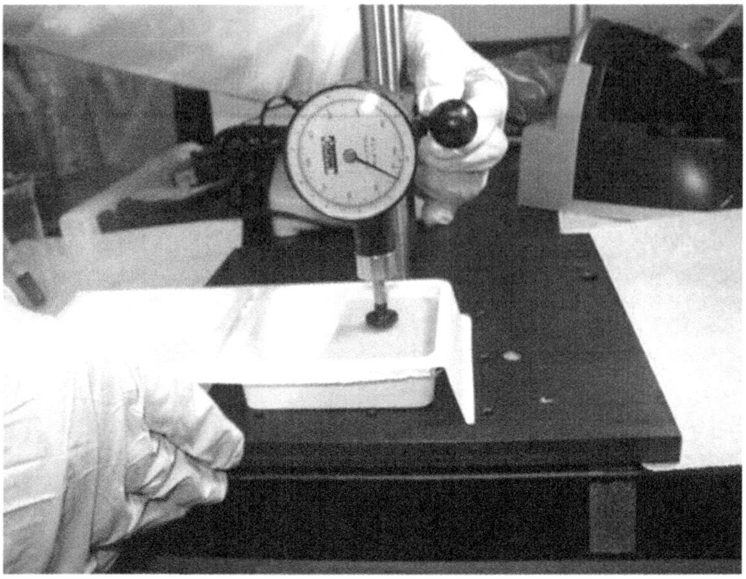

Figure 17.10 A simple penetrometer for testing force required to pierce surface of a gelled diet.

3. Continue the pressure until the surface of the diet was broken and the gage needle remains in place providing an analog reading of the greatest amount of force needed to break the diet's surface.
4. Record the reading, reset the meter, and make the next penetration measurement.
5. Make several measurements in different diet subsamples to help provide a profile of the variability of the diets' gel strength.
6. To make measurements more uniform, reading the surface temperature with a thermal measurement gun (whose readings are instant and noninvasive).
7. Establish a database that will allow you to construct a process control chart.

17.1.20 Enzymatic Activities in Raw Materials or Diets: An Overview

These tests are based on the fact that many diet materials, especially plant-derived components such as soy flour, wheat germ, other grain or cereal products, and most leaves, contain active enzymes—unless they had been destroyed by processing steps such as heating or high pressure. The enzymes are varied, and in a material such as soy flour, they include urease, polyphenol oxidase, catalase (or its similar-acting enzyme family: peroxidases), and lipooxygenase. Several of these enzymes can have deteriorating effects on artificial diets, and proper diet processing will have denatured (destroyed activity) in these enzymes. Also, other proteins that can be harmful to insects if they remain intact in diets include lectins, enzyme inhibitors such as protease inhibitors, and amylase inhibitors. Procedures in food commercial processing often include some kind of heat-based destruction of the enzymes and other harmful proteins, including flash heating, blanching, toasting, or scalding. One of the most common problems that crops up in insect diets is a darkening, which is a result of enzymatic browning from polyphenol oxidase that had somehow escaped the heat destruction process. Other effects resulting from hidden and unwanted enzyme action include rancidity that comes from lipooxygenase that has attacked polyunsaturated fatty acids (discussed in Chapters 3 and 5). Clearly, it would be advantageous to check ingredients that are going into diets to make sure that the potentially destructive enzymes have been denatured once the diet has been synthesized. Fortunately, it is not necessary to check all enzymes that may be present and active, but a good strategy can be to use "sentinel enzymes" in simple tests to detect activity. There are several commercial kits available for the testing several of the enzymes mentioned here including a Sigma-Aldrich kit for determining peroxidase activity (Peroxidase Activity Assay Kit Catalog Number MAK092).

17.1.21 Enzymatic Activities in Raw Materials or Diets: Procedure

17.1.21.1 Urease Determination in Soy Flour

In this procedure, you will be determining the presence of the enzyme *urease*. The enzyme urease catalyzes the following reaction:

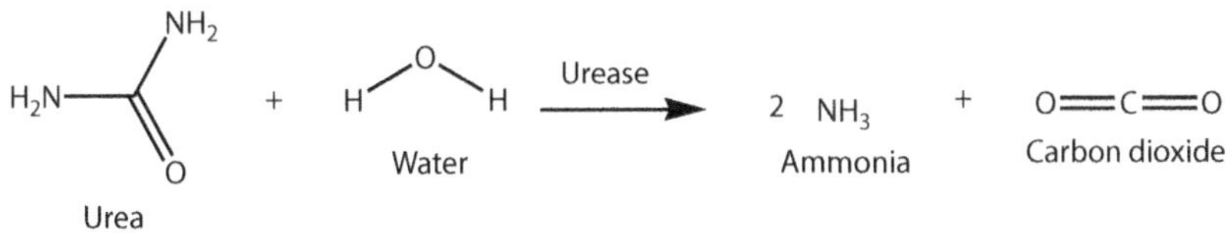

Because this reaction takes place in an aqueous solution, the ammonia molecules generated by urea hydrolysis can join water molecules undergoing the following reaction:

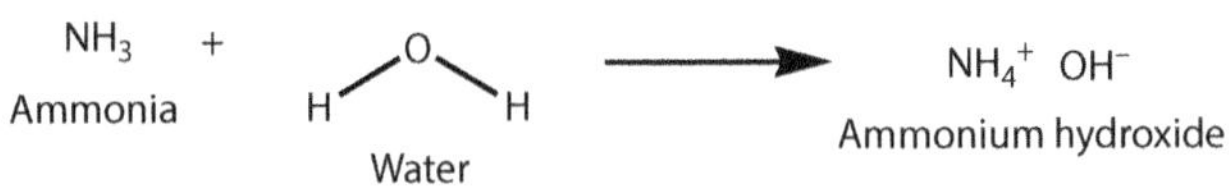

The product, ammonium hydroxide, is a base. Therefore, the progress of the two "tandem" reactions can be detected by changes in pH. In this process, you will set up three beakers with the mixtures specified in Figure 17.1. Once you complete the mixture, you are to monitor the progress of the reaction every minute, using pH paper (or a pH meter). A more quantitative analysis can be done with an instrument known as a spectrophotometer, which measures absorbance of specified wavelengths of light. As the urease reaction progresses, the pH will increase, but if the urease has been properly denatured, there will be little or no increase in pH (which is changed by the release of ammonia from urea and the subsequent dissolution of the ammonia in the water). These are the steps to be followed:

1. Dissolve 30 g of reagent grade urea in 500 mL of deionized water (~500 mM solution).
2. Place 10 g raw soy flour (meal) in beaker 1 (positive control), 10 g of toasted soy flour in beaker 2 (negative control), and 10 g of the soy flour being tested (unknown cooking background) in beaker 3.
3. Add 50 mL of urea solution to the soy flour in each beaker and immediately measure the pH (preferably with a standardized pH meter, but this can be done—less sensitively—with pH paper).
4. Record pH readings in table. Wait for 30 min and remeasure pH.
5. Compare the outcome of test and make decision as to whether the new soy flour has been adequately heat treated to remove active enzymes.

Table for recording urease test results:

	Beaker 1 pH	**Beaker 2 pH**	**Beaker 3 pH**	**Notes**
Color before 30 min				
Color after 30 min				

17.1.22 Determination of Lipid Peroxidation or Rancidity in Wheat Germ, Soy Flour, or Other Diet Components: Using the TBARS Assay: Overview

It should first be noted that more complex analytical methods for the determination of rancidity in diet components are available. For example, Sjovall et al. (2000) did an intricate analysis of the products that are detectable in rancid wheat germ at various stages of rancidity formation. They included head space analysis with gas chromatography and sensory analysis. However, the procedure suggested here has extensive use in the food, medical, and pharmaceutical communities. *Note further*: This is a modification of the procedure used in the Cayman Chemical Company Kit. TBARS is an acronym for thiobarbituric acid–reactive substances. When lipids become rancid (via a peroxidation process), they yield a product called malondialdehyde (illustrated later), and this product is reactive with TBA (also illustrated later). Other lipid peroxidation products are also reactive with TBA, and they form a pink complex that can be quantified by absorbance of 540–550 nm light in a spectrophotometer.

Malondialdehyde

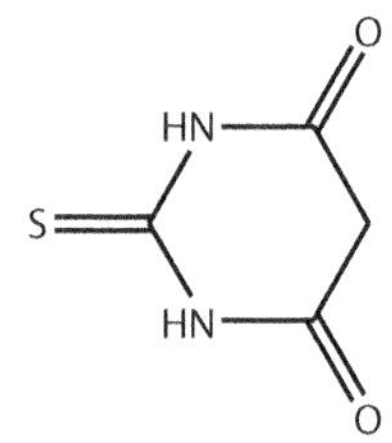

Thiobarbituric acid

Components for the determination of malondialdehyde in foods containing unsaturated fatty acids are provided. Top is the degradation product (malondialdehyde) for oils (fatty acids) that are turning rancid. The second structure (TBA) is the reagent that reacts with malondialdehyde to form a pink complex that is measured with a spectrophotometer.

17.1.23 Determination of Lipid Peroxidation: Procedure

Here is the scheme for preparing your standards.

This gives you an extremely wide range of values.

Tube	MDA (μL)	Water (μL)	MDA Concentration (μM)
A	0	1000	0
B	5	995	0.625
C	10	990	1.25
D	20	980	2.5
E	40	960	5
F	80	920	10
G	200	800	25
H	400	600	50

1. Label vial caps with standard number or sample identification number.
2. Add 100 μL of sample or standard to appropriately labeled 15 mL tube.
3. Add 100 μL of SDS solution to vial and swirl to mix.
4. Add 4 mL of the color reagent forcefully downside of each tube.
5. Cap tubes and place tubes in holder to keep them upright during boiling.

6. Add tubes to vigorously boiling water. After capping tubes, wrap them with a small strip of Parafilm®
   to keep them from popping open. Boil tubes for 1 h.
7. Remove tubes and place in iced bath to stop reaction. Incubate on ice for 10 min.
8. Centrifuge the tubes for 10 min at 1600 × *g*. Tube contents may appear cloudy, but the cloudiness will disappear upon warming to room temperature.
9. Tube contents are stable at room temperature for 30 min.
10. Load 150 μL in duplicate for each sample or standard tube to plate.
11. Read absorbance at 530–540 nm using plate reader.
12. Interpret results for unknowns by fitting absorbance values to standard curve.
13. Divide micromolars of TBARS for each sample by that sample's weight. This is your final value for TBARS assessment of peroxidation. Please remember that the higher the TBARS value, the more peroxidation has occurred in your sample, meaning that the sample with high TBARS values is becoming rancid or defective.

17.2 SANITATION AND CONTAMINATION: DETERMINING THE CLEANLINESS OF FACILITIES IN THE CONTEXT OF SPC

Facilities must be visually inspected to determine an acceptable level of orderliness and cleanliness. Clutter harbors microbes and vermin, and it prevents a thorough cleaning of all surfaces. Visual inspection also reveals spots or layers of dust, dirt, chemical contamination, and debris that can contaminate diets and rearing units. However, diet and rearing facilities that might pass a visual inspection could very well fail a microbial survey as Sikorowski (1984a,b) demonstrates. What is suggested here is that at regular intervals, preferably once per week, a surface assay be taken as recommended by Sikorowski (1984a,b). Two types of surface microbial assessment are the use of swabbing of surfaces and streaking the swabs on a trypticase soy agar (TSA) with lecithin and polysorbate 80 (media available from Difco) or the replicate organism direct agar contact (RODAC) method, also used with TSA. The two methods of testing insectary surfaces are outlined and briefly explained here (Figure 17.11).

17.2.1 Test Procedures for Insectary Surfaces

The following tests can become an excellent part of a process control program, and they can also be used very productively as teaching tools for new insectary workers and for refresher courses in PC/QC for all personnel. However, because microbial growth under the conditions specified can yield very large numbers of potentially harmful microorganisms, the tests should be done cautiously and only if the appropriate equipment is available such as an autoclave to kill cultures and equipment that is to be disposed of and biohazard containers to store the materials that are awaiting sterilization. It is strongly recommended that a specialty manual for microbiology be consulted for details on the techniques. An example of such a manual is Pierce and Leboffe (1999) (Figures 17.11 and 17.12).

1. All equipment and materials should be made sterile prior to testing, preferably by autoclaving.
2. The TSA agar plates that are to be used should be premixed according to the manufacturer's instructions, with the plates used for swab checks poured below the rim of the inner portion of the Petri dish and the plates to be used for RODAC poured deeply enough so that the surface is slightly above the rim of the inner dish. The raised surface of the RODAC plate is to allow contact of the medium with the surface to be tested. It should be noted that the RODAC method is less than 50% as efficient as the swabbing method in capturing microbes on surfaces (Jay et al. 2005). The major advantage of the RODAC method is its simplicity in having fewer steps and less handling of dishes and swabs. It should be noted that the TSA medium that contains lecithin and polysorbate 80 is recommended to neutralize cleaning agents that would carry over antimicrobial activity into the culture plate, thus masking the true contaminant counts.

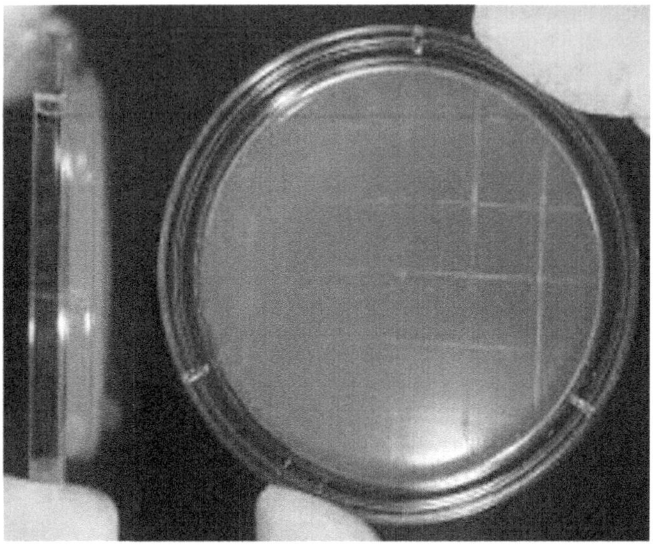

Figure 17.11 RODAC plate: note the raised agar dome where contact is made by firm pressure of plate against flat surface (floors, table tops).

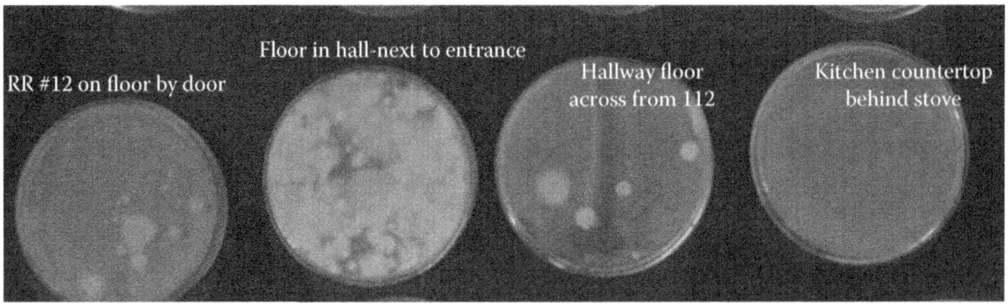

Figure 17.12 Sabouraud agar plates incubated for 48 h after 30 min of exposure to air in an insectary. Note that the colony-forming units in the measurement from the floor next to the entrance are too high to count accurately. This is an indication that corrective action should be taken in the insectary in question.

3. Swab method: The swab is wetted in sterile water and run over the surface to be tested. It is useful to have a premade (sterilized) template such as a 1 cm^2 opening in a flat metal plate that is placed on the surface to be monitored, and the swab is thoroughly wiped within the area. The test swab is then immediately used to inoculate the Petri plate with a zigzag streaking motion done with a twist of the swab to expose the entire surface. During streaking, the lid of the plate should be used as a shield to reduce the possibility of airborne contaminants reaching the plate's surface. In the RODAC method (Figure 17.11), the plate is opened, and the raised medium surface is pressed against the surface to be checked. Then, the plate is closed. In both techniques, once the sampling is done, the plates are incubated for 48 h at 26°C ± 1°C. All tests should be replicated (repeated) three times for each surface being tested (floors, bench tops, laminar flow hoods, equipment).
4. Inspection of plates: The number of colonies that are visible after 48 h is counted. In the refinement of the QC protocol, the QC team may decide to use a magnifier to aid in colony detection, and they may use a colony counter, which is equipped with grids for ease in making counts. Each colony represents what was most likely a single microorganism that was proliferated in the culture process. Therefore, the counts are reported as the number of colony-forming units (CFUs) per unit of surface area.

5. Recording and systematizing data: The number of CFUs per unit surface area should be recorded, preferably on a computer-based spreadsheet. Graphing such a spreadsheet over periods of time, such as over a 6-month interval, can be very telling when the causes of an event are reconstructed such as diagnosis of a large-scale mortality or a drop in fecundity or egg hatch.
6. Interpreting data: Once a clear baseline is established of how many CFUs are to be expected even after high-intensity sanitation efforts, deviations from this norm can be readily diagnosed and linked to events, such as changes in personnel, changes in procedures, or changes in materials. If, for example, a newly colonized species of insect was introduced at a date that corresponds to the onset of higher CFUs, the protocols for rearing the new colony need to be reassessed.

17.2.2 Test Procedures Measuring Cleanliness of Laboratory Air

1. Using three TSA-containing (8 cm diameter = 50 cc^2) Petri plates, remove covers to expose the medium to the laboratory air for 10 min (Sikorowski 1984b) (Figure 17.11). Select places where diet is being made and where insects are cultured to get a good sampling of key points of possible departure from desirable levels of sanitation.
2. Incubate the re-covered plates for 48 h at 36°C ± 1°C.
3. Examine plates to perform colony counts.
4. Record CFUs per plate and enter data into an appropriate spreadsheet for comparison with past data.
5. Interpretation of data when the colony was doing well compared to results of microbial counts when the colony went into a state of decline can help demonstrate which possible events influenced the problems. High colony counts such as more than 10 CFUs per dish indicate that there is an air quality problem, and the management needs to consider improvements in the air filtration system (or adoption of a system, if this has not already been done).

17.2.3 Testing Diets for the Presence of Microbial Contaminants

The issue behind the assessment of environmental contaminants is the cleanliness and wholesomeness of the diet; therefore, it is crucial to assure that the diet is being kept free of excessive amounts of microbial contamination. The protocols for assessing microbial counts in diets are more complex than those used for counts of microbes in the air and on laboratory surfaces. In a most simple assessment, diet is applied to a culture plate and incubated to permit observation of microbes to make possible a nonquantitative estimation of contamination. However, for an efficient system of quality control, a quantitative process is recommended.

17.2.3.1 Level-One Testing: Visual Inspection of Diet—Developing a Strategy of Careful Observation

1. Compare the appearance (and odor) of the freshly made diet or an aliquot of diet that has been refrigerated with a sample of diet that has been used by the target insects at various times through the feeding cycle. For example, with insects such as Heliothis or Helicoverpa, the larval period is about 6–8 days at about 27°C, so inspections should be made at days 3 and 6 or 4 and 8. Under a dissecting microscope, there may be evidence of patches of mold or bacteria, indicating that there is contamination. Also, the moisture of the diets should be noted. This process is the establishment of a baseline.
2. Once the quality control personnel have a clear picture of the visual characteristics of normal healthy growth conditions, deviations will be more readily identified, for example, excessive drying out due to environmental deviations or excessive moisture, which can be a precursor of microbial degradation.
3. What has been stated about the appearance of the diet pertains to the odors of normal diet and healthy insects on that diet vs. the smells that emanate from contaminated or spoiled diet.

17.2.3.2 Level-Two Testing: Using Microbiological Media for Assessing Microbial Contamination of Diet

This test can be performed routinely on all diet components to establish a baseline of microbial contamination of raw materials as they arrive in the insectary and also as a test of storage conditions to determine if there is an inherent problem that leads to contamination (such as poor sanitation by workers when accessing stored materials). Having a baseline of microbial counts will be especially useful if microbial outbreaks suddenly become a problem after a diet-making process has become a reliable, established practice.

The procedure here can be applied to all materials (agar, salt mixtures, vitamin mixtures, sugar, wheat germ, brewer's yeast, etc.). Most (or many) of these materials will have ranges of CFUs for the product on their label, but the tests recommended here are measures of changes that may have taken place during storage and handling in each insectary. It must be remembered that if a material has a high microbial load, it will continue to be a source of diet contamination, and if counts exceed the baseline (determined through SPC methods such as process control charts), it should be discarded and replaced with fresh material, and the source of the contamination should also be determined to prevent further chronic contamination problems.

1. Using sterile instruments (cotton swabs, spatulas, or other transfer devices), apply freshly made diet to a plate containing a general nutrient medium such as nutrient agar or TSA and re-cover plate.
2. Allow plate to incubate for 48 h at 36°C ± 1°C.
3. Inspect plate around insect diet zones of contact with media for the formation of microbial colonies.
4. Perform colony counts to establish acceptable CFUs per unit of diet component weight.
5. Use test routinely to determine if stored or newly purchased components exceed acceptable limits (using a process control chart).

17.3 DIET PROCESSES THAT AFFECT QUALITY OF DIET-BASED REARING OUTCOMES

Several treatments of diet materials and diets themselves have a great impact on the effectiveness of diets, their stability, reliability, and production potential for high-quality insects. This section is intended to guide insectary staff members to consider these factors in developing a comprehensive plan of process control. However, it is not within the scope of the current text to provide detailed step-by-step procedures as was attempted in the earlier sections of this chapter.

17.3.1 Sterilization and Pasteurization

For determining if a material is being heated for pasteurization (i.e., using temperatures below the 121°C and 15 psi of an autoclave), a thermometer that has been surface sterilized can be used. Recording thermometers can give a continuous or periodic readout that can be used for SPC data collection.

For determining if autoclave temperatures have been reached within the diet, there are two commonly used methods: (1) sterilization pellets (purchased from laboratory suppliers such as Cole-Parmer or Fisher) where melting portions of the pellet indicate by color diffusion that the proper temperature has been reached or (2) spores from bacteria within a capsule placed in the material being autoclaved are tested after the heat processing to determine if appropriate temperatures were reached and held for an adequate time interval to kill bacteria and other spore-forming organisms. A modification of this test is where, rather than seeding with known bacteria, aliquots of the diet are transferred by sterile techniques to culture plates and incubated on media such as those described later. Counts of such plates can be made if appropriate dilution techniques are used, again, as described later.

17.3.2 Mixing

Mixing efficiency can be measured with several of the chemical tests described earlier, but it is especially important that the antifungal agents (most prominently, methyl paraben) are measured in extracts made from several subsamples of diet collected from different points of dispensing. The HPLC method readily lends itself to this kind of distribution test because the antimicrobial agents are added in relatively small amounts, which can lead to mixing problems, especially with substances that are poorly soluble (such as methyl paraben).

In diets that have particles that are of different colors than the bulk of the diet, a visual inspection can be useful to determine whether or not mixing took place properly or if settling took place giving the appearance of uneven distribution.

17.3.3 Size Reduction

As discussed in Chapters 3, 4, and 15, particle size can have profound effects on accessibility to insects' mouthparts, diet texture, moisture-holding capacity, and stability. There are several methods that can be as simple as sieving with soil sieves and then weighing size-separated fractions to more sophisticated techniques such as imaging systems or techniques of intermediate sophistication such as sedimentation rate measurements.

17.3.4 Distribution of Diets (Pouring and Packaging)

There are numerous factors that affect the insect/diet interaction once the diet has been fully synthesized. Diets are often "ripened" or allowed to dry out to the extent that puddles or syneretic seepage does not occur. Diets may be sprinkled with an inert material such as covering the surface with corn cob grits or sawdust (to absorb moisture and afford the neonate insect cover) and insulation from harsh light or exposure to other neonates. Diet surfaces are often scored (scratched or gouged) to allow neonate insects to get a better surface for attack with their small mouthparts. Many insectaries store diets that were freshly poured under a laminar flow hood with an added precaution of using an ionizing wavelength of UV light (high UV-C component) to destroy surface microbes.

17.4 OTHER PROCESSES RELATED TO DIETS OR INSECTARY-BASED DIET CONTEXT

As stated in Chapter 1 and briefly in other places throughout this book, rearing insects is a complex process that is made up of subprocesses that impact the quality and economics of the insects we are trying to produce. As explained in Chapter 11 on process and quality control, continuous vigilance in monitoring the processes helps build outcomes of stability, quality products, economic production, and even personnel safety. In this context, several features that are not directly connected to diets are discussed briefly with the goal of offering a reminder that diets are not only complex in their internal (component/component) interactions, but also intricately connected to external features (external to the diets), including environment and the insects themselves. The reader is also encouraged to consult the recent work on insect rearing edited by John Schneider (2013).

17.5 CONTEXT OF THE INSECTARY: ORGANIZATION OF INSECTARY ROOMS, CLIMATE, AND PERSONNEL

The diet does not exist in a vacuum. It is made and used in a context where features such as clean areas, flow of personnel traffic, ventilation systems of diet preparation rooms, storage areas, and rearing rooms, containers and packaging of diets, temperature of storage rooms and rearing areas, humidity, light conditions, and microbial load (based on sanitation) can play a tremendous part of the diet's stability and overall quality. Of course, as has been emphasized in several places in this book, personnel are one of the most important factors. The management of personnel involves training, management style, and the economic resources for hiring the best people possible (e.g., people with good attitudes, strong work ethics, and open-mindedness).

17.5.1 Microclimate in Relationship to Diets

The relationships of microclimate were covered in Chapter 1 and in several other places in this book, but it must be emphasized that as part of a methodology for process control, making good observations and recording the data from these observations can be a key factor of insect stability and problem-solving/problem-prevention techniques. The temperature-measuring equipment in Figure 17.13 offer various degrees of sophistication for measuring diet-related temperature and humidity issues. The data logger is especially useful in establishing—automatically—data collection protocol. Observations of insect-infested diets coupled with data on temperature and humidity close to the diets (in rearing containers) can be a major factor in detecting the system problems that led to mold or bacterial outbreaks.

17.5.2 Sanitation

This topic is covered in considerable detail both in Chapter 13 on microbial contamination and in Section 17.2 of the current chapter. However, it bears repeating that sanitation can be the strongest tool in reducing the microbial load in every insectary or small-scale rearing operations.

Figure 17.13 Standard devices for measuring temperature in the insectary and diet setting.

The components of sanitation begin with personnel, especially their training and the follow-through on expected practices (i.e., are the evaluation standards for personnel written to contain a grade for compliance and initiatives of each staff member toward sanitation and its close partner—safety?) Is the insectary equipped with proper cleanup facilities including soap and warm water, clean towels, clean uniforms, and cleanup equipment that everyone uses in their day-to-day routine? It must further be clarified that sanitation pertains largely to cleanliness with respect to microbial factors, but it also pertains to chemical spills and materials around balances, pH equipment, and other parts of the insectary.

17.6 INSECTS IN RELATION TO DIETS

Although this topic is treated in several places, including Chapter 1 in the discussion of domestication and Chapter 4 in the discussion of selection of diet- and rearing system–adapted tobacco budworms in Yamamoto's laboratory, it is appropriate to comment again on this point. Therefore, it is suggested here that much more research needs to be done on the relationship between artificial diets and the nature of the insects produced on these diets. This raises questions of genetics and indeterminate (possibly genetic, possibly environmental, or even both) characteristics of artificial diet-reared insects.

17.6.1 Genetics

A quotation from Shorey and Hale 1965 states excellently the issue about the genetics of insects reared on artificial diets:

> To determine whether colonies of these various noctuid species could be maintained for successive generations, periodic introduction of new wild specimens into the colonies has not been practiced. However, a great increase in the docility of the adults of each species has been noted, possibly because of continual selection pressures favoring those individuals that survive and reproduce well under confined, crowded conditions. This change in activity patterns may be associated with other unrecognized changes in the physiology of the domesticated individuals. For this reason, if broad inferences are to be made from experiments conducted using these cultivated individuals, it is believed that periodic introduction of wild specimens into colonies is desirable.

The feature of these remarks focused here is that the observed changes in behavior and possibly physiology illustrate the type of interplay between rearing conditions (especially artificial diets) and the target insects, which is of paramount importance. It calls into question how much improvements in rearing outcomes where artificial diets are involved are results of selection of the insects rather than an improvement in the diet. It also raises haunting questions for diet researchers and the community of scientists who use the insect reared on diets about how much their "broad inferences" are to be attributed to the insects. The works of Dr. Alan C. Bartlett (e.g., 1984 and 1994) deal with questions of genetic characteristics of insects subjected to prolonged rearing and domestication, and more recently a group from India, working with silkworms, Ramesha et al. (2012), used conventional and molecular tools to help account for the nutritional genetics involved in rearing silkworms, insects that have been domesticated for about 5000 years. With current molecular techniques being readily available to do detailed analysis of genetic diversity and to pinpoint the genes responsible for nutritionally related attributes, there should be some very useful progress with several of the more prominent insects. It should be noted in closing this section that there is also a considerable body of work with nutrigenic and related environmental questions in several studies of *Drosophila melanogaster* and other congeneric species.

17.6.2 Fitness of the Insect Products

Finally, the basis of quality control programs (discussed in much more detail in Chapter 11) pertains to characteristics of insects that are directly related to the purpose of mass-rearing efforts. Tests of mating capabilities, flight characteristics, survival, biomass, and many other features are routinely analyzed in the various sterile insect facilities for tephritid fruit flies, screwworms, tsetse flies, and also in biological control programs. However, more research is needed to tie the insect quality features to the process of insect production techniques, especially with respect to diet components.

Appendix A: Glossary of Diet and Diet-Related Terms

Absorption: uptake of water or some other solvent.

Acid: chemical matrix with a pH lower than 7.0.

Additive: a chemical added to a diet that improves the shelf life or quality of the diet.

Agglomeration: production of clumps or granules from fine powders (Fellows 2000).

Antibiotic agents: chemicals that kill other organisms; often bacteria are targets.

Antifungal agents: chemicals that kill or suppress the growth or reproduction of fungi (molds, yeasts, rusts, smuts).

Antimicrobial agents: chemicals that kill any or all microbial contaminants.

Antiviral agents: chemicals that kill or render noninfective viruses.

Artificial diet: the food that has been synthesized from one or more ingredients that may be completely defined chemically or that may be partially defined or not defined. An artificial diet and a synthetic diet are essentially synonymous.

Aseptic diet: a diet that is void of living organisms, most notably microorganisms.

Axenic culture: literally without foreign species. Axenic conditions are where the organism being reared is void of living microbes or other living organisms except for the target species itself.

Bacteriocins: naturally occurring peptides that are antimicrobial (Fellows 2000).

Base: chemical matrix with a pH above 7.0.

Blanching: heating diets or diet components to temperatures that are above ambient but below 100°C, usually about 70°C in order to reduce microbial contamination and enzymatic deterioration (Fellows 2000).

Bound moisture: water that is physically and/or chemically bound to a solid component in diet matrix, characterized by having a lower vapor pressure than does free water (Fellows 2000).

Cavitation: production of bubbles by mechanical energy such as ultrasound or high-speed mixing (Fellows 2000).

Centrifugation: the separation of immiscible components by applying centrifugal force via high-speed spinning.

Chelation: formation of a ligand complex between a metal ion and a multidentate molecule forming a ring structure (Miller 2008). Chelating agents may have negative or positive effects on target insects.

Commensalism: a kind of interaction between two species of organisms that live in close proximity with one another with neutral effects of each species upon the other. Contrast with parasitism, competition, and mutualism.

Conditioning: a behavioral response where an insect learns (irreversibly) to respond to a specific diet stimulus. The insect is said to be "tuned" to that stimulus.

Conduction: transfer of heat to or from a diet from solid to solid (see Convection and Radiation) (Fellows 2000).

Continuous phase: the medium that contains the dispersed phase in an emulsion or diet dispersion (see Dispersion and Dispersed Phase) (Fellows 2000).

Convection: transfer of heat to or from a diet by a fluid (gas or liquid) (see Conduction and Radiation) (Fellows 2000).

Critical control point (CCP): a processing factor of which a loss of control would result in unacceptable quality risk (Fellows 2000).

Decimal reduction time: the time needed to destroy 90% of microorganisms (to reduce their number by a factor of 10) (Fellows 2000).

Defined diet: a diet in which all components are chemically pure and of known composition.

Diet: the foods upon which an animal feeds.

Dispersed phase: droplets in an emulsion (Fellows 2000).

Dispersion: a state of matter where nonsoluble materials such as oil droplets or particles are suspended in another material known as the continuous phase (Fellows 2000).

Emulsification: creation of an emulsion by the dispersion of one immiscible liquid (dispersed phase) in the form of small droplets in a second immiscible liquid (continuous phase) (Fellows 2000).

Emulsifying agent: chemical that forms micelles around each droplet in the dispersed phase of an emulsion to reduce interfacial tension and prevent droplets from coalescing (Fellows 2000).

Entropy: a thermodynamic concept of disorder and tendency toward decline in free energy. In diets, it is a law of thermodynamics that predicts that materials will change to states of lower free energy and a higher state of disorder.

Essential nutrient: any substance that serves as a component of metabolism of a target organism and that cannot be manufactured by that organism or spared (replaced) by another substitute nutrient of another chemical species or type.

F-value: the time required to kill a given percentage of microorganisms at a given temperature.

Factitious host: a plant or animal that is not the natural host but is serving as a substitute for the natural host.

Filter cake: solids separated from liquid by filtration.

Filtrate: the liquid remaining after solids are filtered out.

Flash heating: subjecting diet (or food) to a high temperature for a short period of time (e.g., to greater than 72°C for 15 s).

Forming: molding diet or allowing molten diet to gel or solidify into a stable shape.

Friable: easily crumbled into a powder.

Holidic diet (not holistic diet): an artificial diet whose ingredients have been fully defined, chemically.

Host: plant or animal upon which an animal feeds or upon or in which a parasite or parasitoid lives.

Hydrocolloids: gelling agents that increase the viscosity of water by binding with large numbers of water molecules, forming associations that are stable and that immobilize the water.

Hysteresis: the differential in water activity between isotherms of absorption and desorption.

Meridic diet: an artificial diet some of whose ingredients are defined chemically and some not defined. See Holidic Diet and Oligidic Diet for comparison.

Mutualism: a kind of interaction between two species that is beneficial to both; it is a form of symbiosis.

Natural host: the plant or animal upon which an animal feeds under natural conditions.

Nonequilibrium conditions: the unstable state of certain collections of matter, including insect diets. A thermodynamically unlikely state that will tend toward equilibrium and disorder.

Nutrient: any substance that can serve as part of the metabolism of a target organism (including nitrogen sources for building proteins, lipids, carbohydrates, nucleic acids, vitamins, minerals).

Ohmic heating: direct electrical heating of foods or diets (Fellows 2000).

Oligidic diet: generally an artificial diet whose ingredients are not chemically defined or highly purified.

Pathogen: an organism (microbe) that causes disease.

pH: the negative log of the hydrogen ion concentration, a measure of the acidic or basic nature of a diet or diet ingredient.

Phagostimulant: feeding stimulant or substance that elicits a feeding response of a target species.

Phospholipid: any member of the general class of lipids containing phosphate groups and displaying polar characteristics along with fatty acids with nonpolar features.

Preservative: a substance added to a diet to prevent degradation.

Q.S. (Latin *quantum sufficit*): literally as much as will suffice, used when adding a filler to a mixture to bring the weight or volume up to a prescribed level.

Radiation: transfer of heat to or from a diet by electromagnetic energy.

Reactive oxygen species: a molecule that acts strongly as a prooxidant or cause of oxidative stress (often abbreviated ROS).

Rheology: the study of the characteristics of materials under mechanical stresses such as pressures and forces that distort the materials in question. It includes measurements of shear force, viscosity, and flow rates.

Sterol: a cyclic lipid containing an alcohol (OH) group, one of the most universally essential nutrients in Insecta.

Symbiosis: *living together*, an interaction between two species that can be beneficial to both, beneficial to one and harmful to the other, or neutral to both (see Commensalism, Parasitism, and Mutualism).

Syneresis: the release of liquid (usually water) from a gel, considered a gel failure.

Synthetic diet: a synonym for an artificial diet. Often used to connote a defined diet but not limited to that sense.

Token feeding stimulant: any substance that triggers a feeding response but does not have a metabolic role in the target species.

Triacylglycerols: a class of neutral lipids containing three fatty acids connected to a glycerol nucleus via an ester (oxygen) linkage, also known as triglycerides.

Trituration: process of pulverizing a substance such as with a mortar and pestle, often used in diets as a means of mixing two or more solids such as a vitamin present in low concentrations with a sugar present in much higher concentrations.

Appendix B: Historical Landmarks in Insect Diets and Events That Set the Stage for Diet Advancements, History of Artificial Diets for Insects, and Events of Significance to Diet-Related Insect Rearing

Ancient Egypt: Domestication of honeybees (Gillot 1995).
Before 2000 BC in East Asia: Domestication of the silk moth (Gillot 1995).
Circa AD 1000 in East Asia: Cultivation of lac and pela insects (Gillot 1995).
1659: Kircher demonstrated the occurrence of bacteria in milk.
1680: von Leeuwenhoek first described yeast cells (Jay et al. 2005).
1810: Preservation of food by canning patented by Appert in 1810 (Jay et al. 2005).
1842: Food freezing patented (Jay et al. 2005).
1843: Sterilization by steam first attempted by Winslow (Jay et al. 2005).
1853: Autoclaving of food patented by Chevallier-Appert (Jay et al. 2005).
1854: Pasteur began wine-heating studies (Jay et al. 2005).
1855: Powdered milk first made (Jay et al. 2005).
1907: Metchnikoff et al. isolated and named yogurt bacterium (Jay et al. 2005).
1908: Sodium benzoate first given official sanction as preservative for foods (Jay et al. 2005).
1908: Bogdanov first to rear insect from egg to adult (Singh 1977).
1915: Loeb reared *Drosophila* sp. for five generations (Singh 1977).
1917: Guyenot reared *Drosophila ampelophila* on completely artificial diet (Singh 1977).
1932: Michelbacher et al., diet for *Lucilia sericata* (blowflies) where casein was first used in insect diets.
1935: Hobson demonstrates that blowflies require cholesterol (establishing the minimal hypothesis that insects have different nutritional requirements from other animals and the possibility, later confirmed, that insects generally have a sterol requirement in their diets).
1941–1943: Fraenkel reared stored product pests on casein-based diet and began a series of classical studies in the definition of nutritional requirements of insects (Singh 1977).
1942: Bottger was first to rear a phytophagous insect on artificial diet (Singh 1977).
1949: Beck et al. reared phytophagous insects on well-defined diet (Singh 1977).
1949: House began a series of classic works on basic and applied aspects of insect nutrition.
1950: Hagen commenced a series of works on nutrition and dietetics of parasitoids and predators.
1956: Vanderzant began a classical series of studies on insect nutrition including advancement of what is now the most widely used vitamin mixture in insect diets.
1956: Vanderzant and Reiser reared *Pectinophora gossypiella* aseptically on artificial diet with no plant extracts (Singh 1977).
1956: G. R. F. Davis began a series of publications on the mechanistic aspects of insect nutrition and dietetics.
1957: Scheel et al. reared *Oncopeltus fasciatus* on artificial diet—first time rearing of a hemipteran insect on artificial diet.
1957: Dadd begins a series of classic works on the mechanistic aspects of insect nutrition encompassing nutrition and dietetics of numerous species.
1959: Fraenkel advances the plant secondary compound concept in application to insect feeding mechanisms.
1960: Adkisson et al. added wheat germ to the diet of *Pectinophora gossypiella.*
1960: Ito began to communicate a classical series of experiments on basic and applied nutrition of *Bombyx mori.*

1962 and 1963: Mittler and Dadd and Auclair and Cartier reared aphids on artificial diets (Singh 1977).

1964: Chippendale and Beck describe the essentiality of ascorbic acid in *Ostrinia nubilalis* (Singh 1977).

1964: Applebaum advances the concept of antinutrients acting as protease inhibitors (and inhibitors of other digestive enzymes).

1967: Mittler began communicating a classical series of findings on the biochemistry, biophysics, and behavioral aspects of aphid nutrition.

1968: Gordon enunciates the principles of quantitative nutrition in insects.

1968: Waldbauer advances the fundamental concepts of nutritional indices in insects.

1970: Yazgan and House reared a hymenopterous parasite aseptically on artificial diet (Singh 1977).

1977: Singh published a compendium of insect diets.

1982: Debolt published the first artificial diet formulation to support the propagation of continuous generations of plant bugs.

Circa 1982: Twin screw extruder is applied to food technology.

1985: An oligidic diet for entomophagous insects established leading to a series of diets for numerous species of predators and parasitoids.

1997: Edwards et al. use twin screw extruder to process diet for pink bollworms in mass-rearing operations.

Appendix C: Vitamin and Mineral Mixtures Commonly Used in Insect Diets

Comparisons of the two vitamin mixtures, both commonly used in insect diets, show some very important differences. The most major difference between the AIN mixture and the Vanderzant mixture is in the absence of ascorbic acid in the AIN mixture and its abundance in the latter diet (Tables C.1 and C.2). A second important difference is seen in the amount of sugar present in the two mixtures, with about 97% of the AIN being sucrose and about 65% of the Vanderzant mixture being the simple sugar, glucose. Unlike the Vanderzant mixture, the AIN formulation contains vitamin A, vitamin D_3, and vitamin K_2, all known to be required in most vertebrates whose vitamin requirements have been evaluated. However, with the possible exception of vitamin A, whose function is ambiguously interpreted in some insects, these vitamins are not required by insects, in general. Other important differences include the lack of inositol in the AIN formulation and the fact that the α-tocopherol is in its more water-soluble acetate form in the AIN mixture, rather than the less soluble simple α-tocopherol in the Vanderzant formulation.

Likewise, two commonly used mineral (salt) mixtures have strikingly different compositions. Most dramatically, the Wesson mixture has considerably less calcium than does the AIN mixture (Tables C.3 and C.4). It is interesting to note that insects' requirement for calcium is much lower than that of the *higher* vertebrates for which these mineral mixtures were originally designed. It is therefore likely that both mixtures have much more calcium than is required by most insects. Because calcium is strongly inclined to form insoluble compounds and large, cumbersome matrices, it can have detrimental effects on the diet dispersion as a whole, especially if the AIN mixture is used. Other differences include the presence of fluoride in the Wesson mixture but not in the AIN mixture and conversely the presence of zinc, chromium, and selenium in the AIN but not in the Wesson mixture (Table C.5). Possibly one of the most important differences and one that might appear innocuous is the counter ions (anions) for the iron. In the AIN mixture, the iron is in the form of ferric citrate, but in the Wesson mixture, it is ferric phosphate. Insects are known to respond differently to these two forms in terms of efficiency of iron absorption (Keena et al. 1998).

Table C.1 Vanderzant Vitamin Mixture

Vitamin	Amount (g/kg)
α-tocopherol	8
Ascorbic acid	270
Biotin	0.02
Calcium pantothenate	1
Choline chloride	50
Crystalline folic acid	0.25
Inositol	20
Niacinamide	1
Pyridoxine hydrochloride	0.25
Riboflavin	0.50
Thiamine hydrochloride	0.25
Vitamin B_{12} trituration in mannitol	2
Q.S. with glucose	~646.73

Table C.2 AIN Vitamin Mixture 76

Vitamin	Amount (g/kg)
Thiamine hydrochloride	0.6
Riboflavin	0.6
Pyridoxine hydrochloride	0.7
Nicotinic acid	3
D-Calcium pantothenate	1.6
Folic acid	0.2
D-Biotin	0.02
Cyanocobalamin (vitamin B_{12})	0.001
Retinyl palmitate (vitamin A) (250,000 IU/g)	1.6
DL-α-Tocopherol acetate (250 IU/g)	20
Cholecalciferol (vitamin D_3)(400,000 IU/g)	0.25
Menaquinone (vitamin K_2)	0.05
Sucrose (finely powdered)	972.9

Table C.3 Wesson Salt Mixture

Ingredient	Amount (%)
Calcium carbonate	21
Copper sulfate $5H_2O$	0.039
Ferric phosphate	1.470
Manganese sulfate (anhydrous)	0.020
Magnesium sulfate (anhydrous)	9
Potassium aluminum sulfate	0.009
Potassium chloride	12
Potassium dihydrogen phosphate	31
Potassium iodide	0.005
Sodium chloride	10.5
Sodium fluoride	0.057
Tricalcium phosphate	14.9

Table C.4 AIN Mineral Mixture 76

Ingredient	Amount (g/kg) {%}
Calcium phosphate (dibasic)	500 {50}
Cupric carbonate	0.30 {0.03}
Ferric citrate	6.0 {0.6}
Manganese carbonate	3.5 {0.35}
Magnesium oxide	24 {2.4}
Potassium citrate	220 {22.0}
Potassium sulfate	52 {5.2}
Zinc carbonate (70% ZnO)	1.60 {0.16}
Potassium iodate	0.01 {0.001}
Sodium chloride	74 {7.4}
Sodium selenite	0.01 {0.001}
Chromium potassium sulfate	0.55 {0.055}
Sucrose, finely powdered	118 {11.8}

Table C.5 Vanderzant–Adkisson *Special Wheat Germ Diet* as Offered by ICN

Ingredient	Amount (g/kg before Water and Vitamins Are Added)
Vitamin-free casein	28
Sucrose	27.5
Wheat germ	24
Alphacel, nonnutritive bulk	12
Cholesterol U.S.P.	0.05
Linseed oil	0.2
Wesson salt mixture	8.0

Appendix D: Process Control of Environmental Parameters

The team of rearing professionals will keep and regularly examine records of the quality control parameters from each of the categories such as monitoring the temperature, humidity, and light intensity of the rearing rooms (Figure D.1). In the example shown in this figure, it is evident that there is a diurnal cycling of all three parameters, with the lights going off at about 19:00 h (7:00 pm) and coming back on at about 07:00 h (about 7:00 am). It is further evident that the light intensity varied from about 100 to about 250 lm/ft^2. The temperature was correlated with the light intensity and varied by about 1.5°C, and the relative humidity was negatively correlated with the temperature and light intensity. This information was recorded with a Hobo® multievent logger (Onset Computer Corporation). Such a system can be used easily to keep a constant record of major environmental factors including absolute humidity, carbon dioxide content, and oxygen tension, as well as the parameters mentioned here. The recording device is a small unit (pocket sized) whose output can be downloaded to a computer and permanently recorded and readily graphed to detect environmental irregularities.

Devices that simultaneously record temperature and humidity have long been used in insect-rearing laboratories, including the old standby—the hygrothermograph. However, the contemporary data logging devices can monitor more parameters and collect manageable data that is recorded in a digital form. The ease of use of the current generation of data loggers makes these devices suitable for adoption as the heart of a system of environmental monitoring. For example, study of the records of temperature, humidity, and light intensity in the rearing room described by Figure D.1 reveals several potential problems. For example, the air temperature is partly dependent upon the heat generated by the lighting system. The relative humidity is quite variable and can thus have adverse effects on the target insects, especially smaller ones. Finally, the light intensity is extremely variable, probably a result of the lighting source: fluorescent lights. Other factors that can be revealed by this type of data collection can include records of events such as power failures or improper entry into the rearing room as indicated by a sudden increase in light intensity, reflecting entry into the rearing room during the dark phase of the photoperiod.

Should a deterioration of one of the biological parameters be noted, an examination of the environmental profile for the period days or weeks prior to the *deterioration event* had been observed. Indications of temperature, humidity, or light deviations could then be detected and be provisionally related to the biological problem. Even more proactive, if routine surveillance of the environmental data prompted an alert so that those workers who handle the insects or perform the quality control operations on the insects would be encouraged to be especially watchful of potential problems. What is illustrated here is an essential ingredient in a process and quality control system: feedback mechanisms that convey information between responsible parties. *The combination of a good quality control/problem-solving plan, a good, sensitive problem-detection system, and a well-oiled feedback system makes up the heart of high-quality systems operations.* As was the case with detection and actions concerning environmental deviations, the other well-designed, routine quality control measures will help in the systematic detection of problems. This includes routine sampling for determining microbial contamination of equipment, facilities, and personnel, recordkeeping of any deviations from standard operating procedures regarding diet production, ordering ingredients, or storage. As was the case with the anticipation of problems with environmental factors, there is no substitute for proactive preparation for problems by an alert staff that is aware that some change has taken place (a new source of ingredients, changes in the humidifier system, a release of a solvent into the ventilation system, a malfunction of a piece of processing equipment—somehow noticed after the diet had been formulated and presented to most of the colony). Knowing that there may be a problem and what the cause of that problem may be can help a coordinated team of rearing specialists sidetrack a potential disaster.

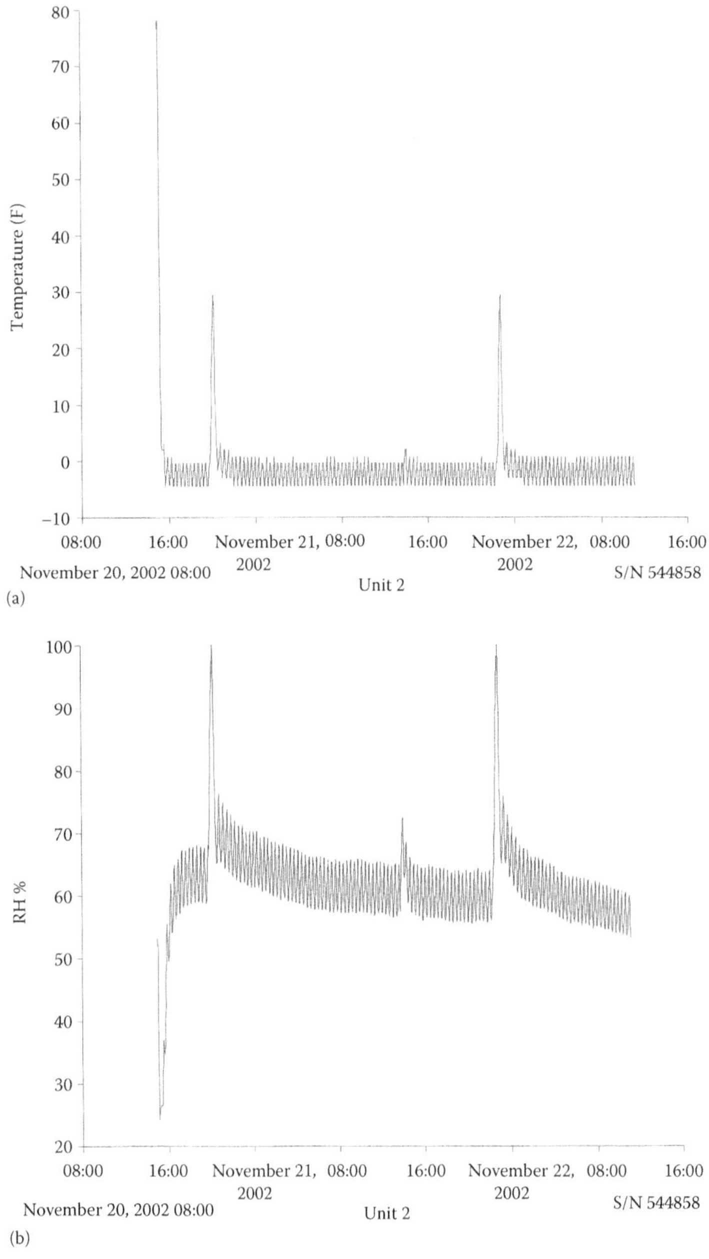

Figure D.1 Outputs from a Hobo® data logger held in a freezer used to store diet components. (a) Temperature (°F) readout and (b) relative humidity readout over a 40 h period.

Appendix E: Explanations of Accuracy and Precision in Measuring Diet Components

Accuracy and precision in QC systems. The concepts of *accuracy* and *precision* are fundamental to understanding QC. The diagrams in Figure E.1 are commonly used to clarify the concepts of accuracy in precision. The degree to which the arrows come close to the bull's eye is the accuracy of the archer. The degree to which the arrows are grouped or clustered is the precision of the archer. The mark (bull's eye) that is being aimed at is the *true* target. Archers who can hit or come very close to this target are considered accurate. Archers who fail to hit the bull's eyes (or even to come close to it) but who can shoot in a tight cluster are considered precise shooters. *Repeatability* is a commonly used synonym for precision.

Examples of accuracy and precision in the insectary setting include the measurement of ingredients of a diet, preparing the diet, and various aspects of insect handling and production. Provided here is a model case, rearing of the predatory big-eyed bug *Geocoris punctipes* with the meat-based diet described in Cohen (1985a). The diet formula calls for specified amounts of components (beef liver, ground beef, sucrose, and water), which are specified in *target* amounts. The formulation calls for 10 mL of a 10% sucrose solution made from 10 g of sucrose in 100 g of water. The insectary worker sets out with these tools, a balance, a weighing boat (or paper), and a clean scoop (or spatula), and starts the weighing (more precisely, the *massing*) process to measure an exact target amount (10 g) of sucrose. If the balance has a precision (readability) of 0.001 g, then the worker can determine the weight of what he/she is putting in the diet to the nearest 0.001 g. Assuming that the worker is very patient, diligent, and skilled at using laboratory implements, he/she adds to or removes amounts of sucrose so that the final reading is 10.000 g. Perfect! Or is it?

What is the accuracy of the balance? Has it been recently calibrated? Does it have a *systematic error* of let's say 0.01 g so that it overreads the 10.000 g, which is really (in the world of absolutes) 9.090 g? Contributing to the question of the reliability of the balance is the haunting problem of balance leveling. The author has witnessed, too many times to count, laboratory workers using balances whose leveling bubbles were so far off center that it often took more than 10 min to get them back in kilter. There is no telling of what this misalignment has done to the measuring of the various diet ingredients!

An effective and practical QC system will be sensitive to true deviations, will give early warnings about deleterious deviations, and will help pinpoint the place in the total process where the deviations are originating. For example, in rearing *G. punctipes* (Cohen 1985a) with four components, it would be useful to know whether or not the dietary lipids from either the liver or the ground beef had deteriorated before it caused a loss of the colony or a decline in adult weight and/or fecundity. The decline in the adult fitness may not express itself until weeks after the original *insult* to the colony had taken place (i.e., the introduction of partially peroxidized fat from stale ground beef). Or as noted by Stewart (1984), the highly variable ingredients such as wheat germ and agar can deviate extremely from the standards that allow production of high-quality diet for pink bollworms. Stewart notes that quality can be substandard in the wheat germ both as a result of new shipments having variations due to a large array of production factors and as this perishable component can also lose its effectiveness due to prolonged or improper storage. It is therefore incumbent upon the insectary staff to adopt standards of quality assessment of such a component to assure that it does not become a source of colony failure or quality loss.

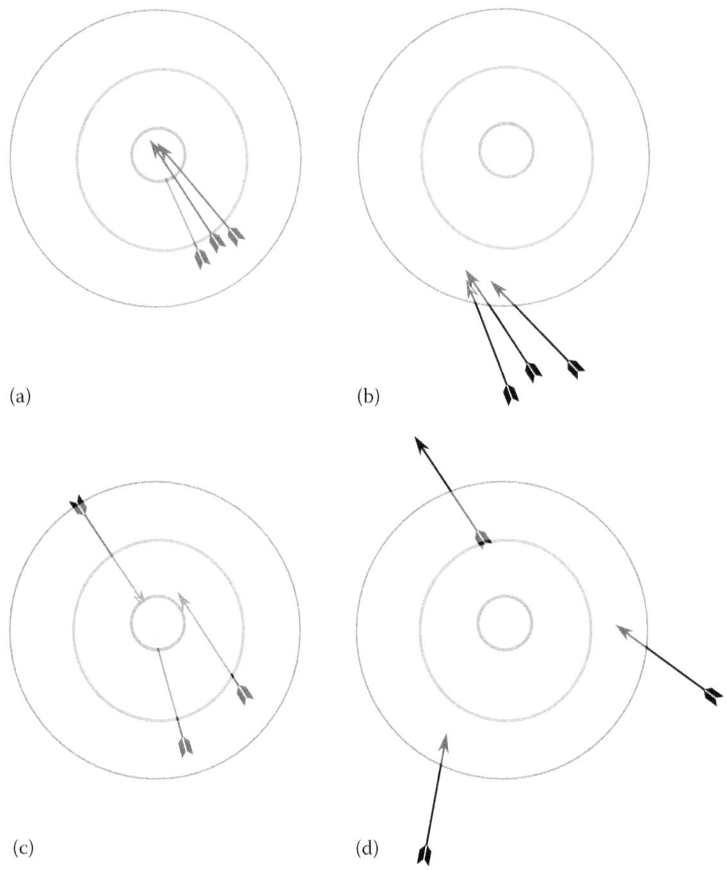

Figure E.1 A diagram illustrating the concepts of accuracy and precision. (a) Arrows indicate accuracy and precision. (b) Arrows show precision but not accuracy. (c) Arrows indicate there is accuracy but less precision than (a) or (b). (d) Arrows show neither accuracy nor precision.

Appendix F: Bioassays in Diet Development, Quality Control, and Testing Effects of Additives

One of the most important tools available to the insect diet professional (and all insect-rearing professionals, for that matter) is the bioassay. This is the case in diet development and improvement settings, as well as in programs of QC/QA and in programs aimed at testing toxins or other biologically active substances. This point has been emphasized throughout this book and cannot be overstated. Bioassays provide answers to the questions, "How are we doing overall? How are we doing with respect to a single, specified variable? How does a given change affect the well-being of the insects that we are targeting?" It is, of course, of paramount importance, whatever other quality tests that we use, that all tests be translated to and designed around the well-being of the insects. The only definitive way of assessing that well-being is through a competently conducted, sensitive, robust bioassay.

A review of the literature reveals that the questions raised in bioassays have included these parameters most commonly: survival, development time, weight of insects at a given stage, linear measurements of given parts of the insects at a given stage, sex ratio, number of eggs laid (fecundity), number of eggs that hatch (fertility), rate or amount of diet consumption, amount of frass (combined fecal and urinary material) produced, diet preferences, other behavior aspects that are specific to the taxon and purpose of the organism being reared (including flight activity, search ability), or physiological factors (such as pheromone production, blood chemistry, oxygen consumption), or finally the appearance of physical abnormalities such as wing deformities.

With human and veterinary medicine, large lists of characteristics have been established for healthy individuals with ranges of normality specified. Measurement of a person's blood pressure, heart rate, electrocardiogram, blood sugar, blood ketones, blood-borne liver products, urine chemistry, body temperature, and numerous other factors can enable a diagnostician to use these measurements for comparisons with well-established normal ranges for these parameters, allowing an objective diagnosis of a disease. Likewise, the normal microbial flora of humans and most domestic animals is well characterized, so that the presence of microbes that deviate from this norm can be detected as an infection that requires therapeutic attention. Unfortunately, no such cadre of normal ranges has been established for any species of insects, let alone for the many species commonly reared in existing programs. There have been a few efforts to establish connections between pathologies such as those stemming from microbial infection (Sikorowski 1984), parasitoid infestation (Dahlman and Vinson 1977), malnutrition (Standifer 1967), or other forms of stress (Cohen and Patana 1982, Cohen 1989), but such background is not comprehensive enough to be used as a robust diagnostic tool. That is to say that measurement of the sugar concentration in the hemolymph, hemolymph protein profiles, or uric acid concentrations in the frass have not been translated into symptoms of a specific malady.

The previously mentioned authors have shown that changes in all these, as well as several other factors, indicate stress of some kind, but they do not pinpoint the kind of problem that the stressed insect is facing. This is in contrast to the state of knowledge of human nutrition, which, though not perfect, provides a reliable set of symptoms for such conditions as protein deficiency (often manifested as kwashiorkor syndrome), iron deficiency (anemia), calcium deficiency (bone loss), thiamine deficiency (beriberi), and ascorbic acid deficiency (scurvy). A competent diagnostician in human or veterinary medicine can use the specific list of symptoms as a basis for diagnosing each of these nutrition-based maladies. There is no such list for any insect malnutrition syndrome. Likewise, the standards of food quality are well developed for human and domestic animal foods in terms of tolerable levels and kinds of microbes, pesticides, various contaminants, and levels of freshness; but again, no such set of standards exists for insect diets or diet components.

This places the burden on the insect-rearing professionals in each insectary to develop their own standards of quality as baselines. The standards adopted should be simple and economical enough to be preformed routinely (e.g., daily or weekly, depending upon circumstances specified later). The standards should also conform to the abilities of insectary personnel or resource people who can be expected regularly to do the testing (e.g., some insectaries have neighboring university laboratories that are capable of performing microbial assessments or biochemical tests that are found to be pertinent to the rearing conditions in question).

The standards should be robust indicators of quality and sensitive enough to locate problems. In this author's experience and in a broad consensus of the literature, survival rates, development rates, and body mass (or size) are the most robust as a general rule. Ideally, a suite of 5–10 standards would be telling of deviations from desirable quality, but most small insectaries do not have adequate staff to conduct these many tests. However, a minimum of three different tests should be used routinely, and meticulous records of data from these tests should be kept. The tests should be selected carefully to assure that each test is not fundamentally a duplicate of the other tests. For example, if weight of last instar larvae, pupal weight, and adult weight are used, there is so much redundancy that the second and third tests add little extra assurance that they are sensitive to what the first test may have missed. Likewise, linear measurements and biomass measurements are so highly correlated that they are redundant. Conversely, development rate, weight, and percent egg hatch are relatively nonrelated and should give substantial resolution to potential problems that are developing (incipient problems).

References

Adkisson, P. L. 1961. Fecundity and longevity of the adult female pink bollworm reared on natural and synthetic diets. *J. Econ. Entomol.* 54: 1224–1227.

Adkisson, P. L., D. L. Bull, and W. E. Allison. 1960a. A comparison of certain artificial diets for laboratory cultures of the pink bollworm. *J. Econ. Entomol.* 53: 791–793.

Adkisson, P. L., E. S. Vanderzant, D. L. Bull, and W. E. Allison. 1960b. A wheat germ medium for rearing the pink bollworm. *J. Econ. Entomol.* 53: 759–762.

Agusti, N. and A. C. Cohen. 2000. *Lygus hesperus* and *L. lineolaris* (Hemiptera: Miridae), phytophages, zoophages, or omnivores: Evidence of feeding adaptations suggested by the salivary and midgut digestive enzymes. *J. Entomol. Sci.* 35: 176–186.

Akey, D. H. and S. D. Beck. 1971. Continuous rearing of the pea aphid, *Acyrthosiphon pisum*, on a holidic diet. *Ann. Entomol. Soc. Am.* 64: 353–356.

Alverson, J. and A. C. Cohen. 2002. Effect of antifungal agents on biological fitness of *Lygus hesperus* Knight (Heteroptera: Miridae). *J. Econ. Entomol.* 95: 256–260.

Anagnostou, C., M. Dorsch, and M. Rohlfs. 2010. Influence of dietary yeasts on *Drosophila melanogaster* life-history traits. *Entomologia Experimentalis et Applicata.* 136: 1–11.

Anand, A. A., S. J. Vennison, S. G. Sankar, D. I. Prabhu, P. T. Vasan, T. Raghuraman, C. J. Geoffrey, and S. E. Vendan. 2010. Isolation and characterization of bacteria from the gut of *Bombyx mori* that degrade cellulose, xylan, pectin and starch and their impact on digestion. *J. Insect. Sci.* 10: 1–20.

Anderson, T. E. and N. C. Leppla. 1992. Preface. In *Advances in Insect Rearing for Research and Pest Management* (T. E. Anderson and N. C. Leppla, Eds.), pp. xi–xii. Westview Press, New Delhi, India.

Anonymous. 1969. National Academy of Sciences. 1969. Control by parasites, predators, and competitors. In *Insects Management and Control*, pp. 100–164. National Academy of Sciences, Washington, DC.

Anonymous. 1999a. A measure of protection. *Pharmaceutical Formulation and Quality*. September/October: 44–45.

Anonymous. 1999b. Product quality control, irradiation and shipping procedures for mass-reared tephritid fruit flies for sterile insect release programmes. Joint FAO/IAEA Programme. Nuclear Techniques in Food and Agriculture. pp. 1–51. Available at http://www.iaea.org/programmes/nafa/d4/public/d4pbl51.html. Accessed November 2014.

Anonymous. 2013. USDA Nutritional database for standard reference, Release 27 (accessed February, 2015).

Anthon, E. S. and S. D. Garret. 1971. Artificial diet and pheromone studies with peach twig borer. Lepidoptera: Gelechiidae. *J. Econ. Entomol.* 64: 259.

Anton, M., V. Beaumal, and G. Gandemer. 2000. Adsorption at the oil–water interface and emulsifying properties of native granules from egg yolk: Effect of aggregated state. *Food Hydrocolloids.* 14: 327–335.

AOAC. A new formulation of Acid Orange 12 reagent solution for the determination of protein, using 16.037-16.041 (Official Methods of Analysis (1957) 12th Ed., AOAC, Washington, DC) was tested for linearity of response and dye binding capacity.

Applebaum, S. W. 1964. Physiological aspects of host specificity in the Brucidae. I. General considerations of developmental compatibility. *J. Insect. Physiol.* 10: 783–788.

Ave, D. A. 1995. Stimulation of feeding: Insect control agents. In *Regulatory Mechanisms in Insect feeding* (R. F. Chapman and G. deBoer, Eds.), pp. 345–363. Chapman & Hall, New York.

Bansal, S. and M. L. Sudha. 2011. Nutritional, microstructural, rheological and quality characteristics of biscuits using processed wheat germ. *Int. J. Food Sci. Nutr.* 62: 474–479.

Barbehenn, R. V. 1999. Non-absorption of ingested lipophilic and amphiphilic allelochemicals by generalist grasshoppers: The role of extractive ultrafiltration by the peritrophic envelope. *Arch. Insect Biochem. Physiol.* 42: 130–137.

Barber, E. J. W. 1992. *Prehistoric Textiles: The Development of Cloth in the Neolithic and Bronze Ages with Special Reference to the Aegean.* Princeton University Press, Princeton, NJ, p. 31.

Bartlett, A. C. 1984. Genetic changes during insect domestication. In *Advances and Challenges in Insect Rearing* (E. G. King and N. C. Leppla, Eds.), pp. 2–8. USDA; ARS, New Orleans, LA.

Bartlett, A. C. 1994. Maintaining genetic diversity in laboratory colonies of parasites and predators. In *Applications of Genetics to Arthropods of Biological Control Significance* (S. K. Narang, A. C. Bartlett, and R. M. Faust, Eds.), pp. 134–145. CRC Press, Boca Raton, FL.

Bass, E. J. 1988. Wheat flour milling. In *Wheat: Chemistry and Technology*, vol. 2 (Y. Pomeranz, Ed.), pp. 1–68. American Association of Cereal Chemists, St. Paul, MN.

Baumberger, J. P. 1917a. The food of *Drosophila melanogaster* Meigen. *Proc. Natl. Acad. Sci. USA* 3: 122–126.

Baumberger, J. P. 1917b. Solid media for rearing *Drosophila. Amer. Nat.* 51(607): 447–448.

Baumberger, J. P. 1919. A nutritional study of insects, with special reference to microorganisms and their substrata. *J. Exp. Zool.* 28: 1–81.

Baumberger, J. P. and R. W. Glaser. 1917. The rearing of *Drosophila ampelophila* loew on solid media. *Science.* 45: 21–22.

BD Biosciences website. http://www.bd.com/ds/technicalCenter/misc/br_3_247.pdf. Accessed February, 2015.

Beale, B. 2002. Probiotics: Their tiny worlds are under scrutiny. *The Scientist.* 16: 20–22.

Beck, S. D. 1972. Nutrition, adaptation, and environment. In *Insect and Mite Nutrition* (J. G. Rodriguez, Ed.), pp. 1–6. North-Holland Publishing Company, Amsterdam, the Netherlands.

Beck, S. D. and J. F. Stauffer. 1950. An aseptic method for rearing European corn borer larvae. *J. Econ. Entomol.* 43: 4–6.

Beckman, H. F., S. M. Bruckart, and R. Reiser. 1953. Laboratory culture of the pink bollworm on chemically defined media. *J. Econ. Entomol.* 46: 627–630.

Bekatorou, C. P. and A. A. Koutinas. 2006. Production of food grade yeasts. *Food Technol. Biotechnol.* 44(3): 407–415.

Bell, J. V., E. G. King, and R. J. Hamale. 1981a. Some microbial contaminants and control agents in a diet and larvae of *Heliothis* spp. *J. Invertebr. Pathol.* 37: 243–248.

Bell, R. A. and F. G. Joachim. 1976. Techniques for rearing laboratory colonies of tobacco hornworms and pink bollworms (Lepidoptera-Sphingidae-Gelechiidae). *Ann. Entomol. Soc. Am.* 69(2): 365–373.

Bell, R. A., D. R. Nelson, T. K. Borg, and D. L. Cardwell. 1975. Wax secretion in non-diapausing and diapausing pupae of the tobacco hornworm, *Manduca sexta. J. Insect. Physiol.* 21: 1725–1729.

Bell, R. A., D. C. Owens, M. Shapiro, and J. R. Tardif. 1981b. Development of mass rearing technology. In *The Gypsy Moth: Research Toward Integrated Pest Management*, Technical Bulletin 1584 (C. C. Doane and M. L. McManus, Eds.), pp. 599–633. USDA, Washington, DC.

BeMiller, J. N. and K. C. Huber. 2008. Carbohydrates. In *Fennema's Food Chemistry*, 4th Ed. (S. Damodaran, K. L. Parkin, and O. R. Fennema, Eds.), pp. 83–154, CRC Press, Boca Raton, FL.

Benzie, I. and J. Strain. 1999. Ferric reducing/antioxidant power assay: Direct measure of total antioxidant activity of biological fluids and modified version for simultaneous measurement of total antioxidant power and ascorbic acid concentration. *Methods Enzymol.* 229: 15–27.

Berger, M., F. Nepveu, and J. Hubert. 2008. Effects of fermentation on the phytochemical composition and antioxidant properties of soy germ. *Food Chem.* 109: 709–721.

Berger, R. S. 1963. *Laboratory Techniques for Rearing Heliothis Species on Artificial Medium*, ARS Series, vols. 33–84, USDA, Beltsville, MD, p. 4.

Bernays, E. A. and R. F. Chapman. 1994. *Host-Plant Selection by Phytophagous Insects.* Chapman & Hall, New York.

Berrios, J. D. J., M. Camara, M. E. Torija, and M. Alonso. 2002. Effect of extrusion cooking and sodium bicarbonate addition on the carbohydrate composition of black bean flours. *J. Food Process. Preserv.* 26: 113–128.

Beuchat, L. R. 1981. Microbial stability as affected by water activity. *Cereal Foods World.* 26(7): 345–349.

Bhatia, A. and N. Khetarpaul. 2002. Effect of fermentation on phytic acid and in vitro availability of calcium and iron of 'Doli ki roti'—An indigenously fermented Indian bread. *Ecol. Food Nutr.* 41: 243–253.

Bidla, G., M. Lindgren, U. Theopold, and M. S. Dushay. 2005. Hemolymph coagulation and phenoloxidase in *Drosophila larvae. Dev. Comp. Immunol.* 29: 669–679.

Blomquist, G. J., C. E. Borgeson, and M. Vundla. 1991. Polyunsaturated fatty acids and eicosanoids in insects. *Insect Biochem.* 21: 99–106.

Blossey, B., D. Eberts, E. Morrison, and T. R. Hunt. 2000. Mass rearing the weevil *Hylobius transversovittatus* (Coleoptera: Curculionidae), biological control agent of *Lythrum salicaria*, on semiartificial diet. *J. Econ. Entomol.* 93: 1644–1656.

Bodenheimer, F. 1951. *Insects as Human Food.* Junk, The Hague, the Netherlands.

Bogdanow, E. A. 1908. Über das Ablängigkeit des Wachstums der Fliegenlarven von. Bacterien und Fermenten un uber Variabilitat und Vererbung bei den Fleischfliegen. *Arch. Anat. Physiol. ABT Suppl:* 173–200.
Boller, E. F. and D. L. Chambers. 1977. Quality control: An idea book for fruit fly workers. *IOBC/WPRS Bull.* 1977/5. 162 p.
Boller, E. F., B. I. Katsoyannos, U. Remund, and D. L. Chambers. 1981. Measuring, monitoring, and improving the quality of mass-reared Mediterranean fruit flies, *Ceratitis capitata* Wied. 1. The RAPID quality control system for early warning. *Z. Angew. Entomol.* 92: 67–83.
Borejszo, Z. and K. Khan. 1992. Reduction of flatulence-causing sugars by high temperature extrusion of pinto bean high starch fractions. *J. Food Sci.* 57: 771–772.
Bowdan, E. and V. G. Dethier. 1986. Coordination of a dual inhibitory system regulating feeding behaviour in the blowfly. *J. Comp. Physiol. A* 158: 713–722.
Brazzel, J. R., C. O. Calkins, D. L. Chambers, and D. B. Gates. Required quality control tests, quality specifications, and shipping procedures for laboratory produces mediterranean fruit flies for sterile insect control programs. APHIS-81-51, United States Department of Agriculture, Hyattsville, MD, September 3, 1986.
Brewer, F. D. 1992. Gel extenders in larval diet of *Cochliomyia homillivorax* (Diptera: Calliphoridae). *J. Econ. Entomol.* 85: 445–450.
Brewer, F. D. and E. G. King. 1979. Consumption and utilization of soyflour-wheat germ diets by *Heliothis* spp. *Ann. Entomol. Soc. Am.* 72: 415–417.
Brewer, F. D. and O. Lindig. 1984. Ingredients for insect diets. Quality assurance, sources, and storage and handling. In *Advances and Challenges in Insect Rearing* (E. G. King and N. C. Leppla, Eds.), pp. 45–50. USDA; ARS, New Orleans, LA.
Brewer, F. D. and D. F. Martin. 1976. Substitutes for agar in a wheat germ diet used to rear the corn earworm and the sugarcane borer. *Ann. Entomol. Soc.* 69: 255–256.
Bridges, C. B. and H. H. Darby. 1933. Culture media for *Drosophila* and the pH of media. *Amer. Nat.* 67: 437–472.
Brinton, F. E., M. D. Proverbs, and B. E. Carty. 1969. Artificial diet for mass production of the codling moth, *Carpocapsa pomonella* (Lepidoptera: Olethreutidae). *Can. Ent.* 101: 577–584.
Brooks, M. A. 1963. The microorganisms of healthy insects. In *Insect Pathology*, vol. 1 (E. A. Steinhaus, Ed.), pp. 215–243. Academic Press, New York.
Brower, L. P. 1969. Ecological chemistry. *Sci. Am.* 20: 22–29.
Brower, L. P. and J. V. Z. Brower. 1964. Birds, butterflies, and plant poisons: A study in ecological chemistry. *Zoologica.* 49: 137–159.
Brun, L. O., V. Gaudichon, J. Wigleyp, and B. Orstom. 1993. An artificial diet for continuous rearing of the Coffee Berry Borer, *Hypothenemus hampei* (Ferrari) (Coleoptera: Scolytidae). *Insect Sci. Appl.* 14: 586–587.
Brust, M. and G. Fraenkel. 1955. The nutritional requirements of the larvae of a blowfly, *Phormia regina* (Meig.). *Physiol. Zool.* 28: 186–204.
Buchanan, B. B., W. Gruissem, and R. L. Jones. 2000. *Biochemistry and Molecular Biology of Plants.* American Society of Plant Physiologists. Rockville, MD.
Buchon, N., N. A. Broderick, and B. Lemaitre. 2013. Gut homeostasis in a microbial world: Insights from *Drosophila melanogaster. Nat. Rev. Microbiol.* 11: 615–626.
Budavari, S. (Ed.). 1996. *The Merck Index*, 12th Ed. Merck & Co., Inc., Whitehouse Station, NJ (discussion of formalin).
Buettner, G. R. 1988. In the absence of catalytic metals ascorbate does not autooxidize at pH 7.0: Ascorbate as a test for catalytic metals. *J. Biochem. Biophys. Methods.* 16: 27–40.
Buettner, G. R. 1993. The pecking order of free radicals and antioxidants: Lipid peroxidation, α-tocopherol, and ascorbate. *Arch. Biochem. Biophys.* 300: 535–543.
Buglione, M. and J. Lozano. 2002. Nonenzymatic browning and chemical changes during grape juice storage. *J. Food Sci.* 67: 1538–1543.
Burton, R. L. and D. J. Schuster. 1986. Meridic diet for rearing tomato pinworm larvae (Lepidoptera: Gelechiidae). *J. Econ. Entomol.* 79: 1143–1146.
Cáceres, C. and P. Rendón. 2008. The FAO/IAEA interactive spreadsheet for design and operating insect mass-rearing facilities. In *Proceedings of the 7th International Symposium on Fruit Flies of Economic Importance* (R. Sugayama, R. Zucchi, S. Ovruski, and J. Sivinski, Eds.), September 10–15, 2006, Bahia, Brazil, pp. 307–312.

Cáceres, C., P. Rendón, and A. Jessup. 2012. The FAO/IAEA spreadsheet for designing and operating insect mass-rearing facilities. FAO Plant Production and Protection Paper 205, Rome, Italy.

Calkins, C. O., K. Bloem, S. Bloem, and D. L. Chambers. 1994. Advances in measuring quality and assuring good field performance in mass reared fruit flies. In *Fruit Flies and the Sterile Insect Technique* (C. O. Calkins, W. Klassen, and P. Liedo, Eds.), pp. 85–96. CRC Press, Boca Raton, FL.

Campbell, B. 1990. On the role of microbial symbiotes in herbivorous insects. In *Insect-Plant Interactions*, vol. I (E. Bernays, Ed.), pp. 1–44. CRC Press, Boca Raton, FL.

Campbell, B. and W. D. Nes. 1983. A reappraisal of sterol biosynthesis and metabolism in aphids. *J. Insect Physiol.* 29: 149–156.

Canavoso, L. E., Z. E. Jouni, K. J. Karnas, J. E. Pennington, and M. A. Wells. 2001. Fat metabolism in insects. *Annu. Rev. Nutr.* 21: 23–46.

Cao, G. and R. L. Prior. 1999. Measurement of oxygen radical absorbance capacity in biological samples. *Methods Enzymol.* 299: 50–62.

Carpenter, J. E. and S. D. Hight. 2012. Rearing the oligophagous *Cactoblastis cactorum* (Lepidoptera: Pyralidae) on meridic diets without host plant materials. *Florida Entomologist.* 95(4): 1132–1141.

Carpenter, K. J., P. K. Weber, M. L. Davisson, J. Pett-Ridge, M. I. Haverty, and P. J. Keeling. 2013. Correlated SEM, FIB-SEM, TEM, and nanoSIMS imaging of microbes from the hindgut of a lower termite: Methods for in situ functional and ecological studies of uncultivable microbes. *Microsc. Microanal.* 19: 1490–1501.

Carrington, D. 2011. Insects could be the key to meeting food needs of growing global population. *The Guardian*, August 1, 2010. Retrieved February 27, 2011.

Carroll, M., A. Hanlon, T. Hanlon, A. R. Zangerl, and M. R. Berenbaum. 1997. Behavioral effects of carotenoid sequestration by the parsnip webworm, *Depressaria pastinacella. J. Chem. Ecol.* 23: 2707–2719.

Caspari, E. 1952. Pleiotropic gene action. *Evolution.* 6: 1–18.

Castrillo, L. A., M. H. Griggs, and J. D. Vandenberg. 2012. Brood production by *Xylosandrus germanus* (Coleoptera: Curculionidae) and growth of its fungal symbiont on artificial diet. *Environ. Entomol.* 41: 822–827.

Cerutti, F. and F. Bigler. 1995. Quality assessment of *Trichogramma brassicae* in the laboratory. *Entomologia Experimentalis et Applicata.* 75(1): 19–26.

Chambers, D. L. 1977. Quality control in mass rearing. *Annu. Rev. Entomol.* 22: 289–308.

Chambers, D. L. and T. R. Ashley. 1984. Putting the control in quality control in insect rearing. In *Advances and Challenges in Insect Rearing* (E. G. King and N. C. Leppla, Eds.), pp. 256–260, U.S. Government Printing Office, Washington, DC.

Chang, C. 2009. Evaluation of yeasts and yeast products in larval and adult diets for the oriental fruit fly, *Bactrocera dorsalis*, and adult diets for the medfly, *Ceratitis capitata*, and the melon fly, *Bactrocera curcurbitae. J. Insect Sci.* 9:23. Available online: insectscience.org/9.23. Last accessed January 2015.

Chang, C. L. and R. Kurashima. 1999. Effects of ascorbic acid-rich bell pepper on development of *Bactorcera latifrons* (Diptera: Tephritidae). *J. Econ. Entomol.* 92: 1108–1112.

Chang, S. K. C. 1998. Protein analysis. In *Food Analysis* (S. Suzanne Nielsen, Ed.), pp. 237–250. Aspen Publishers, Inc., Gaithersburg, MD.

Chang, V. C. S. and L. Jensen. 1972. A diet for studying clonal resistance of sugarcane to the New Guinea sugarcane weevil. *J. Econ. Entomol.* 65: 1197–1199.

Chang, Y. M., D. S. Luthe, F. M. Davis, and W. P. Williams. 2000. Influence of whorl region from resistant and susceptible corn genotypes on fall armyworm (Lepidoptera: Noctuidae) growth and development. *J. Econ. Entomol.* 93: 477–483.

Chapman, R. and G. deBoer. 1995. *Regulatory Mechanisms in Insect Feeding*. Chapman & Hall, New York.

Chapman, R. F. 1995a. Chemosensory regulation of feeding. In *Regulatory Mechanisms in Insect Feeding* (R. F. Chapman and G. de Boer, Eds.), pp. 101–136. Chapman & Hall, New York.

Chapman, R. F. 1995b. Mechanics of food handling by chewing insects. In (R. F. Chapman and G. de Boer, Eds.), pp. 3–31. *Regulatory Mechanisms in Insect Feeding*. Chapman & Hall, New York.

Chapman, R. F. 1998. *The Insects: Structure and Function*. Cambridge, U.K.: Cambridge University Press.

Chapman, R. N. 1924. Nutritional studies on the confused flour beetle, *Tribolium confusum* Duval. *J. Gen. Physiol.* 6: 565–585.

Chaudhury, M. F. and L. A. Alvarez. 1999. A new starch-grafted gelling agent for screwworm (Diptera: Calliphoridae) larval diet. *J. Econ. Entomol.* 92: 1138–1141.

Chaudhury, M. F., L. A. Alvarez, and L. L. Velazquez. 2000. A new meatless diet for adult screwworm (Diptera: Calliphoridae). *J. Econ. Entomol.* 93: 1398–1401.

Chaudhury, M. F., L. A. Alvarez, and J. B. Welch. 1998. An alternative source of blood protein for screwworm (Diptera: Calliphoridae) larval diet. *J. Econ. Entomol.* 91: 1397–1400.
Chaudhury, M. F., S. R. Skoda, and A. Sagel. 2011. Solidifying agent and processing of blood used for the larval diet affect screwworm (Diptera: Calliphoridae) life-history parameters. *J. Econ. Entomol.* 104(3): 1103–1107.
Cherrett, J. M., R. J. Powell, and D. J. Stradling. 1987. The mutualism between leaf-cutting ants and their fungus. In *Insect-Fungus Interactions* (N. Wilding, Ed.), pp. 93–120. Academic Press, New York.
Chippendale, G. M. 1978. The functions of carbohydrates in insect life processes. In *Biochemistry of Insects* (M. Rockstein, Ed.), pp. 2–57. Academic Press, New York.
Chippendale, G. M. and S. D. Beck. 1964. Nutrition of the European corn borer, *Ostrinia nubilalis* (Hubner). V. Ascorbic acid as the corn leaf factor. *Entomol. Exp. Appl.* 7: 241–248.
Chu, E. J. H. 1945. Nutritional requirements of *Drosophila hydei. Tex. Rep. Biol. Med.* 3: 513–527.
Chun, K. H., B. Y. Kim, T. I. Son, and Y. T. Hahm. 1997. The extension of tofu shelf life with water-soluble degraded chitosan as immersion solution. *Korean J. Food Sci. Technol.* 29: 476–481.
Churchill-Stanland, C., R. Stanland, T. T. Y. Wong, N. Tanaka, D. O. McInnis, and R. V. Dowell. 1986. Size as a factor in the mating propensity of Mediterranean fruit flies (*Ceratitis capitata* (Diptera: Tephrididae)), in the laboratory. *J. Econ. Entomol.* 79: 614–619.
Cohen, A. C. 1981. An artificial diet for *Geocoris punctipes* (Say). *Southwest Entomol.* 6: 109–113.
Cohen, A. C. 1983. Improved method of encapsulating artificial diet for rearing predators of harmful insects. *J. Econ. Entomol.* 76: 957–959.
Cohen, A. C. 1984. Food consumption, food utilization and metabolic rates of *Geocoris punctipes.* (Het.: Lygaeidae) fed *Heliothis virescens* (Lep.: Noctuidae) eggs. *Entomophaga.* 29: 361–367.
Cohen, A. C. 1985a. Simple method for rearing the insect predator *Geocoris punctipes* (Heteroptera: Lygaeidae) on a meat diet. *J. Econ. Entomol.* 78: 1173–1175.
Cohen, A. C. 1985b. Metabolic rates of two hemipteran members of a predator-prey complex. *Comp. Biochem. Physiol.* 81A: 833–836.
Cohen, A. C. 1989. Ingestion efficiency and protein consumption by a heteropteran predator. *Ann. Entomol. Soc. Amer.* 82: 495–499.
Cohen, A. C. 1990a. Feeding adaptations of some predaceous Hemiptera. *Ann. Entomol. Soc. Am.* 83: 1215–1223.
Cohen, A. C. 1990b. Fatty acid distribution as related to adult age, sex and diet in the phytophagous heteropteran, *Lygus hesperus. J. Entomol. Sci.* 25: 75–84.
Cohen, A. C. 1992. Using a systematic approach to develop artificial diets for predators. In *Advances in Insect Rearing for Research and Pest Management* (T. E. Anderson and N. C. Leppla, Eds.), pp. 77–91. Westview Press. Boulder, CO.
Cohen, A. C. 1993. Organization of proteinases and preliminary characterization of salivary trypsin-like enzymes in a predaceous heteropteran. *J. Insect Physiol.* 39: 823–829.
Cohen, A. C. 1995. Extra-oral digestion in predatory arthropods. *Ann. Rev. Entomol.* 40: 85–103.
Cohen, A. C. 1996. Plant feeding by predatory Hemiptera: Evolutionary and adaptational aspects of trophic switching. In *Biological Control by Predatory Hemiptera* (R. Wiedenmann, O. Alomar, and D. Livingstone, Eds.), pp. 1–17. Thomas Say Publications in Entomology, Entomological Society of America, Lanham, MD.
Cohen, A. C. 1998a. Biochemical and morphological dynamics and predatory feeding habits in terrestrial Heteroptera. In *Predatory Heteroptera: Their Ecology and Use in Biological Control* (M. Coll and J. R. Ruberson, Eds.), pp. 21–32. Thomas Say Publications in Entomology, Entomological Society of America, Lanham, MD.
Cohen, A. C. 1998b. Solid-to-liquid feeding: The inside(s) story of extra-oral digestion in predaceous Arthropoda. *Am. Entomol.* 44: 103–116.
Cohen, A. C. 1998c. Artificial media for rearing entomophages comprising cooked, whole egg. U.S. Patent #5,834,177. November 10, 1998.
Cohen, A. C. 2000a. Feeding fitness and quality of domesticated and feral predators: Effects of long-term rearing on artificial diet. *Biol. Control.* 17: 50–54.
Cohen, A. C. 2000b. A review of feeding studies of *Lygus* spp. with emphasis on artificial diets. *Southwest. Entomol.* Suppl. 23: 111–119.
Cohen, A. C. 2000c. New oligidic production diet for *Lygus hesperus* Knight and *L. lineolaris* (Palisot de Beauvois). *J. Entomol. Sci.* 35: 301–310.

Cohen, A. C. 2001. Formalizing insect rearing and artificial diet technology. *Am. Entomol.* 47: 198–206.

Cohen, A. C. 2003a. Artificial diet for arthropods. U.S. Patent #6,506,597. January 14, 2003.

Cohen, A. C. 2003b. *Insect Diets: Science and Technology*. CRC Press, Boca Raton, FL.

Cohen, A. C. and C. Cheah. 2011. Development of artificial diets for predators of hemlock woolly adelgids. In *Implementation and Status of Biological Control of the Hemlock Woolly Adelgid*, FHTET-2011-0 (B. Onken and R. Reardon, Eds.), pp. 148–160. U.S. Forest Service Publication, Morgantown, WV.

Cohen, A. C., C. Cheah, K. Kidd, and T. Hodgson. 2011. Defining PC/QC standards for mass-rearing HWA predators. In *Implementation and Status of Biological Control of the Hemlock Woolly Adelgid*, FHTET-2011-0 (B. Onken and R. Reardon, Eds.), pp. 139–147. U.S. Forest Service Publication, Morgantown, WV.

Cohen, A. C., C. C. Chu, T. J. Henneberry, T. Freeman, D. Nelson, J. Buckner, D. Margosan, P. Vail, and L. H. Aung. 1998. Feeding biology of the silverleaf whitefly (Homoptera: Aleyrodidae). *Chin. J. Entomol.* 18: 65–82.

Cohen, A. C. and P. Crittenden. 2004. Deliberately added and "cryptic" antioxidants in three artificial diets for insects. *J. Econ. Entomol.* 97: 265–272.

Cohen, A. C. and J. W. Debolt. 1983. Rearing *Geocoris punctipes* on insect eggs. *Southwest. Entomol.* 8: 61–64.

Cohen, A. C., T. J. Henneberry, and C. C. Chu. 1996. Geometric relationships between whitefly feeding behavior and vascular bundle arrangements. *Entomologia Exp. et Applic.* 78: 135–142.

Cohen, A. C., D. A. Nordlund, and R. A. Smith. 1999. Mass rearing of entomophagous insects and predaceous mites: Are the bottlenecks biological, engineering, economic, or cultural? *Biocontrol News Inform.* 20: 85–90.

Cohen, A. C. and R. Patana. 1982. Ontogenetic and stress-related changes in the hemolymph chemistry of beet armyworms. *Comp. Biochem. Physiol.* 71A: 193–198.

Cohen, A. C. and R. Patana. 1984. Efficiency of food utilization by *Heliothis zea* (Lepidoptera: Noctuidae) fed artificial diets or green beans. *Can. Ento.* 116: 139–146.

Cohen, A. C. and R. Patana. 1985. Chemical composition of tobacco budworm eggs during development. *Comp. Biochem. Physiol.* 81B: 165–169.

Cohen, A. C. and L. Smith. 1998. A novel concept in artificial diets for *Chrysoperla rufilabris*: The efficacy of solid diet. *Biol. Control.* 13: 49–54.

Cohen, A. C. and Staten, R. T. 1994. Long term culturing and quality assessment of predatory big-eyed bugs, *Geocoris punctipes*. In *Applications of Genetics to Arthropods of Biological Control Significance* (S. Karl Narang, A. C. Bartlett, and R. M. Faust, Eds.), pp. 122–132. CRC Press, Boca Raton, FL.

Cohen, A. C. and N. M. Urias. 1988. Nutritional indices of a predator *Geocoris punctipes* (Heteroptera: Lygaeidae) fed artificial diets with rutin. *J. Entomol. Sci.* 23: 174–179.

Cohen, A. C. and A. G. Wheeler. 1998. Role of saliva in the highly destructive four-lined plant bug (Hemiptera: Miridae). *Annals Ent. Soc. Am.* 91: 94–100.

Cohen, A. C., F. Zeng, and P. Crittenden. 2005. Adverse effects of raw soybean extract on survival and growth of *Lygus hesperus*. *J. Entomol. Sci.* 40: 390–400.

Cohen, R. W., S. Friedman, and G. P. Waldbauer. 1988. Physiological control of nutrient self-selection in *Heliothis zea* larvae: The role of serotonin. *J. Insect Physiol.* 34: 935–940.

Cooper, D. 1960. Food preferences of larval and adult *Drosophila*. *Evolution.* 14: 41–55.

Corpas, F. J., J. B. Barroso, and L. A. del Río. 2001. Peroxisomes as a source of reactive oxygen species and nitric oxide signal molecules in plant cells. *Trends Plant Sci.* 6(4): 145–150.

Costa, H. S., D. E. Ullman, M. W. Johnson, and B. E. Tabashnik. 1993. Antibiotic oxytetracycline interferes with *Bemisia tabaci* (Homoptera: Aleyrodidae) oviposition, development, and ability to induce squash silverleaf. *Ann. Entomol. Soc. Am.* 86: 740–748.

Dadd, R. G. 1957. Ascorbic acid and carotene in the nutrition of the desert locust, *Schistocerca gregaria* Forsk. *Nat. London.* 179: 427–428.

Dadd, R. H. 1968. Problems connected with inorganic components of aqueous diets. *Bull. Entomol. Soc. Am.* 14: 22–26.

Dadd, R. H. 1972. Ambiguities in the interpretation of growth experiments with mosquito larvae in semi-synthetic dietary media. In *Insect and Mite Nutrition* (J. G. Rodriguez, Ed.), pp. 199–209. North-Holland Publishing Co., Amsterdam, the Netherlands.

Dadd, R. H. 1985. Nutrition: Organisms. In *Comprehensive Insect Physiology, Biochemistry, and Pharmacology*, vol. 4 (G. A. Kerkut and L. I. Gilbert, Eds.), pp. 313–390. Pergamon Press, Oxford, England.

Dahlman, D. L. and S. B. Vinson. 1977. Effect of calyx fluid from an insect parasitoid on host hemolymph dry weight and trehalose content. *J. Invert. Pathol.* 29: 227–229.

Dalley, S. 2002. *Mari and Karana: Two Old Babylonian Cities*, 2nd Ed. Gorgias Press LLC, Piscataway, NJ, p. 203.

Damodaran, S. 2008. Amino acids, peptides, and proteins. In *Fennema's Food Chemistry*, 4th Ed. (S. Damodaran, K. L. Parkin, and O. R. Fennema, Eds.), pp. 217–330, CRC Press, Boca Raton, FL.

Damodaran, S., K. L. Parkin, and O. R. Fennema (Eds.). 2008. *Fennema's Food Chemistry*, 4th Ed. pp. 1–14, CRC Press, Boca Raton, FL.

Daubert, C. R. and E. A. Foegeding. 1998. Rheological principles for food analysis. In *Food Analysis*, 2nd Ed. (S. Suzanne Nielsen, Ed.), pp. 551–569. Aspen Publishers, Inc., Gaithersburg, MD.

Davis, F. M., G. T. Baker, and W. P. Williams. 1995. Anatomical characteristics of maize resistant to leaf feeding by southwestern corn borer (Lepidoptera: Pyralidae) and fall armyworm (Lepidoptera: Noctuidae). *J. Agric. Entomol.* 12: 55–65.

Davis, F. M. and W. D. Guthrie. 1992. Rearing Lepidoptera for plant resistance research. In *Advances in Insect Rearing for Research and Pest Management* (T. E. Anderson and N. C. Leppla, Eds.), pp. 211–228. Oxford and IBH, New Delhi, India.

Davis, F. M., S. Malone, T. G. Oswalt, and W. C. Jordan. 1990. Medium-sized lepidopterous rearing system using multicellular rearing trays. *J. Econ. Entomol.* 83: 1535–1540.

Debolt, J. 1982. Meridic diet for rearing successive generations of *Lygus hesperus*. *Ann. Entomol. Soc. Am.* 75: 119–122.

Debolt, J. W. and R. Patana. 1985. *Lygus hesperus*. In Singh, P. and R. F. Moore (Eds.), *Handbook of Insect Rearing*, vol. 1, pp. 329–338.

DeFoliart, G. R. 1975. Insects as a source of protein. *Bull. Entomol. Soc. Am.* 21: 161–163.

DeFoliart, G. R. 1999. Insects as food. *Annu. Rev. Entomol.* 44: 21–50.

Delcourt, A. and E. Guyenot. 1910. The possibility of studying certain Diptera in a defined environment. *Comptes rendus hebdomadaires des séances de l'Académie des sciences* (0001-4036). 151: 255–257.

deMan, J. M. 1999. *Principles of Food Chemistry*. Aspen Publishers, Inc., Gaithersburg, MD.

Deming, W. E. 1986. *Out of the Crisis*. Massachusetts Institute of Technology, Cambridge, MA (Originally published in 1982 as *Quality, Productivity, and Competitive Position*).

Dethier, V. G. 1941. Chemical factors determining the choice of food plants by *Papilio* larvae. *Amer. Nat.* 75(756): 61–73.

Dethier, V. G. 1962. *To Know a Fly*. Holden-Day, San Francisco, CA.

Dethier, V. G. 1976. *The Hungry Fly*. Harvard University Press, Cambridge, MA.

Dethier, V. G. and D. Bodenstein. 1958. Hunger in the blowfly. *Z. Tierpsychol.* 15: 129–140.

Dethier, V. G. and E. Bowdan. 1984. Relations between differential threshold and sugar receptor mechanisms in the blowfly. *Behav. Neurosci.* 98:791–803.

Dethier, V. G. and L. E. Chadwick. 1947. Rejection thresholds of the blowfly for a series of aliphatic alcohols. *J. Gen. Physiol.* 30: 247–253.

Dethier, V. G., D. R. Evans, and M. V. Rhoades. 1956. Some factors controlling the ingestion of carbohydrates by the blowfly. *Biol. Bull.* 111: 204–222.

Dethier, V. G. and M. T. Yost. 1979. Oligophagy and absence of food-aversion learning in tobacco hornworms, *Manduca sexta*. *Physiol. Entomol.* 4: 125–130.

DIFCO Laboratories. 1984. *DIFCO Manual*, 10th Ed. DIFCO Laboratories, Detroit, MI.

Dillon, R. J. and V. M. Dillon. 2003. The gut bacterial of insects: Nonpathogenic interactions. *Annu. Rev. Entomol.* 49: 71–92.

Diplock, A., J. Carleux, G. Grozier-Willi, K. Kok, C. Rice-Evans, M. Roberfroid, W. Stahl, and J. Vina-Ribes. 1998. Functional food sciences and defense against reactive oxidative species. *Br. J. Nutr.* 80(Suppl. 1): 77–82.

Dobzhansky, T., D. M. Cooper, H. J. Phaff, E. P. Knapp, and H. L. Carson. 1956. Differential attraction of species of *Drosophila* to different species of yeasts. *Ecology.* 37: 544–550.

Donkor, O. N. and Shah, N. P. 2008. Production of beta-glucosidase and hydrolysis of isoflavone phytoestrogens by *Lactobacillus acidophilus*, *Bifidobacterium lactis*, and *Lactobacillus casei* in soymilk. *J. Food Sci.* 73: 15–20.

Dougherty, E. C. 1959. Introduction to axenic culture of invertebrate metazoan: A goal. *Ann. N. Y. Acad. Sci.* 77: 27–54.

Douglas, A. E. 1998. Nutritional interactions in insect-microbial symbioses: Aphids and their symbiotic bacteria. *Buchnera*. *Annu. Rev. Entomol.* 43: 17–37.

Duffey, S. S. 1980. Sequestration of plant natural products by insects. *Ann. Rev. Entomol.* 31: 321–339.

Dunkel, F. V. and N. R. Read. 1991. Review of the effect of sorbic acid on insect survival in rearing diets with reference to other antimicrobials. *Am. Entomol.* 37: 172–178.

Dunphy, G. B., D. F. Niven, and J. S. Chadwick. 2002. Iron contributes to the antibacterial functions of the haemolymph of *Galleria mellonella. J. Insect Physiol.* 48: 903–914.

Dyck, V. A. 2010. *Rearing Codling Moth for the Sterile Insect Technique.* FAO and IAEA, Rome, Italy.

Economopoulos, A. P. 1992. Adaptation of the mediterranean fruit fly (Diptera: Tephritidae) to artificial rearing. *J. Econ. Entomol.* 85: 753–758.

Edney, E. B. 1977. *Water Balance in Land Arthropods.* Springer-Verlag, Berlin, Germany.

Edwards, R. H., E. Miller, R. Becker, A. P. Mossman, and D. W. Irving. 1996. Twin screw extrusion processing of diet for mass rearing the pink bollworm. *Trans. Am. Soc. Agric. Eng.* 39(5): 1789–1797.

Eichenseer, H., M. C. Mathews, J. L. Bi, J. B. Murphy, and G. W. Felton. 1999. Salivary glucose oxidase: Multifunctional roles for *Helicoverpa zea*? *Arch. Insect Biochem. Physiol.* 42: 99–109.

El-Adawy, T. A., E. H. Rahma, A. A. El-Bedawy, and T. Y. Sobihah. 2000. Effect of soaking process on nutritional quality and protein solubility of some legume seeds. *Food Nahrung*. 44(5): 339–343.

Ellsbury, M. M., E. A. Backus, and D. L. Ullman. 1994. *History, Development, and Application of AC Electronic Insect Feeding Monitors*, Thomas Say Publications in Entomology: Proceedings. Entomological Society of America. Lanham, MD.

El-Tabey, A. S. and E. M. Mrak. 1951. The fate of yeast in the digestive tract of *Drosophila. Amer. Nat.* 85: 381–383.

Ely, M. J. and A. M. Jungreis. 1977a. Radiation-induced inhibition of eclosion in the tobacco hornworm, *Manduca sexta. Biol. Bull.* 152: 169–181.

Ely, M. J. and A. M. Jungreis. 1977b. Effects of x-irradiation on egg hatchability, larval and pupal survival in the tobacco hornworm, *Manduca sexta. J. Insect Physiol.* 23: 95–101.

Endres, J. G. 2001. *Soy Protein Products: Characteristics, Nutritional Aspects, and Utilization.* AOCS Press, Champaign, IL.

Enigl, D. C. 1999. Controlling water activity. *Pharmaceutical Formulation and Quality*. September/October: 29–34.

Erickson, E., A. C. Cohen, D. L. Brummett, and J. Lusby. 1997. Tyrosine nodules in honeybees parasitized by Varroa mites. *J. Invert. Path.* 70: 27–32.

Evans, J. and S. James. 1993. Freezing and meat quality. In Turner, A. (Ed.), *Food Technology International Europe*, pp. 53–56. London, U.K.: Sterling Publications International.

Fanson, B. G., S. Sundaralingam, L. Jiang, B. C. Dominiak, and G. D'Arcy. 2014. A review of 16 years of quality control parameters at a mass-rearing facility producing Queensland fruit fly, *Bactrocera tryoni. Entomologia Experimentalis et Applicata.* 151: 152–159.

FAO/IAEA/USDA. 2003. Manual for product quality control and shipping procedures for sterile mass-reared tephritid fruit flies, Version 5.0, May 2003. International Atomic Energy Agency, Vienna, Austria.

FAO/IAEA/USDA. 2014. Product quality control for sterile mass-reared and released tephritid fruit flies, Version 6.0. International Atomic Energy Agency, Vienna, Austria, 164pp.

Farrell, B. D., A. S. O. Sequeira, B. C. O'Meara, B. B. Normark, J. H. Chung, and B. H. Jorda. 2001. The evolution of agriculture in beetles (Curculionidae: Scolytinae and Platypodinae). *Evolution.* 55: 2011–2027.

Feeny, P. 1970. Seasonal changes in oak leaf tannins and nutrients as a cause of spring feeding by winter moth caterpillars. *Ecology.* 51: 565–581.

Feigenbaum, A. V. 1983. *Total Quality Control*, 3rd Ed. McGraw-Hill, Inc., New York.

Fellows, P. J. 2000. *Food Processing Technology*, 2nd Ed. CRC Press, Boca Raton, FL.

Feng, J., X. Liu, and Z. R. Xu. 2007. Effects of fermented soybean meal on digestive enzyme activities and intestinal morphology in broilers. *Poult. Sci.* 86: 1149–1154.

Fennema, O. R. 1996. Water and ice. In *Fennema's Food Chemistry*, 3rd Ed. (O. R. Fennema, Ed.), Marcel Dekker, Inc. New York.

Fennema, O. R. and S. R. Tannenbaum. 1996. Introduction. In *Food Chemistry* (O. R. Fennema, Ed.), pp. 1–15. Marcel Dekker, Inc., New York.

Finke, M. D. 2002. Complete nutrient composition of commercially raised invertebrates used as food for insectivores. *Zoo Biol.* 21: 269–285.

Finke, M. D. 2005. Nutrient composition of bee brood and its potential as human food. *Ecol. Food Nut.* 44(4): 257–270.

Finke, M. D. 2007. Estimate of chitin in raw whole insects. *Zoo Biol.* 26: 105–115.

Forcada, C., E. Alcacer, M. D. Garcera, A. Tato, and R. Martinez. 1999. Resistance to *Bacillus thuringiensis* Cry1Ac toxin in three strains of *Heliothis virescens*: Proteolytic and SEM study of the larval midgut. *Arch. Insect Biochem. Physiol.* 42: 51–63.

Formalin website. http://stainsfile.info/StainsFile/prepare/fix/agents/terminology.htm.

Fraenkel, G. 1953. The nutritional value of green plants for insects. In *Transactions of the IXth International Congress of Entomology*, vol. 2, pp. 90–100, Amsterdam, the Netherlands.

Fraenkel, G. 1959. The raison d'etre of secondary plant substances. *Science.* 129: 1466–1470.

Fraenkel, G. and M. Blewett. 1942. Biotin as a possible growth factor for insects. *Nat. London.* 149: 301.

Fraenkel, G. and M. Blewett. 1947a. The importance of folic acid and unidentified members of the vitamin-B complex in the nutrition of certain insects. *Biochem. J.* 41: 469–475.

Fraenkel, G. and M. Blewett. 1947b. Folic acid and the growth of insects. *Biochem. J.* 41: 18.

Fraenkel, G. and M. Blewett. 1947c. The requirements of known and unknown members of the vitamin-B complex. *Br. J. Nut.* 1: R1.

Fraenkel, G. and M. Blewett. 1948. BT, a new vitamin of the B-group and its relation to the folic acid group, and other anti-anaemia factors. *Nature.* 161: 981–983.

Francke-Grossmann, H. 1967. Ectosymbiosis in wood inhabiting insects. In: *Symbiosis*, vol. 2 (M. Henry, Ed.), pp.141–205. Academic Press, New York.

Freed, M., S. Brenner, and V. O. Wodicka. 1948. Prediction of thiamine and ascorbic acid stability in canned stored foods. *Food Technol.* 3: 148–151.

Frias, J., Y. S. Song, and C. Martinez-Villaluenga. 2008. Immunoreactivity and amino acid content of fermented soybean products. *J. Agric. Food Chem.* 56: 99–105.

Fukuda, T., M. Suto, and Y. Higuchi. 1960. Silkworm raising on the artificial food. *J. Seric. Sci. Japan.* 29: 1–3 (in Japanese with summary in Singh 1977).

Fukushima, D. 1991. Recent progress of soybean protein foods: Chemistry, technology, and nutrition. *Food Rev. Int.* 7(3): 323–351. Marcel Dekker, Inc.

Funke, B. R. 1983. Mold control for insect-rearing media. *Bull. Entomol. Soc. Am.* 29: 41–44.

Ganter, P. F. 2006. Yeast and invertebrate associations. In *Biodiversity and Ecophysiology of Yeasts* (C. A. Rosa and G. Peter, Eds.), pp. 303–370, Springer, Berlin, Germany.

Ganter, P. F., W. T. Starmer, M.-A. Lachance, and H. J. Phaff. 1986. Yeast communities from host plants and associated Drosophila in southern Arizona: New isolations and analysis of the relative importance of hosts and vectors on community composition. *Oecologia.* 70: 386–392.

Gao, X., L. Bjork, V. Trajkovski, and M. Uggla. 2000. Evaluation of antioxidant activities of rosehip ethanol extracts in different test systems. *J. Sci. Food Agric.* 80: 2021–2027.

Gatehouse, A. M. R., F. M. Dewey, J. Dover, K. A. Fenton, and A. Pusztai. 1984. Effect of seed lectins from *Phaseolus vulgaris* on development of larvae of *Callosobruchus maculates*: Mechanism of toxicity. *J. Sci. Food Agric.* 35: 373–380.

Gatlin, D. M., F. T. Barrows, and P. Brown. 2007. Expanding the utilization of sustainable plant products in aquafeeds: A review. *Aquacult. Res.* 38: 551–579.

GEA Niro site for casein production. http://www.niro.com/niro/cmsdoc.nsf/webdoc/webb7jsdlc.

Gersabeck, E. F., S. D. McCombs, T. E. McGovern, J. Reyes-Flores, and M. de los Santos Ramos. 2010. Fruit fly emergence and release facilities review. Animal and Plant Health Inspection Service United States and Mexico Lower Rio Grande Valley Mexican Fruit Fly Eradication Program Review: Final Report.

Gifawesen, C., B. R. Funke, and F. I. Proshold. 1975. Control of antifungal-resistant strain of *Aspergillus niger* mold contaminants in insect rearing media. *J. Econ. Entomol.* 68: 441–444.

Gillespy, T. G. 1962. The principles of heat sterilization. *Recent Adv. Food Sci.* 2: 93–105.

Gilliam, M. 1975a. How long is Teramycin stable in diets fed to honeybee colonies for disease control? *Am. Bee J.* 115: 230–234.

Gilliam, M. 1975b. Stability of oxytetracycline in diets fed to honeybee colonies for disease control. *J. Invertebr. Pathol.* 26: 383–386.

Gillot, C. 1995. *Entomology*. 2nd Edn. Springer, New York, NY.

Gilmour, D. 1961. *The Biochemistry of Insects*. Academic Press, New York.

Gingrich, R. E. 1972. Nutritional studies: Their bearing on the development of practical oligidic diets for mass rearing larvae of the screwworm, *Cochliomyia hominivorax*. In *Insect and Mite Nutrition* (J. G. Rodriguez, Ed.), pp. 257–268. North-Holland Publishing Company, Amsterdam, the Netherlands.

Gingrich, R. E., A. J. Graham, and B. G. Hightower. 1971. Media containing liquefied nutrients for mass-rearing larvae of the screwworm. *J. Econ. Entomol.* 64: 673–683.

Gladwell, M. 2000. *The Tipping Point: How Little Things Can Make a Big Difference.* Little Brown, New York.

Gleick, J. 1987. *Chaos.* Penguin Books, New York, 352 p.

Gomez, M., J. Gonzalez, and B. Oliete. 2012. Effect of extruded wheat germ on dough rheology and bread quality. *Food Process Technol.* 5: 2409–2418.

Goodenough, J. L. 1984. Materials handling in insect rearing. In *Advances and Challenges in Insect Rearing* (E. G. King and N. C. Leppla, Eds.), pp. 2–8. USDA; ARS, New Orleans, LA.

Gordon, H. T. 1968. Quantitative aspects of insect nutrition. *Am. Zool.* 8: 131–138.

Gould, F. 2008. Broadening the application of evolutionarily based genetic pest management. *Evolution.* 62: 500–510.

Grant, E. L. and R. S. Leavenworth. 1988. *Statistical Quality Control*, 6th Ed. McGraw-Hill, Inc., New York.

Green, E. S., A. R. Zangerl, and M. R. Berenbaum. 2001. Effects of phytic acid and xanthotoxin on growth and detoxification in caterpillars. *J. Chem. Ecol.* 27: 1763–1773.

Gregory, J. F. 2008. Vitamins. In *Fennema's Food Chemistry*, 4th Ed. (S. Damodaran, K. L. Parkin, and O. R. Fennema, Eds.), pp. 439–520, CRC Press, Boca Raton, FL.

Grenier, S., P. Greany, and A. C. Cohen. 1994. Potential for mass release of insect parasitoids and predators through development of artificial culture techniques. In *Pest Management in the Subtropics: Biological Control: A Florida Perspective* (D. Rosen, F. D. Bennett, and J. Capinera, Eds.), pp. 181–295. Intercept Press, Andover, Hants, U.K.

Griffin, J. G. and O. H. Lindig. 1978. Pilot facility for mass rearing of boll weevils. In *Facilities for Insect Research and Production*, Technical Bulletin 1576 (N. C. Leppla and T. R. Ashley, Eds.), pp. 75–78. U.S. Department of Agriculture, Science and Education Administration, Washington, DC.

Griffin, J. G., O. H. Lindig, and R. E. McLaughlin. 1974. Flash sterilizers: Sterilizing diets for insects. *J. Econ. Entomol.* 67: 689.

Gulsen, N., H. D. Umucalilar, and F. Inal. 2007. Impacts of calcium addition and different oil types and levels on in vitro rumen fermentation and digestibility. *Arch. Anim. Nut.* 60: 443–453.

Guss, P. L. and J. L. Krysan. 1973. Maintenance of the southern corn rootworm on a dry diet. *J. Econ. Entomol.* 66: 352–353.

Guyenot, E. 1913a. A biological study of a *Drosophila ampelophila* Low fly I—The possibility of an aseptic life for an individual and the line. *Comptes rendus des séances de la Société de biologie et de ses filiales* (0037-9026). 74: 97–99.

Guyenot, E. 1913b. Biological study in a fly, *Drosophila ampeltophila* Low II—The role of yeast in food. *Comptes Rendus des Seances de la Societe de Biologie et de Ses Filiales.* 74: 178–180.

Guyenot, E. 1913c. Biological studies on a fly, *Drosophila ampelophila* Low V—Nutrition of adults and fertility. *Comptes rendus des séances de la Société de biologie et de ses filiales* (0037-9026). 74: 332–334.

Guyenot, E. 1914a. Biological studies on a fly *Drosophila ampelophila* Low. Need to create a definite feeding environment. *Comptes rendus des séances de la Société de biologie et de ses filiales* (0037-9026). 76: 483–485.

Guyenot, E. 1914b. Primary trials of determination of an artificial nutritive environment for the raising of a fly, *Drosophila ampelophila* Low. *Comptes rendus des séances de la Société de biologie et de ses filiales* (0037-9026). 76: 548–550.

Hagen, K. S. and R. L. Tassan. 1970. Influence of food Wheast and other *Saccharomyces fragilis* yeast products on fecundity of *Chrysopa carnea* (Neuroptera: Chrysopidae). *Can. Entomol.* 102: 806–811.

Hagler, J. R. and A. C. Cohen. 1991. Prey selection by in vitro- and field reared *Geocoris punctipes. Entomol. Exp. Appl.* 59: 201–205.

Harborne, J. B. 1988. *Introduction to Ecological Biochemistry.* 3rd edn. London, U.K.: Academic Press, Inc.

Harper, J. M. 1979. Food extrusion. *CRC Crit. Rev. Food. Sci. Nutr.* February: 155–215.

Harper, J. M. 1981. *Extrusion of Foods*, vol. II. CRC Press, Boca Raton, Fl.

Harrell, E. A. and C. W. Gantt. 1984. In *Advances and Challenges in Insect Rearing* (E. G. King and N. C. Leppla, Eds.) pp. 2–8. USDA; ARS, New Orleans, LA.

Harrell, E. A., A. N. Sparks, W. W. Hare, and W. D. Perkins. 1973. An insect diet filler for an inline form-fill-seal machine. *J. Econ. Entomol.* 66: 1340–1341.

Hartley, S. E. 1998. The chemical composition of plant galls: Are levels of nutrients and secondary compounds controlled by the gall-former? *Oecologia.* 113: 492–501.

Haslam, E. 2001. Plant polyphenols: Old wine in new bottles. *Educ. Chem.* January: 17–20.

Hasnain, S. E., T. K. Taneja, N. K. Sah, M. Mohan, N. Pathak, S. Sahdev, M. Athar, S. M. Totey, and R. Begum. 1999. In vitro cultured *Spodoptera frugiperda* insect cells: Model for oxidative stress-induced apoptosis. *J. Biosci.* 24: 13–19.

Hathaway, D. O., A. E. Clift, and B. A. Butt. 1971. Development and fecundity of codling moths-Lepidoptera-Olethreutidae reared on artificial diets or mature apples. *J. Econ. Entomol.* 64: 1088.

Hathaway, D. O., L. V. Lydin, B. A. Butt, and L. J. Morton. 1973. Monitoring mass rearing of the codling moth. *J. Econ. Entomol.* 66: 390–393.

Haynes, W. M. 2010. *CRC Handbook of Chemistry and Physics*, 91st edn. Boca Raton, FL: Taylor & Francis.

He, X. P. and X. N. Guo. 2007. Ergosterol production from molasses by genetically modified *Saccharomyces cerevisiae*. *Appl. Microbiol. Biotechnol.* 75: 55–60.

Heath, J. J. and J. O. Stireman III. 2010. Dissecting the association between a gall midge, *Asteromyia carbonifera*, and its symbiotic fungus, *Botryosphaeria dothidea*. *Entomologia Experimentalis et Applicata*. 137: 36–49.

Hedin, P. A., O. H. Lindig, P. P. Sikorowski, and M. Wyat. 1978. Suppressants of gut bacteria in boll weevil from the cotton plant. *J. Econ. Entomol.* 71: 394–396.

Heed, W. B. and H. W. Kircher. 1965. Unique sterol in the ecology and nutrition of *Drosophila pachea*. *Science* 149(3685): 758–761.

Heertje, I. and M. Leunis. 1997. Measurement of shape and size of fat crystals by electron microscopy. *Food Sci. Technol.* 30: 141–146.

Heertje, I., H. van Aalst, J. C. G. Blonk, D. A. Nederlof, A. Don, J. Nederlof, and E. H. Lucassen-Reynders. 1996. Observations on emulsifiers at the interface between oil and water by confocal scanning light microscopy. *Food Sci. Technol.* 29: 217–226.

Heinrich, B. 1996. *The Thermal Warriors*. Harvard University Press, Cambridge, MA.

Hickernell, L. M. 1923. The digestive system of the periodical cicada, *Tibicen septendecim* Linn. III. Morphology of the system in the nymph. *Biol. Bull.* 45(4): 213–229.

Hirabayashi, M., T. Matsui, and H. Yano. 1998. Fermentation of soybean flour with *Aspergillus usamii* improves availabilities of zinc and iron in rats. *J. Nutr. Sci. Vitaminol.* 44: 877–886.

Hiratsuka, Y. 1920. The influence of surrounding temperature on the nutrition of silkworms. *Tokyo Nog Kw H* 1918: 943.

Hobson, R. P. 1935. On a fat-soluble growth factor required by blowfly larvae. II. Identity of the growth factor with cholesterol. *Biochem. J.* 29: 2023–2026.

Hoffman, J. D., F. R. Lawson, and R. T. Yamamoto. 1966. Tobacco hornworms. In *Insect Colonization and Mass Production* (C. N. Smith, Ed.), pp. 479–486. Academic Press, New York.

Hölldobler, B. and E. O. Wilson. 2009. *The Superorganism: The Beauty, Elegance, and Strangeness of Insect Societies*. W. W. Norton & Company, New York.

Hoover, K., J. O. Washburn, and L. E. Volkman. 2000. Midgut-based resistance of *Heliothis virescens* to baculovirus infection mediated by phytochemicals in cotton. *J. Insect Physiol.* 46: 999–1007.

Hopkins, T. L. and M. S. Harper. 2001. Lepidopteran peritrophic membranes and effects of dietary wheat germ agglutinin on their formation and structure. *Arch. Insect Biochem. Physiol.* 47: 100–109.

Hou, R. F. and M. A. Brooks. 1975. Continuous rearing of the aster leafhopper, *Macrosteles fascifrons*, on a chemically defined diet. *J. Insect Physiol.* 21: 1481–1483.

House, H. L. 1974a. Nutrition. In *The Physiology of Insecta*, vol. 5 (M. Rockstein, Ed.), pp. 1–62. Academic Press, New York.

House, H. L. 1974b. Digestion. In *The Physiology of Insecta*, vol. 5 (M. Rockstein, Ed.), pp. 63–117. Academic Press, New York.

Howell, J. F. 1970. Rearing codling moth on an artificial diet. *J. Econ. Entomol.* 63: 1148.

Howell, J. F. 1971. Problems involved in rearing codling moth. Lepidoptera: Olethreutidae on diet in trays. *J. Econ. Entomol.* 64: 631.

Howell, J. F. 1972. Modifications of artificial diet for codling moths, Lepidoptera: Olethreutidae to improve larval acceptance and production moths. *J. Econ. Entomol.* 65: 57.

Hsiao, T. H. and G. Fraenkel. 1968. The role of secondary plant substances in the food specificity of the Colorado potato beetle. *Ann. Entomol. Soc. Am.* 61: 485–493.

Hsiao, T. H. and C. Hsiao. 1974. Feeding requirements and artificial diets for the alfalfa weevil. *Entomol. Exp. Appl.* 17: 83–91.

Huang, J., D. Yang, and T. Wang. 2007. Effects of replacing soy-oil with soy-lecithin on growth performance, nutrient utilization and serum parameters of broilers fed corn-based diets. *Asian-Australas. J. Anim. Sci.* 20: 1880–1886.

Huettel, M. D. 1976. Monitoring the quality of laboratory-reared insects: A biological and behavioral perspective. *Environ. Entomol.* 5: 807–814.

Hughes, P. R. and H. A. Wood. 1998. Production of pharmaceutical and other recombinant proteins in insect larvae. *SIM News.* 48: 105.

Hulcr, J. and A. I. Cognato. 2010. Repeated evolution of crop theft in fungus-farming ambrosia beetles. *Evolution.* 64–11: 3205–3212.

Hunt, D. W. A., G. Lintereur, and K. F. Raffa. 1992. Rearing method for *Hylobius radicis* and *H. pales* (Coleoptera: Curculionidae). *J. Econ. Entomol.* 85: 1873–1877.

Hunter, C. D. 1994. *Suppliers of Beneficial Organisms in North America.* California Environmental Protection Agency, Department of Pesticide Regulation Publication, Sacramento, CA, PM 94:3.

Hurst, M. R. H. and T. A. Jackson. 2002. Use of the green fluorescent protein to monitor the fate of *Serratia entomophila* causing amber disease in the New Zealand grass grub, *Costelytra zealandica. J. Microbiol. Methods.* 50: 1–8.

Hutchinson, G. E. 1965. *The Ecological Theater and the Evolutionary Play.* Yale University Press, New Haven, CT.

Inglis, G. D. and A. C. Cohen. 2004. Influence of antimicrobial agents on the spoilage of a meat-based entomophage diet. *J. Econ. Entomol.* 97: 235–250.

Ito, T. 1961a. Nutrition of the silkworm, *Bombyx mori.* I. Nutritive effects of soybean oil for the larva. *Bull. Seric. Exp. Stn. Jpn.* 17: 91–117.

Ito, T. 1961b. Sterol requirements of the silkworm *Bombyx mori. Nat. London.* 191: 882–883.

Ito, T., Y. Horie, and S. Nakasone. 1975. Deterrent effect of soybean meal on feeding of the silkworm, *Bombyx mori. J. Insect Physiol.* 21: 995–1006.

Ito, T., Y. Yorie, K. Watanabe, K. Takamika, M. Furuyama, M. Miyabaya, K. Yamamoto, and M. Nagashim. 1974. Rearing of larvae of silkworm, *Bombyx mori*, entirely on semisynthetic diets. *J. Agric. Chem. Soc. Jpn.* 48: 403–407.

Jacobsen, C., M. Timm, and A. S. Meyer. 2001. Oxidation in fish oil enriched mayonnaise: Ascorbic acid and low pH increase oxidative deterioration. *J. Agric. Food Chem.* 49: 3947–3956.

Jang, M. Y., J. Y. Cho, and J. I. Cho. 2006. Isolation of compounds with antioxidative activity from quickly fermented soy-based foods. *Food Sci. Biotechnol.* 15: 214–219.

Jay, J. M., M. J. Loessner, and D. A. Golden. 2005. *Modern Food Microbiology*, 7th Ed. Springer, New York.

Johnson, K. S. and G. W. Felton. 2000. Digestive protease activity in corn earworm (*Helicoverpa zea*) after molting and in response to lowered redox potential. *Arch. Insect Biochem. Physiol.* 44: 151–161.

Johnson, K. S. and G. W. Felton. 2001. Plant phenolics as dietary antioxidants for herbivorous insects: A test with genetically modified tobacco. *J. Chem. Ecol.* 27: 2579–2597.

Jovani, M., R. Barbera, and R. Farre. 2001. Review: Effect of some components of milk-and soy-based infant formulas on mineral availability. *Food Sci. Technol. Inter.* 7: 191–198.

Kahneman, D. 2011. *Thinking Fast and Slow.* Farrar, Straus, and Giroux, New York.

Kahneman, D., A. Tversky, and P. Slovic (Eds.). 1982. *Judgment under Uncertainty: Heuristics & Biases.* Cambridge University Press, Cambridge, U.K. ISBN 0-521-28414-7.

Karr-Lilienthal, L. K., C. T. Kadzere, and C. M. Grieshop. 2005. Chemical and nutritional properties of soybean carbohydrates as related to nonruminants: A review. *Livest. Prod. Sci.* 97: 1–12.

Kaspari, M. 1990. Prey preparation and the determinants of handling time. *Anim. Behav.* 40: 118–126.

Kasting, R. and A. J. McGinnis. 1965. Measuring consumption of a food by an insect using carbon-14 labelled compounds. *J. Insect Physiol.* 11: 1253–1260.

Keena, M. A., T. M. ODell, and J. A. Tanner. 1998. Environmentally based maternal effects are the primary factor in determining the developmental response of gypsy moth (Lepidoptera: Lymantriidae) to dietary iron deficiency. *Ann. Entomol. Soc. Am.* 91: 710–718.

Kennedy, J. S. 1953. Host plant selection in Aphididae. In *Transactions of the IXth International Congress of Entomology*, vol. 2, pp. 106–113, Amsterdam, the Netherlands.

Kfir, R. 1992. A simple artificial diet for mass rearing the stem borer *Chilo partellus* (Swinhoe) (Lepidoptera: Pyralidae). *J. Entomol. Soc. South Africa* 55: 283–284.

Khuzhaev, V. U., I. Zhalolov, K. K. Turgunov, B. Tashkhodzhaev, M. G. Levkovich, S. F. Aripova, and A. S. Shashkov. 2004. Alkaloids from *Arundo donax*. XVI: Structure of the new dimeric indole alkaloid arundavine. *Chem. Nat. Comp.* 40: 261–265.

Kim, S. C., A. T. Adesogan, and A. D. Arthington. 2007. Optimizing nitrogen utilization in growing steers fed forage diets supplemented with dried citrus pulp. *J. Anim. Sci.* 85: 2548–2555.

King, E. G. and G. G. Hartley. 1985. *Heliothis virescens*. In *The Handbook of Insect Rearing*, vol. II (P. Singh and R. F. Moore, Eds.), pp. 323–328. Elsevier, New York.

King, E. G., K. R. Hopper, and J. E. Powell. 1985. Analysis of systems for biological control of crop arthropod pests in the U.S. by augmentation of predators and parasites. In *Biological Control in Agricultural IPM Systems* (M. A. Hoy and D. C. Herzog, Eds.), pp. 201–227. Academic Press, Inc., Orlando, FL.

King, R. C. 1954. The effect of yeast on phosphorus uptake by *Drosophila*. *Amer. Nat.* 88: 155–158.

Kingsolver, J. G. and T. L. Daniel. 1995. Mechanics of food handling by fluid-feeding insects. In *Regulatory Mechanisms in Insect Feeding* (R. F. Chapman and G. deBoer, Eds.), pp. 32–73. Chapman & Hall, New York.

Klepzig, K. D., J. C. Moser, F. J. Lombardero, R. W. Hofstetter, and M. P. Ayres. 2001. Symbiosis and competition: Complex interactions among beetles, fungi and mites. *Symbiosis.* 30: 83–96.

Klunder, H. C., J. Wolkers-Rooijackers, J. M. Korpela, and M. J. R. Nout. 2012. Microbiological aspects of processing and storage of edible insects. *Food Control.* 26(2): 628.

Knipling, E. F. 1966. Plant protection in the American economy: The entomologist's arsenal. *Bull. Entomol. Soc. Am.* 12: 45–51.

Knipling, E. F. 1979. *The Basic Principles of Insect Population Suppression and Management.* USDA Agriculture Handbook. U.S. Department of Agriculture, Washington, DC, p. 512.

Kobayashi, C., Y. Fukasawa, D. Hirose, and M. Kato. 2008. Contribution of symbiotic mycangial fungi to larval nutrition of a leaf-rolling weevil. *Evol. Ecol.* 22: 711–722.

Koizumi, S., A. Maruyama, and T. Fujio. 1990. Purification and characterization of ascorbic acid phosphorylating enzyme from *Pseudomonas azotocolligans*. *Agric. Biol. Chem.* 54: 3235–3239.

Koyama, K. 1973. Rearing *Inazuma dorsalis* and *Nephotettix cincticeps* on a synthetic diet. Japanese. *J. Appl. Entomol. Zool.* 17: 163–166.

Krautz, R., B. Arefin, and U. Theopold. 2014. Damage signals in the insect immune response. *Front. Plant Sci.* 5: 342.

Kumar, P., P. Coronel, V. D. Truong, J. Simunovic, K. R. Swartzel, K. P. Sandeep, and G. Cartwright. 2008. Overcoming issues associated with the scale-up of a continuous flow microwave system for aseptic processing of vegetable purees. *Food Res. Inter.* 41: 454–461.

Lapointe, S. L., T. J. Evens, and R. P. Niedz. 2008. Insect diets as mixtures: Optimization for a polyphagous weevil. *J. Insect Physiol.* 54: 1157–1167.

Lapointe, S. L., T. J. Evens, R. P. Niedz, and D. G. Hall. 2010a. Artificial diet optimized to produce normative adults of *Diaprepes abbreviatus* (Coleoptera: Curculionidae). *Environ. Entomol.* 39(2): 670–677.

Lapointe, S. L., R. P. Niedz, and T. J. Evens. 2010b. An artificial diet for diaprepes abbreviatus (Coleoptera: Curculionidae) optimized for larval survival. *Florida Entomol.* 93(1): 56–62.

Laturney, M. and J. C. Billeter. 2014. Neurogenetics of female reproductive behaviors in *Drosophila melanogaster* (T. Friedmann, J. C. Dunlap, S. F. Goodwin, Eds.). *Adv. Genet.* 85: 1–108.

Leathem, A. J. and S. A. Brooks. 1998. Light microscopy. In *Lectin Methods and Protocols* (J. M. Rhodes and J. D. Milton, Eds.), pp. 3–20. Humana Press, Totowa, NJ.

Lee, K. P., D. Raubenheimer, and S. J. Simpson. 2004. The effects of nutritional imbalance on compensatory feeding for cellulose-mediated dietary dilution in a generalist caterpillar. *Physiol. Entomol.* 29: 108–117.

Lee, K. P., S. J. Simpson, F. J. Clissold, R. Brooks, J. W. O. Ballard, P. W. Taylor, N. Soran, and D. Raubenheimer. 2008. Lifespan and reproduction in Drosophila: New insights from nutritional geometry. *Proc. Natl. Acad. Sci. USA* 105: 2498–2503.

Lee, S. K., J. H. Han, and E. A. Decker. 2002. Antioxidant activity of phosvitin in phosphatidylcholine liposomes and meat model systems. *J. Food Sci.* 67: 37–41.

Lehane, M. J. 1997. Peritrophic matrix structure and formation. *Annu. Rev. Entomol.* 42: 525–550.

Lehninger, A. L., D. L. Nelson, and M. M. Cox. 1993. *Principles of Biochemistry*. Worth Publishers, New York.

Leppla, N. C. and T. R. Ashley. 1989. Quality control in insect mass production: A review and model. *Bull. Entomol. Soc. Am.* 35(4): 33–44.

Leppla, N. C. and W. R. Fisher. 1989. Total quality control in insect mass production for insect pest management. *J. Appl. Ent.* 108: 452–461.

Leppla, N. C., M. D. Huettel, D. L. Chambers, T. R. Ashley, D. H. Miyashita, T. T. Y. Wong, and E. J. Harris. 1983. Strategies for colonization and maintenance of the Mediterranean fruit fly. *Entomol. Exp. Appl.* 33: 89–96.

Levenbook, L. 1985. Insect storage proteins. In *Comprehensive Insect Physiology, Biochemistry and Pharmacology*, vol. 10 (G. A. Kerkut and L. I. Gilbert, Eds.), pp. 307–346. Pergamon Press, Oxford, U.K.

Levine, R. R. 1973. *Pharmacology, Drug Actions, and Reactions*. Little, Brown, and Co., Boston, MA.

Liang, X., Y. Fu, L. Tong, and H. Liu. 2014. Microbial shifts of the silkworm larval gut in response to lettuce leaf feeding. *Appl. Microbiol. Biotechnol.* 98: 3769–3776.

Lindig, O. H. and O. L. Malone. 1973. Oviposition of boll weevils fed diets containing germinated cottonseed puree or cottonseed meats puree. *J. Econ. Entomol.* 66: 566–567.

Lindsay, R. B. 2008. Food additives. In *Fennema's Food Chemistry*, 4th Ed. (S. Damodaran, K. L. Parkin, and O. R. Fennema, Eds.), pp. 639–688, CRC Press, Boca Raton, FL.

Locke, M. and H. Nichol. 1992. Iron economy in insects. *Annu. Rev. Entomol.* 37: 195–215.

Loeb, J. 1915. The simplest constituents required for growth and the completion of the life cycle in an insect (Drosophila). *Science.* 41(1048): 169–170.

Loeb, J. and J. H. Northrop. 1917. On the influence of food and temperature on the duration of life. *J. Biol. Chem.* 32: 103–121.

Longvah, T., K. Mangthya, and P. Ramulu. 2011. Nutrient composition and protein quality evaluation of eri silkworm (*Samia ricinii*) prepupae and pupae. *Food Chem.* 128: 400–403.

Lorraine, H. and N. D. Bruzzone. 1992. Process control: An untapped resource for managers of sterile insect programs. *Technol. Forecast. Soc. Change.* 41: 57–69.

Lund, D. 1988. Effects of heat processing on nutrients. In *Nutritional Evaluation of Food Processing*, 3rd Ed. (E. Karmas and R. S. Harris, Eds.), pp. 319–354. Van Nostrand Reinhold, New York.

Ma, W. C. 1972. Dynamics of feeding responses in *Pieris brassicae* Linn. as a function of chemosensory input: A behavioural, ultrastructural and electrophysiological study. *Mededelingen Landbouwhogeschool, Wageningen* 72/11: 1–162.

Mackauer, M. 1972. Genetic aspects of insect production. *Entomophaga.* 17: 27–48.

McGinnis, A. J. and R. Kasting. 1964. Comparison of gravimetric and chromic oxide indicator method for measuring percentage utilization and consumption of food by phytophagous insects. *J. Insect Physiol.* 10: 989–995.

McPhail, M. 1937. Relation of time of day, temperature and evaporation to attractiveness of fermenting sugar solution to Mexican fruit fly. *J. Econ. Entomol.* 30: 793–799.

Meloan, C. E. and Y. Pomeranz. 1980. *Food Analysis Laboratory Experiments*. The AVI Publishing Co., Inc., Westport, CT.

Melvin, R. and R. C. Bushland. 1936. A method of rearing *Chochliomyia Americana* C. & P. on artificial media. U.S. Bureau of Entomology and Plant Quarantine, Report ET-88, 2pp.

Melvin, R. and R. C. Bushland. 1940. The nutritional requirements of screwworm larvae. *J. Econ. Entomol.* 33: 850–852.

Mertz, W. 1981. The essential trace elements. *Science.* 213: 1332–1338.

Michelbacher, A. E., W. M. Hoskins, and W. B. Herms. 1932. The nutrition of flesh fly larvae, Lucilia sericata (Meig.). *J. Exp. Zool.* 64: 109–131.

Miller, D. D. 1998. *Food Chemistry: A Laboratory Manual*. John Wiley & Sons, Inc., New York.

Miller, D. D. 2008. Minerals. In *Fennema's Food Chemistry*, 4th Ed. (S. Damodaran, K. L. Parkin, and O. R. Fennema, Eds.), pp. 523–570, CRC Press, Boca Raton, FL.

Min, D. B. and D. F. Steenson. 1998. Crude fat analysis. In *Food Analysis* (S. Suzanne Nielsen, Ed.), pp. 201–216. Aspen Publishers, Inc., Gaithersburg, MD.

Minnich, D. E. 1921. An experimental study of the tarsal chemoreceptors of two nymphalid butterflies. *J. Exp. Zool.* 33: 173–203.

Minno, M. C. and T. C. Holler. 1991. *Procedures Manual for Mass-Rearing the Caribbean Fruit Fly, Anastrepha suspensa (Loew) (Diptera: Tephritidae)*. Sterile Fly Laboratory, Florida Department of Agriculture and Consumer Services, Division of Plant Industry, Gainesville, FL.

Mishra, N., N. C. Hazarika, K. Narain, and J. Mahanta. 2003. Nutritive value of non-mulberry and mulberry silkworm pupae and consumption pattern in Assam, *India. Nut. Res.* 23: 1303–1311.

Mittler, T. E. 1972. Interactions between dietary components. In *Insect and Mite Nutrition* (J. G. Rodriguez, Ed.), pp. 211–223. North-Holland Publishing Company, Amsterdam, the Netherlands.

Mittler, T. E. and R. H. Dadd. 1963a. Studies on the artificial feeding of the aphid *Myzus persicae* (Sulzer) I. Relative uptake of water and sucrose solutions. *J. Insect Physiol.* 9: 623–645.

Mittler, T. E. and R. H. Dadd. 1963b. Studies on the artificial feeding of the aphid *Myzus persicae* (Sulzer) II. Relative survival development, and larva position on different diets. *J. Insect Physiol.* 9: 741–757.

Montgomery, D. C. 1991. *Introduction to Statistical Quality Control*, 2nd Ed. John Wiley & Sons, New York.

Montgomery, D. C. 2009. *Introduction to Statistical Quality Control*, 6th Ed. John Wiley & Sons, New York.

Moran, E. T., J. D. Summers, and E. J. Bass. 1968. Heat-processing of wheat germ meal and its effect on utilization and protein quality for growing chick-toasting and autoclaving. *Cereal Chem.* 45: 304.

Moran, P. J., J. A. Goolsby, A. E. Racelis, A. C. Cohen, M. A. Ciomperlik, K. R. Summy, D. P. R. Sands, and A. Kirk. 2013. Mass-rearing of the stem-galling wasp, *Tetramesa romana*, a biological control agent of the invasive weed, *Arundo donax*. In *Mass Production of Beneficial Organisms: Invertebrates and Entomopathogens* (J. A. Morales-Ramos, M. Guadalupe Rojas, D. I. Shapiro-Ilan, Eds.), pp. 163–192. Elsevier, London, U.K.

Musgrave, A. J. 1964. Mycetomes in sitophilus. *Can. Entomol.* 96: 146.

Musser, R. O., S. M. Hum-Musser, H. Eichenseer, M. Peiffer, G. Ervin, J. B. Murphy, and G. Felton. 2002. Caterpillar saliva beats plant defenses. *Nature.* 416: 599–600.

NASA (National Aeronautics and Space Administration) SP-5076. 1969. *Contamination Control Handbook.* Office of Technology Utilization, NASA, Washington, DC.

Nawar, W. W. 1996. Lipids. In *Food Chemistry* (O. R. Fennema, Ed. 3rd Edition), pp. 225–320. Marcel Dekker, Inc., New York.

Naylor, A. F. 1964. Possible value of casein, gluten, egg albumin, or fibrin as whole proteins in the diet of two strains of the flour beetle, *Tribolium confusum* (Tenebrionidae). *Can. J. Zool.* 42: 1–9.

Nichol, H., J. H. Law, and J. J. Winzerling. 2002. Iron metabolism in insects. *Annu. Rev. Entomol.* 47: 535–559.

Nielsen, J. and R. Molich. 1990. Heuristic evaluation of user interfaces. In *Proceedings of ACM CHI'90 Conference*, April 1–5, Seattle, WA, pp. 249–256.

Nielsen, J. H., G. H. Kristiansen, and H. J. Andersen. 2000. Ascorbic acid and ascorbic acid 6-palmitate induced oxidation in egg yolk dispersions. *J. Agric. Food Chem.* 48: 1564–1568.

Nielsen, S. 1998. *Food Analysis.* Aspen Publishers, Inc. Gaithersburg, MD.

No, H. K., N. Y. Park, S. H. Lee, H. J. Hwang, and S. P. Meyers. 2002. Antibacterial activities of chitosans and chitosan oligomers with different molecular weights on spoilage bacteria isolated from tofu. *J. Food Sci.* 67: 1511–1514.

Nogrady, T. 1998. Numerical phenetic taxonomy and its heuristic aspects. *Hydrobiologia.* 387/388: 97–103.

Nordlund, D. A., A. C. Cohen, and R. A. Smith. 2001. Mass-rearing, release techniques, and augmentation. In *Lacewings in Biological Control* (P. K. McEwen, Ed.), pp. 303–319. Cambridge University Press, Cambridge, U.K.

Nordlund, D. A. and S. M. Greenberg. 1994. Facilities and automation for the mass production of arthropod predators and parasitoids. *Biocontrol. News Inform.* 4: 45–49.

Norris, D. M. and J. M. Baker. 1968. A minimum nutritional substrate required by *Fusarium solani* to fulfill its mutualistic relationship with *Xyleborus ferrugineus*. *Ann. Entomol. Soc. Am.* 61: 1473–1475.

Norris, D. M. and H. M. Chu. 1985. *Xyleborus ferrugineus*. In *Handbook of Insect Rearing*, vol. 1 (P. Singh and R. F. Moore, Eds.), pp. 303–315. Elsevier, Amsterdam, the Netherlands.

ODell, T. M. 1992. Straggling in gypsy moth production strains: A problem analysis for developing research priorities. In *Advances in Insect Rearing for Research and Pest Management* (T. A. Anderson and N. C. Leppla, Eds.), pp. 325–350. Westview, Boulder, CO.

ODell, T. M., R. A. Bell, V. C. Mastro, J. A. Tanner, and J. F. Kennedy. 1984. Production of gypsy moth, *Lymantria dispar*, for research and biological control. In *Advances and Challenges in Insect Rearing* (E. G. King and N. C. Leppla, Eds.), pp. 155–166. U.S. Department of Agriculture, Agricultural Research Service, New Orleans, LA.

ODell, T. M., M. A. Keena, and R. B. Willis. 1997. Dietary influence of iron formulation on the development of gypsy moth (Lepidoptera: Lymantriidae) in laboratory colonies. *Ann. Entomol. Soc. Am.* 90: 149–154.

Ohkuma, M. and A. Brune. 2011. Diversity, structure, and evolution of the termite gut microbial community. In *Biology of Termites: A Modern Synthesis* (D. E. Bignell, Y. Roisin, and N. Lo, Eds.), pp. 413–438. Springer, London, U.K.

Ohkuma, M. and T. Kudo. 1996. Phylogenetic diversity of the intestinal bacterial community in the termite *Reticulitermes speratus*. *Appl. Environ. Microbiol.* 62: 461–468.

Ojeda-Avila, T., H. A. Woods, and R. A. Raguso. 2003. Effects of dietary variation on growth, composition, and maturation of *Manduca sexta* (Sphingidae: Lepidoptera). *J. Insect Physiol.* 49: 293–306.

Oliver, B. F., F. G. Maxwell, and J. N. Jenkins. 1970. Utilization of glanded and glandless cotton diets by the bollworm. *J. Econ. Entomol.* 63: 1965–1966.

Olli Sjovall, O., T. Virtalaine, A. Lapvetelainen, and H. Kallio. 2000. Development of rancidity in wheat germ analyzed by headspace gas chromatography and sensory analysis. *J. Agric. Food Chem.* 48: 3522–3527.

OSHA. Occupational Safety and Health Administration. 2000. OSHA regulations (standards-29 CFR), substance technical guidelines for formalin. U.S. Department of Labor. http://www.osha.slc.gov/OshStd_data/1910_1048_APP_A.html. Accessed March 2014.

Oten, K. L. F., A. C. Cohen, and F. P. Hain. 2014. Stylet bundle morphology and trophically related enzymes of the hemlock woolly adelgid (Hemiptera: Adelgidae). *Ann. Entomol. Soc. Am.* 107(3): 680–690.

Owens, C. D. 1984. Controlled environments for insects and personnel in insect-rearing facilities. In *Advances and Challenges in Insect Rearing* (E. G. King and N. C. Leppla, Eds.), pp. 58–63. USDA; ARS, New Orleans, LA.

Paes, M. C., M. B. Oliveira, and P. L. Oliveira. 2001. Hydrogen peroxide detoxification in the midgut of the blood-sucking insect, *Rhodnius prolixus*. *Arch. Insect Biochem. Physiol.* 48: 63–71.

Paine, T. D., K. F. Raffa, and T. C. Harrington. 1997. Interactions between scolytid bark beetles, their associated fungi and live host conifers. *Annu. Rev. Entomol.* 42: 179–206.

Palanca, L., A. C. Gaskett, C. S. Günther, R. D. Newcomb, and M. R. Goddard. 2013. Quantifying variation in the ability of yeasts to attract *Drosophila melanogaster*. *PLoS ONE* 8(9): e75332.

Park, T. 1932. Studies in population physiology: The relation of numbers to initial population growth in the flour beetle *Tribolium confusum*. *Duval. Ecol.* 13(2): 172–181.

Park, S. K. and D. H. Bae. 2006. Film-forming properties of protein aceous fibrous material produced from soybean fermented by Bacillusnatto. *J. Microbiol. Biotechnol.* 16: 1053–1059.

Parker, A. G. 2005. Mass-rearing for sterile insect release. In *Sterile Insect Technique. Principles and Practice in Area-Wide Integrated Pest Management* (V. A. Dyck, J. Hendrichs, and A. S. Robinson, Eds.), pp. 209–232. Springer, Dordrecht, the Netherlands.

Parrott, W. L., J. N. Jenkins, and J. C. McCarty, Jr. 1983. Feeding behavior of first-stage tobacco budworm larvae on three cotton cultivars. *Ann. Entomol. Soc. Am.* 76: 167.

Pasteur, L. 1870. *Etudes sur la maladie des vers a soie*, Tom I et II. Gauthier-Villars, Paris, France.

Patana, R. 1969. Rearing cotton insects in the laboratory. U.S. Department of Agriculture Product Research Report 108, Washington, DC.

Patana, R. 1977. *Rearing Selected Western Cotton Insects in the Laboratory*, vol. 51 of ARS-W. United States Agricultural Research Service, Western Region, Department of Agriculture, Berkeley, CA, 8pp.

Patana, R. and W. C. McAda. 1973. Tobacco budworms: Use of dry diet flakes in rearing. *J. Econ. Entomol.* 66: 817–818.

Pechan, T., A. C. Cohen, W. P. Williams, and D. S. Luthe. 2002. Insect feeding mobilizes a unique plant defense protease that disrupts the peritrophic matrix of caterpillars. *Proc. Natl. Acad. Sci. USA.* 99: 13319–13323.

Pechan, T., L. J. Ye, Y. M. Chang, A. Mitra, A. Lin, F. M. Davis, W. P. Williams, and D. S. Luthe. 2000. A unique 33-kD cysteine proteinase accumulates in response to larval feeding in maize genotypes resistant to fall armyworm and other Lepidoptera. *Plant Cell.* 12: 1031–1040.

Pellegrini, N., R. Re, M. Yang, and C. Rice-Evans. 1999. Screening of dietary carotenoids and carotenoid-rich fruit extracts for antioxidant activities applying 2,2′-azinobis(3-ethyenebenzothiazoline-6-sulfonic acid) radical cation decolorization assay. *Methods Enzmol.* 299: 379–389.

Pham, T. T. and N. P. Shah. 2008a. Effect of lactulose on biotransformation of isoflavone glycosides to aglycones in soymilk by lactobacilli. *J. Food Sci.* 73: 158–165.

Pham, T. T. and N. P. Shah. 2008b. Effects of lactulose supplementation on the growth of bifidobacteria and biotransformation of isoflavone glycosides to isoflavone aglycones in soymilk. *J. Agric. Food Chem.* 56: 4703–4709.

Pierce, B. E. and M. J. Leboffe. 1999. *Exercises for the Microbiology Laboratory*. Morton Publishing Co., Englewood, CO.

Pike, O. A. 1998. Fat characterization. In *Food Analysis* (S. Suzanne Nielsen, Ed.), pp. 217–236. Aspen Publishers, Inc., Gaithersburg, MD.

Poinar, G. O., Jr. and G. M. Thomas. 1978. *Diagnostic Manual for the Identification of Insect Pathogens*. Plenum Press, New York.

Pólya, G. 1945. *How to Solve It: A New Aspect of Mathematical Method*. Princeton University Press, Princeton, NJ.

Pusztai, A. and G. Grant. 1998. Assessment of lectin inactivation by heat and digestion. In *Lectin Methods and Protocols* (J. M. Rhodes and J. D. Milton, Eds.), pp. 505–514. Humana Press, Totowa, NJ.

Rajeshwara, A. N. and V. Prakash. 1994. Structural stability of lipase from wheat germ. *Int. J. Pept. Protein Res.* 44: 435–440.

Ramadan, M. F., H. E. S. Showky, and A. E. R. M. Sulleman. 2008. Comparison between the effect of gamma-irradiation and roasting on the profile and antioxidant activity of wheat germ lipids. *Grasas y Aceites*. 59: 166–173.

Ramesha, C., H. Lakshmi, S. S. Kumari, C. M. Anuradha, and C. S. Kumar. 2012. Nutrigenetic screening strains of the mulberry silkworm, *Bombyx mori*, for nutritional efficiency. *J. Insect Sci.* 12: 15. Available online: insectscience.org/12.15. Last accessed August 2014.

Rao, C. G., S. V. Seshgiri, C. Ramesh, B. K. Ibrahim, H. Nagaraju, and H. Chandrashekaraiah. 2006. Evaluation of genetic potential of the polyvoltine silkworm (*Bombyx mori* L.) germplasm and identification of parents for breeding programme. *J. Zhejiang Univ. Sci. B*.7: 215–220.

Raubenheimer, D. 2011. Toward a quantitative nutritional ecology: The right-angled mixture triangle. *Ecol. Monogr.* 81(3): 407–427.

Raulston, J. R. and T. N. Shaver. 1970. A low agar casein-wheat germ diet for rearing tobacco budworm. *J. Econ. Entomol.* 63: 1743–1744.

Reed, G. and T. W. Nagodawithana. 1991. *Yeast Technology*, 2nd Ed. Van Nostrand Reinhold, New York.

Ribeiro, J. M. C. 1995. Insect saliva: Function, biochemistry, and physiology. In *Regulatory Mechanisms in Insect Feeding* (R. F. Chapman and G. deBoer, Eds.), pp. 74–98. Chapman & Hall, New York.

Rock, G. C. and E. Hodgson. 1971. Dietary amino acid requirements for *Heliothis zea* determined by dietary deletion and radiometric techniques. *J. Insect Physiol.* 17: 1087–1097.

Rock, G. C. and K. W. King. 1966. Amino acid composition in hydrolysates of the red-banded leaf roller, *Argyrotaenia velutinana* (Lepidoptera: Tortricidae) during development. *Ann. Entomol. Soc. Am.* 59: 273–275.

Rogers, D. J. and C. J. Winks. 1993. Quality control in laboratory-reared codling moth at Mt Albert Research Centre, Auckland, New Zealand. In *Procedings of the 7th Workshop of the IOBC Global Working Group "Quality Control of Mass Reared Arthropods"* (G. Nicoli, M. Benuzzi, and N. C. Leppla, Eds.), pp. 13–21. September 13–16, 1993, Rimini, Italy. Available at http://users.ugent.be/~padclerc/AMRQC/proceedings.html. Accessed November 2014.

Romoser, W. S. and J. G. Stoffolano, Jr. 1998. *The Science of Entomology*. McGraw-Hill, Boston, MA.

Rosenthal, G. A. and M. R. Berenbaum (Eds.). 1992. *Herbivores: Their Interactions with Secondary Metabolites*, vol. 1. The Chemical Participants, Academic Press, New York.

Rothschild, M. 1972. Secondary plant substances and warning colouration in insects. In *Insect/Plant Relationships* (H. F. van Emden, Ed.), pp. 59–83. Royal Entomological Society (London), Symposium 6. Blackwell Scientific, London, U.K.

Roush, D. J. and Y. F. Lu. 2008. Advances in primary recovery: Centrifugation and membrane technology. *Biotechnol. Prog.* 24: 488–495.

Ryan, T. P. 1989. *Statistical Methods for Quality Improvement*. John Wiley & Sons, New York.

Sandberg, AS. 2002. Bioavailability of minerals in legumes. *Br. J. Nut.* 88(Suppl. 3): S281–S285.

SAS. 1999. Statview® Reference. SAS Institute Inc. Cary, NC.

SAS Institute. JMP®, Version 11. SAS Institute Inc., Cary, NC, pp. 1989–2007.

Saunders, J. L. and J. K. Knoke. 1967. Diets for rearing the ambrosia beetle, *Xyleborus ferruginous* (Fabr.) in vitro. *Science*. 157: 460–463.

Savage, D. C. 1977. Microbial ecology of the gastrointestinal tract. *Annu. Rev. Microbiol.* 31: 107–133.

Schneider, J. C. (Ed.). 2013. *Principles and Procedures for Rearing High Quality Insects*. Mississippi State University Press, Jackson, MS.

Schoener, T. W. 1971. Theory of feeding strategies. *Annu. Rev. Ecol. Syst.* 2: 369–404.

Schoonhoven, L. M. 1972. Some aspects of host selection and feeding in phytophagous insects. In *Insect and Mite Nutrition* (J. G. Rodriguez, Ed.), pp. 557–566. North-Holland Publishing Company, Amsterdam, the Netherlands.
Schonrogge, K., L. J. Harper, and C. P. Lichtenstein. 2000. The protein content of tissues in cynipid galls (Hymenoptera: Cynipidae): Similarities between cynipid galls and seeds. *Plant Cell Environ.* 23: 215–222.
Schultz, T. R. and S. G. Brady. 2008. Major evolutionary transitions in ant agriculture. *Proc. Natl. Acad. Sci. USA.* 105(14): 5435–5440.
Schwarz, A. J., A. Zambada, D. H. S. Orozco, J. L. Zavala, and C. O. Calkins. 1985. Mass production of the Mediterranean fruit fly at Metapa. *Mexico. Fla. Entomol.* 68: 467–477.
Scriber, J. M. and F. Slansky, Jr. 1981. The nutritional ecology of immature insects. *Ann. Rev. Entomol.* 26: 183–211.
Scruggs, C. G. 1978. The origin of the screwworm control program. In *The Screwworm Problem—Evolution of Resistance to Biological Control* (R. H. Richardson, Ed.), pp. 11–18. University of Texas Press, Austin, TX.
Serafini, M., M. F. Testa, D. Villaño, M. Pecorari, K. van Wieren, E. Azzini, A. Brambilla, and G. Maiani, 2009. Antioxidant activity of blueberry fruit is impaired by association with milk. *Free Rad. Biol. Med.* 46(6): 769–774.
Shapiro, J. P., J. H. Law, and M. A. Wells. 1988. Lipid transport in insects. *Annu. Rev. Entomol.* 33: 297–318.
Shapiro, M. 1984. Microorganisms as contaminants and pathogens in insect rearing. In *Advances and Challenges in Insect Rearing* (E. G. King and N. C. Leppla, Eds.), pp. 130–142. USDA—ARS, U.S. Government Printing Office, Washington, DC.
Sharma, O. P. and T. K. Bhat. 2009. DPPH antioxidant assay revisited. *Food Chem.* 113: 1202–1205.
Sharp, J. L., E. F. Boller, and D. L. Chambers. 1983. Selection for flight propensity of laboratory and wild strains of *Anastrepha suspensa* and *Ceratitis capitata* (Diptera: Tephritidae). *J. Econ. Entomol.* 76: 302–305.
Sharp, J. L., D. L. Chambers, and F. H. Haramoto. 1975. Flight mill and stroboscopic studies of oriental fruit flies and melon flies, including observations on Mediterranean fruit flies. *Proc. Hawaii. Entomol. Soc.* 22: 137–144.
Shorey, H. H. 1963. A simple artificial rearing medium for the cabbage looper. *J. Econ. Entomol.* 56: 536–537.
Shorey, H. H. and R. L. Hale. 1965. Mass-rearing of the larvae of nine noctuid species on a simple artificial medium. *J. Econ. Entomol.* 58: 522–524.
Sikorowski, P. 1984a. Microbial contamination in insectaries: Occurrence, prevention, and control. In *Advances and Challenges in Insect Rearing* (E. G. King and N. C. Leppla, Eds.), pp. 143–153. USDA; ARS, New Orleans, LA.
Sikorowski, P. P. 1984b. Pathogens and microbial contaminants: Their occurrence and control. In *Boll Weevil Mass Rearing Technology* (P. P. Sikorowski, J. G. Griffin, J. Roberson, and O. H. Lindig, Eds.), pp. 115–169. University Press of Mississippi, Jackson, MS.
Sikorowski, P. and R. H. Goodwin. 1985. Contaminant control and disease recognition in laboratory colonies. In *Handbook of Insect Rearing*, vol. 1. (P. Singh and R. F. Moore, Eds.), pp. 85–105. Elsevier, Amsterdam, the Netherlands.
Sikorowski, P., A. D. Kent, O. H. Lindig, G. Wiygul, and J. Roberson. 1980. Laboratory and insectary studies on the use of antibiotics and antimicrobial agents in mass-rearing of boll weevils. *J. Econ. Entomol.* 73: 106–110.
Sikorowski, P. and A. Lawrence. 1994. Microbial contamination and insect rearing. *Am. Entomol.* 40: 240–253.
Sikorowski, P. and A. Lawrence. 1997. *Major Diseases of Heliothis virescens and Helicoverpa zea in Mississippi Fields and Insectaries.* Technical Bulletin 218. Mississippi Agricultural and Forestry Experiment Station, Starkville, MS.
Sikorowski, P. P, J. G. Griffin, J. R. Roberson, and O. H. Lindig. 1984. *Boll Weevil Mass Rearing Technology.* University Press of Mississippi, Jackson, MS.
Sikorski, Z. E. J. Pokomy, and S. Damodaran. 2008. Physical and chemical interactions of components of food systems. In Damodaran, S., K. L. Parkin, and O. R. Fennema (Eds.), *Fennema's Food Chemistry*, 4th edn., pp. 849–884. Boca Raton, FL: CRC Press.
Silverman, J. and D. N. Bieman. 1993. Glucose aversion in the German cockroach, *Blattella germanica. J. Insect Physiol.* 39: 925–933.
Simpson, S. J. and D. Raubenheimer. 1995. The geometric analysis of feeding and nutrition: A user's guide. *J. Insect Physiol.* 7: 545–553.
Simpson, S. J. and D. Raubenheimer. 2001. The geometric analysis of nutrient–allelochemical interactions: A case study using locusts. *Ecology.* 82: 422–439.
Singh, P. 1977. *Artificial Diets for Insects, Mites, and Spiders.* Plenum, New York.

Singh, P. and H. L. House. 1970a. Antimicrobials: "Safe" levels in a synthetic diet of an insect, *Agria affinis*. *J. Insect Physiol.* 16: 1769–1782.

Singh, P. and H. L. House. 1970b. Antimicrobial agents: Their detrimental effects on size of an insect, *Agria affinis*. *Can. Entomol.* 102: 1340–1344.

Singh, P. and W. L. Howe. 1971. Feeding longevity, and fecundity of the adult western corn rootworm fed artificial diets. *J. Econ. Entomol.* 64: 1136–1137.

Singh, R. P. and D. R. Heldman. 2014. *Introduction to Food Engineering*, 5th Ed. Academic Press, Amsterdam, the Netherlands.

Singh, P. and R. F. Moore (Eds.). 1985. *The Handbook of Insect Rearing*, vols. I and II. New York: Elsevier.

Singkhornart, S., B.-J. Gu, and G. H. Ryu. 2013. Physicochemical properties of extruded germinated wheat and barley as modified by CO_2 injection and difference extrusion conditions. *Int. J. Food Sci. Technol.* 48: 290–299.

Sjovall, O., T. Virtalaine, A. Lapvetelainen, and H. Kallio. 2000. Development of rancidity in wheat germ analyzed by headspace gas chromatography and sensory analysis. *J. Agric. Food Chem.* 48: 3522–3527.

Slansky, F. and M. Schriber. 1981. The nutritional ecology of immature insects. *Annu. Rev. Entomol.* 26: 193–211.

Smith, A. E. and S. D. Garret. 1971. Artificial diet and pheromone studies with peach twig borer. Lepidoptera: Gelechiidae. *J. Econ. Entomol.* 64: 259.

Snodgrass, R. E. 1935. *Principles of Insect Morphology*. New York: McGraw-Hill Book Company.

Solaiman, D. K. Y., R. D. Ashby, and J. A. Zerkowski. 2007. Simplified soy molasses-based medium for reduced-cost production of sophorolipids by *Candida bombicola*. *Biotechnol. Lett.* 29: 1341–1347.

Song, D. and S. K. C. Chang. 2005. Enzymatic degradation of oligosaccharides in pinto bean flour. *J. Agric. Food Chem.* 54: 1296–1301.

Sparks, A. N. and E. A. Harrell. 1976. *Corn Earworm Rearing Mechanization*. USDA; ARS Technical Bulletin. No. 1554, U.S. Department of Agriculture, Agricultural Research Service, Washington, DC, 11pp.

Spencer, N. R. 2000. *X International Symposium on Biological Control of Weeds*. Montana State University Press, Bozeman, MT.

SPSS. 1996. *SYSTAT 6.0 for Windows: Graphics*. SPSS, Inc., Chicago, IL.

Stamps, J. A., L. H. Yang, V. M. Morales, and K. L. Boundy-Mills. 2012. *Drosophila* regulate yeast density and increase yeast community similarity in a natural substrate. *PLoS ONE* 7(7): e42238.

Standifer, L. N. 1967. A comparison of the protein quality of pollens for growth-stimulation of the hypopharyngeal glands and longevity of honey bees, *Apis mellifera* L. (Hymenoptera: Apidae). *Insectes Sociaux.* 14: 415–426.

Standifer, L. N. 1980. Honey bee nutrition and supplemental feeding. In *Beekeeping in the United States*. Agriculture Handbook Number 335. Science and Education Administration, Carl Hayden Center for Bee Research, Tucson, AZ, pp. 39–45. http://www.beesource.com/resources/usda/honey-bee-nutrition-and-supplemental-feeding/. Accessed March 2015.

Starmer, W. T., J. S. F. Barker, H. J. Phaff, and J. C. Fogleman. 1986. Adaptations of *Drosophila* and yeasts: Their interactions with the volatile 2-propanol in the cactus-microorganism-*Drosophila* model system. *Aust. J. Biol. Sci.* 39: 67–77.

Starmer, W. T. and J. C. Fogleman. 1986. Coadaptation of *Drosophila* and yeasts in their natural habitat. *J. Chem. Ecol.* 12: 1035–1053.

Starmer, W. T., W. Heed, M. Miranda, M. W. Miller, and H. J. Phaff. 1976. The ecology of yeast flora associated with cactiphilic Drosophila and their host plants in the Sonoran desert. *Microbial Ecol.* 3: 11–30.

Starmer, W. T., W. B. Heed, and E. S. Rockwood-Sluss. 1977. Extension of longevity in *Drosophila mojavensis* by environmental ethanol: Differences between subraces. *Proc. Natl. Acad. Sci. USA.* 74: 387–391.

Starmer, W. T., M. A. Lachance, H. J. Phaff, and W. B. Heed. 1990. The biogeography of yeasts associated with decaying cactus tissue in North America, the Caribbean, and northern Venezuela. In *Evolutionary Biology* (M. K. Hecht, B. Wallace, R. J. MacIntyre, Eds.), pp. 253–296. Plenum, New York.

StatView. 1999. *StatView Reference*, 3rd Ed. SAS Institute, Inc., Cary, NC.

Steffe, J. F. 1996. *Rheological Methods in Food Process Engineering*, 2nd Ed. Freeman Press, East Lansing, MI.

Steinhaus, E. A. 1940. The microbiology of insects—With special reference to the biologic relationships between bacteria and insects. *Bacteriol. Rev.* 4: 17–57.

Sterling, W. L. and P. L. Adkisson. 1966. An artificial diet for laboratory cultures of boll weevil larvae and adults. *J. Econ. Entomol.* 59: 1074–1077.

Stevens, C. E. and I. D. Hume. 1995. *Comparative Physiology of the Vertebrate Digestive System*. Cambridge University Press, Cambridge, U.K.

Stewart, F. D. 1984. Mass rearing the pink bollworm, *Pectinophora gossypiella*. In *Advances and Challenges in Insect Rearing* (E. G. King and N. C. Leppla, Eds.), pp. 176–187. USDA; ARS, New Orleans, LA.

Stoltz, D. B. and S. B. Vinson. 1979. Viruses and parasitism in insects. *Adv. Virus Res.* 24: 125–171.

Stone, C. W. 2006. Yeast products in the feed industry yeast products in the feed industry: A practical guide for feed professionals. http://en.engormix.com/MA-feed-machinery/formulation/articles/yeast-products-feed-industry-t243/800-p0.htm. Accessed March 9, 2015.

Stukey, B. N. 1972. Antioxidants as food stabilizers. In *Handbook of Food Additives* (T. E. Furia, Ed.), pp. 185–224. CRC Press, Cleveland, OH.

Svoboda, J. A. 1984. Insect steroids: Metabolism and function. In *Isopentenoids in Plants: Biochemistry and Function* (W. D. Nes, G. Fuller, and L. L. Tsai, Eds.), pp. 367–388. Marcel Dekker, Inc., New York.

Svoboda, J. A. and W. R. Lusby. 1986. Sterols of phytophagous and omnivorous species of Hymenoptera. *Arch. Insect Biochem. Physiol.* 3: 13–18.

Svoboda, J. A., M. J. Thompson, W. E. Robbins, and T. C. Elden. 1975. Unique pathways of sterol metabolism in the Mexican bean beetle, a plant-feeding insect. *Lipids.* 10: 524.

Swaisgood, H. E. 2008. Characteristics of milk. In *Fennema's Food Chemistry* (S. Damodaran, K. L. Parkin, and O. R. Fennema, Eds.), pp. 885–922, CRC Press, Boca Raton, FL.

Tamaki, G. and R. E. Weeks. 1972. Biology and ecology of two predators *Geocoris punctipes* Say and *G. bullatus* Stal. *USDA Tech. Bull.* 1446: 46.

Tan, X. L., S. Wang, and F. Zhang. 2013. Optimization an optimal artificial diet for the predatory bug *Orius sauteri* (Hemiptera: Anthocoridae). *PLoS ONE* 8(4): e61129.

Tanada, Y. and H. K. Kaya. 1993. *Insect Pathology*. Academic Press, Inc., San Diego, CA.

Tanaka, N., L. F. Steiner, K. Ohinata, and R. Okamoto. 1965. Low-cost larval rearing medium for mass production of oriental and Mediterranean fruit flies. *J. Econ. Entomol.* 62: 967–968.

TAPPI Test Methods. 1991. *Bursting Strength of Paperboard and Liner Board*, vol. 2. TAO7OM-87. TAPPI Press, Peachtree Corners, GA.

Taoukis, P. and T. P. Labuza. 1996. Summary: Integrative concepts. In Fennema, O. R. (Ed.), *Food Chemistry*, 3rd edn., pp. 1013–1042. New York: Marcel & Dekker.

Taylor, D. B. 1992. Rearing systems for screwworm mass production. In Anderson, T. E. and N. C. Leppla (Eds.), *Advances in Insect Rearing for Research and Pest Management*, pp. 393–403. Boulder, CO: Westview Press.

Taylor, R. 1973. *Butterflies in My Stomach or Insects in Human Nutrition*. Woodbridge, Santa Barbara, CA.

Terra, W. R. 1990. Evolution of digestive systems in insects. *Annu. Rev. Entomol.* 35: 181–200.

Thompson, S. N. 1975. Defined meridic and holidic diets and aseptic feeding procedures for artificially rearing the ectoparasitoid *Exeristes roborator* (Fabricius). *Ann. Entomol. Soc. Am.* 68: 220–226.

Thompson, S. N. 1999. Nutrition and culture of entomophagous insects. *Ann. Rev. Entomol.* 44: 561–592.

Thompson, S. N. and J. S. Barlow. 1983. Metabolic determination and regulation of fatty acid composition in parasitic Hymenoptera and other animals. In *Metabolic Aspects of Lipid Nutrition in Insects* (T. E. Mittler and R. H. Dadd, Eds.), pp. 73–106, Westview Press, Boulder, CO.

Thomsen, M. K., C. Jacobsen, and L. H. Skibsted. 2000. Mechanism of initiation of oxidation in mayonnaise enriched with fish oil as studied by electron spin resonance spectroscopy. *Eur. Food Res. Technol.* 211: 381–386.

Tillman, P. G., G. McKibben, S. Malone, and D. Harsh. 1997. Form-fill-seal machine for mass-rearing noctuid species. In Remy, K. H. (Ed.), Office of Agricultural Communications, Division of Agriculture, Forestry, and Veterinary Medicine, Mississippi State University, Technical Bulletin 213, pp. 1–4.

Timmermann, S. E., A. R. Zangerl, and M. R. Berenbaum. 1999. Ascorbic and uric acid responses to xanthotoxin ingestion in a generalist and a specialist caterpillar. *Arch. Insect Biochem. Physiol.* 42: 26–36.

Toivo, J., K. Philips, A. M. Lampi, and V. Piironen. 2001. Determination of sterols in foods: Recovery of free, esterified, and glycosidic sterols. *J. Food Comp. Anal.* 14: 631–643.

Tomic-Carruthers, N. 2007. Development of a meridic diet for *Hylobius transversovittatus* (Coleoptera: Curculionidae) and the role of carbohydrates in feeding, growth, and survival of larvae. *J. Econ. Entomol.* 100(4): 1062–1070.

Tomic-Carruthers, N. 2009. Rearing *Hylobius transversovittatus* and *Cyphocleonus achetes* larvae on artificial diets (Coleoptera: Curculionidae). *Florida Entomol.* 92(4): 656–657.

Turunen, S. 1979. Digestion and absorption of lipids in insects. *Comp. Biochem. Physiol.* 63A: 455–460.

Turunen, S. 1983. Absorption and utilization of essential fatty acids in lepidopterous larvae: Metabolic implications. In *Metabolic Aspects of Lipid Nutrition in Insects* (T. E. Mittler and R. H. Dadd, Eds.), pp. 57–71. Westview Press, Boulder, CO.

Turunen, S. 1985. Absorption. In *Comprehensive Insect Physiology, Biochemistry, and Pharmacology*, vol. 4 (G. A. Kerkut and L. I. Gilbert, Eds.), pp. 241–277. Pergamon Press, New York.

Udy. Principles of protein measurement. http://www.udyone.com/udydoc/udysys2.shtml. Accessed 2015.

U.S. Forest Service: http://www.na.fs.fed.us/fhp/alb/.

USDA. November 27, 2013. USDA National Nutrient Database for Standard Reference Release 26. Full Report (All Nutrients) 16415, Soy flour, full-fat, raw, crude protein basis (N x 6.25).

Vacek, D. C., W. T. Starmer, and W. B. Heed. 1979. Relevance of the ecology of citrus yeasts to the diet of *Drosophila. Micro. Ecol.* 5: 43–49.

van Huis, A., J. Van Itterbeeck, H. Klunder, E. Mertens, A. Halloran, G. Muir, and P. Vantomme. 2013. Edible insects: Future prospects for food and feed security. FAO Forestry Paper 171. Wageningen UR, the Netherlands.

van Kempen, T. A. T. G., E. van Heugten, and A. J. Moeser. 2006. Selecting soybean meal characteristics preferred for swine nutrition. *J. Anim. Sci.* 84: 1387–1395.

Vanderzant, E. S. 1959. Inositol: An indispensable dietary requirement of the pink bollworm. *J. Econ. Entomol.* 59: 1018–1019.

Vanderzant, E. S. 1963. Nutrition of the boll weevil larva. *J. Econ. Entomol.* 56: 357–62.

Vanderzant, E. S. 1967. Wheat-germ diets for insects: Rearing the boll weevill and the salt-marsh caterpillar. *Ann. Ent. Soc. Amer.* 60: 1062–1066.

Vanderzant, E. S. and T. B. Davich. 1961. Artificial diet for the adult boll weevil and techniques for obtaining eggs. *J. Econ. Entomol.* 54: 923–928.

Vanderzant, E. S., D. Kerur, and R. Reiser. 1957. The role of dietary fatty acids in the development of the pink bollworm. *J. Econ. Entomol.* 50: 606–608.

Vanderzant, E. S., M. C. Pool, and C. D. Richardson. 1962a. The role of ascorbic acid in the nutrition of three cotton insects. *J. Insect Physiol.* 8: 287–297.

Vanderzant, E. S. and C. D. Richardson. 1963. Ascorbic acid in the nutrition of plant-feeding insects. *Science.* 140: 989–991.

Vanderzant, E. S., C. D. Richardson, and T. B. Davich. 1959. Feeding and oviposition by the boll weevil on artificial diets. *J. Econ. Entomol.* 52: 1138–1142.

Vanderzant, E. S., C. D. Richardson, and S. W. Fort, Jr. 1962b. Rearing of the bollworm on artificial diet. *J. Econ. Entomol.* 55: 140.

Vanderzant, E. S. and R. Reiser. 1956. Studies of the nutrition of the pink bollworm using purified casein media. *J. Econ. Entomol.* 49: 454–458.

Vargas, R. I. 1989. Mass production of tephritid fruit flies. In *Fruit Flies, Their Biology, Natural Enemies, and Control.* World crop pests, 3B (G. Hooper and W. Robinson, Eds.), pp. 141–152. Elsevier, Amsterdam, the Netherlands.

Vargas, R. I., H. Chang, and D. L. Williamson. 1983. Evaluation of a sugarcane bagasse larval diet for mass production of the Mediterranean fruit fly (Diptera: Tephritidae) in Hawaii. *J. Econ. Entomol.* 76: 1360–1362.

Vargas, R. I. and S. Mitchell. 1987. Two artificial larval diets for rearing *Dacus latifrons* (Diptera: Tephritidae). *J. Econ. Entomol.* 80: 1337–1339.

Versoi, P. L. and L. K. French. 1992. The establishment of commercial insectaries. In *Advances in Insect Rearing for Research and Pest Management* (T. E. Anderson and N. C. Leppla, Eds.), pp. 457–463. Westview Press, New Delhi, India.

von Elbe, J. H. and S. J. Schwartz. 1996. Colorants. pp. 651–720. In *Fennema's Food Chemistry*, 3rd Ed. (O. R. Fennema, Eds.) Marcel Dekker. New York.

Wada-Katsumata, A., J. Silverman, and C. Schal. 2013. Changes in taste neurons support the emergence of an adaptive behavior in cockroaches. *Science.* 340: 972–975.

Waldbauer, G. P. 1962. The growth and reproduction of maxillectomized tobacco hornworms feeding on normally rejected non-solanaceous plants. *Entomol. Exp. Applicata.* 5: 147–168.

Waldbauer, G. P. 1968. The consumption and utilization of foods by insects. *Adv. Insect Physiol.* 5: 229–288.

Waldbauer, G. P. and A. K. Bhattacharya. 1973. Self-selection of an optimum diet from a mixture of wheat fractions by the larvae of *Tribolium confusum*. *J. Insect Physiol.* 19: 407–418.

Waldbauer, G. P., R. W. Cohen, and S. Friedman. 1984. Self-selection of optimal mix from defined diets by larvae of the corn earworm, *Heliothis zea* (Boddie). *Physiol. Zool.* 57: 590–597.

Walstra, P. and T. von Vliet. 2008. Dispersed systems: Basic considerations. In Damodaran, S., K. L. Parkin, and O. R. Fennema (Eds.), *Fennema's Food Chemistry*, 4th edn., pp. 743–847. Boca Raton, FL: CRC Press.

Werren, J. H. 1997. Biology of *Wolbachia*. *Annu. Rev. Entomol.* 42: 587–609.

Western Electric Co., Inc. 1956. *Statistical Quality Control Handbook*. Western Electric Co., Inc. Newark, NJ.

Wheeler, A. G. 2001. *Biology of the Plant Bugs*. Ithaca, New York: Cornell University Press.

Whitaker, J. R. 1996. Enzymes. In *Food Chemistry* (O. R. Fennema, Ed. 3rd Edition), pp. 331–430. Marcel Dekker, Inc., New York.

Wigglesworth, V. B. 1974. *The Principles of Insect Physiology*, 7th edn. London, U.K.: Chapman & Hall.

Willis, R. B. and P. R. Allen. 1999. Measurement of amorphous ferric phosphate to assess iron bioavailability in diets and diet ingredients. *Analyst.* 124: 425–430.

Windholz, M. (Ed.). 1983. *Merck Index*. Merck and Co., Inc., Rahway, NJ.

Wolf, W. A. 1984. Controlling respiratory hazards in insectaries. In *Advances and Challenges in Insect Rearing* (E. G. King and N. C. Leppla, Eds.), pp. 64–69. USDA; ARS, New Orleans, LA.

Woods, A. E. and L. W. Aurand. 1977. *Laboratory Manual in Food Chemistry*. The AVI Publishing Co., Inc. Westport, CT.

World Health Organization. 1977. Wholesomeness of irradiated foods. Report of the Joint FAO-IAEA-WHO Expert Committee. WHO Technical Report Series No. 604. HMSO, London, U.K.

World Health Organization. 1981. Wholesomeness of irradiated foods. Report of the Joint FAO–IAEA–WHO Expert Committee. WHO Technical Report Series No. 659. HMSO, London, U.K.

World Health Organization. 1994. Review of the safety and nutritional adequacy of irradiated food. Report of a WHO Consultation. Geneva, May 20–22, 1992. WHO. Geneva, Switzerland.

World Health Organization. 2002. Technical report series 935. Protein and amino acid requirements in human nutrition. Report of Joint WHO/FAO/UNU Expert Consultation. WHO. Geneva, Switzerland.

Wu, Z. X., A. C. Cohen, and D. A. Nordlund. 2000. The feeding behavior of *Trichogramma brassicae*: New evidence for selective ingestion of solid food. *Entomologia Experimentalis et Applicata.* 96: 1–8.

Wykes, G. R. 1952. The preference of honeybees for solutions of various sugars which occur in nectar. *J. Exp. Biol.* 29: 511–519.

Xezones, H. and I. J. Hutchings. 1965. Thermal resistance of *Clostridium botulinum* (62A) spores as affected by fundamental food constituents. *Food Technol.* 19: 1003–1005.

Xia, Z., S. Wu, S. Pan, and J. M. Kim. 2012. Nutritional evaluation of protein from *Clanis bilineata* (Lepidoptera), an edible insect. *J. Sci. Food. Agric.* 92: 1479–1482.

Yamamoto, R. T. 1968. Mass rearing of tobacco hornworm: 1 Egg production. *J. Econ. Entomol.* 61(1): 170–174.

Yamamoto, R. T. 1969. Mass rearing of tobacco hornworm: 2 Larval rearing and pupation. *J. Econ. Entomol.* 62: 1427–1431.

Yosinori, T. and H. K. Kaya. 1993. *Insect Pathology*. Academic Press, San Diego, CA.

Young, R. F., K. S. Shields, and G. P. Berlyn. 1995. Hemlock woolly adelgid (Homoptera: Adelgidae): Stylet bundle insertion and feeding sites. *Ann. Entomol. Soc. Am.* 88: 827–835.

Yu, Y., M. Zhang, S. Lin, L. Wang, J. Liu, G. Jones, and H. C. Huang. 2013. Assessment the levels of tartrate-resistant acid phosphatase (TRAP) on mice fed with eggshell calcium citrate malate. *Int. J. Biol. Macromol.* 58: 253–257.

Zabinski, J. 1929. The growth of black beetles and of cockroaches on artificial and on incomplete diets. Part 1. *J. Exp. Biol.* 6: 360–385.

Zeng, F. and A. C. Cohen. 2000. Partial characterization of α-amylase in the salivary glands of *Lygus hesperus* and *L. lineolaris*. *Comp. Biochem. Physiol. B.* 126: 9–16.

Zeng, F., Y. C. Zhu, and A. C. Cohen. 2002a. Molecular cloning and partial characterization of a trypsin-like protein in salivary glands of *Lygus hesperus* (Hemiptera: Miridae). *Insect Biochem. Mol. Biol.* 32: 455–464.

Zeng, F., Y. C. Zhu, and A. C. Cohen. 2002b. Partial characterization of trypsin-like protease and molecular cloning of a trypsin-like precursor cDNA in salivary glands *Lygus lineolaris*. *Compar. Biochem. Physiol Part B* 131: 453–463.

Zha, C. and A. C. Cohen. 2014. Effects of anti-fungal compounds on feeding behavior and nutritional ecology of tobacco budworm and painted lady butterfly larvae. *Entomol. Ornithol. Herpetol.* 3:120.

Zhou, J. and D. Han. 2006. Proximate, amino acid and mineral composition of pupae of the silkworm *Antheraea pernyi* in China. *J. Food Comp. Anal.* 19: 850–853.

Zumreoglu, A., N. Tanaka, and E. J. Harris. 1979. The need for wheat germ in larval diets of the Mediterranean fruit fly (Diptera: Tephritidae) of non-nutritive bulking material. *Turk. Bit. Kor. Derg.* 3: 131–138.

Index

A

E

F

Q

For Product Safety Concerns and Information please contact our EU representative GPSR@taylorandfrancis.com Taylor & Francis Verlag GmbH, Kaufingerstraße 24, 80331 München, Germany

Batch number: 10732714

Printed by Printforce, the Netherlands